AF891980

Mosquitoes of India

This is an up-to-date and comprehensive handbook that presents a wealth of information on the different aspects of one of the largest dipterous family, Culicidae (mosquitoes). India shares more than 10% of the global mosquito fauna and some of the deadliest mosquito species occur in the country, implying that globally, India has always been in the forefront of research and control of such deadly and/or debilitating diseases as malaria, lymphatic filariasis, dengue, chikungunya, Zika, Japanese encephalitis, West Nile Virus, etc. This book illustrates updated information, including chemical, biological, herbal, and genetical, on the control of vectors, and showcases mosquito preponderance in the neighbouring countries. The book additionally deals with the biosafety principles in theory and practice, use of emerging science of artificial intelligence in mosquito identity and control, and, of course, consideration of mosquitoes in human psyche. This authoritative account is a crucial reference source for mosquito-borne disease control and prevention. This book is meant for researchers, university students, medical entomologists, parasitologists, and public health professionals.

Mosquitoes of India

Mosquito-Borne Diseases, Diagnosis and Control

Edited by

Brij Kishore Tyagi

CRC Press
Taylor & Francis Group
Boca Raton London New York

CRC Press is an imprint of the
Taylor & Francis Group, an **informa** business

Designed cover image: Shutterstock

First edition published 2025
by CRC Press
2385 NW Executive Center Drive, Suite 320, Boca Raton FL 33431

and by CRC Press
4 Park Square, Milton Park, Abingdon, Oxon, OX14 4RN

CRC Press is an imprint of Taylor & Francis Group, LLC

© 2025 selection and editorial matter, B.K. Tyagi; individual chapters, the contributors

Reasonable efforts have been made to publish reliable data and information, but the author and publisher cannot assume responsibility for the validity of all materials or the consequences of their use. The authors and publishers have attempted to trace the copyright holders of all material reproduced in this publication and apologize to copyright holders if permission to publish in this form has not been obtained. If any copyright material has not been acknowledged please write and let us know so we may rectify in any future reprint.

Except as permitted under U.S. Copyright Law, no part of this book may be reprinted, reproduced, transmitted, or utilized in any form by any electronic, mechanical, or other means, now known or hereafter invented, including photocopying, microfilming, and recording, or in any information storage or retrieval system, without written permission from the publishers.

For permission to photocopy or use material electronically from this work, access www.copyright.com or contact the Copyright Clearance Center, Inc. (CCC), 222 Rosewood Drive, Danvers, MA 01923, 978-750-8400. For works that are not available on CCC please contact mpkbookspermissions@tandf.co.uk

Trademark notice: Product or corporate names may be trademarks or registered trademarks and are used only for identification and explanation without intent to infringe.

ISBN: 978-1-032-35462-0 (hbk)
ISBN: 978-1-032-35468-2 (pbk)
ISBN: 978-1-003-32702-8 (ebk)

DOI: 10.1201/9781003327028

Typeset in Times
by SPi Technologies India Pvt Ltd (Straive)

Dedication

All mosquito researchers who spent their lives in understanding the strengths and weaknesses of the man's deadliest foe on our planet, Earth,

and

all those mosquito combaters who lived to see the mosquito relinquish her past and don a new role in bettering human lives!

Contents

SECTION II Diagnosis

SECTION III Control

SECTION IV Brief Review of Mosquitoes in Other Countries of the Indian Subcontinent

Preface

Mosquitoes form one of the largest dipterous families, Culicidae, and are among the most important creatures to man on Earth. In fact, no other animal whatsoever has ever directly or indirectly affected man's own survival to the extent that the mosquito is regarded as the deadliest foe of man on the planet! There are over 3,600 species of mosquitoes worldwide. They are present in all regions and counties of the world, save for Antarctica, Iceland, and a handful of French Polynesian micro-islands. They transmit some of the deadliest vector-borne diseases (VBDs)and are the reason for 17% of all communicable diseases globally. Mosquito-borne infections affect almost 60% of the world population, mostly in the tropics and subtropics; malaria alone infects 229 million and kills over 400,000 people annually. Their dangers are on the increase day-by-day due to increased global transcontinental human travel and trade, in addition to the newfound range of distribution of the dominant vectors. Invasion of malaria vector *Anopheles stephensi* in Africa and Sri Lanka (Tyagi, 2023), and of dengue/chikungunya/Zika vector *Aedes albopictus* in Europe and the Indian Ocean region is threatening enough to raise alert signals for future epidemics. The annual losses brought about by mosquitoes to humans are astronomical, with malaria alone levying a phenomenal estimate of disease burden of 53.6 million Disability Adjusted Life Years (DALYs) in 2019.

India shares more than 10% (420 species) of the global fauna. Some of these mosquito species are responsible for transmitting several deadly and/or debilitating diseases, such as malaria, filariasis, dengue, chikungunya, Japanese encephalitis, West Nile, and Zika, exacting death and sickness on millions of people, besides sapping off the country's crucial national resources of economy, health and intelligentsia. Unfortunately, there has never been a single comprehensive book in the past nearly 100 years that offered updated information on various facets of the life of mosquito such as, for example, taxonomy, phylogeny, biogeography, biology, ecology, physiology, and genetics, as well as mosquito-borne diseases and their treatment and control – so important for strategizing their effective control in future for a healthy and prosperous nation! Over the years, many vector mosquitoes have rapidly changed their behaviour both under the continuous and injudicious use of insecticides for their control and the unavoidable impact of climate change and anthropization. Such behavioural changes in vector mosquitoes have rendered them resistant against the insecticides of choice, thereby resulting into unhindered increase in their population. Many other vectors have shifted their resting and feeding behaviour from indoors to outdoors a phenomenon which brings in a big question on the validity of the Indoor Residual Spraying (IRS). Moreover, we have no information on the exotic species sneaked into India, on the one hand, and the status of endangered species (read 'genome'), on the other! Needless to say, with the advancement of scientific and technological tools in recent past, the control of the vector mosquitoes has been thoroughly churned out with an altogether new array of tools. Thus, mosquitoes of India, or for that matter of the world, have altogether a new face to present today, and it is inevitably indispensable to know these highly important animals in their current form of knowledge. Therefore, it is now considered commensurate to be *au fait* about the mosquitoes of India in totality, and the present book, *Mosquitoes of India: Mosquito-Borne Diseases, Diagnosis and Control*, in combination with its first part under the title *Mosquitoes of India: Biology, Phylotaxonomy, Host-Parasite Interaction and Resistance Development*, has been so designed and reinforced with updated knowledge on all important subjects that will make it an important reference work in entirety for not only Indian researchers but also their contemporaries in other parts of the world.

Prof Dr B. K. Tyagi, Jodhpur, India, 2025 Professor of Practice

Acknowledgement

A combo-book of *Mosquitoes of India* in the wake of glittering master pieces as those produced by Christophers (1933), Barraud (1934), and Rao (1984), just to mention a few examples, on one hand, and a nearly complete vacuum of a comprehensive treatment embodying within its folds an updated description of all the important disciplines of Indian Culicidae, on the other, was truly a Himalayan responsibility full of endless challenges, but still hugely gratifying when I thought of it becoming useful to a large number of students and researchers, both in India and abroad! This uphill task was, however, made largely easy by over a hundred dedicated authors engaged in completing as many as 41 chapters for the 2 volumes describing nearly all important facets of life of mosquitoes, especially with reference to India. To all of them I am grateful from the bottom of my heart.

For permission to use figures and/or tables from their publications, my heartfelt gratitude goes to a large number friends and colleagues – namely, Dr Dale E. Greenwalt (research associate, Departments of Paleobiology, National Museum of Natural History, Washington, DC, USA), Dr Derek E. G. Briggs (G. Evelyn Hutchinson Professor of Earth and Planetary Sciences, curator, Yale Peabody Museum, Department of Earth and Planetary Sciences, Yale University, New Haven, Connecticut, USA), Dr Leonardo Puerta (professor, Institute for Immunological Research, University of Cartagena, Cartagena, Colombia), Dr R. Harbach (research entomologist. British Museum Natural History, London), Dr Marc J. Klowden (editor in-chief of the *Journal of Vector Ecology*, Gainesville, USA), and a host of other generous permitters who are all duly acknowledged in appropriate places in the book. Further, I am also deeply indebted to the universal policy of organizations like the World Health Organization, Geneva, Switzerland, the Centers for Disease Control and Prevention, USA, etc., which allows the use of their material with due acknowledgement, which I am respectfully doing herewith. They are all duly cited in appropriate legends in the figure(s), table(s), or other places in the book.

Above all, my cordial remembrances with heartfelt thanks go to the world-famous malariologist and my friend, the late Professor Dr Felix P. Amerasinghe of International Water Management Institute, Colombo, Sri Lanka, who appreciated my malaria research in the Thar Desert, remarking on 7 April 2004, "*Tyagi is best known internationally among malaria and medical entomology circles for his long years of research into mosquitoes, malaria and irrigated agriculture in the Thar Desert of Rajasthan, India.*" His appreciation was my motivation for writing *Mosquitoes of India.*

I am extremely grateful to Dr Rui-De Xue (Anastasia Mosquito Control District of St. Johns County, St. Augustine, Florida, USA) for writing the foreword in the book (see the earlier volume of the combo book, Mosquitoes of India) and giving encouragement to me throughout the course of completing this book.

As always, I wish to thank my wife, Ajita, who gave me full freedom of time and movement so that I could put my whole attention to the accomplishment of this long and meandering journey of the book *Mosquitoes of India.*

Last but not least I duly thank the publisher Taylor & Francis, USA, and their highly competent and able editorial team for their patient guidance throughout the course of processing this book for publication.

Prof Dr B. K. Tyagi, 2025

Editor Bio

Brij Kishore Tyagi currently serves as the Professor of Practice in the Department of Biosciences, University Institute of Biotechnology, Chandigarh University, Mohali (Punjab), India, as well as the Advisor at Vellore Institute Technology, India, and the Consultant at Reckitt Benckiser, Gurugram (Haryana), India. He completed his MSc (zoology) in 1974 and Ph.D. (entomology/zoology) in 1978. He joined the Malaria Research Centre (now the National Institute of Malaria Research) in 1981 on a WHO-TDR (World Health Organization- Special Programme for Research and Training in Tropical Diseases) research project on tribal malaria in south Gujarat, and subsequently the ICMR (Indian Council of Medical Research) mainstream as Senior Research Officer at Vector Control Research Centre, Pondicherry, in 1984, and as Assistant Director and later Deputy Director at Desert Medicine Research Centre, Jodhpur, in 1988. He finally joined as a Scientist 'F' (Joint Director) (2000) and subsequently Scientist 'G' (Director) and Director-in-Charge (2005/07) at the Centre for Research in Medical Entomology, Madurai (Tamil Nadu State), through a 35-years-long and highly distinguished career in diverse medico-entomological research, particularly the various VBDs, in most of the states and UTs (Union Territories) in the country. He retired from the ICMR Service in 2013, with two-year extensions to 2015, with all administrative powers. He was appointed first as the Visiting Fellow, followed by Visiting Professor at Bharathidasan University, Trichy (TN) (2015–16), and subsequently as Visiting Professor at Punjabi University, Patiala (Punjab) (2017–19). He has authored more than 700 scientific titles and 43 books, including a training manual, *Biosafety for Human Health and the Environment in the Context of the Potential Use of Genetically Modified Mosquitoes (GMMs): A Tool for Biosafety Training Based on Courses in Africa, Asia and Latin America, 2008–2011*, published by the World Health Organization (2015).

Section I

Mosquito-Borne Infections

1 Malaria in India

P. K. Srivastava
Ex-National Vector Borne Disease Control Programme, Ministry of Health & Family Welfare, Government of India, Delhi, India

K. Ravi Kumar
Ex-National Vector Borne Disease Control Programme, Bengaluru, India

Devendra Kumar Goyal
Mohanlal Sukhadia University, Udaipur, India

O. Adeogun Adedapo
Center for Vector Management, Nigerian Institute of Medical Research, Lagos, Nigeria

K. Anju Viswan
Ex-State Officer Malaria (SSA, WHO), Raipur, India

A. Usha
District Surveillance Unit, Tumakuru, India

S.K. Ghosh
Ex-ICMR-National Institute of Malaria Research, Nirmal Bhawan, Poojanahalli, Kannamangala, Post Devanahalli, Bengaluru, India

Vas Dev
ICMR-National Institute of Malaria Research, Dwarka, New Delhi, India

Gyan Chand
Ex-ICMR-Regional Medical Research Centre, Jabalpur, India

Anil Prakash
Ex-National Institute for Research in Environmental Health, Bhopal, India

Sridharan Subramanian
Former Chief Entomologist, Public Health Department, State Health & Family Welfare, India

Sagya Singh
National Vector Borne Diseases Control Programme, India

DOI: 10.1201/9781003327028-2

Roop Kumari
Programme Development Officer, WHO SEARO, New Delhi, India

Rajnikant Dixit
ICMR-National Institute of Malaria Research, Dwarka, New Delhi, India

Chaitali Ghosh
Tata Institute for Genetics and Society, Bangalore, India

B.K. Tyagi
Chandigarh University, Mohali, India

1.1 INTRODUCTION

Human malaria is one of the oldest, deadliest, and most-widespread diseases in the world (Manson-Bahr, 1982). The disease is believed to have been associated with man and animals since time immemorial (Bruce-Chwatt, 1985). A mosquito-borne disease, malaria is transmitted to humans by a member of the genus, *Anopheles* (Hehir, 1927; Russell, 1955; Pampana, 1969; Harrison, 1978; Rao, 1984; Spielman and D'Antonio, 2002; Tyagi, 2004, 2020, 2023; Cox Francis, 2010; Dev, 2020; Spooner, 2017). Significantly, malaria has by now been eliminated from some of the endemic countries (Table 1.1), albeit still continuing to sicken over 250 million and kill over 600,000 people annually worldwide with a whopping disease burden of 56,201,200 Disability Adjusted Life Years (GBD, 2017; Tyagi and Tilak, 2022; WHO, 2022b). The estimated economic burden from malaria in India during mid-2010s was close to the US$1,940 million; the major burden sprang from lost earnings (75%), while 24% came from treatment costs (Gupta and Choudhury, 2014). Malaria has not only drastically affected human health in India but also largely impoverished economy and dwarfed intelligentsia (Sinton, 1935; Tyagi, 1992; Winegard, 2019). The nation most recently declared malaria-free in January 2024 is Cabo Verde in Africa.

TABLE 1.1
Countries and Territories Certified Malaria-Free by WHO (1955–2024)

Country/Territory	Countries Certified Malaria-Free[a,b,c,d,e]	Countries Where Malaria Never Existed or Disappeared Without Specific Measures[f]
AFRICA		
Algeria	2019	
Lesotho		2012
Mauritius	1973	
La Réunion (France)	1979	
Seychelles		2012
Cabo Verde	2024	
EASTERN MEDITERRANEAN		
Bahrain		2012
Jordan		2012
Kuwait		1963
Lebanon		2012

(Continued)

TABLE 1.1
(Continued)

Country/Territory	Countries Certified Malaria-Free[a,b,c,d,e]	Countries Where Malaria Never Existed or Disappeared Without Specific Measures[f]
Libya		2012
Morocco	2010	
Qatar		2012
Tunisia		2012
United Arab Emirates	2007	
EUROPE		
Albania		2012
Andorra		2012
Armenia	2011	
Austria		1963
Azerbaijan	2023	
Belarus		2012
Belgium		1963
Bosnia and Herzegovina	1973	
Bulgaria	1965	
Croatia	1973	
Cyprus	1967	
Czechia		1963
Denmark		1963
Estonia		2012
Finland		1963
France (Metropolitan)		2012
Germany		1964
Greece		2012
Hungary	1964	
Iceland		1963
Ireland		1963
Israel		2012
Italy	1970	
Kazakhstan		2012
Kyrgyzstan	2016	
Latvia		2012
Lithuania		2012
Luxembourg		2012
Malta		1963
Monaco		1963
Montenegro	1973	
Netherlands (Kingdom of the)	1970	
Norway		1963
Poland	1967	
Portugal	1973	
Republic of Moldova		2012
Republic of North Macedonia	1973	
Romania	1967	
Russian Federation		2012

(Continued)

TABLE 1.1
(Continued)

Country/Territory	Countries Certified Malaria-Free[a,b,c,d,e]	Countries Where Malaria Never Existed or Disappeared Without Specific Measures[f]
San Marino		1963
Serbia	1973	
Slovakia		1963
Slovenia	1973	
Spain	1964	
Sweden		1963
Switzerland		1963
Tajikistan	2023	
Turkmenistan	2010	
Ukraine		2012
United Kingdom of Great Britain and Northern Ireland		1963
Uzbekistan	2018	
AMERICAS		
Antigua and Barbuda		2012
Argentina	2019	
Bahamas		2012
Barbados		1968
Belize	2023	
Canada		1965
Chile		1968
Cuba	1973	
Dominica	1966	
El Salvador	2021	
Grenada	1962	
Jamaica	1966	
Paraguay	2018	
Saint Kitts and Nevis		2012
Saint Lucia	1962	
Saint Vincent and the Grenadines		2012
Trinidad and Tobago	1965	
United States of America	1970	
Uruguay		2012
SOUTH-EAST ASIA		
Maldives	2015	
Sri Lanka	2016	
WESTERN PACIFIC		
Australia	1981	
Brunei Darussalam	1987	
China	2021	
Cook Islands		1963
Fiji		1963
Japan		2012
Kiribati		2012

(Continued)

TABLE 1.1
(Continued)

Country/Territory	Countries Certified Malaria-Free[a,b,c,d,e]	Countries Where Malaria Never Existed or Disappeared Without Specific Measures[f]
Marshall Islands		1963
Micronesia (Federated States of)		1963
Mongolia		1963
Nauru		1963
New Zealand		1963
Niue		1963
Palau		1963
Samoa		1963
Singapore	1982	
Tonga		1963
Tuvalu		2012

Notes:
[a] Until 1987, the register was known as the "WHO Official Register of Areas Where Malaria Eradication Has Been Achieved."
[b] For the purpose of this publication, reference is made to the official name of WHO Member States as of 11 June 2018.
[c] The Bolivarian Republic of Venezuela (northern) was certified as malaria-free in June 1961.
[d] Taiwan (China) was certified malaria-free in November 1965.
[e] La Réunion is a French overseas region which was certified malaria-free independently from Metropolitan France.
[f] These countries are added to the Supplementary list.

Malaria is a treatable infection; some highly effective antimalarials are available to treat while malaria vaccines such as RTS,S/AS01 (RTS,S) and R21/Matrix-M vaccine, which have been already permitted in a few African nations by the World Health Organization since 2021, are now available to prevent onset of the disease and lessen the burden of the disease. India has strategized its malaria elimination plan to become malaria-free by 2030.

At present, malaria is generally prevalent in tropical and subtropical countries but more prominently in the regions of sub-Saharan Africa, South Asia, and Latin America (Maurice and Pearce, 1987; Greenwood, 1999; Tyagi, 2002, 2003; WHO, 2020; Figure 1.1). Although malaria is a universal malaise, nevertheless, the highest disease prevalence and deaths (90%) occur in the poorest countries (May, 1951). Initially, malaria predominantly remained a rural problem, but with the growth of urbanization, it has of late become equally a serious health problem in the urban agglomerations, on one hand, and, due to extensive anthropization, an emerging threat even in the xeric environment of the Thar Desert (Baeza et al., 2011, 2013; Tyagi, 1994, 2002).

1.1.1 Early History of Malaria

The treatment of malarial fevers was known to the *Ayurved* practitioners or seers several hundred years before the initiation of the Christian Era. '*Suśruta-saṃhitā*,' along with '*Charaka-saṃhitā*', dated 400–500 B.C., is one of the most important surviving ancient treatises on medicine and is considered a foundational text of Ayurveda (Wujastyk, 2003) where all forms of malarial fevers are duly described with their treatment policies. In contrast, the New World's scientific history of malaria dates back only to 1640 A.D. when the Peruvian physician *Juan Del Vega* first used the tincture of the Cinchona tree bark for treating malaria on the Countess of Chinchon of Spain and the wife of Mendoza Chincon, the Viceroy of Peru, South America, and cured her completely (Paidhungat,

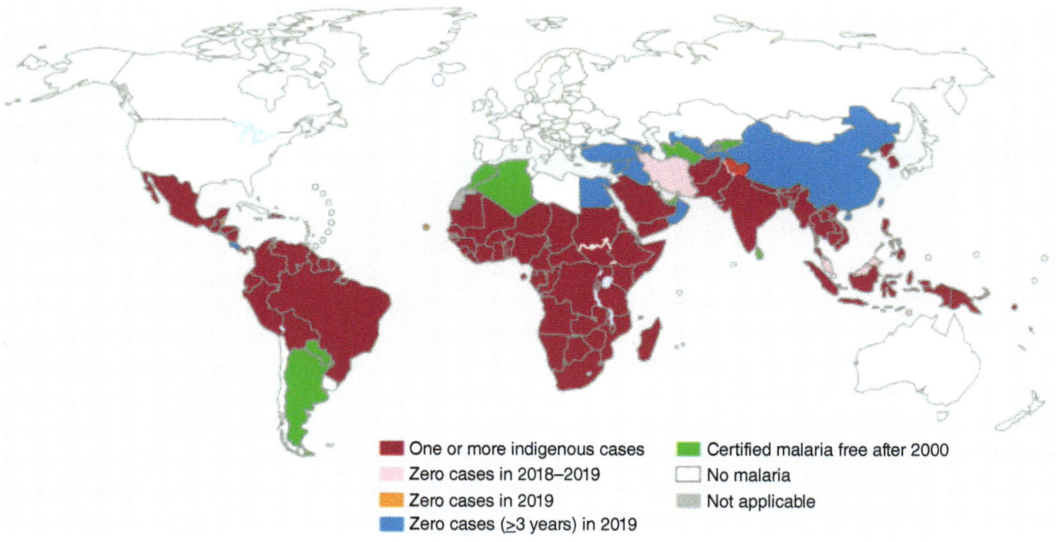

FIGURE 1.1 World map showing countries with indigenous cases in 2000 and their status by 2019. Countries with zero indigenous cases over at least the past three consecutive years are considered to have eliminated malaria. In 2019, China and El Salvador reported zero indigenous cases for the third consecutive year and have applied for WHO certification of malaria elimination; also, the Islamic Republic of Iran, Malaysia, and Timor-Leste reported zero indigenous cases for the second time.

WHO, 2020, auto permitted with due acknowledgement in the book.

2001). As described by Greenwood (1992) and Kaufman et al. (2005) Thompson in 1650 introduced this 'Jesuits' bark' to England: its first recorded use there was by Dr John Metford of Northampton in 1656. Morton in 1696 presented the first detailed description of the clinical picture of malaria and of its treatment with cinchona. In 1816, Gize studied the extraction of crystalline quinine from the cinchona bark, and in 1820 in France, Pelletier and Caventou extracted pure quinine alkaloids, which they named quinine and cinchonine.

1.1.2 The Term: Malaria

The name *malaria*, from the Italian *mala aria*, meaning bad air, comes from the linkage, suggested by Lancisi in 1717, of malaria with the poisonous vapours of swamps (Hempelmann and Krafts, 2013). This species name comes from the Latin *falx* meaning sickle, and *parere* meaning to give birth. In 1847, the German scientist Meckel saw brownish pigments in leukocytes of malaria patients and established its association with the disease. The organism itself was first seen by Laveran on 6 November 1880 at a military hospital in Constantine, Algeria, when he discovered a microgametocyte exflagellating. Manson (1894), who had earlier found out a definitive relationship between lymphatic filariasis worm, *Wuchereria bancrofti*, and the mosquito, *Culex pipiens fatigans*, in China, suggested his hypothesis on the same lines for malaria to Dr Ronald Ross, who during the course of his duties in the British Indian Army Medical Service at Secunderabad (India) demonstrated that mosquitoes transmitted malaria (Ross, 1897). Almost simultaneously, Prof Giovanni Battista Grassi, an Italian zoologist and expert on insect taxonomy, provided undeniable proof of malaria parasite completing sexual cycle of biology in anopheline mosquitoes, *Anopheles sacharovi*, as well as *An. stephensi* (Grassi, 1898). Soon Grassi (1900) also proposed an exoerythrocytic stage in the life cycle of the malaria parasite, later confirmed by Shortt et al. (1948), who found *Plasmodium vivax* in the human liver known as 'hypnozoites.'

1.1.3 Definition

Malaria (məˈlɛːrɪə) is defined as an infectious disease caused by protozoan parasites from the Plasmodiidae family that can be transmitted naturally by the bite of the *Anopheles* mosquito or unnaturally by a contaminated needle or transfusion. Falciparum malaria caused by *P. falciparum* is the deadliest type of the disease. Indian literature of yore has several irrefutable references to both malaria as '*durvāta*' (दुर्वात) or '*sheet-jwar*' (शीत-ज्वर) and various different types of mosquitoes involved in the transmission of malaria and other maladies to man (Satvalekar, 1958; Rao, 1984; Tyagi, 2020). Among the many names for malaria are ague, jungle fever, marsh, or swamp fever, and (in French) *Paludisme*.

1.1.4 Symptoms of Malaria

These include cycles of chills, fever, sweats, muscle aches, and headache that recur every few days. There can also be vomiting, diarrhoea, coughing, and yellowing (jaundice) of the skin and eyes. Persons with severe falciparum malaria can develop bleeding problems, shock, kidney and liver failure, central nervous system problems, coma, and die.

After an infected mosquito bite to a healthy man, signs and symptoms of the malaria disease typically begin within a few weeks. However, in some types of malaria such as vivax and quartan malaria caused by *Plasmodium vivax* and *P. malariae*, respectively, symptoms of malaria may recur after several weeks or even months since the parasite can stay dormant in the liver for a long time until favourable conditions return.

1.1.5 Types of Malaria Parasites

The genus *Plasmodium* (Marchiafava and Celli, 1885), harbours close to 200 species of which as many as 11 have been documented from humans (Igweh, 2012), and yet only five species of *Plasmodium* are responsible for human malaria (Cox Francis, 2010), viz., *P. vivax* (Grassi and Feletti, 1890); *P. falciparum* (Welch, 1897); *P. malariae* (Feletti and Grassi, 1889); *P. ovale* (Stephens, 1922), and *P. knowlesi* (Sinton and Mulligan, 1932). They are concisely defined as follows:

i. *Plasmodium vivax*: This is the most common species and the malaria caused by this species is known as vivax malaria or benign tertian. This kind of malaria is characterized by 'relapse.' This is prevalent in India as the *numero uno* parasite.
ii. *Plasmodium falciparum*: This species is most dangerous and often fatal. It is the second commonest species after *P. vivax*. The malaria due to this species is known as falciparum malaria, malignant tertian, subtertian, quotidian, malignant, pernicious, or tropical aestivo-autumal. This prevails in India and is a major concern for malaria elimination in the country.
iii. *Plasmodium malariae*: This species is scattered throughout the world and responsible for the malaria commonly known as quartan malaria. This is a rare species in India found in the tribal areas.
iv. *Plasmodium ovale*: This is the rarest among all forms of human malaria and for all practical purposes, it is similar to *vivax*. The malaria due to this species is termed as ovale malaria or ovale tertian malaria. Two genetically divergent sub-species *P. ovale curtisi* and *P. ovale wallikeri* are reported from tribal areas in India. The ovale tertian malaria is endemic to tropical Western Africa, although the parasite *P. ovale* has been rarely reported in Odisha (erstwhile Orissa; Jambulingam et al., 1989; Bal et al., 2021). It has also been reported from the tribal area of Chhattisgarh (Krishna et al., 2017).
v. *Plasmodium knowlesi*: Originally known to cause simian malaria, this species is now recognized as the fifth human malarial parasite. This human malaria infection has been reported in nearly all countries, including India, in Southeast Asia.

1.1.6 Classification and Identification of Malaria Parasites

The human malaria parasite is a protozoan. It infects the red blood corpuscles and pass through an asexual reproduction phase in humans. All human malaria parasites are classified as follows:

Kingdom — Protozoa
Subkingdom — Baciliata
Infrakingdom — Alveolata
Phylum — Myzozoa
Subphylum — Apicoplexa
Class — Aconoidasida
Order — Haemosporinia
Family — Plasmodiidae
Genus — Plasmodium
Species — *P. falciparum/P. vivax/P. ovale/P. malariae/P. knowlesi*

These malaria parasites are morphologically distinguishable as depicted in Table 1.2.

TABLE 1.2
Distinguishing Attributes of Four of the Five Human Malaria Parasites

Plasmodium Species	Stages Found in Blood	Appearance of Erythrocytes (RBC)	Appearance of Parasite
P. falciparum	Ring	normal; multiple infection of RBC more common than in other species	delicate cytoplasm; one to two small chromatin dots; occasional applique (accolle) forms
	Trophozoite	norma; rarely "Maurer's" clefts (under certain staining conditions)	seldom seen in peripheral blood; compact cytoplasm; dark pigment
	Schizont	norma; rarely "Maurer's" clefts (under certain staining conditions)	seldom seen in peripheral blood; mature = 8–24 small merozoites; dark pigment, clumped in one mass
	Gametocyte	distorted by parasite	crescent or sausage shape; chromatin in one single mass (microgametocyte) or diffuse (microgametocyte); dark pigment mass
P. vivax	Ring	normal to 1¼ X; round, occasionally fine Schuffner's dots; multiple infection of RBC not uncommon	large cytoplasm with occasional pseudopods; large chromatin dot
	Trophozoite	enlarged 1½–2 X; may be distorted; fine Schuffner's dots	large ameboid cytoplasm; large chromatin; fine, yellowish-brown pigment
	Schizont	enlarged 1½–2 X; may be distorted; fine Schuffner's dots	large, may almost fill RBC; mature =12–24 merozoites; yellowish-brown, coalesced pigment
	Gametocyte	enlarged 1½–2 X; may be distorted; fine Schuffner's dots	round to oval; compact; may almost fill RBC; chromatic compact, eccentric (macrogametocyte) or diffuse (microgametocyte); scattered brown pigment

(*Continued*)

TABLE 1.2
(Continued)

Plasmodium Species	Stages Found in Blood	Appearance of Erythrocytes (RBC)	Appearance of Parasite
P. ovale	Ring	normal to 1¼ X; round to oval; occasionally Schuffner's dots; occasionally fimbriated; multiple infection of RBC not uncommon	sturdy cytoplasm; large chromatin
	Trophozoite	normal to 1¼ X; round to oval; some fimbriated; Schuffner's dots	compact with large chromatin; dark-brown pigment
	Schizont	normal to 1¼ X; round to oval; some fimbriated; Schuffner's dots	mature = 6–14 merozoites with large nuclei, clustered around mass of dark-brown pigment
	Gametocyte	normal to 1¼ X; round to oval; some fimbriated; Schuffner's dots	round to oval; compact; may almost fill RBC; chromatic compact, eccentric (macrogametocyte) or more diffuse (microgametocyte); scattered brown pigment
P. malariae	Ring	normal to ¾ X	sturdy cytoplasm; large chromatin
	Trophozoite	normal to ¾ X; rarely, Ziemann's stippling (under certain staining conditions)	compact cytoplasm; large chromatin; occasional band forms coarse dark-brown pigment
	Schizont	normal to ¾ X; rarely, Ziemann's stippling (under certain staining conditions)	mature = 6–12 merozoites with large nuclei; clustered around mass of dark-brown pigment; occasional rosettes
	Gametocyte	normal to ¾ X; rarely, Ziemann's stippling (under certain staining conditions)	round to oval; compact; may almost fill RMC; chromatin compact, eccentric (macrogametocyte) or more diffuse (microgametocyte); scattered brown pigment

Source: CDC, https://www.cdc.gov/dpdx/resources/pdf/benchaids/malaria/malaria_comparison_p1-2.pdf.

1.2 LIFE CYCLE OF MALARIA PARASITES

The disease is transmitted by the bite of an infective female anopheline mosquito. When inoculated, the parasites are transferred from man to man via the vector mosquito. Male mosquitoes cannot bite or feed on blood. To affect the transmission of malaria, the parasite undergoes two distinct cycles of development: (a) the human cycle (Schizogony): Parasites multiply exponentially, through asexual reproduction, in the liver and then in infected red blood cells. Man represents the 'intermediate host', and (b) the mosquito cycle (Sporogony): Mosquitoes ingest parasites with a blood meal, in whose gut the parasites undergo another reproductive phase, through sexual reproduction, before being passed on to another human. The mosquito acts as the 'definitive host' (Figure 1.2).

The two distinct phases of reproductive cycles are defined briefly as follows:

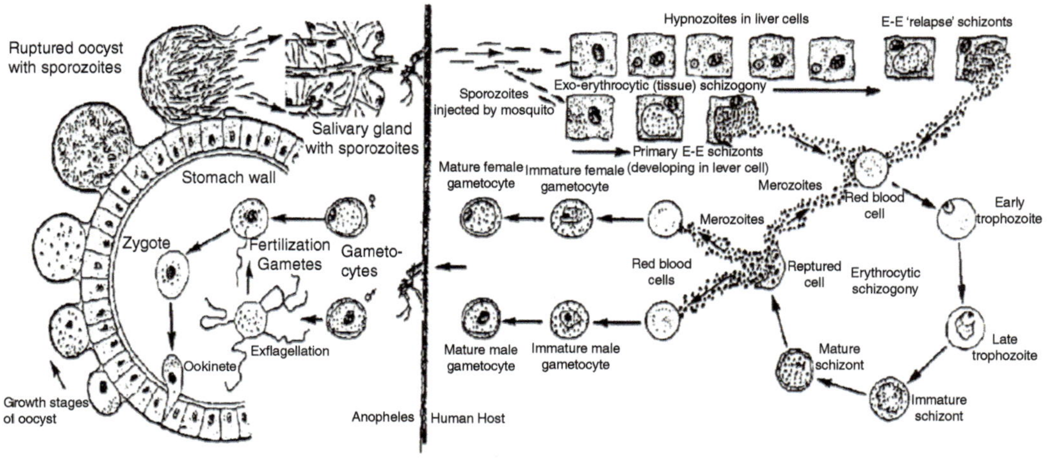

FIGURE 1.2 The life cycle of malaria parasites in the mosquito (sexual phase) and in the human host (asexual phase) according to present views on the exo-erythrocytic schizogony.

Wikipedia.

1.2.1 The Human Cycle

Human is the intermediate or secondary host where the asexual cycle of parasites in human is initiated when an infected mosquito bites a person and injects sporozoites. These tiny, motile forms of the parasite circulate in the blood stream for less than an hour before disappearing from the peripheral blood. While some of the sporozoites are able to make it to the liver cells (hepatocytes), many are destroyed through phagocytosis by the white blood cells. In the liver, following one to two weeks of development (depending upon a particular malarial parasite species), they develop to form hepatic schizonts. On rupture, the schizonts release thousands of very tiny merozoites (a single schizont of *P. falciparum* may result in 40,000 merozoites, 3–20 times higher than in other species). The human liver offers a safe sanctuary for malaria parasites. The schizonts burst out to release merozoites almost synchronously in *P. falciparum*, without leaving behind any persistent tissue phase or the exoerythrocytic phase, as found in all other species as a result of nonsynchronous rupturing of the hepatic schizonts, which later develop into exoerythrocytic schizonts after undergoing pre-erythrocytic schizogony. This phenomenon carries tremendous pathological significance, as they release merozoites into the blood stream, causing relapses of these infections (in case of *P. vivax* and *P. ovale*, these relapses may last up to 2–3 years, while *P. malariae* may persist up to 10–20 years or more!). Relapses in *P. malariae* are said to be due to its persistence in the blood rather than in the liver. However, once the parasites enter the red blood cells (RBCs), they do not reinvade the liver at all. This phase of development of the malarial parasite associated with the liver is called as the 'hepatic phase.'

Again, many merozoites become victim to the waiting phagocytes, but still quite many are able to infest the RBC by attaching to specific receptor sites. At this junction, following a very intricate and interesting mechanism involving dissolution of the cell wall of the RBC, the merozoites penetrate the host RBC. Once inside the merozoites develop to transform into various different forms called mainly the trophozoite and schizont. The latter, under pressure of growth and development of enclosed merozoites, burst to liberate the tiny merozoites to infect afresh the RBCs. This phase of development of the malarial parasite associated with the RBCs is called as the 'erythrocytic phase.' The cycle is repeated over and over again until it is slowed down by the immune response of the host or the drugs or the death of the host or parasites. Each species of malarial parasite is characterized by a definite duration of the erythrocytic cycle; *P. falciparum, P. vivax*, and *P. ovale* have 48 hrs., while *P. malariae* has 72 hrs.

Having undergone erythrocytic schizogony for some time, some of the merozoites, instead of dividing and further developing into trophozoites/schizonts, actually become male and female gametocytes – the sexual forms of the parasite which are infective to mosquito. This phase of development of parasite into gametocytes is called as 'gametogony.' Peripheral blood reveals only the mature gametocytes. Gametogony is completed in four days and, therefore, a person carrying these gametocytes is often referred to as the 'carrier' because by blood feeding on such a carrier, only a mosquito becomes infected.

1.2.2 The Mosquito Cycle

The mosquito is the definitive host where sexual cycle, or sporogony occurs. The cycle begins with the ingestion of the gametocytes with the blood meal by the vector mosquito. The gametocytes continue further development in the body of the mosquito. The gametogony, with formation of gametocytes, starts about ten days, on an average, after the malarial paroxysm has started in man. The male and female gametocytes are called microgametocyte and macrogametocyte, respectively. The presence of at least 12 gametocytes, i.e., both mature male and female sexual parasites, with more macrogametocytes than their counterparts, per cubic millimetre of blood is absolutely imperative for further development of the parasite's sexual cycle in the mosquitoes. The first event to take place in the gut of the mosquito is 'exflagellation' of the microgametocytes, whereby four to eight thread-like processes called microgametes are formed. This phenomenon of exflagellation can also be observed outside the mosquito host in a moist chamber preparation of peripheral blood. The macrogametocyte, on the other hand, undergoes a simple process of maturation to transform into a female gamete or macrogamete. Through a process of chemotaxis, microgametes, which, after detachment from the parent cell, actively wander around, are attracted to the female gamete. As a rule, only one microgamete penetrates and resultantly fertilizes the female gamete, forming a zygote. The whole process of development is completed within 20 minutes, depending on ambient temperature, etc. The resultant zygote is at first a motionless body but soon acquires a change in its form to become oblong 'ookinete' and begins moving within 18–24 hrs. The ookinete dissolves the inner wall of the stomach and penetrates through to reach the outer surface of the stomach where it develops into oocyst. Each oocyst measures about 6–12 μm in diameter containing a single vesicular nucleus and pigment granules of the macrogamete. With further development, oocyst increases to a size of about 60 μm in diameter by the successive nuclear divisions containing several hundreds or thousands of sickle-shaped sporozoites. The number of oocysts formed on a gut may vary from 20 to 30. When mature, the oocyst bursts and liberates sporozoites into the body cavity (haemocoel) of the mosquito. Many of the sporozoites migrate to the salivary glands of the mosquito, and the mosquito is now infective to man. More than one malarial parasite can simultaneously develop in the mosquito body, thereby giving ample possibilities of mixed infections, such as often observed in case of mixed *P. vivax–P. falciparum* infection. The period of time required for the development of the parasite from the gametocyte to sporozoite in the body of the mosquito is about 8–21 days (*P. vivax* 8–10 days at 25°C, *P. ovale* 14–18 days at 25°C, *P. falciparum* 10–12 days at 30°C, and *P. malariae* 18–21 days at 30°C), depending of course on favourable conditions of atmospheric temperature and humidity. This period is also referred to as the 'extrinsic incubation period.'

1.3 VECTOR MOSQUITOES

Only the female vector mosquitoes suck the blood of animals and humans, the males feed on plant juices. The *Anopheles* mosquitoes are characterized mainly by spotted wings and their sitting posture making approximately 45° angle from the surface. There are ten species of anophelines which are known to transmit malaria in India and are called malaria vectors. Out of these, six are primary vectors and four are secondary. The primary vectors are *An. culicifacies, An. stephensi, An. fluviatilis, An. minimus, An. Bimaii*, and *An. sundaicus* (now known as *An. epiroticus*). The secondary vectors are *An. philipinensis, An. varuna, An. Annularis*, and *An. nivipes* (Christophers, 1933; Wattal and Kalra, 1961; Rao, 1984; Nagpal and Sharma, 1995; Subbarao et al., 2019; Tyagi, 2020, 2023).

Major malaria vector species are annotated below.

1.3.1 *Anopheles culicifacies*

Anopheles culicifacies is widely distributed in India, except on Andaman and Nicobar Islands and Lakshadweep. It is predominantly a major vector in rural and peri-urban areas. A complex of species, it comprises five sibling species, viz., A, B, C, D, and E. Species A is more anthropophagic than others, and therefore it is considered a relatively efficient vector. Species B is known to be refractory to *P. vivax* completely but partially to *P. falciparum*. Species B may play a role as a vector of *P. falciparum* in areas with less cattle population (Ghosh et al., 2008). Species E is found in certain locations with low cattle population and behaviourally has propensity for human biting.

The breeding sites of *An. culicifacies* are innumerable, viz., rainwater pools, puddles, borrow pits, riverbed pools, irrigation channels, seepages, rice fields, wells, pond margins, streams, etc. The breeding is generally high following monsoon rains. The species prefers to rest in human dwellings and cattle sheds during daytime. It tends to bite throughout the night, but the peak biting time recorded is between 19:00 and 04:00. It is mainly zoophilic but feeds on humans when in high density. The flight range recorded for *An. culicifacies* is about 1–3 km. It is resistant to DDT in many states but continues to be susceptible to malathion and synthetic pyrethroids.

In India, *An. culicifacies* is the most important and dominant rural vector of malaria, though predominantly incriminated in the plains of north-western and peninsular India, including the southern states of Tamil Nadu, Andhra Pradesh, and Karnataka. The species is responsible for nearly 65% of malaria transmission in India.

1.3.2 *Anopheles stephensi*

Anopheles stephensi is distributed throughout India, except at higher altitudes, but in the north-east, it is found sporadically. It causes malaria in urban and industrial areas. It has three biological variants, viz. (i) type, (ii) intermediate, and (iii) mysorensis based on egg-float ridge numbers. The type found in urban areas; intermediate form in urban and semi-urban localities and *mysorensis* form in rural areas. Both 'type' and 'intermediate' forms act as vectors, whereas the '*mysorensis*' form is not a vector in India. Tyagi et al. (1991) proposed the presence of a possible new sub-species or even a sibling species of *An. stephensi* in the Thar Desert. It is noteworthy that Firooziyan et al. (2018) have recently speculated on the possibility for introducing *An. stephensi* as a species complex based on odorant binding protein 1 intron I sequence. Most importantly, *An. stephensi* has invaded Sri Lanka and several African countries, and raised alarm for future outbreaks of malaria (Sinka, 2013; Sinka et al., 2020; WHO, 2022a; Tyagi, 2023).

An. stephensi breeds in overhead and ground level water tanks, as well as all kinds of manmade potable systems like cisterns, peri-domestic containers, wells, rainwater collections in roof gutters, and underground water storage tanks. In the Thar Desert of Rajasthan, it predominantly breeds in underground 'tanka' and 'beri' (Tyagi and Yadav, 1996; Tyagi, 2002, 2023). The mosquito is both anthrophagic and zoophagic, as it rests in both houses and cattle sheds. Inside houses, it may rest on hanging objects, behind curtains, grooves in roofs, etc. The orientation time for biting of *An. stephensi* varies from area to area and according to the prevailing season, but peak biting activity is generally from 22:00 to 24:00. The species has limited flight range in the urban areas, but in rural areas the flight range may be up to 3 km. Though adults are partially susceptible to DDT, the larvae are susceptible to organophosphorus larvicides (e.g., temephos) used under the Urban Malaria Scheme in India.

An. stephensi is the most important vector of urban malaria the world over, while transmission may also be supplemented by *An. culicifacies* in peri-urban (semirural) areas. It is also a vector of rural malaria in Rajasthan. It has been responsible for malaria epidemics in urban areas and focal outbreaks in project areas and construction sites. Antilarval measures can control the vector population effectively.

1.3.3 *Anopheles fluviatilis*

This species is widely distributed in the foothill areas of India. It plays the role of the main vector in hilly, forested, and forest fringe areas. It is a complex of four sibling species known as S, T, U, and V. Among all the species, S is highly anthropophagic (peak biting periods being 21:00–23:00) and thus is an efficient vector. Though characteristically endophilic and anthropophilic, yet the species is also found to be zoophilic and zoophagic in northern India, implying that *An. fluviatilis* indiscriminately rests in both human dwellings and cattle sheds alike. The species prefers to breed in slow running streams, seepages, and irrigation channels; however, it has also been reported in rice fields and shallow wells.

Aopheles fluviatilis is highly susceptible to DDT, malathion, and synthetic pyrethroids. Being a main vector of malaria in foothills areas, this species is responsible for extended transmission in conjunction with *An. culicifacies*.

1.3.4 *Anopheles minimus*

The species has a restricted distribution in the north-eastern states, although it was nearly eliminated during the 1950s and 1960s due to large-scale spraying of insecticides throughout the country. However, it reappeared during late 1970s and became an efficient vector in the foothill areas of the north-east states. It prefers to breed in shaded slow running streams with grassy margins, swamps, ditches, and shallow wells. It has also been found to breed in borrow pits, rice fields, and slow running seepage water.

Anopheles minimus rests in both houses and cattle sheds, and prefers to rest on the lower part of the wall. A highly anthropophilic species, and consequently a very efficient vector of malaria, it generally orients to bite and blood-feed outdoors (18:00–19:00), as well as indoors (24:00–02:00), although the biting time may vary from locality to locality and seasonality. The species can normally fly for 0.5 km but, by taking help of the wind, it can disperse up to 2 km from the original locality. The species is highly susceptible to DDT and other insecticides.

Anopheles minimus is a major vector of malaria in north-east India and is responsible for maintaining perennial transmission with other vectors.

1.3.5 *Anopheles baimaii*

Anopheles baimaii, a member of the multiple-vector system in north-east states maintaining perennial malaria transmission, is a primary malaria vector in the forest and forest fringe areas. It mainly has a distribution in the forested areas of the north-eastern states where it breeds in pools, stream margins with decaying organic matter, and rainwater collections, besides animal footprints during high monsoon. The species enters human dwellings just for biting and rest for a short period but tends to leave houses soon after blood meal is over. The peak biting activity is from 22:00 to 02:00. It has high preference for human blood but also bites monkeys and other primates, as well as cattle. It is well known for its exophilic behaviour, with a flight range varying from 1 to 2.5 km in forests.

An. baimaii is susceptible to DDT and other insecticides used in the control programme.

1.3.6 *Anopheles sundaicus* (Now Known as *An. epiroticus*)

Mostly prevalent on the Andaman and Nicobar Islands where it is responsible for malaria transmission, *An. sundaicus* has also been reported now and then from the coastal mainland India, e.g., Odisha, Andhra Pradesh, and West Bengal. In Odisha, it was reported characteristically breeding in the brackish water of the Chilka Lake, as on the Andaman and Nicobar Islands. It breeds in the streaks with brackish water pools rife with algae, margins of mangroves, and lagoons, as well as swampy water collections. It can tolerate salinity of 0.08– 2.6% and pH 7.7–8.5.

It rests indoors in human dwellings, cattle sheds, and mixed dwellings. An opportunistic feeder, it prefers to bite man as the biting goes on throughout the night, but peak biting is from 20:00 to 02:00. Its flight range is short and is restricted to about 1–3 km. The species is susceptible to DDT and synthetic pyrethroids.

1.3.7 *Anopheles annularis*

Anopheles annularis is widely reported from India, except on Andaman and Nicobar and Lakshadweep Islands. This species characteristically breeds in a variety of stagnant water bodies with abundant vegetation, and also breeds in wells, moats, tanks, borrow pits, rice fields and other water bodies such as lakes and stream margins with vegetation. It usually rests in houses, cattle sheds, and mixed dwellings, but it is also found in small numbers resting outdoors. The peak biting activity of the species takes place between 22:00 and 24:00, and though mainly a zoophilic mosquito, it is also infrequently noticed feeding on humans. Its flight range is small, up to 1 km. The species is DDT resistant, but susceptible to malathion and synthetic pyrethroids. *An. annularis* plays a role of secondary malaria vector in the plains of Odisha, Jharkhand, and West Bengal.

1.3.8 *Anopheles varuna*

It is widely distributed in north-east plains, central peninsular India, and Lakshadweep islands where breeding is commonly observed in rainwater pools, ponds, tanks, rice fields, drains, irrigation channels, wells, and slow-moving streams with shade. An endophilic species rests indiscriminately in all human dwellings, cattle sheds, and mixed dwellings. However, it has also been found resting outdoors near banks of streams. The species is observed to bite throughout the night, but its peak biting period is between 24:00 and 02:00. It can fly up to 1 km. The species is susceptible to DDT, malathion, and synthetic pyrethroids. It is also a secondary vector of malaria in Andhra Pradesh, Jharkhand, and Odisha.

1.3.9 *Anopheles philippinesis*

An important malaria vector mostly in deltaic West Bengal and north-east states of India. *An. philippinensis* is distributed in the state of West Bengal, north-eastern states, and Andaman and Nicobar Islands. Its common breeding places include tanks, swamps, ditches, rice fields, pools, leaf axils, shaded lake margins, inundated drains, and water bodies with generally good growth of vegetation. Adults characteristically rest in cattle sheds and human dwellings during daytime. Predominantly a zoophagic species, it prefers to bite both outdoors and indoors throughout the night, with two biting peaks from 20:00 to 22:00 and 02:00 to 04:00. It can fly up to 0.8 km. This species is susceptible to all insecticides in use in public health.

1.4 ESTABLISHMENT OF ISO-FEMALE LINES IN VECTOR MOSQUITOES

Several vector mosquito species exhibit either sibling species or biological variants. For example, *An. culicifacies* has five sibling species A, B, C, D, and E; *An. fluviatilis* has four sibling species S, T, U, and V. On the other hand, *An. stephensi* has three biological variants – namely, type, intermediate, and *mysorensis*. It is important to identify each sibling species and variants since they have different bionomics, vector competence, and insecticide resistance. Generally, sibling species is identified consisting of analysis of the polytene chromosomes in ovarian nurse cells. The biological variants of *An. stephensi* is identified based on the ridge numbers on the egg-floats. Such methods are very cumbersome and sampling complex. Isofemale lines of type and intermediate forms were developed for the first time to obtain more homogenized population from which the reference genome was established (Ghosh et al., 2023; Thakare et al., 2022; Srinivasan et al., 2022). To overcome these, molecular techniques, including environmental DNA (eDNA), have been developed, but such methods face many uncertainties to identify with utmost perfection.

1.5 MALARIA PARADIGMS

The malaria situation differs in different eco-environmental situations due to diverse complexity among the behaviour of the population. With malaria being *local* and *focal* in character, control programmes need to stratify their malaria problem into a number of smaller units, keeping in mind the variability of malaria. Such stratification may be based on the epidemiology of malaria or on its determinants, such as ecology and disease transmission capacity of vectors (Schapira and Boutsika, 2012). An ecotypic or paradigmatic classification comprising eight major malaria paradigms to categorize typical disease transmission settings was developed by the World Health Organization around 1990 (WHO, 1993a, 1993b), which was recently modified by adding a new paradigm, 'Desert Malaria' (Tyagi, 2023). The entire list is as follows:

1. Malaria of the African savannah
2. Forest malaria
3. Malaria associated with irrigated malaria
4. Highland fringe malaria
5. Desert malaria
6. Urban malaria
7. Plains malaria
8. Seashore malaria

1.6 MALARIA RESEARCH AND IMPLEMENTATION IN INDIA

India contributed 1.7% of malaria cases and 1.2% deaths globally in the year 2021 (WMR, 2021). India accounted for 79% of malaria cases in the Southeast Asia Region (SEAR). The case load, though steady around two million cases annually in the late '90s, has shown a declining trend since 2002. The reported total number of malaria cases in 2021 was 161,753, including 90 deaths. The reported *Plasmodium falciparum* (*Pf*) cases declined from 1,133,005 in 1995 to 100,442 cases in 2021. The *Pf*% has gradually increased from 39% in 1995 to 63.10% in 2021. Overall, 86.2% decline in malaria cases and 76.6% decline in deaths were noted in 2021 compared to 2015 (NCVBDC Annual Malaria Report, 2021).

1.6.1 History of the National Malaria Programme in India

In 1947, at India's independence, 22% of the country's population was estimated to have suffered from malaria, with 75 million cases and 0.8 million deaths annually. To combat the devastating effects of malaria, the National Malaria Control Programme (NMCP) was launched in 1953 and built around three key activities—insecticidal residual spray (IRS) with DDT, monitoring and surveillance of cases, and treatment of patients. Malaria-related morbidity and mortality in India were rapidly brought down within a few years. Encouraged by the programme's success, it was converted to the National Malaria Eradication Programme (NMEP) in 1958. Following a massive resurgence of malaria in 1976, the Modified Plan of Operations (MPO) was launched in 1977 with a three-pronged strategy: early diagnosis and prompt treatment, vector control, and IEC/BCC (Information, Education and Communication/Behaviour Change Communication) with community participation.

To combat malaria in high transmission areas of the country, the Enhanced Malaria Control Project (EMCP) was launched, with additional support from the World Bank in 1997, and the Intensified Malaria Control Project (IMCP) was launched, with support of the Global Fund to fight AIDS, Tuberculosis, and Malaria (GFATM), in 2005. The malaria control programme and other vector-borne diseases control programmes – namely, kala-azar, dengue, lymphatic filariasis, Japanese encephalitis and chikungunya – were integrated under the umbrella programme for prevention and control of vector-borne diseases. The programme was named the National Vector Borne Disease Control Programme (NVBDCP) in December 2003. Since 2005, NVBDCP has been implemented under the overarching umbrella of the National Rural Health Mission (NRHM), which has now been subsumed under the National Health Mission (NHM), incorporating the National Urban Health Mission (NUHM) as well. In 2021, NVBDCP was renamed the National Centre for Vector Borne Diseases Control (NCVBDC). The Government of India also has 19 Regional Offices for Health and Family Welfare (ROHFW), located in 19 states covering one or more states under their jurisdiction. These ROHFWs play a critical role in monitoring of NCVBDC activities in the respective states. Every state has a Vector-Borne Diseases Control Unit under its State Department of Health and Family Welfare headed by the state programme officer. Each state has a State Health Society and District Health Societies at the state and district levels through which funds are disbursed. At the district level, the vector-borne disease programme, including malaria, is managed by the district vector-borne disease control officer (DVBDCO).

1.6.2 Incidence of Malaria in Recent Years

Incidences of total malaria cases and *Pf* cases from the year 2000 to 2022 are shown in Figure 1.3. Malaria epidemiological data from 2000 to 2022 are given in Table 1.3. There was a drastic reduction in API from 2.09 to 0.12 during the years 2000 to 2021. Death due to malaria also reduced from 932 to 90 over the same years.

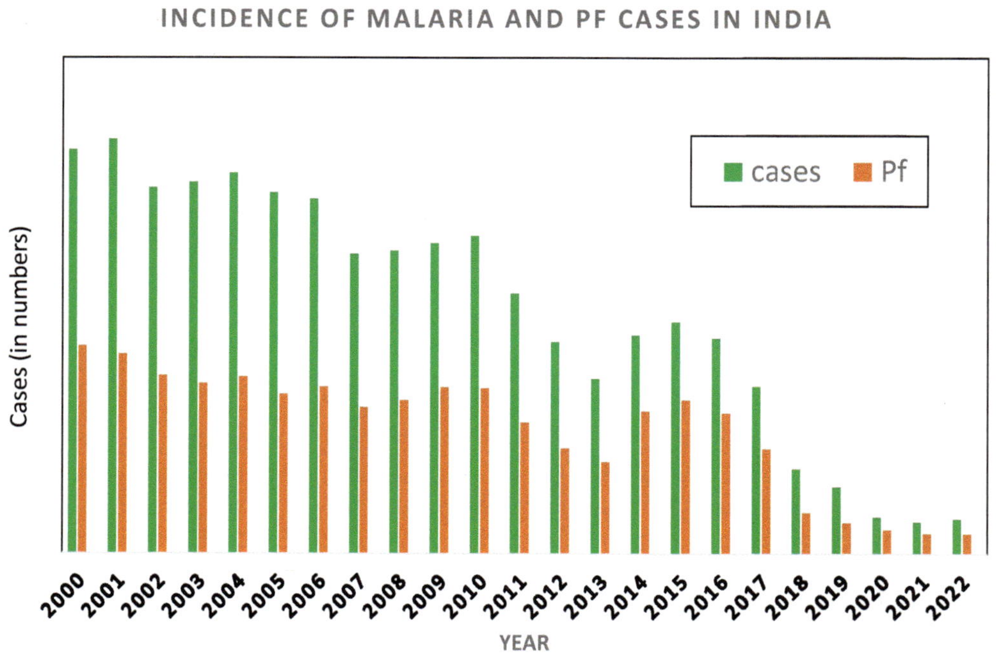

FIGURE 1.3 Incidences of malaria and *Pf* cases.

Developed from NCVBDC website.

TABLE 1.3
Malaria Epidemiological Parameters (2000–2022)

Year	Population (in '000)	Blood Smear Examined	Positive Cases	Pf Cases	Pf%	ABER	API	SPR	AFI	SFR	Deaths
2000	972,148	86,790,375	2,031,790	1,047,218	51.54	8.94	2.09	2.34	1.08	1.21	932
2001	984,579	90,389,019	2,085,484	1,005,236	48.20	9.18	2.12	2.31	1.02	1.11	1,005
2002	1,013,942	91,617,725	1,841,229	897,446	48.74	9.04	1.82	2.01	0.89	0.98	973
2003	1,027,157	99,136,143	1,869,403	857,101	45.85	9.65	1.82	1.89	0.83	0.86	1,006
2004	1,040,939	97,111,526	1,915,363	890,152	46.47	9.33	1.84	1.97	0.86	0.92	949
2005	1,082,882	10,414,3806	1,816,569	805,077	44.32	9.62	1.68	1.74	0.74	0.77	963
2006	1,072,713	106,725,851	1,785,129	840,360	47.08	9.95	1.66	1.67	0.78	0.79	1,707
2007	1,087,582	94,928,090	1,508,927	741,076	49.11	8.73	1.39	1.59	0.68	0.78	1,311
2008	1,119,624	97,316,158	1,526,210	775,523	50.81	8.69	1.36	1.57	0.69	0.8	1,055
2009	1,150,113	103,396,076	1,563,574	839,877	53.72	8.99	1.36	1.51	0.73	0.81	1,144
2010	1,167,360	106,040,223	1,495,817	779,549	52.12	9.21	1.37	1.41	0.67	0.74	1,018
2011	1,194,901	109,313,294	1,310,656	665,004	50.74	9.12	1.1	1.20	0.56	0.61	754
2012	1,211,580	109,048,884	1,067,824	533,695	49.98	9.00	0.88	0.98	0.44	0.49	519
2013	1,221,640	113,445,106	881,730	463,846	52.61	9.26	0.72	0.78	0.38	0.41	440
2014	1,234,995	124,066,331	1,102,205	722,546	65.55	10.05	0.89	0.89	0.59	0.58	562
2015	1,265,173	121,141,970	1,169,261	778,821	66.61	9.58	0.92	0.97	0.62	0.64	384
2016	1,283,303	124,933,348	1,085,823	711,502	65.53	9.74	0.85	0.87	0.55	0.57	331
2017	1,315,092	125,977,799	844,558	529,530	62.70	9.58	0.64	0.67	0.40	0.42	194
2018	1,337,617	124,475,724	429,928	207,198	48.19	9.31	0.32	0.35	0.15	0.17	96
2019	1,349,006	134,230,349	338,494	156,940	46.36	9.95	0.25	0.25	0.12	0.12	77
2020	1,372,316	97,177,024	186,532	119,088	63.84	7.08	0.14	0.19	0.09	0.12	93
2021	1,385,500	114,391,977	161,753	101,566	62.79	8.26	0.12	0.14	0.07	0.09	90
2022	1,396,937	152,083,001	176,522	101,068	57.26	10.89	0.13	0.12	0.07	0.07	83

Source: NCVBDC website.

As a strategy to make India malaria-free, an improvised methodology is being followed to reduced incidence of malaria with a substantial reduction in deaths. As a result, between 2000 and 2021, the annual parasite incidence (API) reduced from 2.09 to 0.13 and deaths from 932 to 83. Admirably, the slide positivity rate (SPR) and annual falciparum incidence (AFI) depleted from 2.34 and 1.08 to 0.12 and 0.07, respectively. The only reason to worry was an escalation in the fatal proportion of Pf% that ascended from 51.54% to 57.26% (Table 1.3), which might be attributed to the significant reduction of overall cases, which is the denominator.

1.7 NATIONAL PROGRAMME FOR CONTROL/ELIMINATION OF MALARIA

Malaria control in India started with antilarval measures way back in 1900. Organochlorine insecticide, DDT, was first introduced into the disease control programme in the 1940s in army camps (Singh, 1962) with considerable success. These sprays were later extended to civilian areas. In the 1940s, large-scale malaria control operations were commissioned in different parts of the country, with spectacular success in the control of malaria, and DDT indoor sprays became the ultimate choice for vector control in the anti-malaria campaign. Prompted by this success, control efforts were extended to the national level by launching a five-year NMCP in 1953 and the targeted population rose to 165 million at the end of the programme, and based on the success of this five-year programme, in the year 1958, the programme was converted to a structured nationwide eradication programme (NMEP presently NCVBDC). During this period, reports of the failure of DDT to effectively control the major vector species *An. culicifacies* started appearing, and later, two insecticides were introduced in 1950s, dieldrin and BHC. BHC was proposed as a replacement insecticide of DDT based on the comparative studies with these insecticides. Malathion was introduced into public health sprays in 1969, and within four years of its introduction, *An. culicifacies* developed resistance to malathion. Resistance to deltamethrin has also been reported within five to six years of its introduction. NMEP was launched with the primary objective to eradicate malaria within a decade before the onset of widespread resistance in this species; this could not be accomplished due to various reasons. In the post-DDT era, due to many administrative, technical, and operational reasons, the success remained short-lived. The cases started increasing in 1965, and the programme witnessed a resurgence of malaria and a record number of cases, 6.45 million, were recorded in 1976, the highest ever recorded in the post-DDT era. Of the many reasons responsible for the resurgence of malaria, insecticide resistance was found to be one of the technical reasons. Later, with the implementation of MPO since 1977, the incidence was brought down to ~2.0 million by 1980 and has been maintained below 2 million (Choudhury, 1985). During the period from 2000 to 2015, cases declined to 1.13 million. Malaria is targeted for elimination by 2030 in the country (Ghosh and Rahi, 2019).

A National Framework for Malaria Elimination (NFME) 2016–30 was launched in February 2016. Subsequently, the National Strategic Plan for Malaria Elimination (2017–22) was developed and disseminated to states because the implementation is done by states and districts. Besides parasite control and cross-cutting interventions, vector control has been the main pillar of malaria control since the beginning (Ghosh and Rahi, 2019).

Malaria control/elimination is now being dealt with through the use of many tools, such as insecticides, biolarvicides, insect growth regulators (IGRs), and long- lasting insecticidal nets (LLINs). The vector control using insecticidal spray in India mainly by indoor residual spray has been the mainstay in the country. The commonly used insecticides are organochlorine insecticide, viz., DDT (now under phasing out), and organophosphate insecticide, viz., malathion and synthetic pyrethroids – namely, deltamethrin, cyfluthrin, lambda cyhalothrin, alphacypermethrin, and bifenthrin. Some of these synthetic pyrethroids are used for impregnation of mosquito nets – insecticide treated nets (ITN) which are used selectively at the community level in some parts of the country as personal protection measures. The ITNs have been replaced with LLINs where the insecticides are impregnated during the manufacturing process and their efficacy lasts through 20 washes of the nets. The major vector control strategies in urban areas are source reduction and larviciding of breeding habitats with temephos, biolarvicides, and IGRs. This is known as larval source management (LSM).

An. culicifacies — the major vector of malaria has developed widespread resistance to the insecticides used in IRS (Yadav et al., 2020) due to continued use of residual insecticides in public health. While other vector species are mostly susceptible to the commonly used insecticides in public health, there are stray reports of resistance in *An. fluviatilis* from different states (Yadav et al., 2020). The resistance status in *An. culicifacies*, the most important vector of malaria in India, is slowly but steadily reaching a point where chemical control with a single insecticide will no longer be a feasible proposition, and the only choice left will be to manage the resistance to increase the useful life of the existing insecticides. The strategy of managing insecticide resistance in disease vectors is also important, as the available insecticides for vector control are limited.

1.8 CONCLUSION

At present, India stands in the final phase of malaria elimination. This means that there are many challenges to achieving this goal by 2030. Reports state that malaria cases were reduced by approximately 86% between 2015 and 2021. Around 0.15 million malaria cases are reported per year, and over 90% are concentrated in 24 districts in 8 states which have significant tribal populations. While these tribal districts contribute only 8.6% of India's population, they account for 50% of malaria cases (mostly *P. falciparum*) and 30% of all malaria-related deaths. On the other hand, *P. vivax* cases are found outside these tribal districts. Compared to *P. falciparum* malaria, *P. vivax* exhibits a complex human malaria due to its relapse-causing hypnozoites in the liver. Primaquine – an eight aminoquinoline compound – is used to kill these hypnozoites. Several policy questions remain on the usage and dosage of primaquine, as well as the use of another anti-relapse drug, tefenoquine. At the same time, keeping in mind the huge impact made by LLINs, the use and replacement of old LLINs is proving another challenge to tackle (Ghosh, 2023).

REFERENCES

Baeza, A., Bouma, M.J., Dhiman, R.C., Baskervillea, E.B., Ceccatod, P., Yadav, R.S. and Pascuala, M. (2013). Long-lasting transition toward sustainable elimination of desert malaria under irrigation development. *Proceedings of the National Academy of Sciences, USA*, 110(37): 15157–15162.

Baeza, A., Bouma, M.J., Dobson, A.P., Dhiman, R., Srivastava, H.C. and Pascual, M. (2011). Climate forcing and desert malaria: The effect of irrigation. *Malaria Journal*, 10: 190. https://doi.org/10.1186/1475-2875-10-190

Bal, M., Rana, R., Das, A., Khuntia, H.K., Somalkar, N., Ghosal, J., Pati, S., Dutta, A. and Ranjit, M. (2021). Neglected malaria parasites in hard-to-reach areas of Odisha, India: Implications in elimination programme. *Malaria Journal*, 20: 482. https://doi.org/10.1186/s12936-021-04010-8

Bruce-Chwatt, L.J. (1985). *Essential Malariology*. 2nd ed., Heinemann: London, 452 pp.

Centre for Disease Control and Prevention, USA (CDC). (n.d.) https://www.cdc.gov/dpdx/resources/pdf/benchAids/malaria/Malaria_Comparison_p1-2.pdf#:~:text=Numerous%3A%20PF%20Multiply%20infected%20RBCs%3A%20PF%20Accessory%20chromatin,PF%20Delicate%3A%20PF%20Trophozoites%20Schizonts%20%28mature%29%20Ameboid%3A%20PV

Choudhury, D.S. (1985). Malaria in India: Past, present and future. *Indian Journal of Pediatrics*, 52: 243–248.

Christophers, S.R. (1933). *The fauna of British India, including Ceylon and Burma. Diptera: Family Culicidae; Tribe Anophelini*. Taylor & Francis, London.

Cox Francis, E.G. (2010). History of the discovery of the malaria parasites and their vectors. *Parasites & Vectors*, 3: 5. http://www.parasitesandvectors.com/content/3/1/5

Dev, V. (2020). Mosquito vectors of human malaria and geographical distribution, pp. 3–8. In: *Vector Biology and Control – An Update for Malaria Elimination Initiative in India*, (Ed.: Vas Dev). National Academy of Sciences, Allahabad, India. http://www.nasi.org.in/BCIL%20Vector%20Biology%20PDF_%2014th%20Aug%202020.pdf

Firooziyan, S., Dinparast, Djadid N. and Gholizadeh, S. (2018). Speculation on the possibility for introducing *Anopheles stephensi* as a species complex: Preliminary evidence based on odorant binding protein 1 intron I sequence. *Malaria Journal*, 17(1): 366. https://doi.org/10.1186/s12936-018-2523-y

GBD (2017). Disease and Injury Incidence and Prevalence Collaborators. Global, regional, and national incidence, prevalence, and years lived with disability for 354 diseases and injuries for 195 countries and territories, 1990–2017: A systematic analysis for the Global Burden of Disease Study 2017. *The Lancet*, 392: 1789–1858.

Ghosh, C., Kumar, N., Kushwah, R.B.S., Joshi, S.G., Ramanjini, C.K., Alalamath, T., Srinivasan, S., Subramani, S., Kumar, S., and Swain, S. (2023). Enrichment of phenotype among biological forms of *Anopheles stephensi* Liston through establishment of isofemale lines. *Parasites and Vectors*, 16(1): 79. https://doi.org/10.1186/s13071-023-05696-2

Ghosh, S.K. (2023). Hammering the last nail on the malaria coffin in India: It's time to address the grey areas. https://www.indiahealthfund.org/hammering-the-last-nail-on-the-malaria-coffin-in-india-its-time-to-address-the-grey-areas/

Ghosh, S.K. and Rahi, M. (2019). Malaria in India – The way forward. *Journal of Vector Borne Diseases*, 56: 32–40.

Ghosh, S.K., Tiwari, S.N., Raghavendra, K., Sathyanarayan, T.S. and Dash, A.P. (2008). Observations on sporozoite detection in naturally infected sibling species of the *Anopheles culicifacies* complex and variant of *Anopheles stephensi* in India. *Journal of Biosciences*, 3: 333–336.

Grassi, B.G. (1898). Rapporti tra la malaria e peculari insetti (zanzaroni e zanzare palustri). *C R Acad Linnei* 7: 163–177.

Grassi, B.G. (1900). *Studi di uno Zoologo Sulla Malaria*. Rome.

Greenwood, B. (1999). Malaria mortality and morbidity in Africa. *Bulletin of the World Health Organization*, 77: 617–618.

Greenwood, D. (1992). The quinine connection. *Journal of Antimicrobial Chemotherapy*, 30(4): 417–427.

Gupta, I. and Choudhury, S. (2014). Economic burden of malaria in India: The need for effective spending. *WHO South-Eat Asia Journal of Public Health*, 3: 95–102.

Harrison, G. (1978). *Mosquitoes, Malaria, and Man: A History of the Hostilities since 1880*. John Murray Publishers Ltd., 320 pp.

Hehir, P. (1927) *Malaria in India*. Humphrey Milford, Oxford University Press, London, 490 pp.

Hempelmann, E. and Krafts, K. (2013). Bad air, amulets and mosquitoes: 2,000 years of changing perspectives on malaria. *Malaria Journal*, 12: 232. https://doi.org/10.1186/1475-2875-12-232

Igweh, J.C. (2012). Biology of Malaria Parasites. (In: O.O. Okwa (Ed.) *Malaria Parasites*. London, UK, IntechOpen [Online]. Available from: https://www.intechopen.com/books/971; https://doi.org/10.5772/1477

Jambulingam, P., Mohapatra, S.S.S., Das, L.K., Das, P.K. and Rajagopalan, P.K. (1989). Detection of *Plasmodium ovale* in Koraput district, Orissa state. *Indian Journal Medical Research*, 89: 115–116.

Kaufman, Teodoro S., Rúveda and Edmundo, A. (2005). The Quest for Quinine: Those Who Won the Battles and Those Who Won the War. *Angewandte Chemie International Edition*, 44(6): 854–885.

Krishna, S., Yadav, A., Bhandari, S., Vishwakarma, A.K., Bharti, P.K., Mandavi, P.L., Bahgel, P., Basak, S., Sharma, R.K. and Singh, N. (2017). Prevalence of malaria in two highly endemic Community Health Centers in the Bastar district, Chhattisgarh showing mixed infections with *Plasmodium* species. *Scientific Reports*, 7(1): 16860. https://doi.org/10.1038/s41598-017-16974-2

Manson, P. (1894). The annual oration, 1894.Malaria and its associated parasite. Hunterian Society Transactions, 76th session (1894-1895), pp. 19–50.

Manson-Bahr, P.H. (1982). *Manson's Tropical Diseases*. The English Language Book Society & Bailliere Tindall, London, 843 pp.

Maurice, J. and Pearce, A.M. (1987). *Tropical Disease Research: A global partnership*. World Health Organization, Geneva.

May, J.M. (1951). Map of the world distribution of malaria vectors. *Geographic Review*, 41: 638–639.

Nagpal, B.N. and Sharma, V.P. (1995). *Indian Anophelines*. Oxford and IBH Publishing Co. Pvt. Ltd., New Delhi.

NCVBDCP (2021). Malaria annual report. https://ncvbdc.mohfw.gov.in/index1.php?lang=1&level=1&sublinkid=5784&lid=3689 (accessed 08.01.2025).

Pampana, E. (1969). *A Textbook of Malaria Eradication*. Oxford University Press, 593 pp.

Paidhungat, J.V. (2001). Medical philately (medical personalities on stamps). *Journal of Association of Physicians of India*, 49: 909.

Rao, T.R. (1984). *The Anophelines of India* (Revised Edition.), Malaria Research Centre, Indian Council of Medical Research, New Delhi, 518 pp.

Ross, R., (1897). On some peculiar pigmented cells found in two mosquitoes fed on malaria blood. *British Medical Journal* 2: 1786–1788.

Russell, P.F. (1955). *Man's Mastery of Malaria*. Geoffrey Cumberlege, Oxford University Press, London, New York and Toronto, 308 pp.

Satvalekar, S.D. (1958). *Atharvaveda*. Swadhaye Mandal, Pardi.

Schapira, A., and Boutsika, K. (2012). Malaria ecotypes and stratification. *Advances in Parasitology*, 78: 97–167. https://doi.org/10.1016/B978-0-12-394303-3.00001-3

Shortt, H.E., Garnham, P.C.C., Covell, G. and Shute, P.G. (1948). The pre-erythrocytic stage of human malaria *Plasmodiun vivax*. *British Medical Journal*, 1: 547. https://doi.org/10.1136/bmj.1.4550.547

Singh, J. (1962). Activities of the Malaria Institute of India during World War II. *Indian Journal of Malariology*, 16: 504–514.

Sinka, M.E. (2013). Global distribution of the dominant vector species of malaria. *IntechOpen*. https://www.intechopen.com/chapters/43624

Sinka, M.E., Pironon, S., Massey, N.C., Longbottom, J., Hemingway, J., Moyes, C.L. and Willis, K.J. (2020). A new malaria vector in Africa: Predicting the expansion range of *Anopheles stephensi* and identifying the urban populations at risk. *Proceedings of National Academy of Sciences, USA*, 117(40): 24900–24908. https://doi.org/10.1073/pnas.2003976117

Sinton, J.A. (1935). *What Malaria Costs India, Nationally, Socially and Economically*. Thacker's Press & Directories, Calcutta, 77 pp.

Sinton, J.A. and Mulligan, H.W. (1932). A critical review of the literature relating to the identification of the malaria parasites recorded from monkeys of the families Cercopithecidae and Colobidae. *Records of Malaria Survey of India*, 3: 357–380.

Spielman, A. and D'Antonio, M. (2002). Mosquito: The story Old Man's Deadliest Foe. *Transactions of Royal Society of Tropical Medicine & Hygiene*, 96(1), 95.

Spooner, J. (2017). *Comprehensive Guide to Anopheles Mosquitoes*. Foster Academics, 492 pp.
Srinivasan, S., Ghosh, C., Das, S., Thakare, A., Singh, S., Ganesh, A., Mahawar, H., Jaisimha, A., Krishna, M., Chattopadhyay, A., Borah, R., Singh, V., Kumar, N., Kumar, S., Swain, S. and Subramani, S. (2022). Identification of a TNF-TNFR-like system in malaria vectors (*Anopheles stephensi*) likely to influence *Plasmodium* resistance. *Scientific Reports*, 12(1): 19079. https://doi.org/10.1038/s41598-022-23780-y
Subbarao, S.K., Nanda, N., Rahi, M., Raghavendra, K. (2019). Biology and bionomics of malaria vectors in India: Existing information and what more needs to be known for strategizing elimination of malaria. *Malaria Journal*, 18(1): 396. https://doi.org/10.1186/s12936-019-3011-8
Thakare, A., Ghosh, C., Alalamath, T., Kumar, N., Narang, H., Whadgar, S., Paul, K., Shrotri, S., Kumar, S., Soumya, M., Rao, R., Chakraborty, M., Choudhary, B., Ghosh, S.K., Subramani, S., Swain, S. and Srinivasan, S. (2022). The genome trilogy of *Anopheles stephensi*, an urban malaria vector, reveals structure of a locus associated with adaptation to environmental heterogeneity. *Scientific Reports*, 12(1): 3610. https://doi.org/10.1038/s41598-022-07462-3
Tyagi, B.K. (1992). Control of malaria vectors in India. *Indian Review of Life Sciences*, 12: 211–238.
Tyagi, B.K. (1994). Distribution of arthropod vector-borne communicable diseases and control of their vectors in India. *Indian Review of Life Sciences*, 14: 223–243.
Tyagi, B.K. and Yadav, S.P. (1996). Malariological and sociological significance of 'tanka' and 'beri' in the Thar Desert, Western Rajasthan, India. *Journal of Arid Environments*, 33(4): 497–501.
Tyagi, B.K. (2002). *Malaria in the Thar Desert: Facts, Figures and Future*. Agrobios (India), p. 162.
Tyagi, B.K. (2003). *Medical Entomology: A Handbook of Medically Important Insects and Other Arthropods*. Scientific Publishers (India). p. 265.
Tyagi, B.K. (2004). *The Invincible Deadly Mosquitoes: India's Health and Economy Enemy No. 1*. Scientific Publishers, India, p. 266.
Tyagi, B.K. (2020). *Mosquito Hunters: A History of Hostilities Against Man's Deadliest Foe – The Mosquito – Since 1881*. Scientific Publishers (India), p. 474.
Tyagi, B.K. (2023). *Desert Malaria: An Emerging Malaria Paradigm and Its Global Impact on Disease Elimination*. Springer-Nature, Singapore, p. 416.
Tyagi, B.K., Bansal, S. K. and Verma, K.V.S. (1991). Intraspecific morphological variations in two ecotypes of *Anopheles stephensi* Liston in northwestern Rajasthan (Diptera: Culicidae). *Annals of Entomology*, 9: 85–89.
Tyagi, B.K. and Tilak, R. (2022). Which is the world's deadliest animal: A lion, tiger, shark, snake or mosquito? *Journal of Medical Arthropodology & Public Health*, 2(1): 1–7.
Wattal, B.L. and Kalra, N.L. (1961). Region wise keys to female Indian Anopheles. *Bulletin of National Society of India for Malaria & other Mosquito Borne Diseases*, 10: 55–138.
Winegard, T.E. (2019). *The Mosquito: A Human History of Our Deadliest Predator*. Penguin Publishing Group, p. 496.
WHO (2020). *World Malaria Report, 2020: 20 Years of Global Progress and Challenges*. World Health Organization, Geneva.
WHO (1993a). *A Global Strategy for Malaria Control*. World Health Organization, Geneva, pp. 1–30.
WHO (1993b). *Implementation of the global malaria control strategy*. World Health Organization, Geneva.
WHO (2022a). *WHO Initiative to Stop the Spread of Anopheles stephensi in Africa*. World Health Organization. https://apps.who.int/iris/handle/10665/363318
WHO (2022b). *World Malaria Report 2022*. https://www.who.int/teams/global-malaria-programme/reports/world-malaria-report-2022
World Malaria Report (2021). Geneva: World Health Organization; 2021. Licence: CC BY-NC-SA 3.0 IGO
Wujastyk, D. (2003). *The Roots of Ayurveda Selections From Sanskrit Medical Writings*. Penguin Books, London; New York.
Yadav, R.S., Raghavendra, K., Velamuri, P.S., Verma, V., Uragayala, S. (2020). Insecticide resistance in vectors of malaria in India and its management, pp. 129–148. In: *Vector Biology and Control – An Update for Malaria Elimination Initiative in India*, ed. Vas Dev: National Academy of Sciences, Allahabad, India. http://www.nasi.org.in/BCIL%20Vector%20Biology%20PDF_%2014th%20Aug%202020.pdf

2 Lymphatic Filariasis in India

P. K. Srivastava
Ex-National Centre for Vector Borne Diseases Control, Ministry of Health & Family Welfare, Delhi, India

I.P. Sunish
ICMR-Regional Medical Research Centre, Andaman & Nicobar ils., India

R.K. Hazra
Ex-ICMR-Regional Medical Research Centre, Bhubaneswar, India

A.M. Khan
ICMR-National JALMA Institute for Leprosy, Tajganj, Agra, India

Mohammad Nazrul Islam
Hamdard University Bangladesh (Affiliated by UGC Bangladesh), Gazaria, Bangladesh

S. Sridharan
Public Health Department Tamil Nadu, Chennai, India

Sagya Singh
National Centre for Vector Borne Disease Control, Jharkhand, India

K. Anju Viswan
Ex-State Coordinator Malaria, WHO Raipur Unit, India

A. Usha
District Surveillance Unit, Tumakuru, India

B.K. Tyagi
Chandigarh University, Mohali, India

2.1 INTRODUCTION

Lymphatic filariasis (LF) is a major cause of acute and chronic morbidity and a significant impediment to socioeconomic development in 73 countries in Africa, South-East Asia, the Americas, and the Pacific region. Initially, 81 countries were at risk which was reduced later to 73. The estimates by the World Health Organization (WHO) in 2012 indicated more than 1.4 billion people living at risk of acquiring the infection of human LF (*Wuchereria bancrofti*, *Brugia malayi*, and *Brugia timori*; WHO, 2013). This infection occurs through repeated and long-term exposure to mosquito bites from different genera of anopheline and culicine mosquitoes, particularly *Culex quinquefasciatus*

 DOI: 10.1201/9781003327028-3

Say, 1823. The adult worm lives in the human lymph vessels, where it mates and produces millions of microscopic worms, also known as microfilariae (mf). Microfilariae circulate in the blood and infect the mosquito when it bites an infected person. Microfilariae develop in the mosquito to infective stage. When this infective mosquito bites another person, the larval worms pass from the mosquito into the human skin, and travel to the lymph vessels, where they grow into adult worms, the process takes approximately six months or more. An adult worm lives for about five to seven years. People with microfilariae in their blood can serve as a source of infection to others. Programmes to eliminate lymphatic filariasis are under way in more than 50 countries. Many countries implementing mass drug administration (MDA) have completed more than five consecutive annual treatment rounds, but as per the latest estimate, 17 countries have been validated for having eliminated LF as a public health problem and are under surveillance (WHO, 2008, 2022). Parasite infection rates and infectivity in the human and mosquito populations decrease after several rounds of MDA, but individuals may remain microfilaria and antigenemia-positive even after transmission has been interrupted (Chu et al., 2013).

2.2 GEOGRAPHICAL DISTRIBUTION

Wuchereria bancrofti, most common of all parasites, occurs in sub-Saharan Africa, South-East Asia, the Indian subcontinent, many of the Pacific islands, and focal areas of Latin America and the Caribbean (including Haiti). The next prodigious species, *B. malayi*, occurs mainly in China, India, Malaysia, the Philippines, Indonesia, and various Pacific islands, whereas *B. timori*, the least common of all worms, occurs on the Timor Island (Figure 2.1). Overall, approximately two-thirds of individuals infected with LF are in Asia. More than 120 million in 73 countries throughout the tropics and sub-tropics of Asia, Africa, the Western Pacific, and parts of the Caribbean and South America were estimated to be infected with LF with more than 90% due to *W. bancrofti* and the remainder due to *B. malayi*. Estimates also suggest that more than 40 million infected individuals are seriously incapacitated and disfigured by LF (Ichimori et al., 2014)). Of these, almost 25 million men have genital disease (most commonly hydrocele), and almost 15 million, mostly women, have lymphoedema or elephantiasis of the leg. A study from India, which accounts for nearly 40% of the global prevalence of infection, estimated that more than $840 million (INR 58,000 million)

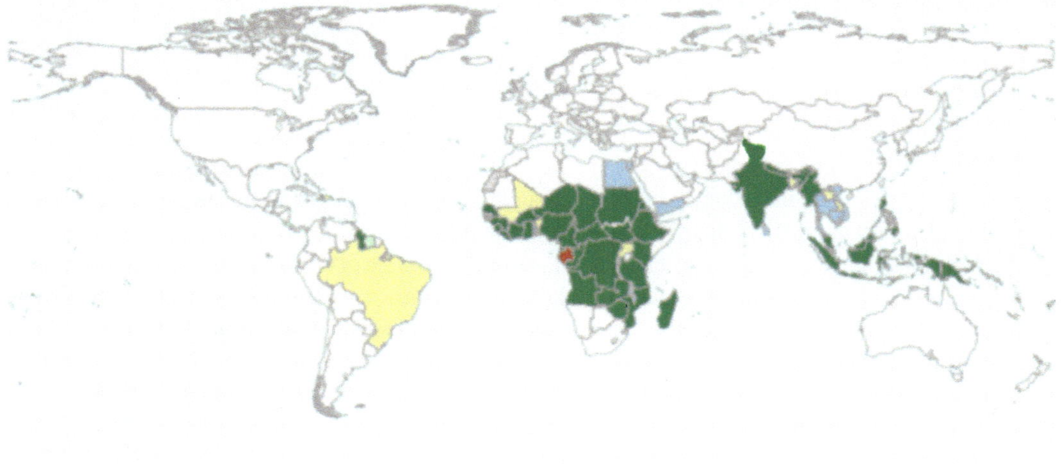

FIGURE 2.1 The areas in dark green indicate the geographic distribution of LF.

Source: WHO, 2022.

are lost each year due to treatment costs and missed working days (Ramaiah et al., 2000). Another study in India noted that patients with chronic filariasis lose around 29 days of work per year due to complications of infection, highlighting the considerable burden the disease places on individuals and on the community (Babu et al., 2006).

In the Americas, only four countries are currently known to be endemic: Haiti, the Dominican Republic, Guyana, and Brazil. Approximately 66% of those at risk of infection live in the WHO South-East Asia region and 33% in the African region.

A district-level endemicity map created for India in 2000 shows that of the 289 districts surveyed up to 1995 (62% of all districts), as many as 257 were found to be endemic (Sabesan et al., 2000). Eighteen states and six Union Territories were identified to be endemic, with about 650 million people exposed to the risk of infection. About 31 million people were estimated to be the carriers of *mf* (microfilariae), and over 23 million suffered from filarial disease manifestations in India (WHO, 2005). The State of Bihar has the highest endemicity (over 17%) followed by Kerala (15.7%), Uttar Pradesh (14.6%), Andhra Pradesh and Tamil Nadu (about 10%), Assam (about 5%), and Madhya Pradesh (above 3%). Goa and Lakshadweep showed the lowest endemicity of about 1%. *B. malayi* was reported to be prevalent in the states of Kerala, Tamil Nadu, Andhra Pradesh, Orissa, Madhya Pradesh, Assam, and West Bengal. The single largest tract of this infection lies along the west coast of Kerala, comprising the districts of Trichur, Ernakulum, Alleppey, Quilon, and Trivandrum, stretching over an area of 1,800 sq km. The infection in the other six states was confined to a few villages. Currently, it is reported only from Alappuzha district of Kerala and a few cases from Balasore district of Odisha.

Both *W. bancrofti* and *B. malayi* infections in mainland of India exhibit nocturnal periodicity of microfilaraemia (Sasa, 1976). In 1974–75, diurnal subperiodic *W. bancrofti* infection was discovered among aborigines, inhabiting Nicobar Group of Andaman and Nicobar Islands. *Ochlerotatus (Finlaya) niveus* mosquito was incriminated as the vector for this infection (Shriram et al., 2008), formerly classified as *Aedes (Finlaya) niveus* (Kalra, 1974), which is now known as *Aedes (Downsiomyia) niveus*.

2.3 TRANSMISSION CYCLE

Humans are the definitive hosts and mosquitoes the intermediate hosts of Bancroftian and Malayan types of filariasis. The adult worms of both *W. bancrofti* and *B. malayi* live in the lymph canals and lymph nodes of humans. The longevity of an adult parasite varies from 5 to 18 years, (Vanamail et al., 1989, 1996), but a maximum lifespan up to 40 years has been reported (Carme & Laigret, 1979). The *mf* is produced from ova in the uterus of the female worm and sheathed *mf* begin to appear in the peripheral blood six months to one year after infection (Mak, 1987). The females are viviparous and give birth to as many as 50,000 *mf* per day, which find their way to blood circulation. The lifespan of *mf* is not exactly known, probably one year or more. The *mf* remains in the arterioles of the lungs during the day and emerges at night (nocturnally periodic). The mosquito cycle begins when *mf* is picked up by a mosquito during blood meal and its sheath is dissolved for further development within the mosquitoes (intermediate hosts) to form the infective larvae.

The following stages of development take place in the vector: (a) *Exsheathing*: In mosquito gut the larva comes out of the sheath in which it is enclosed within 1 to 2 hours of ingestion. (b) *First stage larva* (L_1): after exsheathing the larva is able to penetrate the stomach wall of the mosquito where it grows and develops into a sausage shaped (short thick) form. (c) *Second stage larva* (L_2): the larva increases in length (long thick form) with the development of an alimentary canal but is relatively inactive. (d) *Third stage larva* (L_3): this is the final stage of motile infective larva (long, thin form) which may be found in any part of the mosquito. When it migrates to the proboscis of the mosquito, it is ready to be transmitted to a new host, and the mosquito is said to be infective. The duration of development of *mf* to infective stage varies between species; *W. bancrofti* takes 10–14 days, whereas *B. malayi* takes 7–10 days. When the infective mosquitoes bite, the L_3 larvae escape from the proboscis and actively enter the human host through the wound made by the mosquito for blood meal. The shortest pre-patent period (from the entrance of L_3 to the appearance of *mf* in the peripheral blood) is estimated to be about nine months for *W. bancrofti* and three months for *B. malayi*. The appearance of *mf* depends upon the mating probability of adult worms, the pre-patent period, and the intensity of transmission.

2.4 THE VECTORS

Mosquitoes are the only known vectors of bancroftian and Brugian/Malayan filarial parasites (Omori, 1962). A wide range of mosquitoes can transmit the parasite, depending on the geographic area. In Africa, the most common vector is an *Anopheles* species. In the Americas, it is *Culex quinquefasciatus* Say, 1823, whereas *Aedes* and *Mansonia* species can transmit the infection in the Pacific and in Asia. *Culex quinquefaciatus* is the main vector of bancroftian filariasis in India and other endemic countries. Although *Cx. quinquefaciatus* breeds profusely in dirty/ polluted waters, it can also breed in clear water when denied dirty water.

Mansonoides mosquitoes are the chief vectors of Malayan/Brugian filarial parasites. The major species involved are *Mansonia annulifera, M. uniformis, and M indiana*. The breeding of these mosquitoes is associated with certain aquatic plants such as *Pistia stratiotes*. In the absence of these plants, these mosquitoes cannot breed. Many mosquito bites over several months to years are needed to get LF. People living for a long time in tropical or sub-tropical areas where the disease is common are at greatest risk for infection. Short-term tourists have a very low risk. In some areas, anopheline mosquitoes are responsible for transmitting infection. *Brugian* parasites are confined to areas of east and south Asia, notably India, Indonesia, Malaysia, and the Philippines.

2.5 PATHOGENESIS

The principal pathological changes are the result of inflammatory alterations in the lymphatics, which are typically caused by adult worms and not by microfilariae. Adult worms live in afferent lymphatics or sinuses of lymph nodes and cause lymphatic dilatation and thickening of vessel walls. The infiltration of plasma cells, eosinophils, and macrophages in and around the infected vessels, along with endothelial and connective tissue proliferation, leads to tortuosity of lymphatics leading to damaged or incompetent lymph valves.

Lymphoedema and chronic stasis changes with hard or brawny oedema develop in the overlying skin. These consequences of filariasis are due to both direct effects of worms and inflammatory response of the host to the parasite. Inflammatory responses are believed to cause granulomatous and proliferation processes that precede total lymphatic obstruction. It is thought that the lymphatic vessels remain patent as long as the worms remain viable, and the death of the worms leads to enhanced granulomatous reaction and fibrosis.

2.6 CLINICAL FEATURES

The most common presentations of the LF are asymptomatic (or subclinical) microfilaremia, hydrocele, acute adenolymphangitis (ADL), and chronic lymphatic disease. In areas where *W. bancofti* or *B. malayi* is endemic, the overwhelming majority of infected individuals have few overt clinical manifestations of filarial infections despite a large number of circulating microfilariae in the peripheral blood. Although they may be clinically asymptomatic, virtually all persons with *W. bancrofti* or *B. malayi* microfilaremia have some degree of subclinical disease that includes microscopic haematuria and/or proteinuria, dilated (and tortuous), lymphatic (visualized by imaging), and, in men, lymphangiectasia (detectable by ultrasound). In spite of these findings, the majority of individuals appear to remain clinically asymptomatic for years, and relatively few progress to either acute or chronic disease. Acute ADL is characterized by high fever, lymphatic inflammation (lymphangitis and lymphadenitis), and transient local oedema. The lymphadenitis and lymphangitis can involve both upper and lower extremities in both bancroftian and Brugian filariasis, but involvement of genital lymphatics occurs almost exclusively with bancroftian infection. In endemic areas, another type of acute disease, dermatolymphangioadenitis (DLA), is recognized as a syndrome that includes high fever, chills, myalgias, and headache.

If lymphatic damage progresses, transient lymphoedema can develop into lymphatic obstruction and permanent changes associated with elephantiasis. In bancroftian filariasis, where genital involvement is common, hydrocele may develop. In advanced stages, the condition may evolve into scrotal lymphoedema and scrotal elephantiasis. Furthermore, if there is obstruction of retroperitoneal lymphatics, the increased renal lymphatic pressure leads to rupture of the renal lymphatics and the development of chyluria, which is usually intermittent and most prominent in the morning.

2.7 CONTROL STRATEGY UNDER NATIONAL FILARIA CONTROL PROGRAMME

Under the National Filaria Control Programme (NFCP) of India, the control measures were mainly based on (i) recurrent anti-larval measures at weekly intervals with approved larvicides to control the vector mosquitoes and (ii) anti-parasitic measures by detection and treatment of filaria cases and microfilaria carriers with DEC. NFCP activities through filaria control units were confined to 227 towns and spread over 13 states and 4 Union Territories, with an additional 27 filaria survey units (FSU) and 224 filaria clinics (detection-cum-treatment teams) mainly attached to the control units.

International Task Force for Disease Eradication has identified LF as one of the six infectious diseases to be "eradicable" or "potentially eradicable" (WHO, 1995). In 1997, the World Health Assembly (WHA) adopted resolution WHA 50.29 for the elimination of lymphatic filariasis (ELF) as a global public health problem by 2020 (WHO, 2011a,b; Ottesen et al., 1997). India became a signatory to this resolution, and strategies were revised from control to elimination (Biswas, 1996; Sharma et al., 1995).

2.8 MASS DRUG ADMINISTRATION

Based on various consultative meetings and reviews, it was decided to start nationwide programme to eliminate LF in 2004 (Anon, 2003, NVBDCP, 2005, 2009). The districts were selected as implementation unit, and the basis of inclusion of these districts was historical evidence of filaria endemicity, presence of lymphodema and hydrocele cases, and microfilaria carriers. The Government of India (GoI) in the year 2004, launched the strategy of annual MDA with a single dose of DEC to populations living at risk of filariasis, except children below 2 years, pregnant women, and seriously ill persons. The co-administration of DEC + albendazole was introduced in 2007 (Sunish et al., 2003, 2007). The sensitization, advocacy, commitment of political leaders, and issuance of technical guidelines since 2004 made the programme visible not only in India, but the initiatives and efforts were appreciated on an international platform. The reported coverage of the population showed improvement from average of about 74% to above 90%, however, actual drug compliance remained a challenge, leading to significant gap. The overall microfilaria rate though was reduced from 1.24% in 2004 to about 0.40% in 2017, but the quality of microscopy, especially the time of night survey and examination of slides by trained microscopists, remained a challenge. Triple drug combination therapy with ivermectin, DEC, and albendazole (IDA) was introduced to high endemic districts to interrupt transmission and achieve the elimination within the stipulated time. To narrow down the population at risk, the strategic change was introduced so that the blocks are made as implementation units for MDA instead of districts. The concept was first discussed in a meeting in 2016–17, and some states had already initiated this by excluding part of districts which became non-endemic. Twelve districts which completed more than five rounds of MDA with an effective coverage of > 65% and *mf* prevalence < 1% in the sentinel and spot-check sites were subjected to the Transmission

Assessment Survey (TAS), and all of them qualified for stopping MDA (VCRC, 2014). MDA for LF was being observed on 11 November every year, which has been synchronized since 2017 with deworming programme of department of health, observed on national deworming day every year on 10 February and August on account of a common drug albendazole.

2.8.1 Drug Compliance Assessment

Assessment of MDA was initially done to assess the actual compliance of drug administration, which revealed a large gap between drug coverage and actual drug compliance. This activity was made an inbuilt component of the programme by involving medical colleges and research institutes to give insight into the programme, which also indicated the gap (Kumar et al., 2008; Lahariya and Mishra, 2008; Mukhopadhyay, 2010; Aswathy et al., 2009; Ramaiah et al., 2007; Vanamail et al., 2005; Srivastava & Dhillon, 2008; Srivastava et al., 2013; Srivastava et al., 2014a,b), and accordingly, the programme improved compliance in subsequent years.

2.8.2 Use of DEC-Fortified Salt for LF Elimination

The use of DEC-fortified salt as a unique form of MDA is an alternative supplemental strategy for LF elimination in which a low dose of DEC (0.2% w/w) is added to the cooking salt at the manufacturing facility of iodized salt and consumed by the LF-endemic communities for about two years. Limited community-scale trials in India with DEC-fortified salt at the concentration ranging from 0.1% to 0.4% over 3–12 months with complete coverage led to the reduction of MF rate ranging from 86% to 100%. Implementing DEC-iodine-fortified salt was reported to be a safe, less expensive, more efficient, and prompt approach for achieving the elimination of LF in India (Sabesan et al., 2022). This strategy was implemented successfully to eliminate the single focus of diurnal subperiodic form of LF on Nicobar Islands (Shriram et al., 2020).

2.9 INTEGRATED VECTOR MANAGEMENT IN LYMPHATIC FILARIASIS ELIMINATION

GPELF initially focused on preventive chemotherapy to interrupt the transmission of LF by reducing the microfilaria prevalence at community level. There have been arguments and challenges in integrating vector control with MDA for LF elimination (Burkot and Ichimori, 2002). An integrated vector management (IVM) strategy was implemented from 1981 to 1985 in Pondicherry of South India for the control of the bancroftian filariasis vector *Culex quinquefasciatus*. After 1985, both areas, i.e., IVM and MDA operated areas, were managed conventionally. The switch to conventional strategy resulted in an increase of vector density in both areas. The microfilaremia prevalence in humans showed a general decline ($P < 0 - 05$) from 1986 to 1989 only in the IVM area. There have also been arguments on the cost effectiveness of vector control measures, especially in rural areas with large and innumerable breeding sites. However, since WHO emphasized entomological surveillance and vector control as adjuncts to MDA, they were included as additional pillars in the ELF campaign. They have multiple potential benefits which include (1) the ability to suppress LF transmission without the need to identify all individual 'foci of infection,' (2) minimizing the risk of reestablishment of transmission from imported mf positive individuals, and (3) reducing the spread of any possible DEC or albendazole resistant *W. bancrofti* which might emerge. The reduction of nuisance in addition to vector mosquito biting is likely to enhance community support for ELF programmes. Implementation of vector control strategies as adjuncts to the MDA campaigns will better ensure the success of the elimination efforts and enhance the prospects for sustainable benefits. Mosquito surveillance and control as adjuncts to MDA are already included in filariasis elimination plans of many countries, even though they may not be actively implemented. Among PacELF (the Pacific Program to Eliminate Lymphatic Filariasis) countries, activities commonly mentioned include larval surveys for filariasis vectors, environmental sanitation to reduce mosquito breeding sites, and use of mosquito nets and ultra-low-volume (ULV) spraying against adult mosquitoes (Burkot et al., 2006). We need to support vector control efforts to ensure that the resources available to ministries of health are spent effectively. There are a number of vector control strategies whose efficacy against transmission of LF has been demonstrated. These include the use of insecticide-treated mosquito nets where *Anopheles* are LF vectors (Bogh et al., 1998; Pedersen and Mukoko, 2002) and the use of polystyrene beads alone (Sunish et al., 2007) and in combination with insecticide impregnated curtains with larvivorous fishes (Sunish et al., 2016) for control of *Culex* vectors that breed in pit latrines/soakage pits, unused wells, and drains (Maxwell et al., 1990). Mosquito nets will suppress LF transmission whenever the vectors are night-biting (i.e., *Anopheles, Culex, Mansonia, Ae. vigilax*). These strategies are ready for widespread implementation, but the degree of impact needs to be confirmed in controlled trials before advocating them for implementation at the national level. Minor engineering and closing open drains for breeding source reduction are the permanent measures for reducing mosquito breeding.

REFERENCES

Aswathy, S., Beteena, K., Leelamoni, K. (2009). Mass drug administration against filariasis in India: perceptions and practices in a rural community in Kerala. *Annals of Tropical Medicine & Parasitology*, 103, 617–24.

Biswas, G. (1996). *Revised strategy for the control of lymphatic filariasis in India, New Delhi.* National Institute of Communicable diseases and National Malaria Eradication program, Delhi.

Burkot, T.R., Durrheim, D.N., Melrose, W.D., Speare, R. and Ichimori, K. (2006). The argument for integrating vector control with multiple drug administration campaigns to ensure elimination of lymphatic filariasis. *Filaria Journal*, 5, 10

Burkot, T.R., Ichimori, K. (2002). The Pacific Program for the Elimination of Lymphatic Filariasis: Will mass drug administration be enough? *Trends in Parasitology*, 18, 109–115.

Babu, B.V., Swain, B.K. and Rath, K. (2006). Impact of chronic lymphatic filariasis on quantity and quality of productive work among weavers in an endemic village from India. *Tropical Medicine & International Health*, 11, 712.

Bogh, C., Pedersen, E.M., Mukoko, D.A., Ouma (1998). Permethrin impregnated bednet effects on resting and feeding behaviour of lymphatic filariasis vector mosquitoes in Kenya. *Medical and Veterinary Entomology*, 12, 52–9.

Carme, B. and Laigret, J. (1979). Longevity of *Wuchereria bancrofti* var. *pacifica* and mosquito infection acquired from a patient with low level parasitemia. *American Journal of Tropical Medicine and Hygiene*, 28, 53–5.

Chu, B.K., Deming, M., Biritwum, N.K., Bougma, W.R. et al. (2013). Transmission assessment surveys (TAS) to define endpoints for lymphatic filariasis mass drug administration: a multicenter evaluation. *PLoS Neglected Tropical Diseases*, 7(12), e2584. doi: 10.1371/journal.pntd.0002584

Ichimori, K., King, J. D., Engels, D., Yajima, A., Mikhailov, A., Lammie, P., & Ottesen, E. A. (2014). Global programme to eliminate lymphatic filariasis: the processes underlying programme success. *PLoS neglected tropical diseases*, 8(12), e3328. https://doi.org/10.1371/journal.pntd.000

Kalra, N. L. (1974) Filariasis among aborigines of Andaman and Nicobar islands. 1. Detection of non-periodic Bancroftian filariasis among Nicobarese of Nanncowry group of Nicobar Islands. *Journal of Communicable Diseases*, 6: 40–56.

Kumar, P., Prajapati, P., Saxena, D., Kavishwar, A. B., & Kurian, G. (2008). An evaluation of coverage and compliance of mass drug administration 2006 for elimination of lymphatic filariasis in endemic areas of Gujarat. *Indian Journal of Community Medicine: Official Publication of Indian Association of Preventive & Social Medicine*, 33(1), 38–42.

Lahariya, C. Mishra, A. (2008) Strengthening of mass drug administration implementation is required to eliminate lymphatic filariasis from India: an evaluation study. *Journal of Vector Borne Diseases*, 45, 313–20.

Mak J. W. (1987). Epidemiology of lymphatic filariasis. *Ciba Foundation Symposium, 127*, 5–14. https://doi.org/10.1002/9780470513446.ch2

Maxwell CA, Curtis CF, Haji H, Kisumku S, Thalib AI, Yahya SA (1990). Control of Bancroftian filariasis by integrating therapy with vector control using polystyrene beads in wet pit latrines. *Journal of Vector Borne Diseases*, 84, 709–14.

Mukhopadhyay, A.K. (2010) Lymphatic filariasis in Andhra Pradesh Paper Mill Colony, Rajahmundry, India after nine rounds of MDA programme. *Journal of Vector Borne Diseases*, 47, 55–7.

NVBDCP (2005). Guidelines for elimination of Filariasis. National Vector Borne Disease Control Programme. https://ncvbdc.mohfw.gov.in/WriteReadData/l892s/LF%20manual_final.doc

NVBDCP (2009). Guidelines on Filariasis Control in India and its Elimination. National Vector Borne Disease Control Programme. https://ncvbdc.mohfw.gov.in/WriteReadData/l892s/43461824631532409675.pdf

Omori, N. (1962). A review of the role of mosquitoes in the transmission of Malayan and bancroftian filariasis in Japan. *Bulletin of the World Health Organization*, *27*, 585–594.

Ottesen, E.A. Duke, B.O., Karam, M. and Behbehani, K. (1997). Strategies and tools for the elimination of lymphatic filariasis. *Bulletin of the World Health Organization*, 75, 491–503.

Pedersen, E.M. and Mukoko, D.A. (2002). Impact of insecticide-treated materials on filaria transmission by the various species of vector mosquito in Africa. *Annals of Tropical Medicine & Parasitology*, 96(Suppl. 2), S91–S95.

Ramaiah, K.D., Das, P.K., Michael, E. and Guyatt, H. (2000). The economic burden of lymphatic filariasis in India. *Parasitology Today* 16(6), 251–3.

Ramaiah, K.D., Vanamail, P. and Das P.K. (2007). Changes in *Wuchereria bancrofti* infection in a highly endemic community following 10 rounds of mass administration of diethylcarbamazine. *Royal Society of Tropical Medicine and Hygiene*,101, 250–5.

Sabesan, S., Krishnamoorthy, K., Hoti, S. L., Subramanian, S., Srividya, A., Roy, N., Jain, T., Kumar, A. and Rahi, M. (2022) Diethylcarbamazine citrate-fortified salt for lymphatic filariasis elimination in India. *Indian Journal of Medical Research*, 155, 347–55.

Sabesan S, Palaniyandi M, Das PK, Michael E (2000). Mapping of lymphatic filariasis in India. *Annals of Tropical Medicine & Parasitology*, 94, 591–606.

Sasa, M. (1976). *In*: *Human filariasis*: A global survey of epidemiology and control, University of Tokyo. 819 pp.

Sharma, R.S., Biswas, H., Saxena, N.B.L. (1995). National Filaria Control Programme, Operational Manual. National Malaria Eradication Programme, Delhi.

Shriram, A.N., Krishnamoorthy, K. and Sehgal, S.C. (2008). Transmission dynamics of diurnally subperiodic lymphatic filariasis transmitted by *Ochlerotatus (Finlaya) niveus* in the Andaman & Nicobar Islands. *Indian Journal of Medical Research*, 127, 37–43.

Shriram, A.N., Premkumar, A., Krishnamoorthy, K., De, A., Paul, S.K., Subramanian, S., Vijayachari, P. and Jambulingam, P. (2020). Elimination of diurnally sub-periodic *Wuchereria bancrofti* in Andaman and Nicobar Islands, India, using mass DEC-fortified salt as a supplementary intervention to MDA. *Parasitology Research*, 119, 1467–83.

Srivastava, P.K., Dhariwal, A.C. and Bhattacharjee, J. (2013). Status of Lymphatic Filariasis in India, *Health Action*, 19.

Srivastava, P.K. and Dhillon, G.P. (2008). Elimination of lymphatic filariasis in India--a successful endeavour. *Journal of Indian Medical Association*, 106, 673–4.

Srivastava, P.K., Krishnamoorthy, K., Govenkar, S., Perni, S., Dalvi, S., Subramanain, S., Dhariwal, A.C., Bhattacharjee, J. and Dash, A.P. (2014b). Elimination of Lymphatic Filariasis in Goa: First Successful Transmission Assessment Survey in India. *Journal of Commicable Diseases*, 46 (2), 7–16.

Srivastava, P.K., Sharma, S.N., Bhattacharjee, J., Dhariwal, A.C. and Krishnamoorthy, K. (2014a). A Tool for Monitoring Epidemiological Impact of Mass Drug Administration (MDA) in the Elimination of Lymphatic Filariasis – An Indian Experience. *Journal of Communicable Dis*eases, 46(2), 1–6.

Sunish, I.P., Kalimuthu, M., Ashok Kumar V., Munirathinam, A., Nagaraj, J., Tyagi, B.K., White, G.B. and Arunachalam, N. (2016) Can community-based integrated vector control hasten the process of LF elimination? *Parasitology Research*, 115, 2353–62.

Sunish, I.P., Rajendran, R., Mani, T.R., Gajanana, A., Reuben, R. and Satyanarayana, K. (2003). Long-term population migration: an important aspect to be considered during mass drug administration for elimination of lymphatic filariasis. *Tropical Medicine & International Health*, 8(4), 316–21.

Sunish, I.P., Rajendran, R., Mani, T.R., Munirathinam, A., Dash, A.P., Tyagi, B.K. (2007). Vector control complements mass drug administration against bancroftian filariasis in Tirukoilur, India. *Bulletin of the World Health Organization*, 85, 138–45.

Vanamail, P., Ramaiah, K.D., Pani, S.P., Das, P.K., Grenfell, B.T. and Bundy, D.A. (1996). Estimation of the fecund life span of *Wuchereria bancrofti* in an endemic area. *Transactions of the Royal Society of Tropical Medicine and Hygiene*, 90, 119–21.

Vanamail P, Ramaiah KD, Subramanian S, Pani SP, Yuvaraj J, Das PK (2005) Pattern of community compliance with spaced, single-dose, mass administrations of diethylcarbamazine or ivermectin, for the elimination of lymphatic filariasis from rural areas of southern India. *Annals of Tropical Medicine & Parasitology*, 99, 237–42.

Vanamail, P., Subramanian, S. and Das, P.K. et al (1989) Estimation of age-specific rates of acquisition and loss of *Wuchereria bancrofti* infection. *Transactions of the Royal Society of Tropical Medicine and Hygiene*, 83, 689–93.

VCRC (2014). *Report on independent appraisal of Elimination of Lymphatic Filariasis Programme in India.* Vector Control Research Centre, Indian council of Medical Research, Puducherry.

WHO (1995). *Bridging the gap. The World Health Report 1995.* World Health Organization, Geneva.

WHO (2005). Sixth meeting of the Technical Advisory Group on the Global Elimination of Lymphatic Filariasis, Geneva, Switzerland. *Weekly Epidemiological Records*, 80, 401–8.

WHO (2008). Global programme to eliminate lymphatic filariasis. *Weekly Epidemiology*, 83(37), 333–41.

WHO (2011a) Global Programme to Eliminate Lymphatic Filariasis: progress report on mass drug administration, 2010. *Wkly Epidemiological Records*, 35, 86, 377–88.

WHO (2011b). Monitoring and epidemiological assessment of Mass Drug Administration. A manual for National elimination programmes. WHO/HIM/NTD/PCT/2011.4

WHO (2013). Global programme to eliminate lymphatic filariasis. Progress report for 2012. *Weekly Epidemiological Records*, 88(37), 389–99

WHO (2022). Global programme to eliminate lymphatic filariasis: progress report, 2021. *Weekly Epidemiological Records*, 1(97), 513.

3 Dengue in India

Kalpana Baruah, Amit Katewa and Gavendra Singh
National Center for Vector Borne Diseases Control, Directorate General of Health Services, Ministry of Health & family Welfare, Govt. of India, Delhi, India

Roop Kumari
World Health Organization Country Office for India, New Delhi, India

Mangesh Gokhale
Entomology Group, ICMR-National Institute of Virology, Pune, India

Sudeep Balan
ICMR-iNational Institute of Virology, Microbial Containment Complex, Pashan, India

T. Dilip Kumar
Kerala State Public Health Department, Thiruvananthapuram, India

S. Sridharan
Tamil Nadu Public Health Department, Chennai, India

Jhansi Charles
Rajaji Medical College, Madurai, India

Himmat Singh
ICMR-National Institute of Malaria Research, New Delhi, India

B.K. Tyagi
Chandigarh University, Mohali, India

3.1 INTRODUCTION

Dengue, an arboviral infection with major public health consequences and caused by one of the four serotypes of dengue virus (DENV 1–4), is transmitted by mosquitoes, specifically *Aedes aegypti* and *Ae. albopictus* (Christophers, 1960; WHO-ROSEA, 2011). The disease has been indicated by the World Health Organization (WHO, 2019) as one of the ten threats to global health. Dengue is prevalent worldwide in over 100 tropical and subtropical countries where vector species are also predominantly prevalent.

Continued geographic expansion of dengue viruses and their mosquito vectors has seen the magnitude and frequency of epidemic dengue fever/dengue haemorrhagic fever (DF/DHF) increase dramatically (Gubler et al., 2014). The incidence of dengue has increased by more than 30-fold in the last 50 years (WHO, 2020a) with an estimated 100 million dengue cases and over 390 million

DOI: 10.1201/9781003327028-4

infections worldwide annually (WHO, 2020b). Although the full global burden of the disease estimated to be 2,946,900 (WHO, 2018) is still uncertain due obviously to many serious lacunae and variations in surveillance, case reporting, etc., the patterns are alarming for both human health and the economy. As far as India is concerned, it is believed that as far back as 1780, Madras (now Chennai) had suffered with an epidemic of clinical dengue-like illness but in the absence of any knowledge about the dengue viruses (DENV 1–4), the disease consequences on health or economy were hardly understood meaningfully. Even at present, its impact on both the disease burden and economic burden across India's heterogenous geographical regions remains poorly quantified, hence, largely underdetermined and likely merely symbolic in nature (Shepard et al., 2014; Hariharan et al., 2019; Murhekar et al., 2019; Baruah and Dhariwal, 2011; Baruah et al., 2021). All four serotypes of dengue are prevalent in India.

Based on the transmission dynamics of dengue, WHO has also grouped India in 2011 with Group 'A' countries in South-East Asia Region (SEAR) where dengue is a major public health problem and the leading cause of hospitalization and death among children and hyperendemicity in urban centres, as well as spreading to rural areas with multiple virus serotypes circulating (WHO, 2011).

3.1.1 Definition

Dengue is an inconsequential disease presenting initially with flu-like symptoms, but it may be fatal in its severe form. Ironically, immunity developed against a given virus serotype is limited to this only and does not provide cross-protection against any one of the other three serotypes, and upon contracting a second infection subsequently with a different serotype, a patient with a strain-specific immunity may stand a greater risk (Guzman and Kouri, 2002).

3.1.2 Signs and Symptoms of Dengue

Mild symptoms of dengue can be confused with other illnesses that cause fever, aches and pains, or a rash. Symptoms of dengue typically last two to seven days. Most people will recover after about a week (Figure 3.1). Rarely Gullian Barre Syndrome (GBS) may occur and cause neurological disorder (Sulekha et al., 2004).

The most common symptom of **classical dengue** is fever with any of the following: (i) nausea, vomiting, (ii) rash, (iii) aches and pains (eye pain, typically behind the eyes, muscle, joint, or bone pain), and (iv) any warning sign. However, if unattended, symptoms of dengue can become severe within a few hours. **Severe dengue** is a medical emergency. It can result in shock, internal bleeding, and even death. If one has had dengue in the past, then he/she is more likely to develop severe dengue, and infants and pregnant women are at higher risk for developing severe dengue. It is desired to watch for signs and symptoms of severe dengue, which usually begin in 24–48 hours after fever has gone away. Patient is advised to immediately go to a local clinic or hospital if any of these symptoms appears: (i) stomach pain, tenderness; (ii) vomiting (at least 3 times in 24 hours); (iii) bleeding from the nose or gums; (iv) vomiting blood or blood in the stool; (v) feeling tired, restless, or irritable; or (vi) severe joint pain all over body.

3.1.3 Treatment

There is no specific medicine to treat dengue infection or any vaccine to prevent infection (WHO, 1999). Cases are managed symptomatically. National guidelines for clinical management of dengue fever were developed in 2007 and updated in 2014 and 2023. These guidelines proved helpful in effective case management to minimize the case fatality rate (CFR; deaths per 100 cases) in the country. Self-medication should be avoided, and it is advised not to take aspirin or any NSAID (non-steroidal anti-inflammatory drugs). Dengue patients should drink plenty of fluids to stay hydrated.

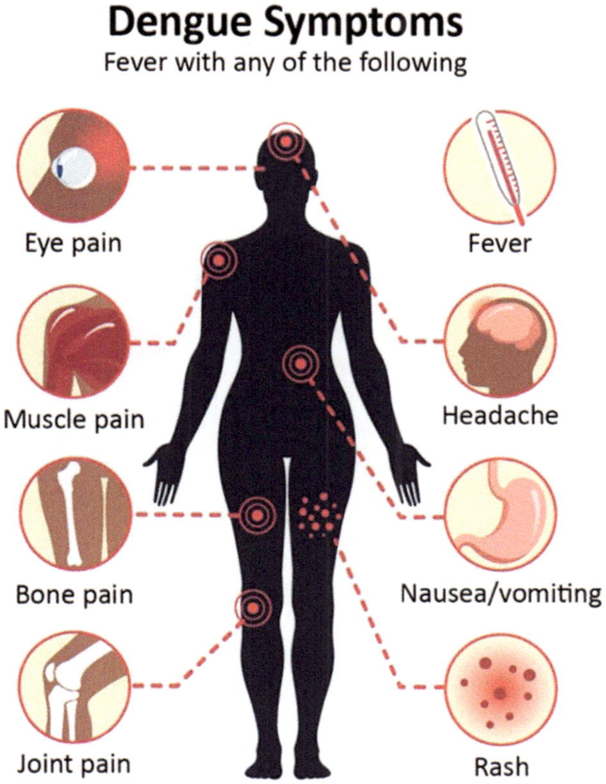

FIGURE 3.1 Common symptoms of dengue.

Source: Centres for Disease Control and Prevention; https://www.cdc.gov/dengue/symptoms/index.html.

3.2 GLOBAL SCENARIO OF DENGUE

Worldwide nearly 1.8 billion (more than 70%) of the 2.5 billion people at risk for dengue live in member states of the WHO SEAR and Western Pacific region, which together bear nearly 75% of the disease burden (Beatty et al., 2011; Bhatt et al., 2013; Gubler, 2002; Murray et al., 2013; WHO, 2009). The dengue scenario during the past two decades has steeply altered globally with the number of cases having increased from 2.2 million in 2010 to over 4.2 million in 2019, which were hitherto also the highest ever reported (WHO, 2019, 2020b). The disease is now endemic in 129 countries in Africa, the Americas, Eastern Mediterranean, South-East Asia, and the Western Pacific (PAHO, 2020). The true numbers are probably far more since severe underreporting and misclassification of dengue cases have been documented by the countries (WHO, 2012). The increasing number of dengue cases imported from endemic countries and autochthonous transmission of dengue in non-endemic areas such as the United States (Hawaii, Florida, and Texas), France, Croatia, and Angola highlight the importance of global collaboration to manage dengue epidemics (Adalja et al., 2012; Effler et al., 2005; Gould et al., 2010; Gjenero-Margan et al., 2011; Sessions et al., 2013). Climate change, the expansion of dengue vectors to new geographic regions, increasing human movement across borders, global trade, and urban migration have collectively changed the scope and scale of dengue fever from a national to a global concern (Bhatt et al., 2013; Murray et al., 2013; Gupta et al., 2012; Rochlin et al., 2013; Schwartz et al., 2008; Wilder-Smith and Gubler, 2008).

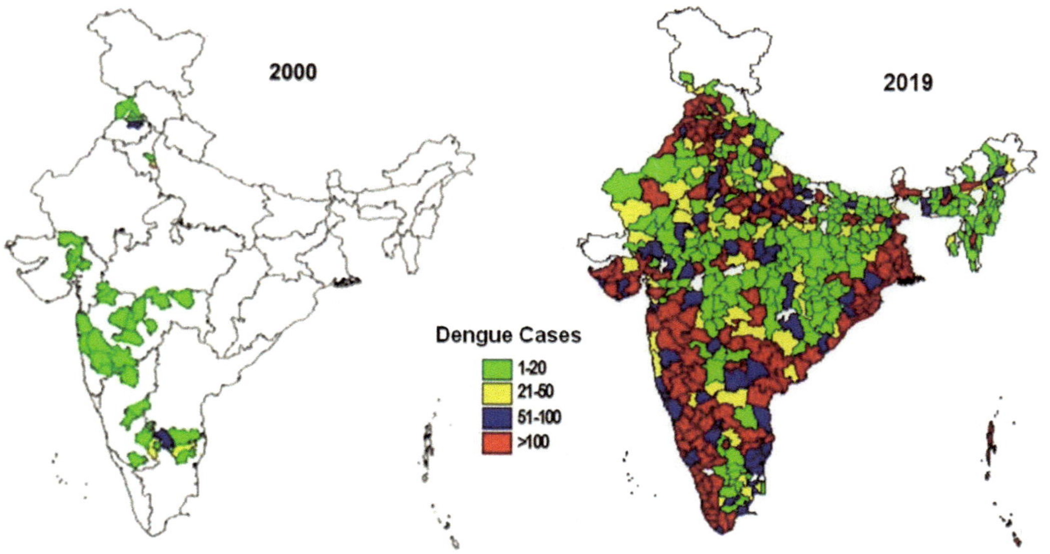

FIGURE 3.2 Geographical spread of dengue over the last two decades (from 2000 to 2019) in India.

Source: Baruah et al., 2021.

3.3 DENGUE DISTRIBUTION IN INDIA

During the last two decades, significant geographical spread of dengue has been observed with an increase in number of reported cases (Tyagi, 2005; Baruah et al., 2020). Currently all states/union territories are endemic for Dengue (except Ladakh). Newer areas are reporting outbreaks every year, including a few from high altitudes (like Bilaspur in Himachal Pradesh, Nainital, Dehradun in Uttarakhand from the foothills of Himalayas).

The geographical expansion of dengue in India in the last two decades from 2000 to 2019 is depicted in Figure 3.2. The number of states/UTs reporting dengue cases increased from 8 in 2000 to 35 at present (except Ladakh). There was an 11-fold increase in the number of dengue cases from 2000 to 2019 (Baruah et al., 2020).

3.4 DENGUE PREVALENCE IN INDIA

In the WHO-SEA region alone, all ten countries are endemic for dengue, save for the Democratic People's Republic of Korea. India, which contributed only 6% of the cases in the region in 2009, had the maximum share of 33.5% in 2018 (see Figure 3.3 and Table 3.1).

It is noteworthy that although the dengue virus was isolated in India for the first time in 1944 and the virologically confirmed epidemic was reported from the Vellore district in Tamil Nadu during 1956, it took nearly two decades for incriminating a vector, *Ae. aegypti*, in the wild population for the first time (Carey et al., 1964). Another vector, *Ae. albopictus* was incriminated in 1980s (Reuben et al., 1988). Following a major dengue outbreak of DHF in Calcutta (now Kolkata), West Bengal in 1963, dengue began to quickly spread to other neighbouring states/UTs with increased number of cases and deaths, and tipped by the massive outbreak in Delhi in 1996, the infection invaded almost the entire country, including those states/UTs which were hitherto a *terra incognito* for the arbovirus (Gupta et al., 2012).

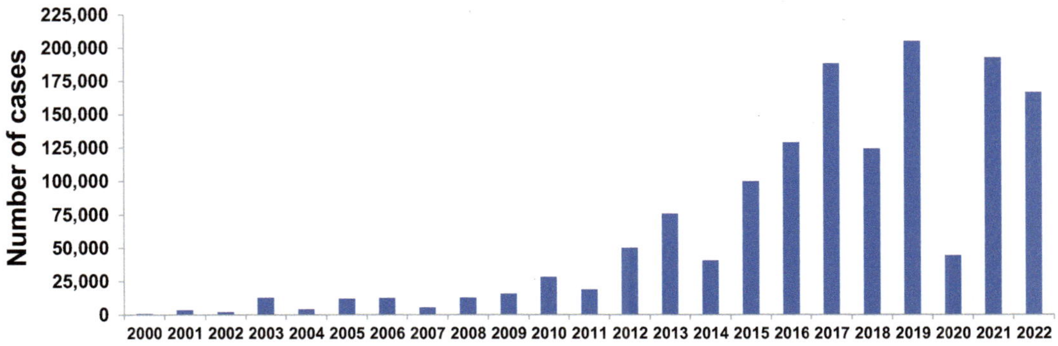

FIGURE 3.3 Dengue morbidity trend in India during 2000–2022.

Source: NCVBDC

TABLE 3.1

Dengue Incidence in India (2017–22)

Sl. No.	State/UT	2017	2018	2019	2020	2021	2022	2017	2018	2019	2020	2021	2022
		C	D	C	D	C	D	C	D	C	D	C	D
1	Andhra Pradesh	4,925	0	4,011	0	5,286	0	925	0	4,760	0	6,391	0
2	Arunachal Pradesh	18	0	1	0	123	0	1	0	7	0	114	0
3	Assam	5,024	1	166	0	196	0	33	0	103	0	1,826	2
4	Bihar	1,854	0	2,142	0	6,712	0	493	2	633	2	13,972	32
5	Chhattisgarh	444	0	2,674	10	722	0	57	0	1,086	0	2,679	10
6	Goa	235	0	335	1	992	0	376	0	649	0	443	1
7	Gujarat	4,753	6	7,579	5	18,219	17	1,564	2	10,983	14	6,682	7
8	Haryana	4,550	0	1,898	0	1,207	0	1,377	0	11,835	13	8,996	18
9	Himachal Pradesh	452	0	4,672	7	344	2	21	0	349	0	3,326	1
10	Jharkhand	710	5	463	1	825	0	79	0	220	1	290	0
11	Karnataka	17,844	10	4,427	4	16,986	13	3,823	0	7,393	7	9,889	9
12	Kerala	19,994	37	4,083	32	4,652	16	4,399	5	3,251	27	4,432	29
13	Madhya Pradesh	2,666	6	4,506	5	4,189	2	806	0	15,592	11	3,318	2
14	Meghalaya	52	0	44	0	82	0	4	0	129	0	26	0
15	Maharashtra	7,829	65	11,011	55	14,907	29	3,356	10	12,720	42	8,578	27
16	Manipur	193	1	14	0	359	0	37	0	203	0	503	4
17	Mizoram	136	0	68	0	42	0	67	0	83	0	1,868	5
18	Nagaland	357	0	369	0	8	0	1	0	24	0	154	0
19	Odisha	4,158	6	5,198	5	3,758	4	496	0	7,548	0	7,063	0
20	Punjab	15,398	18	14,980	9	10,289	14	8,435	22	23,389	55	11,030	41
21	Rajasthan	8,427	14	9,587	10	13,706	17	2,023	7	20,749	96	13,491	10
22	Sikkim	312	0	320	0	444	0	11	0	243	1	264	0
23	Tamil Nadu	23,294	65	4,486	13	8,527	5	2,410	0	6,039	8	6,430	8
24	Tripura	127	0	100	0	114	0	24	0	349	0	56	0
25	Telangana	5,369	0	4,592	2	13,331	7	2,173	0	7,135	0	8,972	0
26	Uttar Pradesh	3,092	28	3,829	4	10,557	26	3,715	6	29,750	29	19,821	33

(Continued)

TABLE 3.1
(Continued)

Sl. No.	State/UT	2017	2018	2019	2020	2021	2022	2017	2018	2019	2020	2021	2022
		C	D	C	D	C	D	C	D	C	D	C	D
27	Uttarakhand	849	0	689	3	10,622	8	76	1	738	2	2,337	0
28	West Bengal	37,746	46	23,301	86	47,928	27	5,166	10	8,264	7	67,271	30
29	A&N Islands	18	0	49	0	168	0	98	0	175	0	1,014	3
30	Chandigarh	1,125	0	301	0	286	0	265	0	1,596	3	910	1
31	Delhi	9,271	10	7,136	4	5,077	0	1,269	0	13,089	23	10,183	9
32	Dadra and Nagar Haveli	2,064	0	493	0	1,491	2	248	0	547	0	685	0
	Daman and Diu	59	0	163	0	625	2	71	0	279	0	228	0
33	Jammu and Kashmir	488	0	214	0	439	0	53	0	1,709	4	8,269	18
34	Ladakh	0	0	0	0	0	0	0	0	0	0	0	0
35	Lakshadweep	0	0	0	0	0	0	0	0	1	0	67	0
36	Puducherry	4,568	7	592	2	2,030	2	633	1	1,625	1	1,673	3
	Total	**188,401**	**325**	**124,493**	**258**	**205,243**	**193**	**44,585**	**66**	**193,245**	**346**	**233,251**	**303**

3.5 DENGUE VECTORS: *AEDES AEGYPTI* AND *AEDES ALBOPICTUS*

3.5.1 Classification

The systematic classification of *Aedes aegypti* (Linnaeus, 1962) and *Ae. albopictus* is presented in Table 3.2 and localizes these two species within the order Diptera, family Culicidae, subfamily Culicinae, tribe Aedini, genus *Aedes*, subgenus *Stegomyia. Aedes aegypti* presents two sub-species or subpopulations:

i. The first sub-species, *Ae. aegypti formosus*, is the ancestor of the domestic form of *Ae. aegypti* still lives in forests and vegetated ecotones in sub-Saharan Africa (Lounibos, 1981). In addition to its attraction to tree holes for breeding habitats and egg laying, it has a preference for non-human blood as sources of blood meals (required by females for egg production) and feeds on wild animals. Morphologically, this form is much darker than the form adapted to human habitats (McClelland, 1974). India does not so far have this sub-species.
ii. The second sub-species, *Ae. aegypti aegypti* (often designated by the shorter name *Ae. aegypti*) is found globally in tropical and subtropical regions, typically in association with humans, but is absent from the interior of Africa south of the Sahara (Moore et al., 2013; Powell and Tabachnick, 2013). In contrast to the first sub-species, *Ae. aegypti aegypti* predominantly breeds in artificial containers provided by humans, it also breeds indoors, and it has a preference for feeding on human blood (Moore et al., 2013). India has only *Ae. aegypti aegypti* (Tyagi, 2020; Tyagi et al., 2015).

A third sub-species was previously thought to exist, *Ae. aegypti queenslandensis*; it was described as a light-coloured form found in the Mediterranean Basin (Mattingly, 1967). However, recent analysis suggests *Ae. aegypti queenslandensis* is genomically identical to the second sub-species *Ae. aegypti aegypti* (Rašić et al., 2016).

TABLE 3.2
Organization of the Two Major Dengue Vectors, *Aedes aegypti* and *Ae. albopictus*, in the Classificatory System

Sl. No.	Taxon	Nomenclature (Authority)
1	Kingdom	Animalia (Margulis and Schwartz, 1998)
2	Subkingdom	Bilateria (Hatschek, 1888)
3	Infrakingdom	Protostomia (Grobben, 1908)
4	Superphylum	Ecdysozoa (Aguinaldo et al., 1997)
5	Phylum	Arthropoda (Latreille, 1829)
6	Subphylum	Hexapoda (Latreille, 1825)
7	Class	Insecta (Linnaeus, 1758)
8	Subclass	Pterygota (Lang, 1888)
9	Infraclass	Neoptera (Martynov, 1923)
10	Superorder	Endopterygota (Sharp, 1898)
11	Order	Diptera (Linnaeus, 1758)
12	Suborder	Nematocera (Berthold, 1927)
13	Infraorder	Culicomorpha (Wood & Borkent, 1989)
14	Family	Culicidae (Stephens, 1829)
15	Subfamily	Culicinae (Meigen, 1818)
16	Tribe	Aedini (Neveu-Lemaire, 1902)
17	Genus	*Aedes (Meigen, 1818)*
18	Subgenus	*Stegomyia (Theobald, 1901)*
19	Species	*Aedes aegypti* (Linnaeus, 1762)

3.5.2 Morphology

3.5.2.1 Larvae (Figure 3.4)

Aedes aegypti larvae resemble other mosquito larvae in their morphology; in general, they have an ovoid head, thorax, and abdomen of nine segments. The posterior segment (anal) has four lobed gills for osmotic regulation and a short barrel-shaped siphon bearing a single pair of subventral tufts for breathing at the water surface (Figure 3.4; Nelson, 1986; Clements, 2000; Service, 2012). Additional morphologic characteristics include at least three pairs of setae in the ventral brush, antennae that are not greatly flattened, and a lack of enormous setae on the thorax. These characteristics are sufficient in distinguishing *Aedes* larvae from most others belonging to the family Culicidae and subfamily Culicinae (Service, 2012).

The resting position at the water surface is also different among the various mosquito species: *Anopheles* larvae lay parallel to the water surface. *Culex* larvae rest at an angle. *Aedes* larvae hang almost vertically. The larvae pass through four instars (I, II, III, and IV respectively) with growth and changes in form and size occurring during their development. The first instar *Ae. aegypti* larva is only about 1 mm in length, whereas in the fourth instar stage, it reaches a length of approximately 8 mm (Schaper and Hernandez-Charvarria, 2006; Bar and Andrew, 2013a). Growth and development of larval instars are temperature dependent, however, complex interactions with other factors such as resource availability and intraspecific density also contribute to variation in development rate (Couret and Benedict, 2014). At cool environmental temperatures (around 15°C), *Ae. aegypti* larvae can remain in a particular instar for months, so long as the water supply is minimally sufficient (Foster and Walker, 2002; Bar and Andrew, 2013a; Brady et al., 2013).

The most distinguishing characteristics facilitating the differentiation of *Ae. aegypti* larvae from many other species of the *Aedes* genus are the 2 lateral spines on each side of the thorax and the

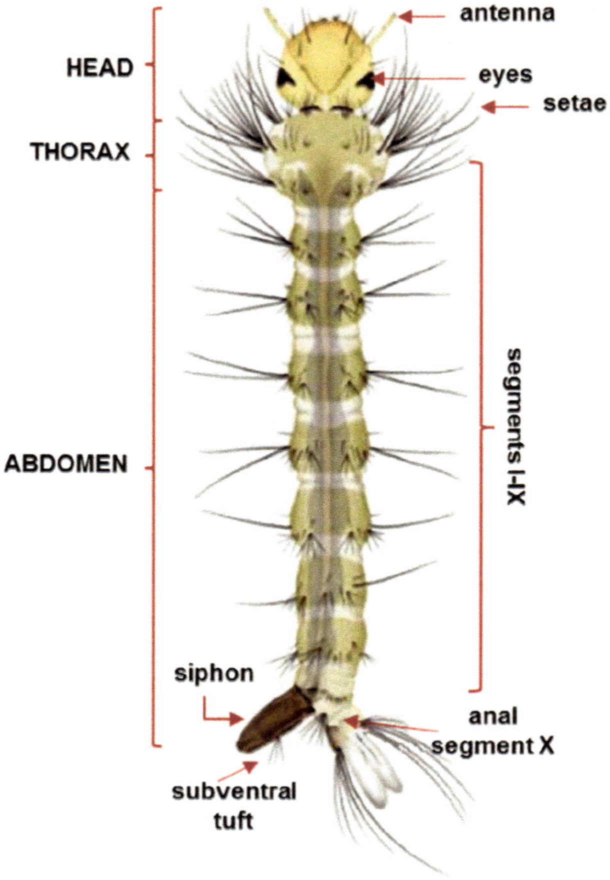

FIGURE 3.4 Dorsal view of *Ae. aegypti* larva.

Source: Rueda, 2004.

straight row of 7 to 12 comb scales on the 8th abdominal segment. *Ae. aegypti* exhibits a medial spine with stout, subapical spines (Figure 3.5, panel A) which are absent in *Ae. albopictus* (Figure 3.5, panel B; Nelson, 1986).

3.5.2.2 Pupae (Sexual Dimorphism)

The pupa is the stage of the life cycle of mosquitoes that follows the last larval instar and precedes the adult stage. Pupae are comma-shaped, composed of two main sections: cephalothorax (head and thorax fused) and abdomen (Nelson, 1986; Service, 2012). At the base of the cephalothorax of the pupa is a pair of breathing tubes or "trumpets" that pierce the water surface to allow breathing (Nelson, 1986). At the tip of the abdomen, there is a pair of oars or paddles used for swimming, which in the female (Figure 3.6) are wider and overlap but in the male (Figure 3.6, panel B) are narrow and separated (Vargas, 1968).

Another morphologic difference between female and male pupae is their overall size, with the female usually being larger than the male. Since the range in body size between female and male pupae overlaps considerably and can be affected by both biotic and abiotic, including environmental factors such as diet, temperature, rearing conditions, overcrowding, it is deemed necessary to select additional sexually dimorphic characteristics such as the differences in paddles in order to determine the sex of pupae (Vargas, 1968).

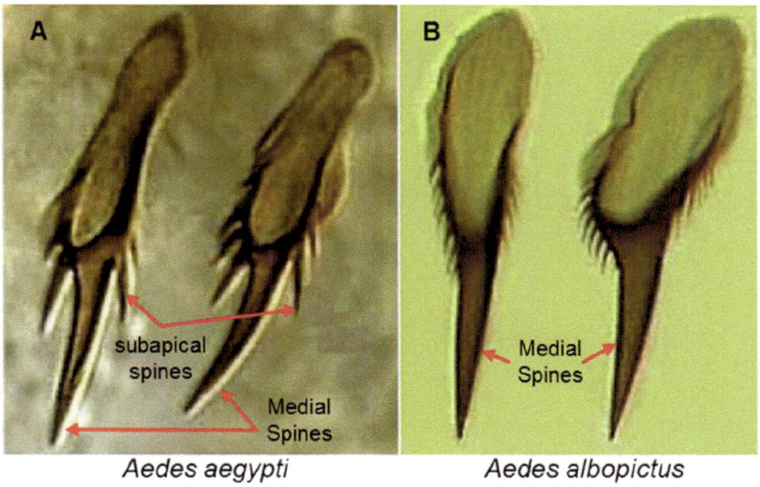

FIGURE 3.5 Comb scales of *Aedes aegypti* (A) exhibiting a median spine with stout, subapical spines and *Ae. albopictus* without subapical spines (B).

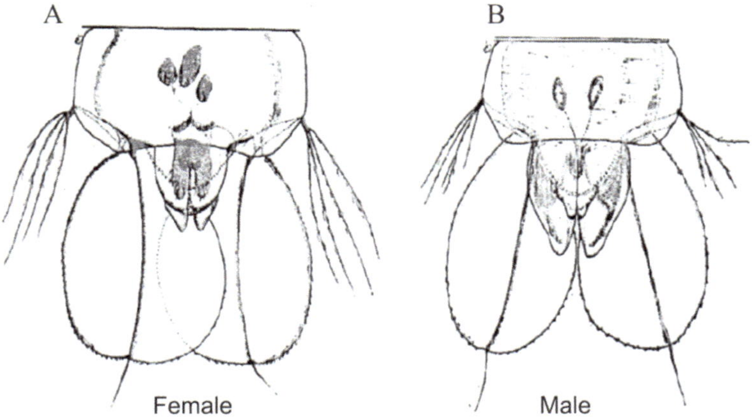

FIGURE 3.6 Anal segments of pupae. Ventral view, showing differences in dimorphism characters between sexes: female (A) and male (B).

Source: Vargas, 1968.

3.5.2.3 Adults (Male and Female)

The body of an adult *Ae. aegypti* mosquito is composed of head, thorax, and abdomen (Figure 3.7). *Ae. aegypti* males and females are similar in appearance except for the differences in size and form of the antennae (males have plumose antennae), maxillary palps (females have shorter palps), abdomen, claws and in scale markings (Bar and Andrew, 2013b). These differences are described in detail in the following sections.

3.5.2.3.1 Head

In both male and female *Ae. aegypti*, dorsally, the head is globular in shape and laterally convex with a vertex that has silvery white flat scales. The female clypeus has two silvery white dots, whereas the male has no dots. Females have a larger head capsule (0.55 ± 0.09 mm) than males (0.53 ± 0.06 mm; Bar and Andrew, 2013b). The head bears several structures critical to the mosquito's ability to feed, as well as to act as a vector of human diseases.

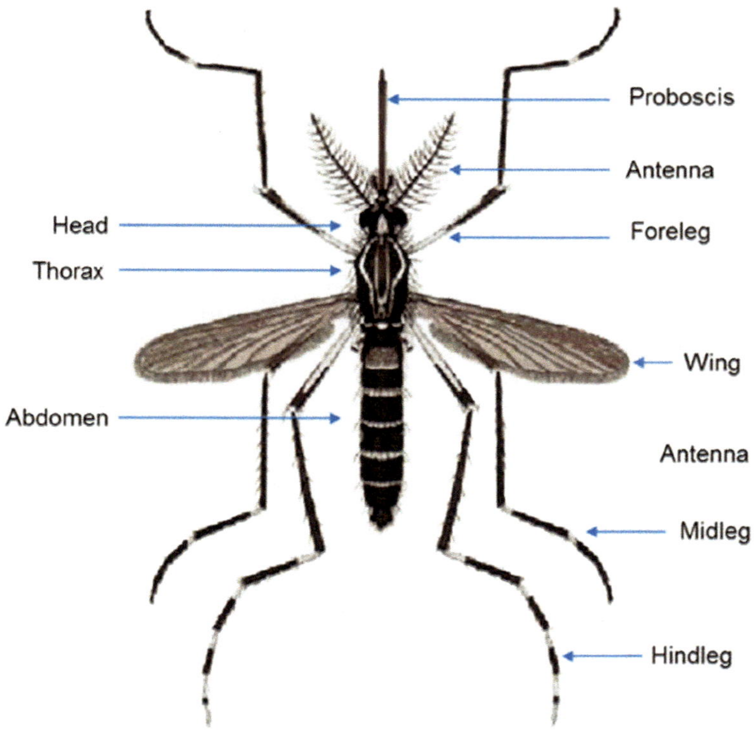

FIGURE 3.7 Dorsal view of the female mosquito *Ae. aegypti.*

Source: Rueda, 2004.

a. *Mouthparts*. The mouthparts in these mosquitoes include a pair of maxillary palps, which have five white scale bands and are longer (0.77 ± 0.06 mm) and more developed in males than in females (0.53 ± 0.06 mm; Bar and Andrew, 2013b). The proboscis is longer in males (0.76 ± 0.04 mm) than in females (0.66 ± 0.03 mm; Bar and Andrew, 2013b). However, in females, this structure is adapted for skin penetration to enable blood feeding, even though they may survive in nature by sucking plant juices. The male proboscis is adapted to feed on nectar and plant juices rich in carbohydrates (Clements, 1992).
b. *Antenna*. Each antenna of *Ae. aegypti* arises from a globular pedicel, has 13 flagellar segments, and a greatly reduced scape. Males have longer antennae (0.57 ± 0.03 mm) than females (0.52 ± 0.07 mm). The antennal hairs are bushy and plumose in males, whereas in females, they are smaller and less dense (Nelson, 1986; Bar and Andrew, 2013b).

3.5.2.3.2 Thorax

Females of *Ae. aegypti* have a larger thorax measuring 0.50 ± 0.08 mm in length and 0.35 ± 0.07 mm in width, while the shorter male thorax is 0.41 ± 0.06 mm in length and 0.29 ± 0.02 mm in width. The thorax of *Ae. aegypti* is black or dark brown coloured and consists of the pro-, meso-, and metathoracic segments, which together bear the wings (one pair), legs (three pairs), and halteres (one pair; Bar and Andrew, 2013b). Many, but not all, *Aedes* adults have conspicuous patterns on the thorax formed by white or silver coloured scales (Service, 2012), and these patterns vary between species.

An example of the difference across species is the case of *Ae. aegypti* with its typical white, lyre-shaped markings (Figure 3.8), compared to *Ae. albopictus* with its median-longitudinal white stripe (Figure 3.8, right panel Nelson, 1986). The scutellum in *Ae. aegypti* is three-lobed, with each lobe

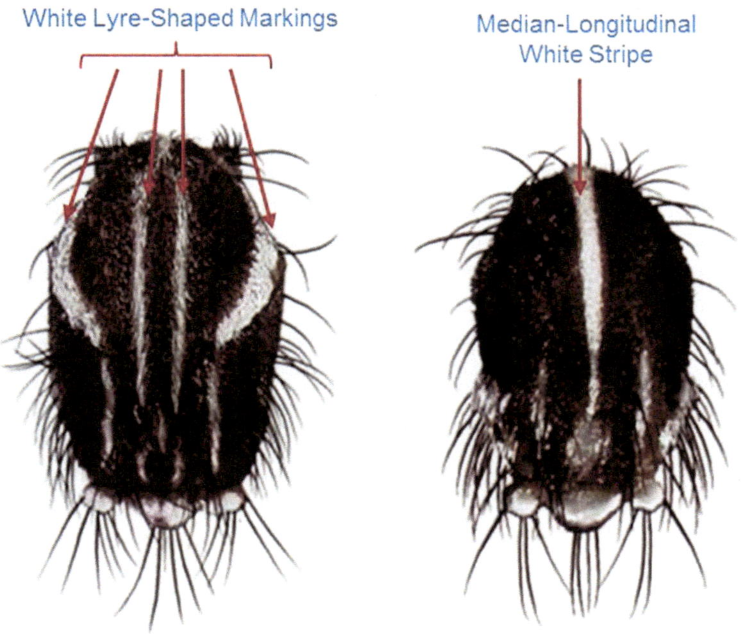

FIGURE 3.8 Comparative dorsal view of thoracic scutum of *Ae. aegypti* and *Ae. albopictus*.

Source: Rueda, 2004.

having silvery white scale patches, and a few dark scales at the apex of the mid-lobe (Bar and Andrew, 2013b).

At the same time, adults of *Aedes* and other Culicinae may be distinguished from adult *Anopheles* mosquitoes by their shorter palps and their resting position, which is more horizontal or parallel to the resting surface (Nelson, 1986; Service, 2012).

3.5.2.3.3 Abdomen

The abdomen consists of eight segments covered with black and white scales forming distinctive patterns in both males and females. In females, the eighth segment is greatly reduced. The tergites (dorsal portion of each abdominal segment) are dark brown in colour, and the first abdominal segment has a patch of pale, median scales. The dorsal side of abdominal segments II through VII has transverse white bands. The size of the abdomen in males is larger (length 3.03 ± 0.18 mm and width 0.51 ± 0.07 mm) than in females (length 2.94 ± 0.20 mm and width 0.41 ± 0.06 mm; Bar and Andrew, 2013b).

The posterior tip of the abdomen is narrow in males, while in females it has a broad round shape. *Ae. aegypti* can be differentiated from most of the other Culicinae by their pointed abdomen and the absence of spiracular bristles (Service, 2012).

With age, the lyre-shaped markings on the thorax may disappear, but the distinctive white scales on the pedicel, clypeus, and tip of the palps, and the pattern of white scales on abdominal sternites (ventral plate on each abdominal segment) III–V, usually remain. These characteristics are essential for the identification of *Ae. aegypti* females with damaged morphological structures and to differentiate them from *Ae. albopictus* females (Nelson, 1986; Savage and Smith, 1995).

3.5.2.4 *Aedes* Breeding Sites

Different micro-environmental conditions affect the suitability of each type of container with regards to volume, material, physical properties, and exposure to sunlight, etc. In the northern parts of the country, from December to March, breeding mainly found in plastic drums and large containers, thereafter from April onwards in coolers and indoor water filled containers. Following pre-monsoon intermittent rains, gradually breeding spills over to overhead/cement tanks etc. which further expands during monsoon and post-monsoon to indoor water storage containers and outdoor rain fed containers.

Ae. albopictus is known to breed in natural habitats. However, in Kerala, breeding is observed in indoor situations like freeze trays, plant pots, tyres, coconut shells, etc. In affluent areas, breeding mainly detected in ornamental breeding habitats. Like any other vector-borne disease, dengue and Chikungunya are also local and focal. The source reduction strategy for one area may not be suitable to another area if specific containers are not mapped. The national programme has developed guidelines to strategize the source reduction activities.

3.6 DENGUE VIRUSES

The dengue virus is a single stranded positive sense RNA (ribonucleic acid) virus of the family Flaviviridae and genus *Flavivirus*. The genome is composed of three structural protein genes encoding the nucleocapsid of core protein (C), a membrane associated protein (M), an envelope protein (E), and seven nonstructural protein (NS) genes. The virion consists of a nucleocapsid with cubic symmetry enclosed in a lipoprotein envelope. The dengue virus genome is approximately 11,000 base pairs in length and is composed of three structural protein genes. Based on antigenic and biological characteristics, dengue fever can be caused by any one of the four serotypes of dengue virus: DEN-1, DEN-2, DEN-3, and DEN-4. The spectrum of clinical illness is broad and ranges in severity from mild symptoms to death. Dengue fever (DF) is rarely fatal, but DHF and dengue shock syndrome (DSS) are serious diseases that lead to death. Dengue virus infection is clinically similar to many other acute febrile illnesses in which laboratory diagnosis plays an important role in early diagnosis and management. The spread of dengue is attributed to expanding geographic distribution of the four serotypes and of their mosquito vectors, the most important of which is the predominantly urban species *Aedes aegypti*.

3.7 DENGUE TRANSMISSION

Dengue is driven by complex interactions among host, vector, and virus that are influenced by climatic factors and anthropization (Baruah et al., 2004, 2008). Temperature and rainfall are important climatic factors in mosquito population and disease transmission dynamics (Randolph and Rogers, 2010). Precipitation is often required to create and maintain breeding sites and consequently has a strong influence on vector distribution (Tseng et al., 2009). Warm temperatures and high humidity favour increased longevity of the adult mosquitoes and shorten the viral incubation period within the vector and its blood-feeding intervals, thus leading to faster virus replication and increased transmission intensity (Arcari et al., 2007). The association between climate and dengue varies across geographical locations and socio-environmental strata (Sharmin et al., 2015). Similar observations were also reported about the correlation of climatic variability and dengue incidence having a strong positive correlation between temperature, humidity, and incidence. Cases increase from July to November and then decline. From 2008 to 2009, transmission continued beyond October. Gradually, the peak of transmission shifted from September to October from 2012 onwards till 2018. In 2019, the peak has further shifted to November with the expansion of transmission window. Consequently, dengue has visibly shifted its peak transmission from September to October over nearly a decade (2012–19; Figure 3.9).

Quarterly analysis of the data over two decades (2000–2009 and 2010–19) at the national level reveals that in 2000–2009, during September to November, 65.6% of the total cases were reported, followed by 19.2% during June to August and lowest numbers of cases (4.1%) were reported during the months of March, April and May. Almost similar trend was observed during 2010–19. A total of 68% cases were reported during September to November, 17.7% during June to August, and the

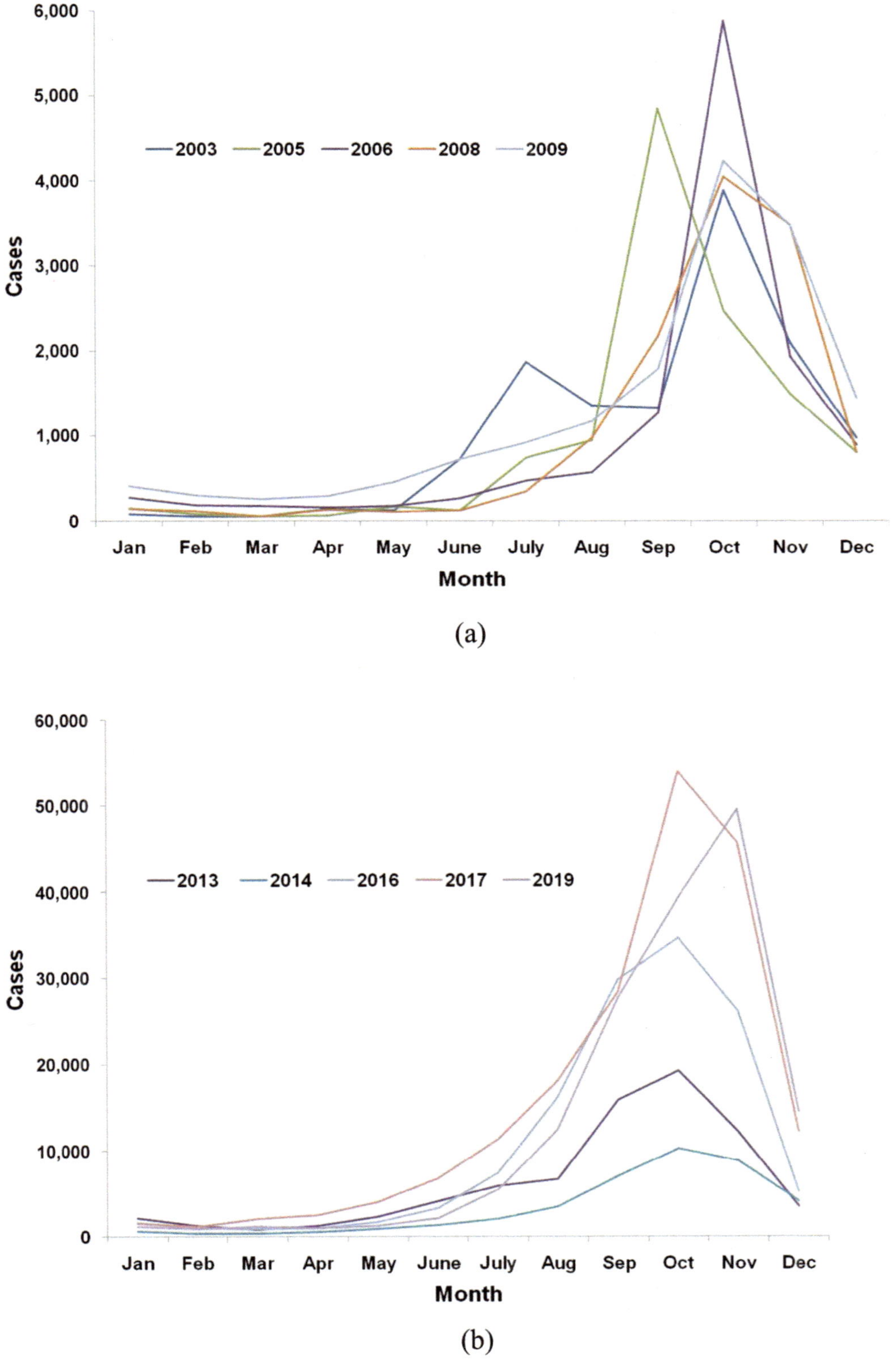

FIGURE 3.9 A clear shift in the peak in transmission of dengue from (a) September (2000–2019) to (b) October (2010–19).

Source: Baruah et al., 2021.

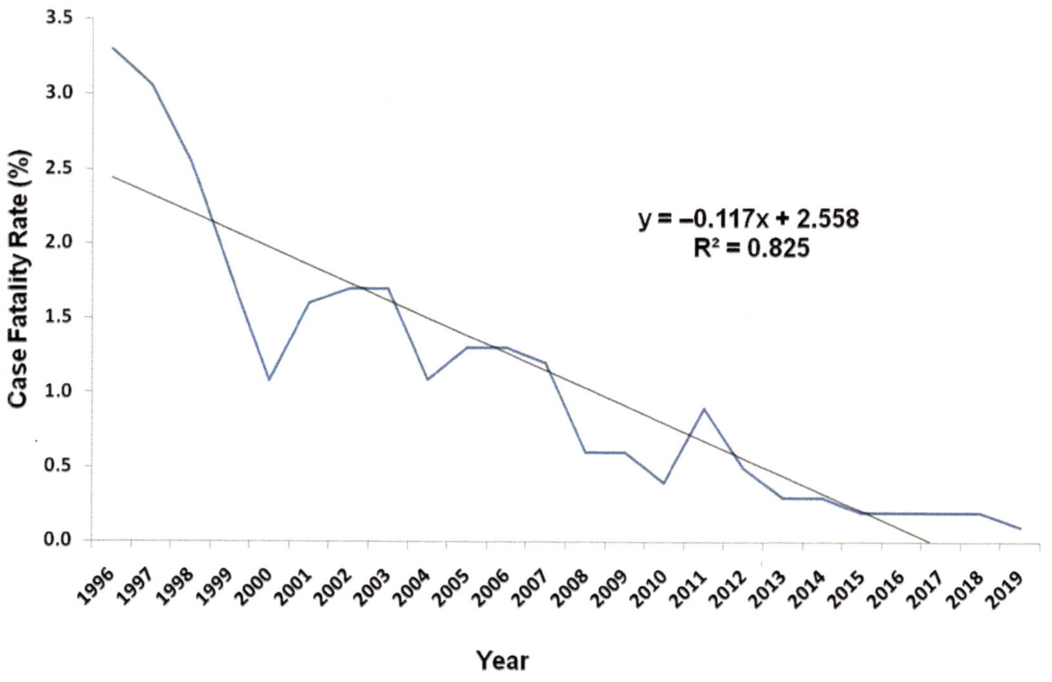

FIGURE 3.10 Dengue CFR in India during 1996–2019.

Source: Baruah et al., 2021.

lowest numbers of cases, i.e., 4.1%, were reported during the months of March, April, and May. Similarly, during 2022, 2.89% of total cases were reported in January–March, whereas, 3.90% in April–June, 28.46% during July-September and 64.75% in October–December. In northern and eastern states of the country, dengue cases start to increase with the onset of monsoon, whereas transmission is perennial in south-western states from where the cases are reported throughout the year.

Though the incidence of dengue has increased, however, the CFR (deaths per 100 cases) has declined from 3.3% in 1996 to 0.4% in 2010, to 0.25% in 2013. As per national target, dengue CFR has to be sustained at < 1% at the national level. After the introduction of first national guidelines for dengue case management and repeated trainings, CFR was brought down to < 1% during 2008 (0.6%). But it increased in 2011 (0.9%), when Orissa (now Odisha) reported its' first dengue outbreak with 1,816 cases and 33 deaths. Punjab and Maharashtra also reported 33 and 25 deaths, respectively, in 2011. The CFR was further reduced after updating the national guidelines for dengue case management in 2014. It was sustained at 0.2% from 2015 to 2018 and brought down to 0.1% in 2019. Early diagnosis and timely referral also played a pivotal role in bringing down the CFR. The CFR for dengue in India was steadily declining from 1996 to 2019 (R2 = 020.825), which is depicted in Figure 3.10. Since 2019, the CFR at the national level has been maintained at 0.1%.

3.8 PREVENTION AND CONTROL OF DENGUE

As of this date, no specific anti-viral treatment or vaccine against dengue is available. The key options for preventing and controlling dengue are as follows:

- To control larval habitats in and around homes, workplaces, unclaimed empty spaces, dump sites, and public areas, such as places of worship/recreational centres, roadsides, playgrounds, cemeteries
- To reduce human-vector contact through mechanical barriers

- To ensure prompt diagnosis of cases of fever and appropriate clinical management as per national guidelines

***Aedes* Control Methods:** A range of *Aedes* control methods now exist, many of which have been tried and proven for different settings as follows:

- Environmental measures to reduce mosquito breeding sites (source reduction), such as physical management of water containers (e.g., mosquito-proof covers for water storage containers, polystyrene beads in water tanks), better designed and reliable water supplies, and recycling of solid waste such as discarded tyres, bottles, and cans
- Biological methods (e.g., larvivorous fish that feed on mosquito larvae) to kill or reduce larval mosquito populations in large water containers
- Chemical methods against the mosquito's aquatic stages for use in larger water containers (e.g., larvicide, insect growth regulator)
- Chemical methods directed against adult mosquitoes, such as insecticide space sprays
- Personal protection through the use of repellents, vaporizers, mosquito coils, and insecticide-treated screens, curtains, bed nets (for daytime use against *Aedes*)

3.9 NATIONAL PROGRAMME FOR DENGUE

Under the Ministry of Health and Family Welfare, the Government of India, the National Center for Vector Borne Diseases Control (NCVBDC – earlier National Vector Borne Disease Control (NVBDCP)) is mandated for prevention and control of dengue, along with other vector borne diseases. Technical guidance is provided to the states and UTs for prevention and containment of the disease. To augment free diagnostic services for dengue across the country, a network of laboratories has been established.

3.10 CONCLUSION

In the past few decades, Dengue has emerged as the most important public health problem in the world where it is currently prevalent in approximately 141 countries, mostly in tropical and subtropical regions (Bhatt et al., 2013; Shepard et al., 2014). According to WHO, dengue fever is one of the top ten global health threats. It's also the most rapidly spreading; there has been a 30-fold increase in global incidence over the past 50 years. One of the deadliest systemic arboviral infections, dengue is transmitted between humans by *Aedes* mosquitoes (*Aedes aegypti* or *Ae. albopictus*). Despite certain candidate vaccines in the final stage of testing, there are as yet no licensed vaccines or specific therapeutics available for public administration. Therefore, the only remedy in tackling dengue is substantial vector control which has not stopped its rapid emergence and global spread. In 2023, and as of 27 July, over 3 million cases and over 1,500 dengue-related deaths have been reported globally.

Dengue has become a highly serious public health issue with enormous disease burden (Gupta et al., 2012; Shepard et al., 2014; Murhekar et al., 2019). Climate change, escalated urbanization, swift and widespread transportation, man-made mosquitogenic conditions through behavioural shift (anthropization), and, above all, vector mosquitoes' reproductive plasticity as well susceptibility to dengue viruses (DENV 1–4) are the main challenges before humanity (Tyagi, 2005). Dengue is a climate-sensitive disease, and models show that climate can predict mosquito-borne disease dynamics in India (Tyagi et al., 2006; Bal and Sodoudi, 2020; Kakarla et al., 2020).

Dengue is currently prevalent in all states and UTs of India, except Ladakh, and is on the rise nationally due to various factors (Baruah et al., 2021). Initially, only *Aedes aegypti* (Figure 3.11) was regarded as the sole vector of major public health importance (Sarkar et al., 1964; Carey et al., 1964), but Tyagi et al. (2006) demonstrated that *Ae. albopictus* could act not only as a dominant vector but also transmit dengue viruses as the primary vector in the absence of *Ae. aegypti*. Even, a

FIGURE 3.11 *Aedes aegypti*, the most dangerous dengue vector in the world.

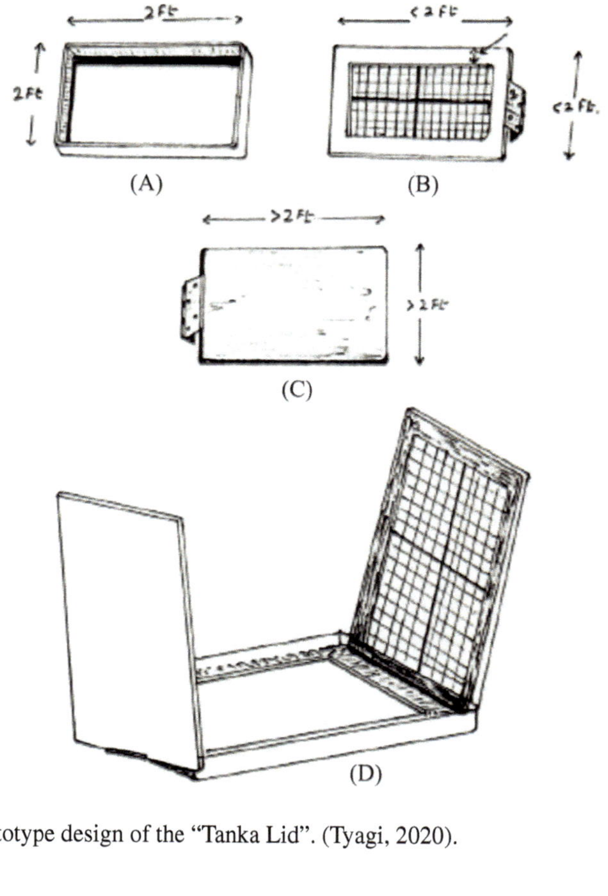

FIGURE 3.12 A prototype design of the "Tanka Lid". (Tyagi, 2020).

study carried out in Delhi by Kumari et al. (2011) shows that *Ae. albopictus* adapted to breeding in a variety of man-made containers in urban areas in addition to its natural breeding sites.

To overcome challenges associated with dengue, a strategic approach is needed at various levels with involvement of the non-health sector and community until an effective drug or vaccine is available.

In the recent past, WHO offered a useful slogan: "*Think globally, act locally.*" Working from this tenet, Tyagi and Hiriyan (2004), Tyagi (2020) invented the "Tanka Lid" in the year 2000 to prevent malaria (*Anopheles stephensi*) and dengue vector (*Aedes aegypti*) from breeding in thousands of tankas, the underground water reservoirs, in the Thar Desert (Figure 3.12).

REFERENCES

Adalja, A.A., Sell, T.K., Bouri, N. and Franco, C. (2012). Lessons learned during dengue outbreaks in the United States, 2001–2011. *Emerging Infectious Diseases*, 18, 608–614.

Arcari, P., Tapper, N. and Pfueller, S. (2007). Regional variability in relationships between climate and dengue/ DHF in Indonesia. Singapore. *Journal of Tropical Geography*, 28, 251–272.

Bal, S. and Sodoudi, S. (2020). Modelling and prediction of dengue occurrences in Kolkata, India, based on climate factors. *International Journal of Biometeorology*, 64(8),1379–1391. doi: 10.1007/s00484-020-01918-9

Bar, A. and Andrew, J. (2013a). Morphology and morphometry of *Aedes aegypti* larvae. *Annual Review and Research in Biology*, 3, 1–21.

Bar, A. and Andrew, J. (2013b). Morphology and morphometry of *Aedes aegypti* adult mosquito. *Annual Review and Research in Biology*, 3, 52–69.

Baruah, K., Arora, N., Sharma H and Katewa A. (2021). Dengue in India: Temporal and spatial expansion in last two decades. *Journal of Medical Arthropodology and Public Health*, 1(1), 15–32.

Baruah, K. and Dhariwal, A.C. (2011). Epidemiology of dengue, its prevention and control in India. *Journal of Indian Medical Association*, 109(2), 82–86.

Baruah, K., Katewa, A., Singh, G. and Dhingra, N. (2020). Epidemiological stratification of dengue in India and strategic challenges. *Dengue Bulletin*, 2020. 41-149-165.

Baruah, K., Kumar, A. and Meena, V.R. (2004). Entomological investigations for DF/DHF in Alwar district, Rajasthan, India. *Dengue Bulletin*, 28, 213–215.

Baruah, K., Singh, I. K., Agrawal, C.M. and Dhillon, G.P.S. (2008). An outbreak of dengue in Moreh: A small rural town in Chandel District, Manipur near Indo-Myanmar border. *Dengue Bulletin*, 32, 219–221.

Beatty, M.E., Beutels, P., Meltzer, M.I., Shepard, D.S., Hombach, J., Hutubessy, R., Dessis, D., Coudeville, L., Dervaux, B., Wichmann, O., Margolis, H.S., Kuritsky, J.N. (2011). Health economics of dengue: A systematic literature review and expert panel's assessment. *American Journal Tropical Medicine and Hygiene*, 84, 473–488.

Bhatt, S., Gething, P.W., Brady, O.J., Messina, J. P., Farlow, A.W., Moyes, C.L., Drake, J.M., Brownstein, J. S., Hoen, A.G., Sankoh, O., Myers, M.F., George, D.B., Jaenisch, T., Wint, G.R., Simmons, C.P., Scott, T.W., Farrar, J.J. and Hay, S.I. (2013). The global distribution and burden of dengue. *Nature*, 496, 504–507. doi: 10.1038/nature12060

Brady, O.J. et al. (2013). Modelling adult *Aedes aegypti* and Aedes albopictus survival at different temperatures in laboratory and field settings. *Parasites and Vectors*, 6, 351.

Carey, D.E., Myers, R. M. and Reuben R. (1964). Dengue Types 1 and 4 viruses in wild-Caught mosquitoes in South India. *Science* 143(3602), 131–132. DOI: 10.1126/science.143.3602.131

Christophers, S.R. (1960). *Aedes aegypti* (L.). *The yellow fever mosquito. Its life history, bionomics and structure.* Cambridge University Press, Cambridge.

Clements, A.N. (1992). *The Biology of Mosquitoes, Volume I. Development, Nutrition and Reproduction.* Chapman and Hall, London.

Clements, A.N. (2000). *The Biology of Mosquitoes, Volume I. Development, Nutrition and Reproduction. Second Edition*, CABI Publishing, Oxford.

Effler, P.V., Pang, L., Kitsutani, P., Vorndam V., Nakata, M., Ayers, T., Elm, J., Tom, T., Reiter, P., Rigau-Perez, J.G., Hayes, J.M., Mills, K., Napier, M., Clark, G.G. and Gubler, D. J. (2005). Hawaii Dengue Outbreak Investigation Team Dengue fever, Hawaii, 2001–2002. *Emerging Infectious Disease*, 11, 742–749.

Foster, W.A. and Walker, E.D. (2002). Mosquitoes (Culicidae). *In*: G. Mullen and L. Durden (eds.), *Medical and Veterinary Entomology*, Academic Press, San Diego, pp. 203–262.

Gjenero-Margan, I., Aleraj, B., Krajcar, D., Lesnikar, V., Klobucar, A., Pem-Novosel, I., Kurecic-Filipovic, S., Komparak, S., Martic, R., Duricic, S., Betica-Radic, L., Okmadzic, J., Vilibic-Cavlek, T., Babic-Erceg, A., Turkovic, B., Avsic-Zupanc, T., Radic, I., Ljubic, M., Sarac, K., Benic, N. and Mlinaric-Galinovic, G. (2011). Autochthonous dengue fever in Croatia, August–September 2010. *Euro Surveillence*, 16, e19805.

Gould, E.A., Gallian, P., De Lamballerie, X. and Charrel, R.N. (2010). First cases of autochthonous dengue fever and chikungunya fever in France: From bad dream to reality! *Clinical Microbiology and Infection*, 16,1702–1704.

Gubler, D.J. (2002). Epidemic dengue/dengue hemorrhagic fever as a public health, social and economic problem in the 21st century. *Trends in Microbiology*, 10, 100–103.

Gubler, D.J., Ooi, E.E., Vasudevan, S. and Farrar, J. (2014). *Dengue and Dengue Hemorrhagic Fever*. Publisher: CABI Publishing, 624 pp.

Gupta, N., Srivastava, S., Jain, A. and Chaturvedi, U.C. (2012). Dengue in India. *Indian Journal of Medical Research*, 136, 373–390.

Guzman, M. G. and Kouri, G (2002). Dengue: An update. *Lancet Infectious Disease*, 2, 33.
Hariharan, D., Das, M.K., Shepard, D.S. and Arora, N.K. (2019). Economic burden of dengue illness in India from 2013 to 2016: A systematic analysis Author links open overlay panel. *International Journal of Infectious Diseases*, 84, 68–73.
Kakarla, S.G., Bhimala, K.R., Kadiri, M.R., Kumaraswamy, S and Mutheneni, S.R. (2020). Dengue situation in India: Suitability and transmission potential model for present and projected climate change scenarios. *Science Total Environment*, 15,739:140336. doi: 10.1016/j.scitotenv.2020.140336
Kumari, R., Kumar, K. and Chauhan, L.S. (2011). First dengue virus detection in *Aedes albopictus* from Delhi, India: Its breeding ecology and role in dengue transmission. *Tropical Medicine and International Health*, 16(8), 949–954. doi: 10.1111/j.1365-3156.2011.02789
Lounibos, L.P. (1981). Habitat segregation among African tree hole mosquitoes. *Ecological Entomology*, 6, 129–154.
Mattingly, P.F. (1967). Taxonomy of *Aedes aegypti* and related species. *Bulletin of the World Health Organization*, 36, 552–554.
McClelland, G.A.H. (1974). A worldwide survey of variation in scale pattern of the abdominal tergum of *Aedes aegypti* (L.) (Diptera: Culicidae). *Transactions of the Royal Entomological Society, London*, 126, 239–259.
Moore, M. et al. (2013). Dual African origins of global *Aedes aegypti* s.l. populations revealed by mitochondrial DNA. *PLoS Neglected Tropical Diseases*, 7(4), e2175.
Murhekar, M. V., Kamaraj, P., Kumar, M.S., Khan, S.A., Allam R.R., Barde, P., Dwibedi, B., Kanungo, S., Mohan, U., Mohanty, S.S., Roy, S., Sagar, V., Savargaonkar, D., Tandale, B.V., Topno, R.K., Sapkal, G., Kumar, G.C. P., Sabarinathan, R., Kumar, V.S., Bitragunta, S., Grover, G.S., Lakshmi, P.V.M., Mishra, C.M., Sadhukhan, P., Sahoo, P. K., Singh, S. K., Yadav, C.P., Bhagat, A., Srivastava, R., Dinesh, E.R., Karunakaran, T., Govindhasamy, C., Rajasekar, T.D., Jeyakumar, A., Suresh, A., Augustine, D., Ashok Kumar, P., Kumar, R., Dutta, S., Toteja, G. S., Gupta, N. and Mehendale, S.M. (2019). Burden of dengue infection in India, 2017: A cross-sectional population based serosurvey. *The Lancet Global Health* 7(8), E1065–E1073. doi: 10.1016/S2214-109X(19)30250-5
Murray, N.E., Quam, M.B. and Wilder-Smith, A. (2013). Epidemiology of dengue: Past, present and future prospects. *Clinical Epidemiology*, 5, 299–309.
Nelson, M.J. (1986). *Aedes aegypti*: Biology and Ecology. *Pan American Health Organization, Washington, DC*, PNSP/86-63, pp. 50.
Pan American Health Organization (PAHO) (2020). Dengue prevention and control during COVID-19 pandemic. Available from: https://www.paho.org/en/documents/dengue-prevention-and-control-during-covid-19-pandemic, accessed on November 17, 2020.
Powell, J.R. and Tabachnick, W.J. (2013), History of domestication and spread of *Aedes aegypti* – A review. *Memórias do Instituto Oswaldo Cruz*, 108, 11–17.
Rašić, G. et al. (2016). The *queenslandensis* and the type form of the dengue fever mosquito (*Aedes aegypti* L.) are genomically indistinguishable. *PLoS Neglected Tropical Diseases*, 10(11), e0005096.
Reuben, R., Kaul, H.N., and Soman, R.S. (1988). Mosquitoes of arboviral importance in India. *Mosquito-borne Disease Bulletin*, 5, 48.
Rochlin, I., Ninivaggi, D.V., Hutchinson, M.L. and Farajollahi, A. (2013). Climate change and range expansion of the Asian tiger mosquito (*Aedes albopictus*) in northeastern USA: Implications for public health practitioners. *PLoS ONE*, 8, e60874.
Rueda, L.M. (2004). Pictorial keys for the identification of mosquitoes (Diptera: Culicidae) associated with dengue virus transmission. *Zootaxa* 589, Magnolia Press, 60. pp.
Rueda, L.M. et al. (1990). Temperature-dependent development and survival rates of *Culex quinquefasciatus* and *Aedes aegypti* (Diptera. Culicidae). *Journal of Medical Entomology*, 27(5), 892-898. doi: 10.1093/jmedent/27.5.892
Sarkar, J.K., Chatterjee, S.N. and Chakravarty, S.K. (1964). Haemorrhagic fever in Calcutta: Some epidemiological observations. *Indian Journal of Medical Research*, 52, 651–659.
Savage, H.M. and Smith G.C. (1995). *Aedes albopictus* y *Aedes aegypti* en las Américas: Implicaciones para la transmisión de arbovirus e identificación de hembras adultas dañadas [Aedes albopictus and Aedes aegypti in the Americas: Implications for the transmission of arboviruses and identification of damaged adult females], *Boletin de la Oficina Sanitaria Panamericana*, 118, 473–487.
Schaper, S. and Hernandez-Charvarria, F. (2006). Scanning electron microscopy of the four larval instars of the Dengue fever vector *Aedes aegypti* (Diptera: Culicidae). *Revista de Biología Tropical*, 54, 847–852.
Schwartz, E., Weld, L.H., Wilder-Smith, A., von Sonnenburg, F., Keystone, J.S., Kain, K.C., Torresi, J. and Freedman, D.O. (2008). GeoSentinel Surveillance Network Seasonality, annual trends, and characteristics of dengue among ill returned travellers, 1997–2006. *Emerging Infectious Disease*, 14, 1081–1088.

Service, M. (2012). *Medical Entomology for Students*, 5th Ed. Cambridge University Press, New York, 303 pp.

Sessions, O.M., Khan, K., Hou, Y., Meltzer, E., Quam, M., Schwartz, E., Gubler, D.J. and Wilder-Smith, A. (2013). Exploring the origin and potential for spread of the 2013 dengue outbreak in Luanda, Angola. *Global Health Action*, 6, 21822.

Sharmin, S., Glass, K., Viennet, E. and Harley, D. (2015). Interaction of mean temperature and daily fluctuation influences dengue incidence in Dhaka, Bangladesh. *PLoS Neglected Tropical Diseases*, doi:10.1371/journal.pntd.0003901

Shepard, D.S., Halasa, Y.A., Tyagi, B.K., Adhish, S.V., Nandan, D., Karthiga, K.S., Chellaswamy, V., Gaba, M. and Arora, N.K. (2014). The Inclen Study Group. Economic and disease burden of dengue illness in India. *American Journal of Tropical Medicine and Hygiene*, 91(6), 1235–1242. doi: 10.4269/ajtmh.14-0002

Sulekha, C., Kumar, S. and Philip, J. (2004). Gullian-Barre syndrome following dengue fever. *Indian Paediatrics*, 41, 948.

Tseng, W.C., Chen, C.C., Chang, C.C. and Chu, Y.H. (2009). Estimating the economic impacts of climate change on infectious diseases: A case study on dengue fever in Taiwan. *Climatic Change*, 92, 123–140.

Tyagi, B.K. (2005). Thar Desert's changing climate and emergence/exacerbation of vector-borne diseases: Past, present and future. (*In*: *Changing faunal ecology in the Thar Desert*. Eds. B.K. Tyagi and Q.H. Baqri). Scientific Publishers, India, Jodhpur, pp. 323–367.

Tyagi, B.K. (2020). *Mosquito Hunters: A history of hostilities against man's deadliest Foe – the mosquito – since 1881*. Scientific Publishers, pp. 474.

Tyagi, B.K. and Hiriyan, J. (2004). Breeding of dengue vector, *Aedes aegypti* (Linnaeus) in the rural Thar Desert, north western Rajasthan, India. *Dengue Bulletin*, 28, 220–222.

Tyagi, B.K., Hiriyan, J. P. Samuel, P., Tewari, S.C. and Paramasivan, R. (2006). Dengue in Kerala: A critical review. *ICMR Bulletin*, 36(4–5), 13–22.

Tyagi, B.K., Munirathinam A. and Venkatesh, A. (2015). A catalogue of Indian mosquitoes. *International Journal of Mosquito Research*, 2(2), 50–97.

Vargas, V.M. (1968), Sexual dimorphism of larvae and pupae of *Aedes aegypti* (Linn.). *Mosquito News*, 28, 374–379.

WHO-ROSEA (2011). Comprehensive guidelines for prevention and control of dengue and dengue haemorrhagic fever. *SEARO Technical Publication Series* No. 60, 196 pp.

Wilder-Smith, A. and Gubler, D.J. (2008). Geographic expansion of dengue: The impact of international travel. *Medical Clinics of North America* 92,1377–1390.

World Health Organization (1999). Prevention and Control of dengue and Dengue Haemorrhagic Fever comprehensive guidelines WHO Regional Publications SEARO No. 29, New Delhi. Available from: https://apps.who.int/iris/bitstream/handle/10665/204894/B4751.pdf?sequence=1&isAllowed=y, accessed on November 13, 2020.

World Health Organization (2009). *Dengue: Guidelines for Diagnosis, Treatment, Prevention and Control.* Geneva: World Health Organization, Geneva.

World Health Organization (2011). Comprehensive Guidelines for prevention and control of dengue and dengue Haemorrhagic fever – WHO, SEARO, Revised and expanded edition. Available from https://apps.who.int/iris/bitstream/handle/10665/204894/B4751.pdf?sequence=1&isAllowed=y, accessed on November 16, 2020.

World Health Organization (2012). Global strategy for dengue prevention and control - 2012–2020. WHO Geneva. Available from: https://www.who.int/denguecontrol/9789241504034/en/, accessed on November 16, 2020.

World Health Organization (2018). A Toolkit for national dengue burden estimation, World Health Organization. Licence: CC BY-NC-SA 3.0 IGO.

World Health Organization (2019). Ten threats to global health in 2019. https://www.who.int/news-room/spotlight/ten-threats-to-global-health-in-2019

World Health Organization (2020a). Ten threats to global health in 2019. Available from: https://www.who.int/news-room/feature-stories/ten-threats-to-global-health-in, accessed on November 17, 2019.

World Health Organization (2020b). Dengue and severe dengue. 2020. https://www.who.int/news-room/fact-sheets/detail/dengue-and-severe-dengue, accessed on November 17, 2020.

4 Chikungunya in India

Kalpana Baruah, Gavendra Singh and Amit Katewa
National Center for Vector Borne Diseases Control, Directorate General of Health Services, Ministry of Health & family Welfare, Govt.of India, Delhi, India

Roop Kumari
WHO Country Office India, New Delhi, India

Mangesh Gokhale
Entomology Group, ICMR-National Institute of Virology, Pune, India

A.B. Sudeep
ICMR-National Institute of Virology, Microbial Containment Complex, Pashan, Pune, India

T. Dilip Kumar
Kerala State Public Health Department, Thiruvananthapuram, India

R. Balasubramanian
ICMR-National Institute of Virology-Kerala Unit, Dept. of Health Research, Ministry of Health & Family Welfare, Govt. T. D. Medical College Hospital, Vandanam, India

Ayushman Ghosh
Brainware University, Kolkata, India

S. Sridharan
Tamil Nadu Public Health Department, Chennai, India

Jhansi Charles
Rajaji Medical College, Madurai, India

Himmat Singh
ICMR-National Institute of Malaria Research, Sector 8, Dwarka, New Delhi, India

B.K. Tyagi
Chandigarh University, Mohali, India

 DOI: 10.1201/9781003327028-5

4.1 INTRODUCTION

Chikungunya, an important *Aedes* mosquito-borne arboviral disease, is caused by chikungunya virus (CHIKV), an *Alphavirus* of the *Togaviridae* family (Staples et al., 2009). It was described in 1952 in the Makonde Plateau, south-eastern Tanzania (Lumsden, 1955; Weinbren et al., 1958), and invaded India in 1963 when a major epidemic of chikungunya fever was reported in Kolkata (West Bengal), soon followed by a wave of epidemics concurrently in 1965 in Pondicherry (now Puducherry), Chennai (Tamil Nadu), Rajahmundry, Visakhapatnam, Kakinada (Andhra Pradesh), Sagar (Madhya Pradesh), and Nagpur (Maharashtra). In 1973, it was reported in Barsi (Maharashtra), although thereafter, sporadic cases continued to be recorded, especially in Maharashtra state during 1983 (Sudeep and Parashar, 2008; TRC-CVI, 2021). Since the re-emergence of virus in 2006, chikungunya outbreaks were reported from almost all regions of India (Dinkar et al., 2018). The recent re-emergence and travel-related spread of chikungunya infection to Europe and the United States has drawn global attention; in fact, the international travel, together with the rapid geographical dissemination of the vector *Aedes albopictus*, stands out as one of the major risk factors for the rapid global spread of the disease (Simon et al., 2023). An invasive *Ae. albopictus* induced outbreak of CHIKV, with more than 200 cases, was reported in Italy in 2007, also for the first time in Europe (Watson, 2007). Later, of course, a few autochthonous cases linked to imported cases were reported in France in 2010 and 2014. Currently, *Ae. albopictus* is established in many European countries, and it is feared that chikungunya will spread in the temperate European region under the impact of climate change and global warming. For the first time in the American region, an outbreak of chikungunya fever was reported in the Caribbean from the French part of the island of Saint Martin on 6 December 2013. Since then, autochthonous transmission of chikungunya fever has spread in the Caribbean and quickly in Central, North and South America.

The transmission involves a human-mosquito-human cycle commencing with an infective mosquito bite (Tyagi et al., 2014). Following two to six days of incubation, symptoms develop with high fever and severe arthralgia or joint bone pain. Generally, the disease tends to be self-limiting and, if no complications occur, usually subsides within two weeks. However, if infected person is not properly taken care of medically, the chronic disease can last up to several months or even years with persistent polyarthralgia with several severe systemic complications. The disease develops the stooped posture due to the rheumatologic manifestations describing the word 'chikungunya.' Symptoms may often overlap with dengue, and co-infection may be found with other pathogens (Venkatasubramani et al., 2015; Mukherjee et al., 2017); co-infection of dengue virus with CHKV and/or Japanese encephalitis virus has been frequently reported in recent years (Thenmozhi et al., 2013; Jain et al., 2017). Diagnosis and treatment of the chikungunya disease pose many challenges since several mutations have been identified in circulating strains of the virus resulting in improved adaptations or increased fitness in the vector(s), effective transmission, and disease severity. Interestingly, CHIKV is transmitted by the same vectors which spread dengue, i.e., *Aedes aegypti* and *Ae. albopictus*.

After keeping quiescent for more than three decades, CHIKV re-emerged as a massive epidemic in India in 2006. Currently, chikungunya is also reported in all states/Union Territories (UTs) of the country (except Ladakh), resulting in huge losses of economy and productivity. CHIKV evolution evidenced by quick mutations to form new and more viraemic strains has been a significant driver for epidemics in India. Early diagnosis and disease management are keys to successful disease curtailment (Lee et al., 2012). A live attenuated chikungunya vaccine "IXCHIQ", manufactured by Valneva has been approved by the Food and Drug Administration (FDA) in November 2023 and licensed in the United States (CDC, 2024).

4.1.1 Origin of 'Chikungunya' Term

The term "chikungunya" often refers to both the virus (CHIKV) and the illness or fever (CHIKF) caused by this virus. It was derived from the African dialect Swahili or Makonde and translates as "*to be bent over.*"

4.1.2 Chikungunya Virus

Chikungunya is caused by the CHIKV, which belongs to the genus *Alphavirus* under the family *Togaviridae*.

4.1.3 Vectors

Aedes aegypti and *Ae. albopictus* are the two vectors of chikungunya which are as shown in Figure 4.1 (for more details see Chapter 3).

Both *Ae. aegypti* and *Ae. albopictus* are involved in chikungunya transmission in India. *Aedes albopictus* has posed serious threats of chikungunya transmission in certain geographical regions such as Kerala and other southern states (Kaur et al., 2008, Kumar et al., 2011).

Breeding habitats of both the vectors varies in different parts of the country. *Ae. aegypti* breeding habitats are domestic and peri-domestic while *Ae. albopictus* breeds in vegetation, such as rubber plantation and other wild habitats like Cocoa pod (*Theobroma cacao*; Tyagi et al., 2006, Sumodan, 2003, 2019; Hiriyan and Tyagi, 2004; Eapen et al., 2010; Viswan et al., 2020).

4.1.4 Chikungunya Virus Transmission

In India, chikungunya is spread by the bite of an *Aedes* mosquito, primarily by *Ae. aegypti* or *Ae. Albopictus* like other endemic countries in Asia. Humans are thought to be the major source or reservoir of CHIKV for mosquitoes. Therefore, the mosquito usually transmits the disease by biting an infected person and then biting someone else a few days later. *Aedes* vector mosquitoes bite during the daytime. This type of transmission, in view of the historical occurrence in urban agglomerations, is generally referred to as 'urban cycle (Figure 4.2B),' but in modern times, the disease, with this modified type of transmission cycle, is no longer restricted to urban areas and is severely affecting rural populations (Figure 4.2A).

However, in the forests of Africa, the aetiological chikungunya (CHIKV) evolved in non-human primates (NHPs) and forest (sylvatic) mosquitoes, like other arboviruses, such as the dengue virus (DENV) and Zika virus (ZIKV; Valentine et al., 2019). In the forests, primatophilic mosquitoes (e.g., *Ae. furcifer-taylori, Ae. luteocephalus*, and *Ae. dalzieli*) maintain the viruses in sylvatic Non-Human Primates (NHP)-mosquito-NHP transmission cycles (Diallo et al., 1999). The characteristics of the sylvatic transmission cycle are a circulation periodicity with silent intervals.

FIGURE 4.1 Chikungunya vectors in India: (a) *Aedes aegypti* Linn. 1762, and (b) *Aedes albopictus* Skuse, 1894.

Source: Free Wikipedia.

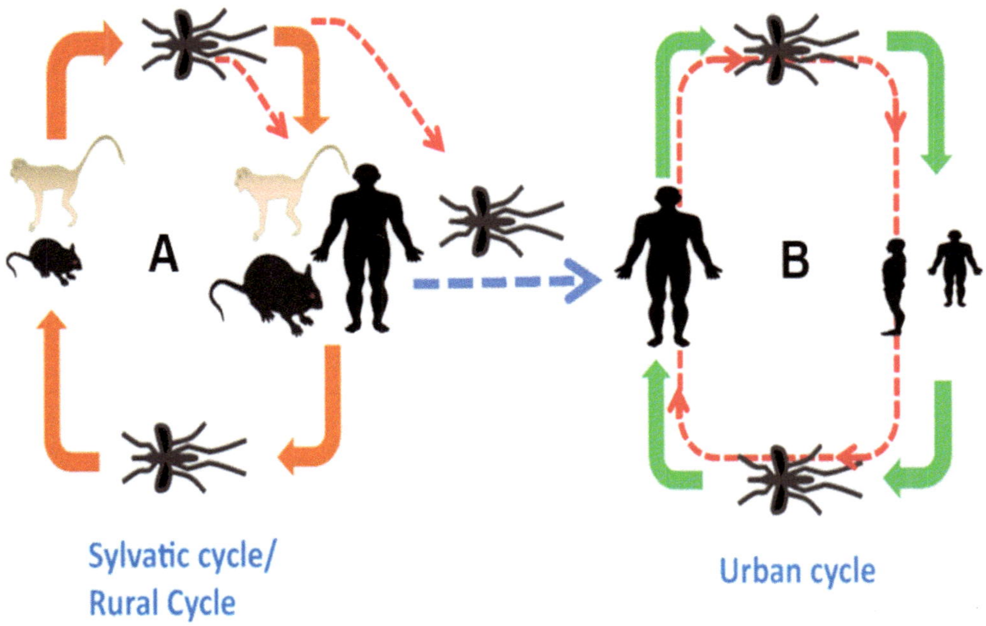

FIGURE 4.2 Graphic presentation of two different transmission cycles of CHIKV. (A) Sylvatic cycle that primarily occurs in Africa; the virus is maintained between non-human primates, bats, rodents, and other vertebrates, with mosquitoes of various genera as conveyors. (B) Urban cycle that primarily occurs in Asia/India, where the transmitted virus directly circulates in human hosts through the human-mosquito-human transmission cycle, with mosquitoes belonging to the genus *Aedes*.

4.1.5 Distribution

Globally, Chikungunya is present in 115 countries and causes severe and joint-disabling arthritis. Currently, besides Asia and Africa, chikungunya has spread to Europe and the Americas.

Presently, in India, all states/UTs in the country (except Ladakh) are affected by chikungunya. Since its re-emergence in 2006, every year, cases are reported in varying numbers from these endemic areas. Till date, no deaths directly attributable to chikungunya have been reported by any state/UT (NCVBDC, 2024).

4.2 EPIDEMIOLOGY AND TREND OF PREVALENCE

CHIKV is responsible for periodic and explosive outbreaks of a febrile disease that is characterized by severe and sometimes prolonged polyarthritis (Kannan et al., 2009). Chikungunya is transmitted principally by *Ae. aegypti*, whereas *Ae. albopictus* plays a key role in places where *Ae. aegypti* is absent or in low density. Chikungunya re-emerged in an epidemic form in 2006 in India after 32 years. Phylogenic analysis based on partial sequences of NS4 and E1 genes showed that all earlier isolates (1963–73) were of Asian genotype, whereas the current and the Yawat isolates were of African genotype (Yergolkar et al. 2006), which is comparatively more pathogenic. The most pathogenic mutant of CHIKV, i.e., A226V, was also isolated from Kerala (Kumar et al., 2008). The reported number of clinically suspected chikungunya cases in the country in 2006 was 1.39 million, thereafter keeping a low ebb in the following years. Starting with urban outbreaks, CHIKV transmission has invaded rural areas, too, in the country (Tyagi, 2007, 2008; Tyagi and Hiriyan, 2004). The epidemiological data for chikungunya cases in India for the period of 2017–22 are shown in Table 4.1 (NCVBDC, 2024).

TABLE 4.1
State-Wise Epidemiological Data for clinically suspected Chikungunya Cases in India for the Period of 2017–22

Sl. No.	Name of the State	2017	2018	2019	2020	2021	2022
1	Andhra Pradesh	1,162	622	832	318	753	1,611
2	Arunachal Pradesh	133	507	332	0	0	0
3	Assam	41	3	0	0	2	0
4	Bihar	1,251	156	594	38	40	67
5	Chhattisgarh	0	0	0	0	0	252
6	Goa	509	455	867	64	114	868
7	Gujarat	7,953	10,601	8,084	8,120	32,372	20,855
8	Haryana	220	62	0	930	765	2425
9	Himachal Pradesh	0	0	0	0	0	253
10	Jharkhand	269	3,405	1,691	627	1,064	2,113
11	Karnataka	32,831	20,411	43,698	16,111	40,134	65,340
12	Kerala	78	77	109	2302	3030	1511
13	Madhya Pradesh	2,477	3,211	2,749	1,871	3,822	42,52
14	Meghalaya	236	44	114	3	0	0
15	Maharashtra	8,110	9,884	5,158	4,258	21,680	14,785
16	Mizoram	0	93	0	34	0	195
17	Manipur	0	2	40	0	0	0
18	Odisha	0	0	242	15	0	0
19	Punjab	3,251	736	372	25	644	1,087
20	Rajasthan	1,612	254	365	1,015	1,157	212
21	Sikkim	130	384	1,126	253	457	1,863
22	Tamil Nadu	131	284	623	1461	3654	4365
23	Telangana	1,277	1,954	5,352	364	220	6,608
24	Tripura	574	683	1,105	454	948	885
25	Uttar Pradesh	103	58	72	53	70	200
26	Uttarakhand	0	29	1	0	3	2,000
27	West Bengal	2,103	52	0	391	154	1,533
28	A&N Islands	93	205	702	610	187	65
29	Chandigarh	1,810	357	82	0	397	3,600
30	Delhi	940	407	520	126	112	61
31	D&N Haveli	0	0	0	0	0	1
32	Jammu and Kashmir	0	1	0	1	7	37
33	Lakshadweep	0	0	0	0	0	103
34	Puducherry	475	2,876	7,084	3,980	7,284	11,440
	Total	**67,769**	**57,813**	**81,914**	**43,424**	**119,070**	**148,587**

Chikungunya in India has been escalating in quantum and spatial distribution since its inception in the country. Cases have increased substantially particularly after 2015, although in several states, the trend was discernible many years before. Many states have become susceptible to cyclic outbreaks on annual basis. The year-wise chikungunya cases reported from India during 2010–2022 are shown in Figure 4.3.

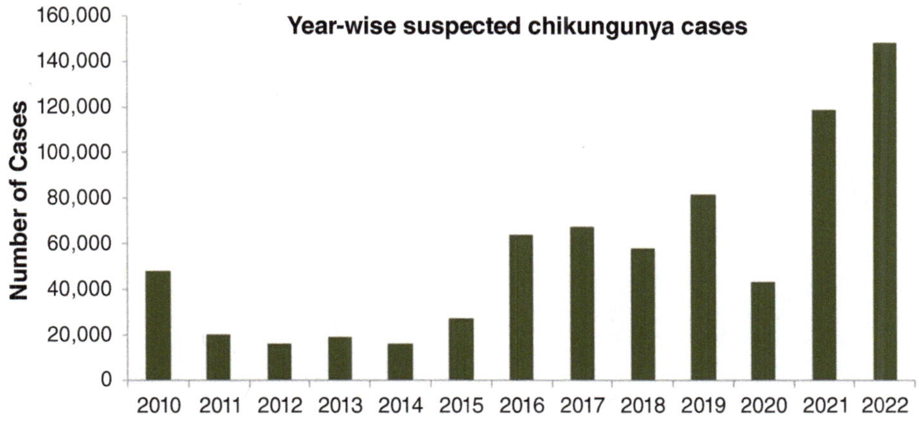

FIGURE 4.3 Year-wise clinically suspected chikungunya cases (2010–22).

4.3 DIAGNOSIS

Clinical manifestation of chikungunya (CHIK) fever resembles those of dengue and other arthropod-borne viruses of the *Alphavirus* genus (Kannan et al., 2009). Laboratory diagnosis including molecular and, wherever applicable, immunochromatography, for rapid detection of CHIKV antigen, is essential to establish the cause of fever for individual case management and initiate specific public health response (Reshma et al., 2021; Paramasivan et al., 2009; Jain et al., 2018). Genetic diversity in CHKV is assuming enormous significance in both diagnosis and treatment of the disease (Yergolkar et al., 2006; Kumar et al., 2008; Grivard et al. 2007; Álvarez-Argüelles et al. 2019).

4.4 SUSCEPTIBILITY OF *AEDES* VECTORS TO CHIKUNGUNYA VIRUS

Although having immense clinical relevance, only abysmally low information is so far available from the investigations which targeted to understand CHIKV susceptibility and growth in *Aedes* populations in India. Gokhale et al. (2015) investigated CHIKV susceptibility and growth kinetics in *Ae. aegypti*, along with the genetic heterogeneity of *Ae. aegypti* populations, and showed no selective advantage for a particular strain of CHIKV in *Ae. aegypti*. At even 100% infection rate, five geographic *Ae. aegypti* populations showed differences in dissemination to three CHIKV strains. Morphometric studies revealed phenotypic variation in all the studied populations. The allelic frequencies and genetic identity values showed that genetic differences between the populations were small, but significant. Ghosh et al. (2019) investigated the role of receptor proteins in midguts contributing to the permissiveness of CHIKV to *Aedes* spp. mosquitoes.

The CHIKV initially infects the anterior and posterior regions of *Ae. aegypti* midguts (white arrows in Figure 4.4) and then progressively infects the entire midgut. CHIKV infection was observed in the entire midgut of *Ae. albopictus* from two days PI (post-infection) and not localized to any specific region. All five *Ae. aegypti* and *Ae. albopictus* mosquitoes at each time point (100%; 5/5) were positive for CHIKV antigen (apple green fluorescence). The fluorescence intensity observed is represented as '+' (low fluorescence), '++' (moderate fluorescence), and '+++' (high fluorescence). [Colour figure can be viewed at wileyonlinelibrary.com] (*Source*: Ghosh et al., 2019)

CHIKV infection of *Ae. aegypti* and *Ae. albopictus* midguts and dissemination to heads was observed to be similar. A plaque assay performed with infected mosquito homogenates revealed that CHIKV replication dynamics were similar in *Aedes* sp. mosquitoes until 28 days post-infection (Figure 4.4). VOPBA (Virus Overlay Protein Binding Assay) performed with mosquito midgut membrane proteins revealed that prohibitin could serve as a putative CHIKV receptor on *Aedes* mosquito midguts.

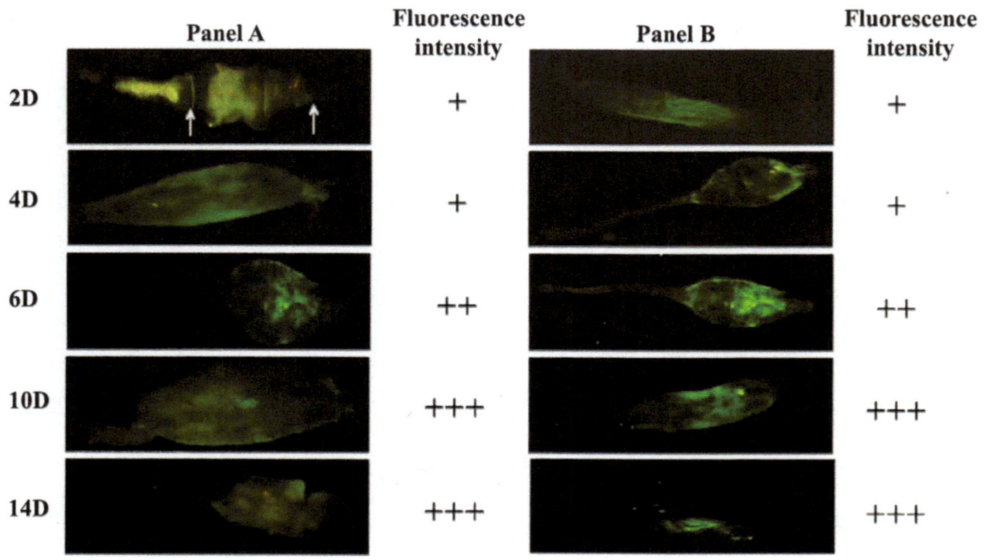

FIGURE 4.4 Localization of CHIKV in infected *Aedes* sp. mosquito midguts by an immunofluorescence assay. (A) *Aedes aegypti* midgut squashes. (B) *Aedes albopictus* midgut squashes.

4.5 PREVENTION AND CONTROL

Prevention of infection by avoiding mosquito bites is the best protection. Patients suspected for CHIKV infection should avoid getting mosquito bites during the first week of illness to prevent further transmission to mosquitoes, who may in turn infect other people. The main method to reduce transmission of CHIKV is through control of the mosquito vectors. This requires mobilization of communities, which are critical in reducing mosquito breeding sites through emptying and cleaning containers that contain water on a weekly basis, disposing of waste and supporting local mosquito control programmes. During outbreaks, insecticides may be sprayed (fogging) to kill flying adult mosquitoes, applied to surfaces in and around containers where the mosquitoes land, and used to treat water in containers to kill immature larvae. For protection against chikungunya, full sleeved clothing which minimizes skin exposure to the day-biting vectors is advised. Window and door screens should be used to prevent mosquitoes from entering homes.

Repellents can be applied to exposed skin or to clothing in strict accordance with product label instructions. Repellents available locally should be used (Kandasamy, 2021). Mosquito nets should be used by persons who sleep during the daytime, for example young children, sick patients, or older people. People travelling to the areas of active CHIKV transmission should apply basic precautions, including the use of insect repellents, wearing long sleeves and pants, and ensuring rooms are fitted with screens to prevent mosquitoes from entering.

4.6 CONCLUSION

There is no specific antiviral drug for treatment of CHIKV infection. Cases are managed symptomatically. National guidelines for clinical management of CHIKV were developed in 2016, which were updated in 2023. These guidelines have proved helpful in effective case management. There is no commercial vaccine available to protect against CHIKV infection.

Chikungunya has become a public health problem in India and various other endemic countries. The persistence of CHIKV and the regular occurrence of outbreaks is a major cause of concern. Besides the efforts by health departments, effective intersectoral convergence and community participation are crucial for timely containment of any chikungunya outbreak. Until the availability of a specific drug or effective vaccine, vector control will remain the key component for the prevention and control of chikungunya fever.

REFERENCES

Álvarez-Argüelles, M.E., Rojo Alba, S., Pérez, M.R., Riveiro, J.A.B. and García, S.M. (2019). Diagnosis and molecular characterization of chikungunya virus infections. Current topics in neglected tropical diseases. *IntechOpen*. https://doi.org/10.5772/intechopen.86957

Center for Disease Control and Prevention (CDC) (2024). Chikungunya virus. https://www.cdc.gov/chikungunya/hcp/vaccine/index.html (accessed on 12 September 2024).

Dinkar, A., Singh, J., Prakash, P., Das, A., Nath, G. (2018). Hidden burden of chikungunya in North India: A prospective study in a tertiary care centre. *Journal of Infection and Public Health* 11(4): 586–591. https://doi.org/10.1016/j.jiph.2017.09.008

Eapen, A., Ravindran, K.J., Dash, A.P. (2010). Breeding potential of *Aedes albopictus* (Skuse, 1895) in chikungunya affected areas of Kerala, India. *Indian Journal of Medical Research*, 132(6): 733–735.

Ghosh, A., Mullapudi, T., Bomanna, S., Tyagi, B.K., Ravi, V., Desai, A, (2019). Understanding the mechanism of Chikungunya virus vector competence in three species of mosquitoes. *Medical and Veterinary Entomology* 33(3): 375–387. https://doi.org/10.1111/mve.12376

Gokhale, M.D., Paingankar, M.S., Sudeep, A.B. and Parashar, D. (2015). Chikungunya virus susceptibility and variation in populations of *Aedes aegypti* (Diptera: Culicidae) mosquito from India. *Indian Journal of Medical Research* 142(1): S33–43. https://doi.org/10.4103/0971-5916.176614

Grivard, P., Le Roux, K., Laurent, P., Fianu, A., Perrau, J., Gigan, J., Hoarau, G., Grondin, N., Staikowsky, F., Favier, F. and Michault, A. (2007). Molecular and serological diagnosis of Chikungunya virus infection. *Pathology & Biology* 55(10):490–494. https://doi.org/10.1016/j.patbio.2007.07.002

Hiriyan, J. and Tyagi, B.K. (2004). Cocoa pod (Theobroma cacao): a potential breeding habitat of *Aedes albopictus* (Skuse) in dengue sensitive Kerala State, India. *Journal of American Mosquito Control Association* 20 (3): 323–325.

Jain, J., Okabayashi, T., Kaur, N., Nakayama, E., Shioda, T., Gaind, R., Kurosu, T. and Sunil, S. (2018). Evaluation of an immunochromatography rapid diagnosis kit for detection of chikungunya virus antigen in India, a dengue-endemic country. *Virology Journal*, 15, 84. https://doi.org/10.1186/s12985-018-1000-0

Kandasamy, C. (2021). DEPA (N, N-Diethyl Phenylacetamide): Commercial Journey of India's First indigenous mosquito repellent with novel properties. In: Tyagi, B.K. (eds) *Genetically Modified and other Innovative Vector Control Technologies*. Springer, Singapore. https://doi.org/10.1007/978-981-16-2964-8_19

Kannan, M., Rajendran, R., Sunish, I.P., Balasubramaniam, R., Arunachalam, N., Paramasivan, R., Tewari, S.C., Philip Samuel, P. and Tyagi, B.K.(2009). A study on chikungunya outbreak during 2007 in Kerala, South India. *Indian Journal of Medical Research*, 129: 311–315.

Kaur, P., Ponniah, M., Murhekar, M.V., Ramachandran, V., Ramachandran, R., Raju, H.K., Perumal, V., Mishra, A.C. and Gupte, M.D. (2008). Chikungunya outbreak, South India, 2006. *Emerging Infectious Disease* 14(10): 1623–1625. https://doi.org/10.3201/eid1410.070569

Kumar, N.P., Joseph, R., Kamaraj, T. and Jambulingam, P. (2008). A226V mutation in virus during the 2007 chikungunya outbreak in Kerala, India. *Journal of General Virology* 89: 1945–1948. https://doi.org/10.1099/vir.0.83628-0

Kumar, N.P., Suresh, A., Vanamail, P., Sabesan, S., Krishnamoorthy, K.G., Mathew, J., Jose, V.T. and Jambulingam, P. (2011). Chikungunya virus outbreak in Kerala, India, 2007: a seroprevalence study. *Memórias do Instituto Oswaldo Cruz* 106(8):912–916.

Lee, V.J., Chow, A., Zheng, X., Carrasco, L.R., Cook, A.R., Lye, D.C. et al. (2012). Simple clinical and laboratory predictors of Chikungunya versus dengue infections in adults. *PLoS Neglected Tropical Disease*, 6, e1786. https://doi.org/10.1371/journal.pntd.0001786

Lumsden, W.H. (1955). An epidemic of virus disease in Southern Province, Tanganyika Territory, in 1952-53. II. General description and epidemiology. *Transactions of the Royal Society of Tropical Medicine & Hygiene*, 49(1), 33–57. https://doi.org/10.1016/0035-9203(55)90081-x

National Center for Vector Borne Diseases Control (NCVBDC) (2024). Chikungunya situation in India. https://ncvbdc.mohfw.gov.in (accessed on 12 September 2024).

Sudeep, A.B. and Parashar, D. (2008). Chikungunya: an overview. *Journal of Biosciences*, 33(4): 443–449. https://doi.org/10.1007/s12038-008-0063-2

Mukherjee, S., Dutta, S.K., Sengupta, S. and Tripathi, A. (2017). Evidence of dengue and chikungunya virus co-infection and circulation of multiple dengue serotypes in a recent Indian outbreak. *European Journal of Clinical Microbiology and Infectious Diseases* 36(11): 2273–2279. https://doi.org/10.1007/s10096-017-3061-1

Paramasivan, R., Philip Samuel, P., Thenmozhi, V., Rajendran, R., Victor, S., Leo, J., Dhananjeyan, K.J., Krishnamoorthi, R., Arunachalam, N. and Tyagi, B.K. (2009). Chikungunya virus isolated in Lakshadweep Islands in the Indian Ocean: Evidence of the Central/East African Genotype. *Japanese Journal of Infectious Diseases* 62(1): 67–69.

Reshma, R.S., Meenu, M.S., Saini, P., Ajithlal, P.M., Sukumaran, M., Chintha, S., Anish, T.S., Ashwani Kumar and Kumar, N.P. (2021). Clinical and virological characteristics of patients during the outbreak of chikungunya in Thiruvananthapuram, Kerala, India, 2019–2020. *Journal of Clinical Virology Plus* 1(4): 100043. https://doi.org/10.1016/j.jcvp.2021.100043

Simon, F., Caumes, E., Jelinek, T., Lopez-Velez, R., Steffen, R. and Chen, L.H. (2023). Chikungunya: risks for travellers. *Journal of Travel Medicine* 30(2): taad008. https://doi.org/10.1093/jtm/taad008

Staples, J.E., Breiman, R.F. and Powers, A.M. (2009). Chikungunya fever: an epidemiological review of a re-emerging infectious disease. *Clinical Infectious Disease* 49(6): 942–948. https://doi.org/10.1086/605496

Sumodan, P.K. (2003). Potential of rubber plantations as breeding source for *Aedes albopictus* in Kerala, India. *Dengue Bulletin*, 27, 197–198.

Sumodan, P.K. (2019). Unconventional breeding habitats of *Aedes (Stegomyia) albopictus* (Skuse) related to agriculture crops in Kerala, India: A review. *International Journal of Mosquito Research* 6(6): 113–115.

Thenmozhi, V., Paramasivan, R., Samuel, P.P., Kamaraj, T., Balaji, T., Dhananjeyan, K.J., Venkatasubramani, K. and Tyagi, B.K. (2013). Dual infection in human by Japanese encephalitis virus & chikungunya virus in Alappuzha district, Kerala, India. *Indian Journal of Medical Research* 138(3):362–363.

Venkatasubramani, K., Paramasivan, R., Thenmozhi, V., Dhananjeyan, K.J., Balaji, T. and Leo, S. (2015). Diagnosis of chikungunya dominated co-infection with dengue during an outbreak in south India (2010 and 2012). *Tropical Doctor* 45. https://doi.org/10.1177/0049475515585476

Translational Research Consortia for chikungunya virus in India (TRC-CVI) (2021). Current Status of chikungunya in India. *Frontiers in Microbiology* 12: 695173. https://doi.org/10.3389/fmicb.2021.695173

Tyagi, B.K., Hiriyan, J., Philip Samuel, P., Tewari, S.C. and Paramasivan, R. (2006). Dengue in Kerala: A critical review. *ICMR Bulletin* 36(4–5): 13–22.

Tyagi, B.K. (2007). Chikungunya in India. *Bulletin De La Societe De Pathologie Exotique* 100 (5): 321.

Tyagi, B.K. (2008). *Vector-borne diseases: Epidemiology and control.* Scientific Publishers (India), 298 pp.

Tyagi, B.K. and Hiriyan, J. (2004). Breeding of dengue vector, *Aedes aegypti* (Linnaeus) in the rural Thar Desert, north western Rajasthan, India. *Dengue Bulletin* 28: 220–222.

Tyagi, B.K., Thenmozhi, V. and Selvi K.S. (2014). Transmission dynamics of Japanese encephalitis, with emphasis on gaps in understanding and priority areas on Japanese encephalitis and other acute encephalitis syndromes in India. *Journal of Communicable Diseases* 1(1): 1–10.

Valentine, M.J., Murdock, C.C. and Kelly, P.J. (2019). Sylvatic cycles of arboviruses in non-human primates. *Parasites & Vectors* 12: 463.

Viswan, A.K., Dileep Kumar, T., Meenakshi, V. and Srivastava, P.K. (2020). An observation on breeding habitats of *Aedes* mosquitoes in Kozhikode district, Kerala. *International Journal of Mosquito Research* 7(2): 57–60.

Yergolkar, P.N., Tandale, B.V., Arankalle, V.A., Sathe, P.S., Sudeep, A.B., Gandhe, S.S., Gokhale, M.D., George, J.P., Hundekar, S.L. and Mishra, A.C. (2006). Chikungunya outbreaks caused by African Genotype, India. *Emerging Infectious Diseases* 12(10): 1580–1583. https://doi.org/10.3201/eid1210.060529

Watson, R. (2007). Europe witnesses first local transmission of chikungunya fever in Italy. *British Medical Journal* 335(7619):532–533. https://doi.org/10.1136/bmj. 39332.708738.DB

Weinbren, M.P., Haddow, A.J. and Williams, M.C. (1958). The occurrence of chikungunya virus in Uganda I. Isolation from mosquitoes, *Transactions of the Royal Society of Tropical Medicine and Hygiene* 52(3): 253–262. https://doi.org/10.1016/0035-9203(58)90084-1

5 Zika in India

Himmat Singh and Nasreen Akhtar
ICMR-National Institute of Malaria Research, Sector 8, Dwarka, Delhi, India

R. Balasubramanian
National Institute of Virology, Kerala Unit, Govt. T. D. Medical College Hospital, Alappuzha, India

T. Dilip Kumar
Regional Office for Health and Family Welfare, Thiruvananthapuram, India

Kalpana Baruah
Ministry of Health & Family Welfare, Government of India, Delhi, India

B.K. Tyagi
Chandigarh University, Mohali, India

5.1 INTRODUCTION

Zika, an arboviral infection, is transmitted mainly by *Aedes* mosquitoes, notwithstanding several other modes of transmission, such as, for example, vector transmission, mother-to-child transmission (transplacental, perinatal, and breastfeeding), sexual transmission (male to female and male to male), and blood transfusion (Runge-Ranzinger et al., 2019). Though rare, other routes of transmission have been also reported – namely, monkey bites, needle stick injuries, transplants, and haemodialysis. Over two billion people are estimated to live in areas with potential Zika transmission (Messina et al., 2016). Zika was first diagnosed in 1947 in Uganda where the Zika virus (ZIKV) was isolated from a rhesus monkey from the Zika forest (Dick et al., 1952). Later, Simpson (1964) carried out some serological experiments on human serum collected in Uganda and showed the presence of neutralizing antibodies against the virus that the ZIKV, thereby providing the first evidence that the virus could infect humans. Subsequently, ZIKV-driven outbreaks were reported in Indonesia and other Asian countries (Olson et al., 1981) and Africa (Wikan and Smith, 2016).

5.1.1 Key Facts

i. ZIKV is transmitted primarily by *Aedes* mosquitoes, which bite mostly during the day.
ii. Most people with ZIKV infection do not develop symptoms; those who do typically have symptoms including rash, fever, conjunctivitis, muscle and joint pain, malaise, and headache that last for two to seven days.

DOI: 10.1201/9781003327028-6

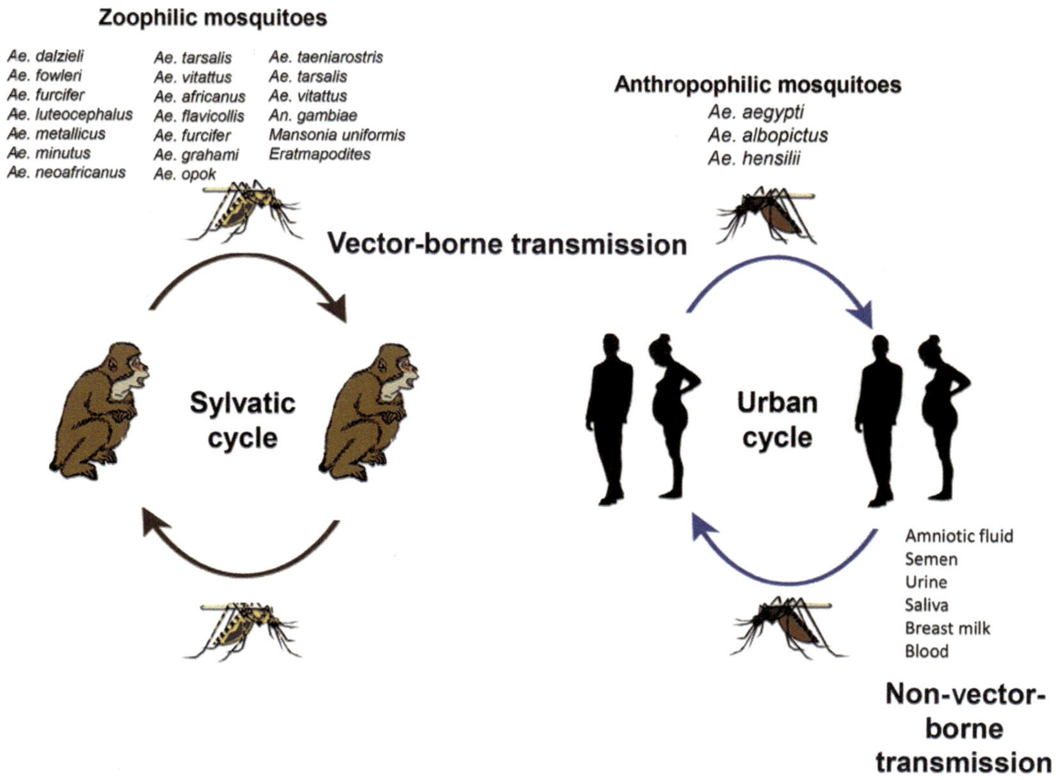

FIGURE 5.1 Vector and non-vector-borne transmissions of ZIKV. Vector-borne cycles comprise a sylvatic cycle with a viral circulation between non-human primates and zoophilic mosquitoes, and an urban cycle with humans as reservoir and amplification hosts and anthropophilic mosquitoes as vectors. Non-vector-borne mode involves direct human-to-human transmission through contact with fluids.

Source: Boyer et al., 2018.

iii. ZIKV infection during pregnancy can cause infants to be born with microcephaly and other congenital malformations, as well as preterm birth and miscarriage.
iv. ZIKV infection is associated with Guillain-Barré syndrome (GBS), neuropathy, and myelitis in adults and children.
v. In February 2016, WHO declared Zika-related microcephaly a Public Health Emergency of International Concern (PHEIC), and the causal link between the ZIKV and congenital malformations was confirmed.
vi. Although cases of ZIKV disease declined from 2017 onwards globally, transmission persists at low levels in several countries in the Americas and other endemic regions.

5.1.2 Definition

Zika is an infectious disease caused by a virus, similar to dengue fever, yellow fever, and West Nile virus. The virus is carried by infected mosquitoes, including the *Aedes aegypti* and *Aedes albopictus* species, which are found throughout the world (Figure 5.1). The infection is associated with a birth defect called microcephaly, which can affect babies born to people who become infected with Zika while pregnant.

The infection itself (sometimes called Zika fever) is usually mild and goes away on its own. However, the impact of the virus has most significantly been seen in pregnant women and their foetuses.

5.1.3 Symptoms

Most people infected with ZIKV do not develop symptoms. Among those who do, they typically start 3–14 days after infection, are generally mild including rash, fever, conjunctivitis, muscle and joint pain, malaise, and headache, and usually last for 2–7 days.

These symptoms are common to other arboviral and non-arboviral diseases; thus, the diagnosis of ZIKV infection requires laboratory confirmation. Zika can be diagnosed through a blood test. There is currently no vaccine or antiviral treatment for Zika. Treatment addresses symptoms, and cases clear up on their own.

5.1.4 Complications

i. ZIKV infection during pregnancy is a cause of microcephaly and other congenital malformations in the infant, including limb contractures, high muscle tone, eye abnormalities, and hearing loss. These clinical features are collectively referred to as congenital Zika syndrome.
ii. The risk of congenital malformations following infection in pregnancy remains unknown; an estimated 5%–15% of infants born to women infected with ZIKV during pregnancy have evidence of Zika-related complications. Congenital malformations occur following both symptomatic and asymptomatic infection. Zika infection in pregnancy can also cause complications such as foetal loss, stillbirth, and preterm birth.
iii. ZIKV infection can also cause GBS, neuropathy and myelitis, particularly in adults and older children.

5.2 ZIKA VIRUS

ZIKV is an arbovirus (arthropod-borne virus) classified into the flavivirus genus within the *Flaviviridae* family. It is spread by daytime-active *Aedes* mosquitoes, such as *Ae. aegypti* and *Ae. albopictus*. Its name comes from the Zika Forest of Uganda, where the virus was first isolated in 1947. ZIKV shares a genus with the dengue, yellow fever, Japanese encephalitis, and West Nile viruses. Since the 1950s, it has been known to occur within a narrow equatorial belt from Africa to Asia. From 2007 to 2016, the virus spread eastward across the Pacific Ocean to the Americas, leading to the 2015–16 ZIKV epidemic. ZIKV is an icosahedral, enveloped, single-stranded positive-sense RNA virus (Leier et al., 2018; Scaramozzino et al., 2001; Wikan and Smith, 2016). Whole genome sequence was obtained from the National Center for Biotechnology Information (NCBI) and analyses using Molecular Evolutionary Genetic Analysis (MEGA) software (Haddow et al., 2012; Singh et al., 2019).

Phylogenetic analyses clearly indicated that ZIKV can be grouped into two distinct lineages – the Asian lineage and the African lineage – based on their complete genome sequences obtained from the NCBI and analyses using MEGA software (Haddow et al. 2012; Singh et al. 2019). The virus is related to the following flaviviruses: West Nile virus, yellow fever virus (YFV), dengue virus (DENV), tick-borne encephalitis, St Louis encephalitis, Langat virus, the Modoc virus, Rio Bravo virus, the Powassan virus, and the Japanese encephalitis virus. Their genomic organisation and size are also highly related (Abrams et al., 2017; Ali et al., 2017; Leier et al., 2018). ZIKV has a genome of approximately 10.7 kb in length. It has a single polyprotein and codes for about 10.2 kb. The genome of ZIKV has been processed into two major segments: the three structural proteins, Capsid-C, pre-membrane (membrane), PrM/M, and the Envelope-E, and the seven non-structural proteins (Tan et al., 2020; Rasmussen et al., 2016; Shi and Gao, 2017). The ZIKV genome is bordered by the 5′ and 3′ untranslated regions (Charrel et al., 2017; Duffy et al., 2009; Lanciotti et al., 2008). The virus is more related to DENVs and YFVs in terms of vector transmission.

5.2.1 Vectors for Zika

Besides the main vectors, viz., *Aedes aegypti* and *Ae. albopictus*, ZIKV has so far been isolated from three additional species – namely, *Anopheles gambiae, Mansonia uniformis*, and *Eretmapodites* (Grard et al., 2014). For disease surveillance, it is interesting to see that other genera are also capable of vectorizing Zika, though evidence has thus far supported the major roles played by *Aedes* spp. as the primary vectors. Although, other genera are minor players and dependent on locality where they are present, it is important to note that *Aedes* is not the only competent vector species (Figure 5.2). As such, the routes of dissemination of Zika across the various continents are also highly diversified (Figure 5.3).

A summary of various mosquito species so far incriminated by the ZIKV is as follows:

1. ***Aedes* spp.** (*Ae. aegypti, Ae. africanus, Ae. apicoargenteus, Ae. dalzieli, Ae. flavicollis, Ae. fowleri, Ae. furcifer, Ae. grahami, Ae. hirsutus, Ae. jamoti, Ae. luteocephalus, Ae. metallicus, Ae. neoafricanus, Ae. opok, Ae. taylori, Ae. taeniarostris, Ae. unilineatus, Ae.* vittatus)
2. ***Anopheles* spp.** (*An. coustani, An. gambiae*)
3. ***Culex* sp.** (*Cx. perfuscus*)
4. ***Mansonia* sp.** (*Ma. uniformis*)
5. ***Eretmapodites* spp.** (*Er. quinquevittatus, Er. inornatus*)

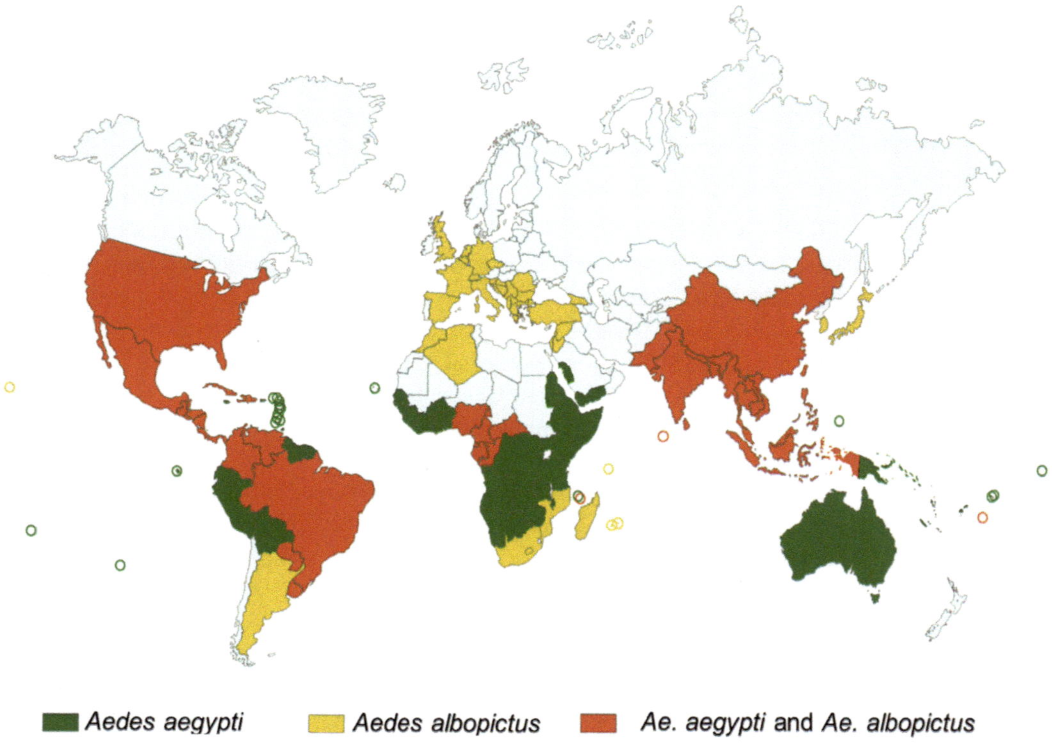

FIGURE 5.2 Geographic distribution of *Ae. aegypti* and *Ae. albopictus.*

Source: Boyer et al., 2018.

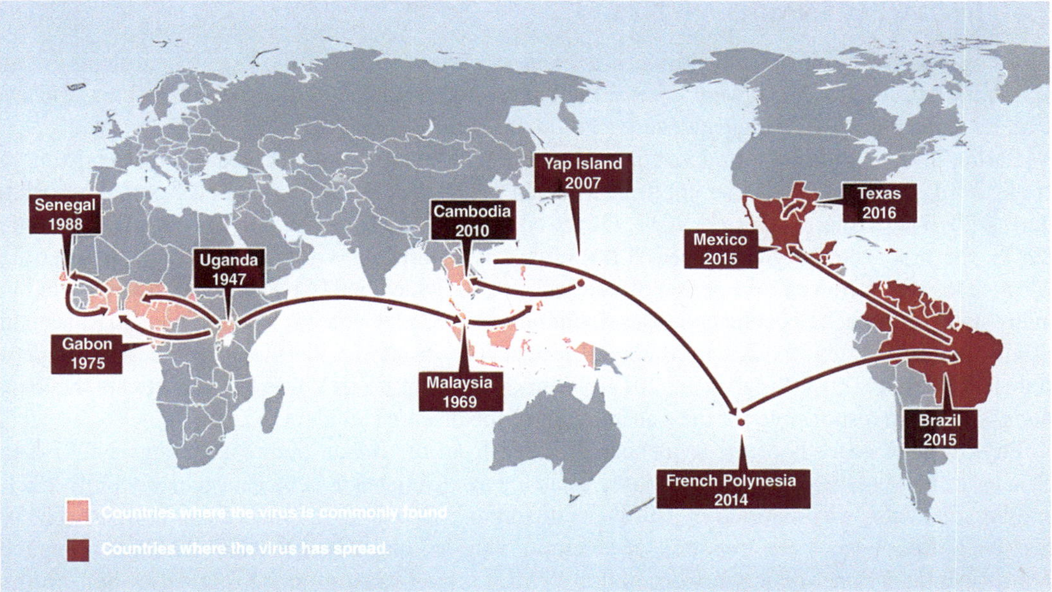

FIGURE 5.3 Pathway of spread of ZIKV from its cradle in Africa to all over the world.

Source: Adapted from Tyagi, 2020.

5.3 WORLDWIDE SCENARIO OF ZIKV

The World Health Organization (WHO, 2017) reported 84 countries, territories, or subnational areas with evidence of vector-borne ZIKV transmission; 64 countries, territories, or subnational areas where the competent vector is established but with no documented past or current on ZIKV transmission; 13 countries have reported evidence of person-to-person transmission of ZIKV (of those five are in the Americas, seven in Europe, and one in the Western Pacific); 31 countries or territories have reported microcephaly and other central nervous system (CNS) malformations potentially associated with ZIKV infection or suggestive of congenital infection; and 23 countries or territories have reported an increased incidence of GBS and/or laboratory confirmation of a ZIKV infection among GBS cases. In the 21st century, the first major epidemic of Zika fever was reported in 2007 in the Yap Islands of the Federated States of Micronesia, in which 185 cases with illness characterized by rash, conjunctivitis, fever, arthralgia, and arthritis were confirmed. But there was no severe morbidity or mortality found (Duffy et al., 2009). After that first virus was isolated from an *Aedes africanus* mosquito the year after (Lanciotti et al., 2008), in 2013, another outbreak occurred in other Pacific islands, viz., French Polynesia, the Easter Islands, the Cook Islands, and New Caledonia (Tognarelli et al., 2016). Later studies revealed the presence of infection in other regions of Africa, as well as in Asian countries (Wikan and Smith, 2016). Since the discovery of ZIKV, several outbreaks have been reported in tropical Africa and some areas of South-East Asia (Wiwanitkit, 2016).

After a streak of ZIKV outbreaks worldwide, the WHO declared Zika a PHEIC on 1 February 2016, and called for active surveillance in countries where no outbreaks were reported initially, including India (Vythilingam et al., 2016). The ZIKV has the potential for endemic transmission. To date, a total of 86 countries and territories have reported evidence of mosquito-transmitted Zika infection.

5.4 INDIAN SCENARIO OF ZIKV

ZIKV was first serologically confirmed in India as early as 1954, following which neutralizing antibodies specific to ZIKV were detected in human sera (16.8%) obtained from various districts without a virologically documented case (Smithburn et al., 1954).

Earlier investigations reported that the Asian strain of the virus had emerged from India (Mourya et al., 2016). While the first four confirmed human cases of ZIKV were reported from Tamil Nadu during 2016–17 (Bhardwaj et al., 2017; Gupta et al., 2019; Yadav et al., 2019; CDC, 2016; Anon, 2017), the first laboratory-confirmed ZIKV case was reported in Gujarat in 2018 (Sapkal et al., 2018). This was followed by a period of quiescence after which a massive outbreak was detected in many states of India; 159 confirmed cases (including 63 pregnant women) were reported in Rajasthan (Yadav et al., 2019), 130 cases in Madhya Pradesh (Biswas et al., 2020), and one case in Gujarat state in December 2018 (Singh et al., 2019; Saxena et al., 2019). ZIKV was also detected in the local *Aedes* mosquito responsible for the transmission of the disease (Singh et al., 2019).

Phylogenetic analysis of the sequences obtained from the Indian outbreak through NCBI data base indicated that the virus belonged to an Asian lineage (Singh et al., 2019). Again, on 8 July 2021, another Zika case was reported in Thiruvananthapuram, Kerala state. Since then, 83 more ZIKV-positive cases have been reported in Kerala, with a majority of cases hailing from the Thiruvananthapuram Municipal Corporation (TMC) region (Sasi et al., 2021). Elsewhere in the country, new cases continued to surface, as a ZIKV-positive case was first detected in Kanpur, Uttar Pradesh, on 22 October 2021, followed by 100 more ZIKV cases, including 17 children (Scroll.in, 2021; Akhtar et al., 2022). (Figure 5.4, Table 5.1).

A multidisciplinary team was rushed to control and contain the disease (Kumar and Singh, 2022). Another study on ZIKV infection by Yadav et al. (2022) stated that ZIKV-positive cases were also detected in two states, Telangana and Jharkhand (one case from each state), and three more cases of co-infection of ZIKV with other flaviviruses, viz., DENV and ZIKV, were also confirmed by qRT-PCR in which there was one case each from Aligarh (Uttar Pradesh) and Thiruvananthapuram (Kerala). At the same time, one case (51 years/M) from Amritsar (Punjab) was confirmed to have co-infection with all three major circulating arboviruses – namely, DENV, ZIKV, and CHIKV, and the conclusion of the study was that there was no significant difference in the profile of clinical symptoms and duration of illness between the patients suffering from stand-alone ZIKV and those with co-infections (CHIKV and DENV).

Due to the high cross-reactivity of ZIKV with DENV and other flaviviruses, it was difficult to confirm Zika virus infection in India based on serology. In the context of ZIKV epidemiology, there are still some major concerns in the affected countries, such as the travel-related ZIKV introductions; specific symptoms, risks, preventions, and clinical management of the individuals; clinical aspects of the affected pregnant women; and associated microcephaly. However, after confirming many of the ZIKV cases, outbreak control strategies were initiated by the coordinated efforts of the Ministry of Health and Family Welfare (MoHFW), Integrated Disease Surveillance Programme (IDSP), Indian Council of Medical Research (ICMR), Rashtriya Bal Swasthya Karyakram (RBSK), National Institute of Virology (NIV) Pune, National Centre for Disease Control, and the National Center for Vector Borne Disease Control (NCVBDC). As a preparedness measure, the ICMR, the apex body of medical research in India, initiated nationwide vector surveillance of ZIKV in the high-risk zones comprising 49 districts in 14 states from 2016 to 2019. Initially, surveillance was started in 3 of the 14 states/UTs, viz., Kerala, Tamil Nadu, and Pondicherry, but was later extended to 11 other states as well, viz., Karnataka, Odisha, Madhya Pradesh, Gujarat, Rajasthan, New Delhi, Assam, Arunachal Pradesh, Nagaland, Meghalaya, and the Andaman and Nicobar Islands. The programme was undertaken on a multicentre-basis, run by seven ICMR institutions, with the ICMR-Vector Control Research Centre as the coordinating agency. A uniform standard operating protocol (SOP) was followed for the investigation. This vector surveillance programme also envisaged screening *Aedes* species sampled across the country for DENV to further understand the

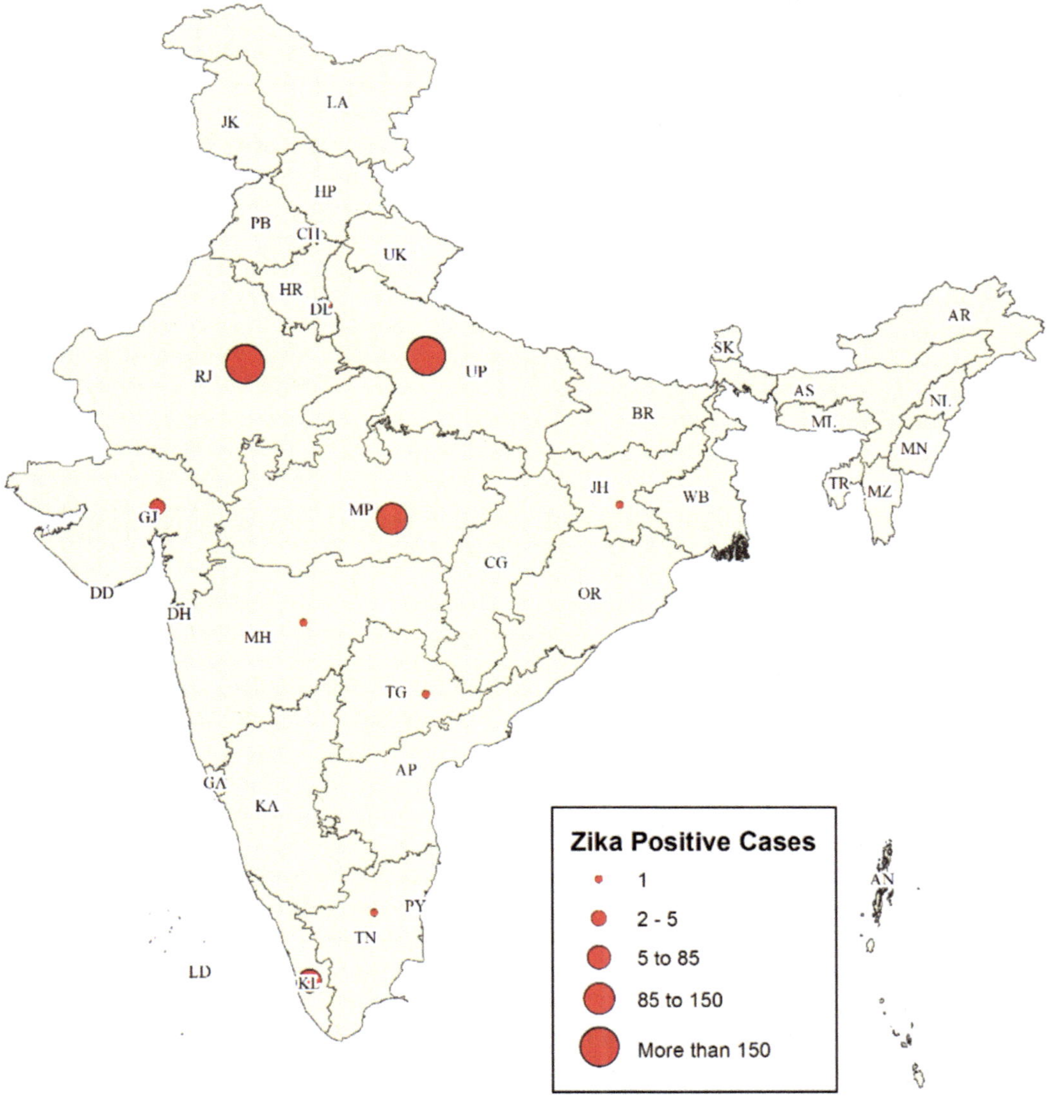

FIGURE 5.4 The map showing ZIKV affected states in India during ZIKV outbreaks (2017–21).

Source: The author; GIS Based Map of study areas.

TABLE 5.1
Year-wise Zika Outbreaks in States of India from 2017 to 2021

Sr. No.	Year	ZIKV-Positive Area/District	Name of the State	No of ZIKV-Positive Cases
1.	2017	Ahmedabad	Gujarat	3
2.	2017	Krishnagiri	Tamil Nadu	1
3.	2018	Jaipur	Rajasthan	159
4.	2018	Vidisha, Bhopal, Sehore, Sagar, Raisen, Hoshangabad	Madhya Pradesh	130
5.	2018	Ahmedabad	Gujarat	1

(Continued)

TABLE 5.1
(Continued)

Sr. No.	Year	ZIKV-Positive Area/District	Name of the State	No of ZIKV-Positive Cases
6.	2021	Thiruvananthapuram	Kerala	83
7.	2021	Kanpur	Uttar Pradesh	149
8.	2021	Palghar	Maharashtra	1
9.	2021	Shadipur (North-east Delhi)	Delhi	1
10.	2021	Hyderabad	Telangana	1
11.	2021	East Singhbhum	Jharkhand	1
12.	2021	Belsar (Pune)	Maharashtra	1

Source : the main author.

transmission pattern of this endemic arboviral disease in India and assess the potential linkage between transmission of the two viruses, i.e., DENV and ZIKV.

5.5 DIAGNOSIS OF ZIKA VIRUS

There is relatively limited literature describing the performance characteristics of diagnostic tests for ZIKV. This section describes the methods that have been used to identify ZIKV infections in human specimens and adult *Aedes* mosquitoes.

5.5.1 Viral Culture and Detection

Culture-based methods for ZIKV detection are used in research laboratories for research purposes only for the development of antibodies against ZIKV but are not generally available for clinical purposes. The ZIKV can be cultured in both mosquito and mammalian cells. Primary dermal fibroblasts, immature dendritic cells, cortical neural progenitor cells, and Vero cells are susceptible to Zika viral infection (Baronti et al., 2014; Hamel et al., 2015; Tang et al., 2016). The reference method developed for the isolation of ZIKV and other arboviruses is intracerebral mouse inoculation (Dick et al., 1952; Digoutte et al., 1992). ZIKV is also culturable in several cell lines, including African green monkey (Vero) and rhesus monkey kidney (LLC-MK2), as well as *Aedes pseudoscutellaris* (MOS61 or AP-61) and *Aedes albopictus* (C6/36; Digoutte et al., 1992; Haddow et al., 2012).

5.5.2 Antibody Detection

Evidence of human ZIKV infection was obtained primarily via the identification of ZIKV neutralizing antibodies in human serum (Dick et al., 1952; Hammon et al., 1958; Macnamara et al., 1959; Olson et al., 1983; Smithburn, 1954; Simpson, 1964). Serologic surveys conducted in Nigeria (Macnamara et al., 1959) noted that sera capable of neutralizing ZIKV in mouse protection experiments were also strongly associated with neutralization of DENV and YFV, suggesting serologic overlap or cross-reactivity.

As ZIKV spreads throughout the areas of the world where DENV infection is endemic, detection of DENV IgM antibodies in patients with a dengue-like illness, particularly in the absence of DENV non-structural protein 1 (NS1) or DENV RNA, should prompt consideration of the possibility of ZIKV infection. While DENV serologic tests are widely used, assays for ZIKV antibodies are currently available only through public health laboratories such as the Centers for Disease Control and Prevention (CDC) laboratories. Furthermore, the development of commercial ZIKV serologic tests for use in clinical laboratories and at the point of care will require careful evaluation of test specificity, particularly in populations with a high prevalence of DENV and of infections with other flaviviruses.

5.5.3 Antigen Detection

Unlike in the DENV diagnostics field, where NS1 antigen assays are widely available, antigen assays for the diagnosis of acute ZIKV infection have not yet been developed.

5.5.4 Viral Genome Detection

Laboratory diagnosis of acute ZIKV infection currently depends upon the detection of ZIKV RNA in patient specimens (serum, plasma, urine, saliva, and amniotic fluid) and adult *Aedes* mosquitoes. Perhaps the best-studied ZIKV real-time reverse transcription PCR assay (qRT-PCR) assay was designed by the CDC using sequences derived from the 2007 Yap Islands epidemic (Lanciotti et al., 2008) using the Super Script III One-Step RT-PCR System with Platinum Taq High Fidelity DNA Polymerase (Invitrogen Inc, USA).

5.6 CLINICAL MANIFESTATION

The clinical presentation of patients with acute ZIKV infection typically includes a combination of fever, headache, retro-orbital pain, conjunctivitis, a maculopapular rash, myalgias, conjunctivitis, and/or arthralgias (Duffy et al., 2009; Roos, 2016; Zanluca et al., 2015). Fever is often low grade (38°C), though cases with fevers of up to 40°C have been reported (Duffy et al., 2009; Levy-Blitchtein and Del Valle-Mendoza, 2016; Olson et al., 1981; Centers for Disease Control and Prevention on ZIKV, 2019; Zanluca et al., 2015). Diagnosing ZIKV can only be confirmed through laboratory tests on blood or other body fluids (urine or semen) and adult mosquitoes also. A causal link has been established between in-utero ZIKV infection and CZS (congenital Zika syndrome; Moore et al., 2017). CZS reflects the spectrum of symptoms observed in affected infants (Moore et al., 2017). Early reports focused on findings of microcephaly. However, we now know that CZS has variable presentation, severity, and prognosis. A recent review concluded that CZS differs from other congenital infections because of five distinct features: (1) severe microcephaly with a partially collapsed skull, (2) thin cerebral cortices with subcortical calcifications, (3) macular scarring and focal pigmentary retinal mottling, (4) congenital contractures, and (5) marked early hypertonia and symptoms of extrapyramidal involvement (Moore et al., 2017).

5.7 DISEASE TRANSMISSION

The most common mode of transmission of Zika and other flaviviruses (chikungunya and dengue) is through *Aed*es mosquitoes such *as Aedes aegypti* (Diallo et al., 2014; Ferreira-de-Brito et al., 2016; Marchette et al., 1969) and *Aedes albopictus* (Boyer et al., 2018; Grard et al., 2014). *Ae. aegypti* is one of the most capable vectors: it feeds primarily on humans, frequently bites several times in a single meal, has an almost imperceptible bite, and lives very close to humans (Gubler, 2002). However, it is geographically restricted because it does not winter well in cold climates. *Ae. albopictus* has a wider geographical distribution (it can be found in subtropical and temperate climates), is resilient and aggressive, can survive in rural as well as urban environments, is relatively long-lived (four to nine weeks), and is more able to survive through cold winters (Medlock and Leach, 2015). Both *Ae. aegypti* and *Ae. albopictus* bite primarily during the daytime, limiting the usefulness of insecticide-impregnated bed nets. Overall, the ability of a vector species to transmit a pathogen depends on location and time.

However, other routes of transmission (congenital, sexual, transplacental, and by contact with body fluids) were also reported (Rodriguez-Morales et al., 2016). A pregnant woman can pass the ZIKV to her foetus during pregnancy and cause microcephaly and other severe foetal brain defects. The ZIKV has been detected in human milk also, but the risk of transmission through breastfeeding is unknown. Several factors can influence the dynamics of transmission, such as environmental and climate factors, the interaction between host and pathogen, and the development of immunity in the population (WHO, 2009).

5.8 TREATMENT

No specific antiviral treatment is available for ZIKV disease. Treatment is generally supportive and can include rest, fluids, and the use of analgesics and antipyretics. Because aspirin and other non-steroidal anti-inflammatory drugs (NSAIDs) can increase the risk of haemorrhage in patients with dengue, avoid the use of these medications until dengue can be ruled out. People infected with Zika, dengue, or chikungunya viruses should be protected from further mosquito exposure during the first week of illness to decrease the possibility of local transmission. Carefully evaluate pregnant women with laboratory evidence of ZIKV infection; closely manage these cases for possible adverse pregnancy outcomes. Guidance for the diagnosis, evaluation, and management of infants with possible congenital ZIKV infections is available from the Centers for Disease Control and Prevention (2022).

5.9 VACCINE ADVANCES

According to the FDA (Food and Drug Administration, 2020), there are no approved vaccines or other specific treatments for ZIKV infection (FDA data on vaccine development). Several organizations, however, are actively developing vaccines using a variety of platforms and technologies. While many of these are still in the early stages of development, many are based on previously approved platforms and designs against dengue and other infectious disease agents. In March 2016, the WHO (2016) announced a list of organizations describing approaches toward Zika vaccine development. An ideal Zika vaccine needs to be safe for at-risk populations, including women of childbearing age, pregnant women, young children, the elderly, and immune-compromised individuals. To achieve such a broad population spectrum, multiple vaccine platforms may be required. In addition, a Zika vaccine should not induce Guillain-Barré-like neurological side effects or congenital malformations in fetuses. Most of the Zika vaccine candidates currently in human trials have been shown to protect foetuses against ZIKV challenges and prevent neurological, placental, and testicular damage from it in mice and non-human primate models, but if this will hold true in humans still needs to be determined in clinical trials (Castanha and Marques, 2020).

There are multiple challenges in developing vaccines against Zika. First, there is currently no ideal small animal model for pre-clinical testing. Wild-type (WT) mice are poorly susceptible to Zika infection and develop low-level, short-lived viremia without apparent pathology. So far, non-human primates (NHP), one of the natural reservoirs of the virus, have been the best pre-clinical models (Lee and Ng, 2018). The cost and maintenance of testing vaccines on NHP are extremely high and pose a hurdle in vaccine development. As ZIKV is linked to congenital malformations, a major hurdle to overcome is the need for a vaccine that is safe for pregnant women and their offspring. Another key challenge in designing a vaccine against ZIKV is the possible risk of antibody-dependent enhancement of heterologous *flavivirus* infections such as dengue due to the sequence and antigenic similarity between them.

5.10 PREVENTION AND CONTROL

No vaccine or preventive drug is available. However, several candidate vaccines are in various stages of development. All travellers to areas with ZIKV transmission should take steps to avoid mosquito bites to prevent the risk of Zika virus and other vector-borne infections. People who may have been exposed to the ZIKV and want to reduce their risk of sexual transmission to an uninfected partner should use protection. Blood donations in the United States are now routinely screened for ZIKV RNA to reduce the risk of transfusion-associated transmission. Pregnant women should avoid travelling to any area with a risk of ZIKV transmission. Pregnant women who travel to one of these areas should talk to their health-care provider before travelling and strictly follow steps to avoid mosquito bites or sexual transmission during travel. There is no restriction on the use of insect repellents by pregnant women if they are used in accordance with the instructions on the product label. Mothers are encouraged to breastfeed infants even after possible ZIKV exposure, as available evidence indicates the benefits of breastfeeding outweigh the theoretical risks associated with ZIKV infection transmission through breast milk (Centers for Disease Control and Prevention, 2022).

ACKNOWLEDGEMENT

The authors are thankful to the ICMR, New Delhi, for providing samples from other states. The authors are also thankful to Mrs. Poonam Saroha from ICMR-National Institute of Malaria Research, New Delhi, for making the village map of the study. Thanks are also due to the field staff of NIMR and MCD, New Delhi, for conducting the surveys.

REFERENCES

Abrams, R. P. M., Solis, J., & Nath, A. (2017). Therapeutic Approaches for Zika Virus Infection of the Nervous System. *Neurotherapeutics: The Journal of the American Society for Experimental NeuroTherapeutics*, 14(4), 1027–1048. 10.1007/s13311-017-0575-2

Akhtar, N., Gupta, S.K., & Singh, H. (2022). Surveillance of Zika and Dengue viruses in field-collected *Aedes aegypti* mosquitoes from different states of India. *Virology*, 574, 96–101. 10.1016/j.virol.2022.07.004

Ali, A., Wahid, B., Rafique, S., & Idrees, M. (2017). Advances in research on Zika virus. *Asian Pacific Journal of Tropical Medicine*, 10 (4), 321–331. 10.1016/j.apjtm.2017.03.020

Anon. (2017). Tamil Nadu reports first case of Zika. Available at: https://www.thehindu.com/news/national/tamil-nadu/tamil-nadureports-first-case-of-zika-virus/article19254690.ece (accessed on 30 September 2017).

Baronti, C., Piorkowski, G., Charrel, R. N., Boubis, L., Leparc-Goffart, I., & de Lamballerie, X. (2014). Complete coding sequence of zika virus from a French polynesia outbreak in 2013. *Genome Announcements*, 2(3), e00500–14. 10.1128/genomeA.00500-14

Bhardwaj, S., Gokhale, M.D., & Mourya, D.T. (2017). Zika virus: Current concerns in India. *The Indian Journal of Medical Research*, 146(5), 572. 10.4103/ijmr.IJMR_1160_17

Biswas, A., Kodan, P., Gupta, N., Soneja, M., Baruah, K., Sharma, K.K. and Meena, S. (2020). Zika outbreak in India in 2018. *Journal of Travel Medicine*, 27 (4), taaa001, 10.1093/jtm/taaa001

Boyer, S., Calvez, E., Chouin-Carneiro, T., Diallo, D., & Failloux, A.B. (2018). An overview of mosquito vectors of Zika virus. *Microbes and infection*, 20(11–12), 646–660. 10.1016/j.micinf.2018.01.006

Castanha, P. M. S., a&nd Marques, E. T. A. (2020). A glimmer of hope: Recent updates and future challenges in Zika Vaccine Development. *Viruses*, *12*(12), 1371. 10.3390/v12121371

Centers for Disease Control and Prevention (CDC) (2016). *Zika virus disease and Zika virus infection 2016 case definition* (accessed on June 2016).

Centers for Disease Control and Prevention (CDC) (2019). Zika virus: For health care providers: Clinical evaluation & disease. *Centers for Disease Control and Prevention*. Available at: https://www.cdc.gov/zika/hc-providers/clinicalevaluation.html (accessed on 28 January 2019).

Centers for Disease Control and Prevention (CDC) (2022). Evaluation & Testing for Zika virus. Available at: www.cdc.gov/pregnancy/zika/testing-follow-up/evaluation-testing.html (accessed on 14 April 2022).

Charrel, R., Mögling, R., Pas, S., Papa, A., Baronti, C., Koopmans, M., Zeller, H., Leparc-Goffart, I. and Reusken, C.B. (2017). Variable sensitivity in molecular detection of Zika virus in European expert laboratories: external quality assessment. *Journal of Clinical Microbiology*, 55 (11), pp. 3219–3226. 10.1128/JCM.00987-17

Diallo, D., Sall, A.A., Diagne, C.T., Faye, O., Faye, O., Ba, Y., Hanley, K.A., Buenemann, M., Weaver, S.C., & Diallo, M. (2014). Zika virus emergence in mosquitoes in southeastern Senegal, 2011. *PloS one*, 9(10), 109442. 10.1371/journal.pone.0109442

Dick, G.W., Kitchen, S.F., & Haddow, A.J. (September 1952). Zika virus. I. Isolations and serological specificity. *Transactions of the Royal Society of Tropical Medicine and Hygiene*, 46(5), 509–520. 10.1016/0035-9203(52)90042-4

Digoutte, J.P., Calvo-Wilson, M.A., Mondo, M., Traore-Lamizana, M., & Adam, F. (1992). Continuous cell lines and immune ascitic fluid pools in arbovirus detection. *Research in Virology*, 143, 417–422. 10.1016/S0923-2516(06)80135-4

Duffy, M.R., Chen, T.H., Hancock, W.T., Powers, A.M., Kool, J.L., Lanciotti, R.S., Pretrick, M., Marfel, M., Holzbauer, S., Dubray, C., & Guillaumot, L. (2009). Zika virus outbreak on Yap Island, federated states of Micronesia. *New England Journal of Medicine*, 360(24), 2536–2543. 10.1056/NEJMoa0805715

Ferreira-de-Brito, A., Ribeiro, I.P., Miranda, R.M.D., Fernandes, R.S., Campos, S.S., Silva, K.A.B.D., Castro, M.G.D., Bonaldo, M.C., Brasil, P., &Lourenço-de-Oliveira, R. (2016). First detection of natural infection of *Aedes aegypti* with Zika virus in Brazil and throughout South America. *Memórias do Instituto Oswaldo Cruz*, 111, 655–658. 10.1590/0074-02760160332

Food & Drug Administration (FDA) (2020). Zika Virus Response Updates from FDA, Available at: https://www.fda.gov/emergency-preparedness-and-response/mcm-issues/zika-virus-response-updates-fda (accessed on 2 May 2020).

Grard, G., Caron, M., Mombo, I.M., Nkoghe, D., Mboui Ondo, S., Jiolle, D., Fontenille, D., Paupy, C., &Leroy, E.M. (2014). Zika virus in Gabon (Central Africa)- 2007: a new threat from *Aedes albopictus?. PLoS Neglected Tropical Diseases*, 8(2), 2681. 10.1371/journal.pntd.0002681

Gubler, D.J. (2002). Epidemic dengue/dengue hemorrhagic fever as a public health, social and economic problem in the 21st century. *Trends in Microbiology*, 10(2), 100–103. 10.1016/S0966-842X(01)02288-0

Gupta, N., Kodan, P., Baruah, K., Soneja, M., & Biswas, A. (2019). Zika virus in India: past, present and future. *QJM: Monthly Journal of the Association of Physicians*, hcz273. Advance online publication. 10.1093/qjmed/hcz273

Haddow, A.D., Schuh, A.J., Yasuda, C.Y., Kasper, M.R., Heang, V., Huy, R., Guzman, H., Tesh, R.B., &Weaver, S.C. (2012). Genetic characterization of Zika virus strains: geographic expansion of the Asian lineage. *PLoS Neglected Tropical Diseases*, 6(2), 1477. 10.1371/journal.pntd.0001477

Hammon, W., Schrack Jr, W.D., & Sather, G.E. (1958). Serological Survey for Arthropod-Borne Virus Infections in the Philippines. *The American Journal of Tropical Medicine and Hygiene*, *7*(3), 323–328. 10.4269/ajtmh.1958.7.323

Hamel, R., Dejarnac, O., Wichit, S., Ekchariyawat, P., Neyret, A., Luplertlop, N., Perera-Lecoin, M., Surasombatpattana, P., Talignani, L., Thomas, F., Cao-Lormeau, V. M., Choumet, V., Briant, L., Desprès, P., Amara, A., Yssel, H., & Missé, D. (2015). Biology of Zika Virus Infection in Human Skin Cells. *Journal of Virology*, 89(17), 8880–8896. 10.1128/JVI.00354-15

Kumar, S. & Singh, T. (2022). Zika virus in India: A Cause for Concern. *Indian Journal of Preventive & Social Medicine*, 53(1), 1–5. Retrieved from https://ijpsm.co.in/index.php/ijpsm/article/view/14

Lanciotti, R.S., Kosoy, O.L., Laven, J.J., Velez, J.O., Lambert, A.J., Johnson, A.J., Stanfield, S.M., & Duffy, M.R. (2008). Genetic and serologic properties of Zika virus associated with an epidemic, Yap State, Micronesia, 2007. *Emerging Infectious Diseases*, 14(8), 1232–1239. 10.3201/eid1408.080287

Lee, C.Y., & Ng, L.F.P. (2018). Zika Virus: from an obscurity to a priority. *Microbes Infection*, 20(11–12), 635–645. 10.1016/j.micinf.2018.02.009

Levy-Blitchtein, S., & Del Valle-Mendoza, J. (2016). Zika virus is arriving at the American continent. *Asian Pacific Journal of Tropical Medicine*, 9(10), 1019–1021. 10.1016/j.apjtm.2016.07.030

Leier, H.C., Messer, W.B., & Tafesse, F.G. (2018). Lipids and pathogenic flaviviruses: An intimate union. *PLoS Pathogens*, 14 (5), pp. e1006952. 10.1371/journal.ppat.1006952

Macnamara, F.N., Horn, D.W., & Porterfield, J.S. (1959). Yellow fever and other arthropod-borne viruses: a consideration of two serological surveys made in South Western Nigeria. *Transactions of the Royal Society of Tropical Medicine and Hygiene*, 53(2), 202–212. 10.1016/0035-9203(59)90072-0

Marchette, N.J., Garcia, R., & Rudnick, A. (1969). Isolation of Zika virus from *Aedes aegypti* mosquitoes in Malaysia. *American Journal of Tropical Medicine and Hygiene*, 18(3), 411–415. 10.4269/ajtmh.1969.18.411

Medlock, J.M., & Leach, S.A. (2015). Effect of climate change on vector-borne disease risk in the UK. *The Lancet Infectious Diseases*, 15(6), 721–730. 10.1016/S1473-3099(15)70091-5

Messina, J. P., Kraemer, M. U., Brady, O. J., Pigott, D. M., Shearer, F. M., Weiss, D. J., Golding, N., Ruktanonchai, C. W., Gething, P. W., Cohn, E., Brownstein, J. S., Khan, K., Tatem, A. J., Jaenisch, T., Murray, C. J., Marinho, F., Scott, T. W., & Hay, S. I. (2016). Mapping global environmental suitability for Zika virus. *Elife*, 5, e15272. 10.7554/eLife.15272

Moore, C.A., Staples, J.E., Dobyns, W.B., Pessoa, A., Ventura, C.V., Da Fonseca, E.B., Ribeiro, E.M., Ventura, L.O., Neto, N.N., Arena, J.F. and Rasmussen, S.A. (2017). Characterizing the pattern of anomalies in congenital Zika syndrome for pediatric clinicians. *JAMA Pediatrics*, 171(3), 288–295. 10.1001/jamapediatrics.2016.3982

Mourya, D.T., Shil, P., Sapkal, G.N., & Yadav, P.D. (2016). Zika virus: Indian perspectives. *The Indian Journal of Medical Research*, 143(5), 553–564. 10.4103/0971-5916.187103

Olson, J.G., Ksiazek, T.G., Gubler, D.J., Lubis, S.I., Simanjuntak, G., Lee, V.H., Nalim, S., Juslis, K., & See, R. (1983). A survey for arboviral antibodies in sera of humans and animals in Lombok, Republic of Indonesia. *Annals of Tropical Medicine & Parasitology*, 77(2), 131–137. 10.1080/00034983.1983.11811687

Olson, J. G., Ksiazek, T. G., Suhandiman, & Triwibowo (1981). Zika virus, a cause of fever in Central Java, Indonesia. *Transactions of the Royal Society of Tropical Medicine and Hygiene*, 75(3), 389–393. 10.1016/0035-9203(81)90100-0

Rasmussen, S.A., Jamieson, D.J., Honein, M.A. and Petersen, L.R. (2016). Zika virus and birth defects-reviewing the evidence for causality. *New England Journal of Medicine*, 374(20), 1981–1987. 10.1056/NEJMsr1604338

Rodriguez-Morales, A.J., Bandeira, A.C., & Franco-Paredes, C. (2016). The expanding spectrum of modes of transmission of Zika virus: a global concern. *Annals of Clinical Microbiology and Antimicrobials*, 15(1), 1–2. 10.1186/s12941-016-0128-2

Roos, R. P. (2016). Zika Virus-A Public Health Emergency of International Concern. *JAMA Neurology*, 73(12), 1395–1396. 10.1001/jamaneurol.2016.3677

Runge-Ranzinger, S., Morrison, A. C., Manrique-Saide, P., & Horstick, O. (2019). Zika transmission patterns: a meta-review. *Tropical Medicine and International Health*, 24(5): 523–529. 10.1111/tmi.13216

Sapkal, G.N., Yadav, P.D., Vegad, M.M., Viswanathan, R., Gupta, N., & Mourya, D.T. (2018). First laboratory confirmation on the existence of Zika virus disease in India. *Journal of Infection*, 76(3), 314–317. 10.1016/j.jinf.2017.09.020

Sasi, M.S., Rajendran, R., Meenakshy, V., Suresh, T., Pillai, R.H., Kumar, T.D., Sugathan, A. and Regu, K. (2021). Study on Vector Dynamics of Zika Virus Outbreak in Thiruvananthapuram, Kerala, India. *International Journal of Current Microbiology and Applied Sciences*, 10(12), 54–71. 10.20546/ijcmas.2021.1012.008

Saxena, S.K., Kumar, S., Sharma, R., Maurya, V.K., Dandu, H.R. and Bhatt, M.L. (2019). Zika virus disease in India-update October 2018. *Travel Medicine and Infectious Disease*, 27, 121–122. 10.1016/j.tmaid.2018.10.022

Scaramozzino, N., Crance, J.M., Jouan, A., DeBriel, D.A., Stoll, F., & Garin, D. (2001). Comparison of flavivirus universal primer pairs and development of a rapid, highly sensitive heminested reverse transcription-PCR assay for detection of flaviviruses targeted to a conserved region of the NS5 gene sequences. *Journal of Clinical Microbiology*, 39(5), 1922–1927. 10.1128/JCM.39.5.1922-1927.2001

Scroll.in (2021). Inside Uttar Pradesh's Zika Outbreak: Can India's Most Populous State Contain the Virus Spread?'. Available at: https://scroll.in/article/1010555/inside-uttar-pradeshs-zika-outbreak-can-indias-most-populous-state-contain-thevirus-spread (accessed on 30 November 2021).

Shi, Y., & Gao, G.F. (2017). Structural biology of the Zika virus. *Trends in Biochemical Sciences*, 42(6), 443–456. 10.1016/j.tibs.2017.02.009

Simpson, D.I. (1964). Zika virus infection in man. *Transactions of the Royal Society of Tropical Medicine and Hygiene*, 58 (4), 335–338. 10.1016/0035-9203(64)90200-7

Singh, H., Singh, O.P., Akhtar, N., Sharma, G., Sindhania, A., Gupta, N., & Valecha, N. (2019). First report on the transmission of Zika virus by *Aedes (Stegomyia) aegypti* (L.) (Diptera: Culicidae) during the 2018 Zika outbreak in India. *Acta tropica*, 199, 105–114. 10.1016/j.actatropica.2019.105114

Smithburn, K.C. (1954). Neutralizing antibodies against arthropod-borne viruses in the sera of long-time residents of Malaya and Borneo. *American Journal of Hygiene*, 59(2), 157–163. 10.1093/oxfordjournals.aje.a119630

Smithburn, K.C., Kerr, J.A., & Gatne, P.B. (1954). Neutralizing antibodies against certain viruses in the sera of residents of India. *The Journal of Immunology*, 72(4), 248–257.

Tan, T. Y., Fibriansah, G., Kostyuchenko, V. A., Ng, T. S., Lim, X. X., Zhang, S., Lim, X. N., Wang, J., Shi, J., Morais, M. C., Corti, D., & Lok, S. M. (2020). Capsid protein structure in Zika virus reveals the flavivirus assembly process. *Nature Communications*, 11(1), 895. 10.1038/s41467-020-14647-9

Tang, H., Hammack, C., Ogden, S.C., Wen, Z., Qian, X., Li, Y., Yao, B., Shin, J., Zhang, F., Lee, E.M., & Christian, K.M. (2016). Zika virus infects human cortical neural progenitors and attenuates their growth. *Cell Stem Cell*, 18(5), 587–590. 10.1016/j.stem.2016.02.016

Tognarelli, J., Ulloa, S., Villagra, E., Lagos, J., Aguayo, C., Fasce, R., Parra, B., Mora, J., Becerra, N., Lagos, N., Vera, L., Olivares, B., Vilches, M., & Fernández, J. (2016). A report on the outbreak of Zika virus on Easter Island, South Pacific, 2014. *Archives of Virology*, 161(3), 665–668. 10.1007/s00705-015-2695-5

Tyagi, B.K. (2020). *Mosquito Hunters: A history of hostilities against man's deadliest Foe – the mosquito – since 1881*. Scientific Publishers, India, 474 pp.

Vythilingam, I., Sam, J. I., Chan, Y. F., Khaw, L. T., & Sulaiman, W. Y. (2016). New Paradigms for Virus Detection, Surveillance and Control of Zika Virus Vectors in the Settings of Southeast Asia. *Frontiers in Microbiology*, 7, 1452. 10.3389/fmicb.2016.01452

Wikan, N., and Smith, D.R. (2016). Zika virus: history of a newly emerging arbovirus. *The Lancet Infectious Diseases*, 16 (7), e119–e126. 10.1016/S1473-3099(16)30010-X

Wiwanitkit, V. (2016). The Current Status of Zika Virus in Southeast Asia. *Epidemiology and Health*, 38, 1–5. doi: 10.4178/epih.e2016026

WHO (2009). *Epidemic and Pandemic Alert. Dengue: guidelines for diagnosis, treatment, prevention and control*. World Health Organization, Geneva.

WHO (2016). *Director General Summarizes the Outcome of the Emergency Committee Regarding Clusters of Microcephaly and Guillain-Barré Syndrome*. World Health Organization, Geneva.

WHO (2017). Emergencies Preparedness, Response, Zika Virus Infection-India. World Health Organization, Geneva. Available at: https://www.who.int/csr/don/26-may-2017-zika-ind/en/ (accessed on 16 June 2017).

Yadav, P.D., Kaur, H., Gupta, N., Sahay, R.R., Sapkal, G.N., Shete, A.M., Deshpande, G.R., Mohandas, S., Majumdar, T., Patil, S., & Pandit, P. (2022). Zika a Vector Borne Disease Detected in Newer States of India Amidst the COVID-19 Pandemic. *Frontiers in Microbiology*, 13. 10.3389/fmicb.2022.888195

Yadav, P.D., Malhotra, B., Sapkal, G., Nyayanit, D.A., Deshpande, G., Gupta, N., Padinjaremattathil, U.T., Sharma, H., Sahay, R.R., Sharma, P., & Mourya, D.T. (2019). Zika virus outbreak in Rajasthan, India in 2018 was caused by a virus endemic to Asia. *Infection, Genetics and Evolution*, 69, 199–202. 10.1016/j.meegid.2019.01.026

Zanluca, C., Melo, V.C., Mosimann, A.L., Santos, G.I., Santos, C.N., Luz, K. (2015). First report of autochthonous transmission of Zika virus in Brazil. *Memórias do Instituto Oswaldo Cruz*, 110(4), 569–572. 10.1590/0074-02760150192

6 Japanese Encephalitis in India

R. Balasubramanian
ICMR-National Institute of Virology-Kerala Unit, Dept. of Health Research, Vandanam, Alappuzha, India

Roop Kumari
National Professional Officer (Malaria and VBDs), WHO Country Office India, Nirman Bhawan, New Delhi, India

Surajita Banerjee
State Entomologist, Integrated Disease Surveillance Programme, Kolkata, India

Sunil Mishra
ICMR-Model Rural Health Research Unit, CHC Bhunga, Hoshiarpur, India

B.K. Tyagi
University Institute of Biotechnology, Chandigarh University, Mohali, India

6.1 INTRODUCTION

Japanese encephalitis (JE) is a focal disease. The prevalence of JE parallels the degree of its connection with the local environmental conditions. The virus is maintained in nature by a complex cycle involving pigs as amplifying hosts, ardeid birds such as egrets and herons as reservoirs, and *Culex* mosquitoes (usually many in number) acting as vectors (Tyagi et al., 2014). Humans, having no role, are the dead end in the Japanese encephalitis virus (JEV) transmission cycle, as they do not develop adequate viremia to infect the mosquitoes. Twenty-four countries in the World Health Organization (WHO) South-East Asia (Tiwari et al., 2012; Kumar et al., 2018; Wuryadi and Suroso, 1989) and Western Pacific regions have endemic JEV transmission, exposing more than three billion people to risks of infection (Figure 6.1). In India, JE is a major public health problem in many parts, particularly in southern (Gajanana, 1998; NVBDCP, 2006; Tyagi et al., 2011) and northern India (Kumari and Joshi, 2012; Kumari et al., 2013), as well as the northeastern states (Dev et al., 2015; Phukan et al., 2004). The clinical manifestation of the disease ranges from the onset of acute febrile illness to neurological manifestations. Diagnosis of JE is mainly by virus isolation or by various serological detection methods. The common method for laboratory confirmation of a JEV infection is the detection of the JE virus-specific IgM antibody in a single sample of cerebrospinal fluid (CSF) or serum by ELISA assay specifically for JEV. Molecular diagnostic methods such as RT–PCR and real time PCR are also available.

JEV is currently endemic in Australia (Torres Strait Islands) and South-East and East Asia, including the temperate zone of north-eastern China, Japan, and Korea (Spickler, 2016), and exists

DOI: 10.1201/9781003327028-7

FIGURE 6.1 Distribution of Japanese encephalitis virus (JEV) in the world.

Source: Wikipedia.

in five different genotypes. The JEV exists as five distinguishable genotypes – G-I, G-II, G-III, G-IV, and G-V – based on nucleotide homology in the E protein gene; three subtypes (a, b and c) are also distinguishable within G-I. Genotypes one (G-I), two (G-II), and three (G-III) are found throughout Asia, genotype four (G-IV) in Indonesia, and genotype five (G-V) in Malaysia, China, and Korea (Le Flohic et al. 2013). G-III was the predominant genotype in Japan and Korea up to the 1990s (Schuh et al., 2010). A shift towards the dominance of G-I strains has, however, been recorded since 1995 (Nga et al., 2004). G-III strains have also been detected outside of their endemic areas, e.g., in Italy and Angola (Faizah et al., 2020). As far as India is concerned, the genotype G-III was the only strain occurring in the country before 2006, and Fulmali et al. (2011) confirmed G-I occurring simultaneously with G-III in JE patients in eastern Uttar Pradesh, India.

6.1.1 Definition

Japanese encephalitis is an arboviral infection caused by JEV, a flavivirus related to dengue, yellow fever, and West Nile viruses, and is spread by mosquitoes, mostly among children below 15 years of age who are also impacted the hardest by the disease. Although symptomatic JE is rare, the case fatality rate among those with encephalitis can be as high as 30%. Permanent neurologic or psychiatric sequelae can occur in 30%–50% of those with encephalitis.

6.1.2 Signs and Symptoms

Most JEV infections are mild (fever and headache) or without apparent symptoms, but approximately 1 in 250 infections results in severe clinical illness. The incubation period is between 4 and 14 days. In children, gastrointestinal pain and vomiting may be the dominant initial symptoms. Severe disease is characterized by rapid onset of high fever, headache, neck stiffness, disorientation, coma, seizures, spastic paralysis, and, ultimately, death. The case fatality rate can be as high as 30% among those with disease symptoms.

Of those who survive, 20%–30% suffer permanent intellectual, behavioral, or neurological sequelae, such as paralysis, recurrent seizures, or the inability to speak.

6.1.3 Cure and Treatment

There is no evidence that vector control of vectors, which breed in extensive open fields, especially under paddy cultivation, has any appreciable impact on disease burden. Since there is no cure for the disease, the treatment is focused on relieving severe clinical signs and supporting the patient to overcome the infection. Safe and effective JE vaccines are available to prevent disease, including inactivated mouse brain-derived vaccines, inactivated Vero cell-derived vaccines, live-attenuated vaccines, and live recombinant vaccines. Thorough the knowledge of the transmission of disease is very essential in control/prevention programms.

6.2 VIRUS STRUCTURE

JE is an enveloped, spherical virus of approximately 50 nm in diameter (Figure 6.2). The virus has an approximately 11-kb-long, single-stranded (3,800 kD), positive-sense RNA genome, which is capped at the 5’ end but is without a poly (A) tail at the 3’ end. The genome contains two noncoding regions, with one each at the 5’ and the 3’ ends, which play a major role in viral genome replication. The viral genome codes for an approximately 3,400 amino acid polyprotein. It contains a single open reading frame (ORF) encoding a polyprotein, which is co- and post-translationally cleaved into three structural proteins and seven non-structural proteins. Structural protein consists of a capsid [C], pre-membrane [prM], and envelope [E], which encapsulates the viral RNA (Lindenbach and Rice, 2003), and the non-structural proteins include NS1, NS2A, NS2B, NS3, NS4A, NS4B, and NS5 (Mohan Babu et al., 2010).

The viral C protein plays a significant role in virus assembly. The E protein forms an icosahedral protein shell and is responsible for cellular attachment and fusion of the virus envelope with the plasma membrane (Heinz et al., 2004), which plays various important roles in infection, such as receptor binding, membrane fusion, and cell entry. The E protein is the principal neutralization antigen that induces JEV-neutralizing antibodies in a virus-infected host. The prM protein contains a pre-signal peptide at its N-terminus that helps to facilitate the intracellular maturation and transport of E protein during virus particle morphogenesis (Heinz and Allison, 2003). Also, prM undergoes

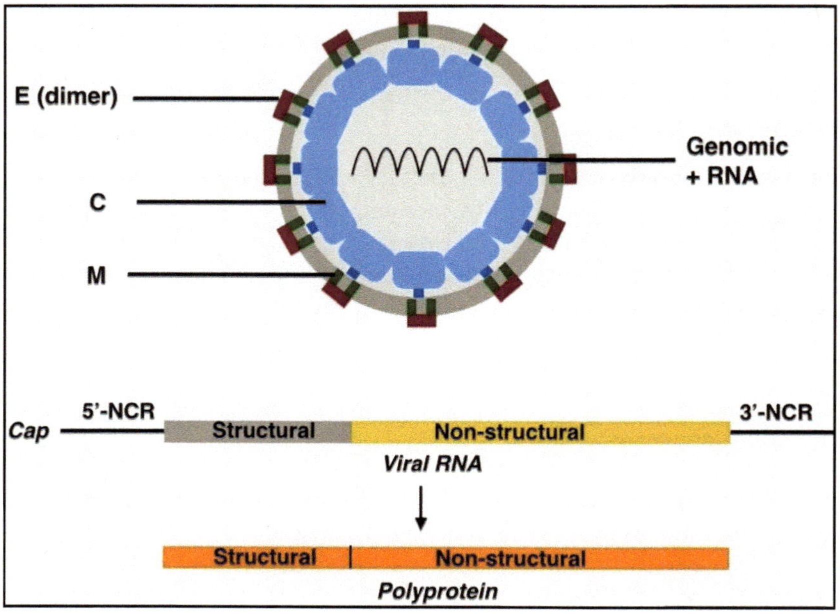

FIGURE 6.2 JEV.

proteolytic cleavage – an event required to produce a fully infectious virus, leaving the M protein in the mature viral envelope (Elshuber et al., 2003; Heinz and Allison, 2003).The external shell of the virus or the viral envelope is formed by the endoplasmic reticulum (ER)-derived lipid bilayers inserted with the mature M proteins and E protein. The JEV exists as five distinct genotypes in Asia that contain G-I, G-II, G-III, G-IV, and G-V (Uchil and Satchidanandam, 2001), based on nucleotide homology in the E protein gene (Table 6.1). Genotype I (G-I) includes strains isolated in South-Eeast Asia, Cambodia, Korea, Japan, India, Australia, and northern Thailand. Genotype II (G-II) has strains isolated in Australia, southern Thailand, Indonesia, and Malaysia (Le Flohic and Gonzalez, 2011). Genotype III was the most frequently isolated genotype throughout most of Asia from 1935, and genotype I was isolated from pigs (Wang and Liang, 2015; Duong et al., 2017). G-IV is found only in Indonesia, and G-V includes a single strain isolated from Singapore.

Three subtypes (a, b, and c) are present within G-I. However, all JEV strains belong to a single serotype. But antigenic variation has been recognized using polyclonal and monoclonal antibodies, for example, Nakayama and the Jagar-01 are two examples of subtypes of JEV (Okuno et al., 1968).

6.3 CLINICAL STUDIES

The clinical disease in horses is similar to that in humans, as they also develop severe encephalitis (Ellis et al., 2000). In serological surveys, pigs display higher geometric mean titres than other domestic or wild animals. They act as the source of blood meals, providing a sensitive indicator of virus transmission in endemic areas for mosquitoes (Burke et al., 1985).

JE infection in humans can range from asymptomatic to mildly febrile symptomatic illness to life-threatening disease affecting the central nervous system (CNS). Though only a tiny fraction of persons infected with the JEV develop clinical manifestations of encephalitis, those manifestations are so severe that the disease is greatly feared. It affects both adults and children; globally, nearly 67,900 JE cases occur each year, and 51,000 are among children aged 0–14 years (Campbell et al., 2011). JE causes the loss of more disability-adjusted life years than any other arthropod-borne virus due to frequent neurological sequelae of the condition (Turtle and Solomon, 2018).

TABLE 6.1
Summary of Strains of JEV Known so Far

Sl. No.	Strain	Year	Location	Source	GenBank Accession No.
1	Beijing-1	1949	China	Human	L48961
2	P3	1950	China	Mosquito	U47032
3	Muar	1952	Malaysia	Human	Hasegawa et al., 1994
4	SA14	1954	China	Mosquito	U14163
5	P20778	1958	India	Human	AF080251 (E)
6	HVI	1958	Taiwan	Human	AF098736
7	JaGAr01	1959	Japan	Mosquito	AF069076
8	Saigon	1962	Vietnam	Information Unavailable	U03696 (E)
9	Chiang Mai	1964	Thailand	Human	U70393 (E)
10	Ling	1965	Taiwan	Human	Tom Solomon et al., 2003
11	Sarawak	1965	Malaysia	Mosquito	AY184214
12	M859	1967	Cambodia	Mosquito	U70410 (E)
13	JE827	1968	Malaysia	Mosquito	D00994
14	691004	1969	Sri Lanka	Human	Z34097 (E)
15	WTP-70-22	1970	Malaysia	Mosquito	U70421 (E)
16	GP78	1978	India	Human	AF075723
17	JKT1724	1979	Indonesia	Mosquito	U70404 (E)
18	JKT5441	1981	Indonesia	Mosquito	U70406 (E)
19	KPP034-35CT	1982	Thailand	Mosquito	U03693 (E)
20	JaOArS982	1982	Japan	Mosquito	NC001437
21	B1065/63	1983	Thailand	Pig	U70388
22	PhAn1242	1984	Philippines	Pig	U70417 (E)
23	B-2524	1985	Nepal	Human	U70392 (E)
24	RP-9	1985	Taiwan	Mosquito	AF014161
2	K87P39	1987	Korea	Mosquito	U34927 (E)
26	CH1392	1990	Taiwan	Mosquito	AF254452
27	ThCMAr4492	1992	Thailand	Mosquito	D45362 (E)
28	K94P05	1994	Korea	Mosquito	AF045551
29	CH2195LA	1994	Taiwan	Mosquito	AF221499
30	FU	1995	Australia	Human	AF217620
31	T1P1	1997	Taiwan	Mosquito	W.-J. Chen
32	Ishikawa	1998	Japan	Mosquito	AB051292
33	CNS138-11	1999	Malaysia	Human	AY184213
34	Torres Strait 2000	2000	Australia	Pig	AF289818 (E)
35	Indonesia	Information Unavailable	Indonesia	Mosquito	U03692 (E)
36	TC	Information Unavailable	Taiwan	Information Unavailable	AF098736

Children with JE are more likely to have dystonia and poor outcomes at six months than adults based on immunological factors, CNS maturation, and neuronal plasticity (Kalita et al., 2003). When an infected mosquito ingests blood from a human, the virus enters the host. After an incubation period of 5 to 15 days, the virus enters the bloodstream and travels to various glands. And it multiplies, transfers, settles in the brain, and attacks the CNS. When the infection reaches secondary lymphoid tissue, viremia spreads to visceral organs, e.g., liver, kidney, and spleen. If the virus escapes from the blood, the virus avoids targeting the immune system and is transported to the brain and spinal cord through a hematogenous route. The virus crosses the blood-brain barrier by passive transport after active replication in epithelial cells. Mainly, the virus causes acute encephalitic syndrome and severely infects the brain and the spinal cord (Ayukawa et al., 2004). The symptoms include high fever, headache, nausea, dizziness, myalgia, convulsions, focal neurological signs, and decreased consciousness. JE is generally asymptomatic, and clinical manifestations are seen only in approximately 0.1%–1% of JEV-infected persons. Failure to make the proper and timely diagnosis of the JEV can lead to a high infant mortality rate. The fatality rate can reach 30%–50% of survivors who may continue to suffer definitive neurological or psychiatric sequelae (Fischer et al., 2008). Therefore, early recognition of clinical features of JE disease helps in disease management and cuts down the infant mortality rate (Gunasekaran et al., 2012).

The preferred breeding sites of the vector for JE have irrigated rice paddies and large pools densely filled with masses of algae which also attract migratory wading birds that help the continuation of the sylvatic transmission cycle (Jeffries and Walker, 2015). Thus outbreaks of JE infection in these areas are reported often. Human traveling, vector mobility, and bird migration increase the spread of the JEV in non-endemic regions. An increase in temperature and rainfall helps the vector thrive as the surface water covers support mass breeding. In addition, the presence of flood water in rivers, puddles, ponds, and vegetation fields in endemic areas widely favors the survival of vector populations that associate with virus invasion of the human population. As a result of urbanization, a virus has been transmitted to new sites where it has been slowly establishing itself.

6.4 EPIDEMIOLOGICAL STUDIES

JEV causes JE, a complex and intriguing disease of the CNS. Mainly two epidemiological patterns of JE are recognized: epidemic and endemic. Epidemic patterns were observed mainly in northern areas, including Bangladesh, Bhutan, the People's Republic of China, Taiwan, Japan, South Korea, North Korea, Nepal, northern Vietnam, northern India, northern Thailand, Pakistan, and Russia. JEV infection was present in Japan in 1871, but the first outbreak occurred in 1924, involving over 6,125 cases and 3,797 deaths (Rappleye, 1939). Then a large outbreak occurred in 1935 and annual episodes from 1946 to 1952 (Pond and Smadel, 1954).JE was reported in the Republic of China for the first time in the 1940s and became the leading cause of viral encephalitis (Wang et al., 2009). A human JE case was first confirmed in South Korea in 1946 (Yang et al., 2004), and a large outbreak was reported in 1983 with hundreds to thousands of cases (Lee et al., 2012). JE outbreak was reported in Bangladesh in 1977, which resulted in 22 cases, with 7 deaths confirmed by serological testing (Khan et al., 1981). JE cases were first reported in 1978 in Nepal (Bista and Shrestha, 2005), where a total of 2,040 JE cases, including 205 deaths, were reported from 2005 to 2010 after mass immunization in 2006 (Dumre et al., 2013). JE was confirmed in Pakistan by a seroepidemiological survey in the early 1980s (Darwish et al., 1983). An acute encephalitis case was further confirmed during the JE epidemic in 1992 (Igarashi et al., 1994).

Endemic patterns are found in southern areas, including Australia, Burma, Brunei Darussalam, Cambodia, Indonesia, Laos, Malaysia, Papua New Guinea (PNG), Philippines, Singapore, south Vietnam, southern Thailand, Southern India, Sri Lanka, and Timor-Leste. JE has been recognized as a clinical disease in Thailand and the Philippines since 1950. The first serologic evidence of JEV infection in Thailand detected in horses occurred in 1961 (Thongcharoen, 1985), and the first outbreak of Japanese encephalitis was reported in 1969. The serological study on clinical viral encephalitis cases revealed that 85.2% of JE cases were under 15 years old, and JE cases were reported each year in individuals ranging from 2 to 77 years; however, 72.9% were under 17 years of age (Trosper et al., 1980; Barzaga, 1989). JE was first recorded in the Philippines in the early 1950s (Trosper et al., 1980). In Sarawak in East Malaysia and Penang, Perak, Selangor, and Johore in West Malaysia, JE was reported repeatedly in 1952 (Paterson et al., 1952). JE cases were first reported in Singapore in 1952

(Erlanger et al., 2009). Approximately 100 JE cases were reported in the 1970s and early 1980s, and a dozen were reported during 1985–1992 (See et al., 2002; Ting et al., 2004).

JE infection then spread to Australia from 1955 to 1998 (Shield et al., 1996) this resulted in the isolation and identification of a new strain of JEV (genotype II). In Phnom Penh, the capital of Cambodia, two years of sentinel surveillance performed in six hospitals revealed that 95% of JE patients were under 12 years of age, and the estimated incidence of JE was 11.1 per 100,000 in children less than 15 years of age (Touch et al., 2009; Chastel and Rageau, 1966). A serological survey confirmed the existence of JE in Burma in 1968. The first outbreak of JE was reported in 1974 in Shan State, in which five cases and four deaths were reported (Swe et al., 1979). Sri Lanka reported JE cases in 1968, and since then, cases have been reported yearly in pigs. Acute encephalitis accounts for over 1,000 hospital admissions yearly, but only 30 patients in the last 10 years were confirmed serologically.

In Laos, JE was first reported in 1990, and JEV was isolated from human blood in 2009, which belonged to genotype 1 (Aubry et al., 2013). The first serologically confirmed JE cases in Indonesia were reported in 1996. Two hundred and thirty-nine JE cases in pediatric patients were confirmed, which estimated that the JE incidence rate was 8.2 per 100,000 in children under 10 years of age (Kari et al., 2006).

6.5 INDIAN SCENARIO OF JEV

The virus is active almost in every part of India, where significant outbreaks have been reported from 231 districts of 23 states. The most affected states are Andhra Pradesh, Assam, Bihar, Haryana, Karnataka, Kerala, Maharashtra, Manipur, Tamil Nadu, Orissa, Uttar Pradesh, and West Bengal (Reuben and Gajanana,1997; Kulkarni et al., 2018). Significant outbreaks were reported from different parts of the country, predominantly in rural areas. The first human case was reported from the Vellore of North Arcot district in Tamil Nadu, in 1955 (Hanna et al., 1996). The first major JE epidemic was reported in the Burdwan and Bankura districts of West Bengal in 1973, followed by another outbreak in 1976, which resulted in 700 cases and 300 deaths (Saxena and Dhole, 2008).

Eastern Uttar Pradesh faced the first epidemic in October of 1978, reporting 3,000 cases, most of which affected children (Kanojia, 2007: Arunachalam et al., 2008). In 1981, JE was reported in Bellary district in Karnataka and adjoining areas, which involved more than 219 people, of which 61 died (George et al., 1990). In the state of Odisha in eastern India, an outbreak of JE was reported in Rourkela city of the Sundergarh district in 1989 (Vajpayee et al. 1991), followed by another outbreak in July and August 1993. In the north-western part of India, in 1990, the first JE outbreak was reported in Haryana State (Kar and Saxena, 1998). JE, an acute neurologic illness affecting children with high case fatality rates, was written about in the Muzaffarpur district of Bihar in 1995 (Kripasindhu et al., 2022). In Kerala, the first JE outbreak was reported in 1996 in the Alappuzha district; during the outbreak, 48 cases were reported (Dhanda et al., 1997a,b). From 2003 to 2004, JE cases were reported in Maharashtra for the first time.

Then a large outbreak of JE was reported in Gorakhpur and adjoining districts in Uttar Pradesh from 2005 to 2018, there were 47,509 acute encephalitis syndrome (AES) cases reported in Uttar Pradesh among which 8373 patients died. Another outbreak followed these in 2006 with 2,320 cases and 528 deaths, and 3,024 cases and 645 deaths in 2007 (Kanojia, 2007: Arunachalam et al., 2008; UT Wise Cases and Deaths due to JE, 2013–17; 2018). The outbreaks in UP led to a significant decision to introduce the vaccine in high endemic areas in 2006 (National Vector Borne Disease Control Programme, 2014). An outbreak of encephalitis was investigated in Uttar Pradesh to identify and characterize the etiologic agent that showed homology with JEV (Joshi et al., 2012). Recently, India witnessed another large outbreak in Malkangiri, Odisha, in 2012 (Dwibedi et al., 2015).

Recently, JE has emerged as a public health problem in Kerala, where two cases and one death in 2015, and five cases and two deaths in 2018 (Directorate of National Vector Borne Disease Control Programme, State-Wise Number of AES/JE Cases and Deaths from 2015 to 2022). In 2019, another case of JE was reported in Thrissur, Kerala (Times of India, 2019), and 11 cases and 2 deaths were observed (Data on Communicable Diseases, 2019). Earlier, a major outbreak of JE and West Nile virus complex outbreak was reported in 2011 in Kerala.

6.6 VECTORS AND ISOLATION OF JEV FROM MOSQUITOES

Since JE distribution is significantly linked to irrigated rice fields combined with pig rearing, mosquito species belonging to the *Culex vishnui* subgroup of comprising *Culex tritaeniorhynchus* Giles, *Cx. vishnui* Theobald and *Cx. pseudovishnui* Colless have been implicated as major vectors of JEV (Reuben et al., 1988). In India, however, JEV has been isolated from 17 species of mosquitoes, 10 species of Culex, 3 species each of Anopheles and Mansonioides, and 1 species of Armigeres (Tyagi et al., 2014), as shown in Table 6.2.

Notwithstanding JEV isolation from a large number of mosquito species belonging to different genera, *Cx. tritaeniorhynchus* is arguably the most important vector the dissemination of the virus in the whole of Asia. In Kerala, *Mansonia* species have emerged as a major contributor to the JEV transmission, albeit having a presence in many other states as well.

6.7 TRANSMISSION CYCLE OF JE

When a host, upon exposure to the pathogen, becomes infected, it does not always result in disease. The infected host can amplify or carry the pathogen for an extended period. The host is known as a reservoir host when there is an increase in the pathogen load but not for a long duration (amplifying host) or either (dead-end host). JE is spread by mosquito vectors of variety and has a seasonality of occurrence (Lian et al., 2002; Gulati et al., 2012). The virus population is maintained by a cycle of mosquito infection and vertical transmission, infection in water birds and pigs, and further infection in uninfected mosquitoes. In the case of the JEV, the common transmission route happens when

TABLE 6.2
Mosquito Species So Far Incriminated with JEV in India

Sl. No.	Mosquito Species	No. of Isolations	State of India
1	*Anopheles peditaeniatus*	1	Karnataka
2	*An. barbirostris*	1	West Bengal
3	*An. subpictus*	9	Karnataka, Kerala, Tamil Nadu
4	*Culex tritaeniorhynchus*	29	Karnataka, Kerala, Tamil Nadu
5	*Cx. vishnui*	30	Karnataka, Kerala, West Bengal
6	*Cx, pseudovishnui*	8	Karnataka, Goa
7	*Cx. epidesmus*	1	West Bengal
8	*Cx. bitaeniorhynchus*	3	Karnataka, West Bengal
9	*Cx. infula*	6	Tamil Nadu
10	*Cx. gelidus*	8	Tamil Nadu, Karnataka
11	*Cx. whitmorei*	4	Tamil Nadu, Karnataka, Andhra Pradesh, West Bengal
12	*Cx. fuscocephala*	7	Tamil Nadu, Karnataka
13	*Cx. quinquefasciatus*	1	Karnataka
14	*Mansonia annulifera*	2	Kerala, Assam
15	*Ma. indiana*	3	Kerala
16	*Ma. uniformis*	4	Kerala, Karnataka
17	*Armigeres subalbatus*	1 (male)	Uttar Pradesh (Gorakhpur district)

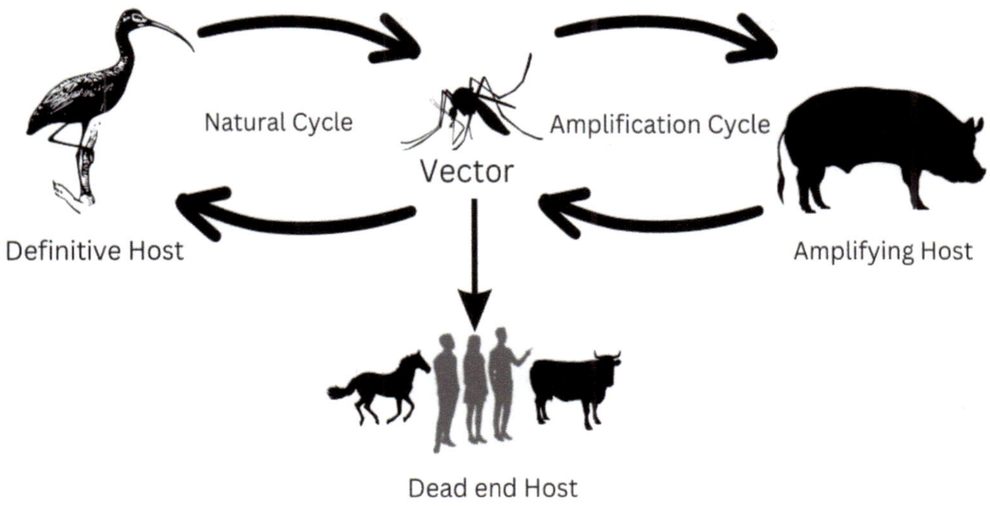

FIGURE 6.3 Life cycle of JEV.

an uninfected mosquito acquires the virus by feeding on an infected host (birds and pigs) and the former becomes infected. The infected mosquito then takes a blood meal from another uninfected host. Thus the host becomes infected. However, humans, horses, and cattle are dead-end hosts. This means that there is no subsequent transmission (Figure 6.3).

In the enzootic cycle, the JEV was maintained and circulated in birds, especially heron and egrets and mosquitoes. Infection of pigs and birds occurs through the bite of infected mosquitoes and further spreads to other non-infected mosquitoes and pigs. Severe febrile illness was noticed in some of the Kyasanur forest disease (KFD) infected monkeys. Humans contact infection mainly through the bite of infected female mosquitoes. Horizontal transmission between humans has not been reported. Persons visiting JE endemic areas for recreation or other purposes will contract infection by accidental mosquito bites. Neutralizing antibodies of JEV have been found in other vertebrate hosts, including domestic vertebrates, donkeys, chickens, ducks, buffalos, cattle, and sheep, as well as vertebrates in nature, including mice, bats, snakes, frogs, and flying foxes (Upadhyay, 2015a; Miller et al., 2012). JEV can replicate in a wide variety of arthropod hosts, too, for example, midges, *Lasiohelea taiwana* (Wu and Wu, 1957), and ticks, *Haemaphysalis japonica* (Lvov, 1978).

The mosquito of the genus Culex, mainly *Culex tritaeniorhynchus*, acts as a major vector for JE. These mosquitoes breed in stagnant water bodies. Widespread distribution of this species of mosquito in paddy fields surrounded villages. JEV has been isolated from various other species of mosquitoes. The mosquito vector was subsequently confirmed by virus isolation from *Culex tritaeniorhynchus* mosquitoes in 1938 (Mitamura et al., 1938). *Cx. tritaeniorhynchus* is considered to be the primary vector in JEV transmission throughout most of Asia (Kramer et al., 2011; Solomon, 2006; Jeffries and Walker, 2015), but in the case of Australia, the primary vector of JE is *Cx. annulirostris* (Hall Mendelin et al., 2012;Van den Hurk et al., 2009). JE has been isolated from *Culex Whitmore, Cx. pseudovishnui, Cx. Vishnu*, and *Aedes subpictus* in India; *Anopheles Hyrcanus* group and *An. barbirostris* in Pakistan; *Cx. tritaeniorhynchus* mosquito species in Cambodia (Chastel and Rageau, 1966); *Ae. chemulpoensis, Cx. pipiens, An. hyracanus, Armigeres obturans*, and *Aeedes albopictus* in China; and *Cx. fuscocephala, Cx. Whitmore*, and *Mansonia uniformis* in Sri Lanka (Huang, 1982; Gajanana et al., 1997). *Aedes spp.* showed less efficiency in JEV transmission under laboratory conditions. However, in India, JEV has been isolated from 17 species of mosquitoes (Tyagi et al., 2014; Philip Samuel, 2000). JEV was isolated from wild-caught *Culex tritaeniorhynchus* Giles, *Mansonia indiana*, and *Ma. uniformis* (Dhanda et al., 1997a,b). Entomological studies were carried out in Kuttanadu, Kerala (1999–2000), Gorakhpur, Uttar Pradesh (2005–06; 2010–15), etc., to determine the seasonal abundance and JEV infection rates in *Cx. tritaeniorhynchus* (Arunachalam et al., 2004; Tyagi et al. 2014).

6.8 INNOVATIVE RESEARCH AND DEVELOPMENT OF TECHNOLOGIES

ICMR's erstwhile Centre for Research in Medical Entomology in Madurai has developed a new technique to isolate JEV from the wild-caught desiccated mosquitoes using a TOXO-IFA system which was transferred to the Department of Public Health, Government of Tamil Nadu (Figure 6.4).

6.9 VACCINES

Prevention by vaccination, as well as mosquito control, is the most effective measure. Several JE vaccines have been developed, including inactivated and live vaccines. The earliest human vaccine tried were crude inactivated antigens prepared from Nakayama strain-infected chicken embryos or mouse brains. The dose could be reduced by one-tenth in adults and two to 7 year-old children. The recommended regimen was a primary series of 2 to 3 doses at days 0, 7, and 30, followed by boosters after 6 to 12 months, and then every three years (Poland et al., 1990).

The only live-attenuated vaccine derived through multiple passages of the wild-type SA14 strain in animals and a cell culture approved for use in humans is the SA14-14-2 vaccine developed by the Institute of Biological Products of Chengdu and has been extensively used in China. The SA14-14-2 strain was completely attenuated for mice and safe and immunogenic in mice, guinea pigs, pigs, and humans, including children (Yu, 2010). It was first licensed for use in 1988 in the People's Republic of China, and currently, the vaccine is licensed in India, South Korea, and Nepal. The vaccine is mainly recommended for children 1 to 2 years of age, with a booster one year later. The only genetically engineered licensed vaccine for JE is JE-CV (ChimeriVax-JE); these are designed to be attenuated. Cell culture-derived inactivated (CCDI) vaccines have now replaced the animal tissue-derived vaccines; the first such vaccine was the PHK cell propagated, formalin-inactivated, wild-type OCT-541 vaccine which showed two to three times higher potency than the OCT-541 LAV vaccine (Darwish and Hammon, 1966).

In 2005, India witnessed a massive outbreak of JE; in response to this outbreak, the Government of India, with the support of the Program for Appropriate Technology in Health (PATH), initiated a pilot project in 2006 to immunize children in hyper-endemic areas. Uttar Pradesh, Assam, West Bengal, and Karnataka were selected for the immunization programm. SA-14-14-2 live JE vaccine manufactured by China was used. It was given to children aged 1 to 15 years old as a single subcutaneous dose. JENVAC is a single-dose, inactivated, Vero cell-derived JE vaccine produced in India for the first time. Mouse brain-derived vaccines (MBDVs) have been used nowadays due to their protective solid efficacy. Currently available vaccines in India are as follows in Table 6.3.

These vaccines have limitations in availability and requirements for multiple doses. Several novel JE vaccines have been developed in recent years, which include cell culture-based inactivated vaccines and live-attenuated vaccines (Daniels, 2003), chimeric virus vaccines, recombinant adenovirus-based vaccines, plasmid DNA-based vaccines, and peptide vaccines (Karen et al., 2017; Igarashi, 2002; Nah et al., 2015). Vaccination of travelers is highly essential because they remain at greater risk of JEV infection (Mohanbabu et al., 2010). High-standard medical facilities, bio-informatics facilities, and proper and timely vaccination are required to reduce the infection rate.

6.10 VECTOR CONTROL

6.10.1 CHEMICAL

Many personal protective measures like mosquito repellents in coils and creams are available on the market. Control of the disease transmission by the vector can be achieved by using pesticides as the primary vector, for the disease resides in the paddy field, which includes insecticides such as

FIGURE 6.4 (Above) Event of transfer of technology by Dr V.M. Katoch, director general, ICMR and secretary of the Department of Health Research, MOH&FW, GOI, to Dr K. Durairaj, addl. director, representative from the Health Directorate, Tamil Nadu, in the presence of (from left to right) Dr D.V.P. Raja, Dr V.M. Katoch, Dr K. Durairaj, Dr B.K. Tyagi, Dr V. Thenmozhi, Mr Sanjeev Dutta, and Dr R. Paramasivan, and (below) the transfer of technology document

Source: Dr B.K. Tyagi, pers. archive.

TABLE 6.3
JE Vaccines Available in India

Sl. No	Vaccine Name	Vaccine Type	Manufacturer(Country)
1	JEEV	JE vaccine (inactivated)	Biological E. Limited (India)
2	JAVAC	JE vaccine (inactivated)	Bharat Biotech (India)
3	Live attenuatedSA 14-14-2	JE vaccine (live attenuated)	Chengdu Institute of Biological Products (People's Republic of China)

DDT, pyrethroids, organophosphates, and carbamates. Insecticides are used by fogging and spraying methods, or they can also be applied directly to mosquito breeding sites or resting sites (indoor and outdoor spray) to kill larvae and adults, respectively. State and local agencies commonly use the organophosphate insecticides Malathion and synthetic pyrethroid insecticides like Prallethrin, Etofenprox, Pyrethrins, Permethrin, Resmethrin, and Sumithrin for adult mosquito control.

Mosquito repellents containing 30% to 50% DEET (N, N-diethyl-m-tolumide) can be used to reduce mosquito bites. Pyrethroid impregnated jute or polypropylene curtains have been used for the control of JE vectors nowadays (Kabilan et al., 2004).

6.10.2 Biological Control

The development of eco-friendly plant-based drugs (Tu, 2011) and plant-synthesized nanoparticles are nowadays used as mosquito control agents (Subramaniam et al., 2016). For example, extracts of *Piper retrofractum* (Piperaceae) and essential oils are used as larvicide and oviposition detergents (Govindarajan et al., 2018). Biological control of mosquitoes using vertebrates has focused chiefly on larvivorous fish that are very effective at reducing mosquito larval populations (Kumar and Hwang, 2006; Chandra et al., 2008). Some omnivorous copepods such as *Cyclops vernalis, Megacyclops formosanus, Mesocyclops aspericornis*, and *Mesocyclops edax* have been reported as active predators of young mosquito instars (Marten et al., 1989; Rawlins et al., 1997). Also, the introduction of composite fish culture for mosquito control in rice fields has been evaluated and proved successful.

Naturally occurring microorganisms like bacteria (e.g., *Wolbachia* Endo symbiotic bacteria; Zug and Hammerstein, 2012) and fungi are also used for the control of mosquito species. Due to cytoplasmic incompatibility, mosquitoes infected with the obligate intracellular bacteria *Wolbachia* are unable to reproduce successfully, so the release of *Wolbachia*-infected males can reduce mosquito spp. Mosquitoes with *Wolbachia* have been successfully used in Mexico, Australia, Singapore, and Thailand (Michael et al., 2014). *Bacillus thuringiensis* var. *israelensis* (Bti) is used as an insecticide spray for adults and larvicides (Ben-Dov et al., 2014) that has been commercialized. Microgel droplet formulations of *Bacillus sphaericus* can be used as a control method for mosquito larvae in rice fields (Keiser et al., 2005). The entomopathogenic fungi, especially *Metarhizium anisopliae* and *Beauveria bassiana*, produce spores that attach and penetrate the cuticle of mosquitoes (Scholte et al., 2004) and also produce several toxins during the infection that are lethal to mosquitoes (Scholte et al., 2007).

Neem products in rice fields can be used to reduce the breeding of mosquito larvae. Using neem products as fertilizer in rice fields not only enhances grain production but also suppresses the breeding Culicinae vectors of JE. Control of mosquitoes through chemical or biological methods could potentially lead to ecological imbalance since they could affect all the mosquitoes.

6.10.3 Mechanical Control

Insecticide-treated mosquito nets are used in some Cambodian rural areas to reduce the seroconversion rate in humans and pigs in India (Dutta et al., 2011). Safe window mesh and regular water storage tank treatment can decrease mosquitoes' bite rate. More attention should be paid to the control of viruses by applying clean drives, livestock and pig farm shifting, waste disposal, and the supply of safe drinking water. A water management system with an intermittent irrigation system can be used to reduce the breeding of mosquito larvae. Water management is nothing but a strategy of alternately drying and wetting water.

6.11 CONCLUSION

JE is an endemic zoonotic viral mosquito-borne disease affecting mainly children and old-age people. Every year in India, hundreds of human cases are reported. Being a mosquito-borne disease, strict measures should be taken to control the mosquito population, both immature and adult. Effective JE vaccines are available to prevent disease. Regular vaccination should be carried out in children and old-age populations with increased area coverage. JEV is still widespread and remains a source of infection for humans. There is an urgent need for effective control strategies so that this type of mosquito-borne infection can be controlled.

REFERENCES

Allison, S. L., Tao, Y. J., O'Riordain, G., Mandl, C. W., Harrison, S. C., Heinz, F. X. (2003). Two distinct size classes of immature and mature subviral particles from tick-borne encephalitis virus. *Journal of Virology*, 77. 10.1128/jvi.77.21.11357-11366.2003

Arunachalam, N., Philip Samuel, P., Paramasivan, R., Balasubramanian, A., Tyagi, B.K. (2008). Japanese encephalitis in Gorakhpur Division, Uttar Pradesh. *Indian Journal of Medical Research*, 128, 775–777.

Arunachalam, N, P. Philip Samuel, J. Hiriyan, V. Thenmozhi, and A. Gajanana Japanese Encephalitis in Kerala, South India: Can Mansonia (Diptera: Culicidae) Play a Supplemental Role in Transmission? *Journal of Medical Entomology*, 41(3): 456Ð461 (2004).

Aubry, F., Vongsouvath, M., Nougairede, A., Phetsouvanh, R., SibounheuangB., Charrel, R., Rattanavong, S., Phommasone, K., Sengvilaipraserth, O., de Lamballerie, X., Newton, P.N., Dubot-Pérès, A. (2013). Complete genome of a genotype I Japanese encephalitis virus isolated from a patient with encephalitis in Vientiane, Lao PDR. *Genome Announcements*, 1(1), e00157–12.

Ayukawa, R., Fujimoto, H., Ayabe, M., Shoji, H., Matsui, R., Iwata, Y., Fukuda, H., Ochi, K., Noda, K., Ono, Y., Sakai, K., Takehisa, Y., Yasui, K. (2004).An unexpected outbreak of Japanese encephalitis in the Chugoku district of Japan, 2002, *Japanese Journal of Infectious Diseases*, 57(2), 63–66.

Mohan Babu, Appaiahgari, and SudhanshuVrati.A Yellow fever virus-based novel Japanese encephalitis vaccine. *Expert Review of Vaccines*, 9(12), 1371–1384 (2010)

Barzaga, N. G. (1989). A review of Japanese encephalitis cases in the Philippines (1972–1985). *The Southeast Asian Journal of Tropical Medicine and Public Health*, 20(4):587–592.

Ben-Dov, E. Bacillus thuringiensis subsp. israelensis and Its Dipteran-Specific Toxins. *Toxins* 2014, 6, 1222–1243.

Bista, M.B., Shrestha, J.M. Epidemiological situation of Japanese encephalitis in Nepal. *JNMA; Journal of the Nepal Medical Association* 2005, 44(158):51–56.

Burke, D.S., Tingpalapong, M., Ward, G. S., Andre, R., Leake, C.J. (1985). Intense transmission of Japanese encephalitis virus to pigs in a region free of epidemic encephalitis. *The Southeast Asian Journal of Tropical Medicine and Public Health*, 16:199–206.

Campbell, G. L., Hills, S. L., Fischer, M., Jacobson, J. A., Hoke, C. H., Hombach, J. M., ... & Ginsburg, A. S. (2011). Estimated global incidence of Japanese encephalitis: A systematic review. *Bulletin of the World Health Organization*, *89*(10), 766–774.

Chandra, G., Bhattacharjee, I., Chatterjee, S.N., Ghosh, A.Mosquito control by larvivorous fish. *The Indian Journal of Medical Research* 2008, 127, 13–27.

Chastel, C., & Rageau, J. (1966). *Isolation of arbovirus in Cambodia from naturally infected mosquitos*.

Daniels, P. T.Arboviruses of Veterinary Significance in The Asia-Western Pacific Region, Such as Japanese Encephalitis Virus, Office International des Epizooties: Paris, France, 2003.

Darwish, M. A., Hammon, W.M. Studies on Japanese B encephalitis virus vaccines from tissue culture. 8. Formalin inactivated OCT-wild strain vaccine. *Journal of Immunology* 1966, 97: 506–11.

Darwish, M. A., Hoogstraal, H., Roberts, T. J., Ahmed, I. P., Omar, F. A seroepidemiological survey for certain arboviruses (Togaviridae) in Pakistan. *Transactions of the Royal Society of Tropical Medicine and Hygiene* 1983, 77(4):442–445.

Dev, V., Sharma, V. P., Barman, K. (2015) Mosquito-borne diseases in Assam, north-east India: current status and key challenges. *WHO South-East Asia Journal of Public Health* 4:20–9. http://www.who-seajph.org/text.asp?2015/4/1/20/206616

Dhanda, V., V. Thenmozhi, N. P. Kumar, J. Hiriyan, N. Arunachalam, A. Balasubramanian, A. Ilango, and A. Gajanana. 1997a. Virus isolation from wild-caught mosquitoes during a Japanese encephalitis outbreak in Kerala in 1996. *The Indian Journal of Medical Research*, 106: 4 Ð 6.

Dhanda,V., V. Thenmozhi, N.P. Kumar, J. Hiriyan, N. Arunachalam, A. Balasubramian, A. Ilango and A. Ganjanan.1997b.Virus isolation from wild-caught mosquitoes during a Japanese encephalitis outbreak in Kerala in 1996. *The Indian Journal of Medical Research*, 106, 4–6.

Dumre, S. P., Shakya, G., Na-Bangchang, K., et al.Dengue virus and Japanese encephalitis virus epidemiological shifts in Nepal: a case of opposing trends. *The American Journal of Tropical Medicine and Hygiene* 2013, 88(4):677–680

Duong, V., Choeung, R., Gorman, C., Laurent, D., Crabol, Y., Mey, C., Peng, B., Di Francesco, J., Hul, V., Sothy, H., et al.Isolation and full-genome sequences of Japanese encephalitis virus genotype I strains from Cambodian human patients, mosquitoes and pigs. *The Journal of General Virology* 2017, 98, 2287–2296.

Dutta, P., Khan, S.A., Khan, A.M., Borah, J., Sarmah, C.K., Mahanta, J. The Effect of Insecticide-Treated Mosquito Nets (ITMNs) on Japanese Encephalitis Virus Seroconversion in Pigs and Humans. *The American Journal of Tropical Medicine and Hygiene* 2011, 84, 466–472.

Dwibedi, B., Sabat, J., Hazra, R. K., Kumar, A., Dinesh, D. S., & Kar, S. K. (2015). Chandipura virus infection causing encephalitis in a tribal population of Odisha in eastern India. *National Medical Journal of India*, *28*(4), 185–187.

Ellis, P., Daniels, P.W., Banks, D.J. Emerging infectious diseases: Japanese encephalitis. *The Veterinary Clinics of North America. Equine Practice* 2000, 16: 565.

Elshuber, S., Allison, S. L., Heinz, F. X., Mandl, C. W. (2003). Cleavage of protein prM is necessary for infection of BHK-21 cells by tick-borne encephalitis virusFN1. *Journal of General Virology*, 84(1):183–91.

Erlanger, T. E., Weiss, S., Keiser, J., Utzinger, J., Wiedenmayer, K. (2009). Past, present, and future of Japanese encephalitis. *Emerging Infectious Diseases*, 15(1): 1–7.

Faizah, A. N., Kobayashi, D., Amoa-Bosompem, M., Higa, Y., Tsuda, Y., Itokawa, K., Miura, K., Hirayama, K., Sawabe, K., Isawa, H. (2020) Evaluating the competence of the primary vector, Culex tritaeniorhynchus, and the invasive mosquito species, Aedes japonicus japonicus, in transmitting three Japanese encephalitis virus genotypes. *PLoS Neglected Tropical Diseases*, 14:e0008986. 10.1371/journal.pntd.0008986

Fischer, M., Hills, S., Staples, E., Johnson, B., Yaich, M., Solomon, T. (2008). Japanese Encephalitis Prevention and Control: Advances, Challenges, and New Initiatives. *Emerging Infectious Diseases*, 8, 93–124.

Fulmali, P. V., Sapkal, G. N., Athawale, S., Gore, M. M., Mishra, A. C., Bondre, V. P. (2011). Introduction of Japanese encephalitis virus genotype I, India. *Emerging Infectious Diseases* 17(2), 319–321. 10.3201/eid1702.100815

Gajanana, A. (1998). Epidemiology ans surveillance of Japanese encephalitis in Tamil Nadu. *ICMR Bulletin*, 28:33–37.

Gajanana, A., Rajendran, R., et al. (1997) Japanese encephalitis in south Arcot district. Tamil Nadu, India: a three-year longitudinal study of vector abundance and infection frequency. *Journal of Medical Entomology*, 34:651–659.

George, S., Yergolker, P. N., Hanumaiah, K.C.S. (1990) Outbreak of encephalitis in Bellary district of Karnataka and adjoining areas of Andhra Pradesh. *The Indian Journal of Medical Research*, 91: 328–30.

Govindarajan, M., Vaseeharan, B., Alharbi, N.S., Kadaikunnan, S., Khaled, J.M., Al-Anbar, M.N., Alyahya, S.A., Maggi, F., Benelli, G. (2018). High efficacy of (Z)-γ-bisabolene from the essential oil of Galinsogaparviflora (Asteraceae) as larvicide and oviposition deterrent against six mosquito vectors. *Environmental Science and Pollution Research*, 25, 10555–10566.

Gulati, B.R., Singha, H., Singh, B.K., Virmani, N., Kumar, S., Singh, R.K. (2012). Isolation and genetic characterization of Japanese encephalitis virus from equines in India. *Journal of Veterinary Science*, 13, 111–118.

Gunasekaran, P., Kaveri, K., Arunagiri, K., Mohana, S., Kiruba, R., Kumar, V. S., Padmapriya, P., Suresh Babu, B. V., Sheriff, A. K. Japanese encephalitis in Tamil Nadu (2007–2009). (2012). *The Indian Journal of Medical Research*, 135(5), pp.680–2.

Hall-Mendelian, S., Jansen, C. C., Cheah, W. Y., Montgomery, B. L., Hall, R. A., Ritchie, S. A., Van den Hurk, A. F. (2012). Culexannulirostris (Diptera: Culicidae) host feeding patterns and Japanese encephalitis virus ecology in northern Australia. *Journal of Medical Entomology*, 49, 371–377.

Hanna, J. N., Ritchie, S. A., Phillips, D. A. et al. (1996) an outbreak of Japanese encephalitis in the Tones Strait, Australia, 1995. *The Medical Journal of Australia*, 165, 256–260.

Huang, C.H. (1982) Studies of Japanese encephalitis in China. *Advances in Virus Research*, 27:71–101.

Igarashi, A. (2002). Control of Japanese encephalitis in Japan: Immunization of humans and animals, and vector control. *Current Topics in Microbiology and Immunology*, 267, 139–152.

Igarashi, A., Tanaka, M., Morita, K. et al. (1994). Detection of west Nile and Japanese encephalitis viral genome sequences in cerebrospinal fluid from acute encephalitis cases in Karachi, Pakistan. *Microbiology and Immunology*, 38(10):827–830.

Jeffries, C. L., Walker, T. (2015). The potential use of Wolbachia-based mosquito biocontrol strategies for Japanese encephalitis. *PLoS Neglected Tropical Diseases*, 9, e0003576.

Joshi, R., Kalantri, S. P., Reingold, A., Colford, J. M. Jr. (2012). Changing landscape of acute encephalitis syndrome in India: A systematic review. *National Medical Journal of India*, 25:212–20.

Kabilan, L., Rajendran, R., Arunachalam, N., Ramesh, S., Srinivasan, S., Samuel, P. P., Dash, A. P. (2004). Japanese encephalitis in India: an overview. *The Indian Journal of Pediatrics*, 71(7):609–15.

Kalita, J., Misra, U. K., Pandey, S., Dooley, T. N. (2003). *Control of Japanese Encephalitis in Japan: Immunization of Humans Animals, and vector control.*

Kanojia P.C. Ecological study on mosquito vectors of Japanese encephalitis virus in Bellary district, Karnataka. *The Indian Journal of Medical Research*, 126, 2007, pp 152–157.

Kar, N. J., Saxena, V. K. (1998) some epidemiological characteristics of Japanese encephalitis in Haryana state of northern India. *Journal of Communicable Diseases*, 30, 129–131.

Kari, K., Liu, W., Gautama, K. et al. (2006). A hospital-based surveillance for Japanese encephalitis in Bali, Indonesia. *BMC Medicine*, 4:8.

Keiser, J., Maltese, M. F., Erlanger, T. E., Bos, R., Tanner, M., Singer, B. H., Utzinger, J. (2005). Effect of irrigated rice agriculture on Japanese encephalitis, including challenges and opportunities for integrated vector management. *Acta Tropica*, 95(1):40–57.

Khan, A. M., Khan, A. Q., Dobrzynski, L., Joshi, G. P., Myat, A. (1981). A Japanese encephalitis focus in Bangladesh. *The Journal of Tropical Medicine and Hygiene*, 84, 41–44.

Kiermayr, S., Kofler, R. M., Mandl, C. W., Messner, P., Heinz, F. X. (2004). Isolation of capsid protein dimers from the tick-borne encephalitis flavivirus and in vitro assembly of capsid-like particles. *Journal of Virology*, 78. 10.1128/jvi.78.15.8078-8084.2004

Kramer, L. D., Chin, P., Cane, R. P., Kauffman, E. B., Mackereth, G., 2011. Vector competence of Zealand mosquitoes for selected arboviruses. *The American Journal of Tropical Medicine and Hygiene*, 85, 182–189.

Kripasindhu, Chatterjee, Bhadra, Atanu, Ray, Swapan Kumar, Mukherjee, Subhasis, Sen, Sukanta (2022). Review of japanese encephalitis vaccines in India. *Asian Pacific Journal of Health Sciences*. 10.21276/apjhs.2022.9.4.52

Kulkarni, R., Sapkal, G.N., Kaushal, H., Mourya, D.T. (2018). Japanese Encephalitis: A Brief Review on Indian Perspectives. *Open Virology Journal*, 12, 121–30.

Kumar, Kiven, Arshad, Siti Suri, Selvarajah, Gayathri Thevi, Abu, Jalila, Toung, Ooi Peck, Abba, Yusuf, Yasmin, A.R., Bande, Faruku, Sharma, Reuben, Ong, Bee Lee (2018) Japanese encephalitis in Malaysia: An overview and timeline. *Acta Tropica*, 185, 219–229.

Kumar, R., Hwang, J.S. (2006). Larvicidal efficiency of aquatic predators: A perspective for mosquito biocontrol. *Zoological Studies*, 45, 447–466.

Kumari, R, Joshi, PL. (2012) A review of Japanese encephalitis in Uttar Pradesh, India. *WHO South-East Asia Journal of Public Health*, 1, 374–95.

Kumari, R, Kumar, K, Rawat, A, Singh, G, Yadav, NK, Chauhan, LS. (2013) First indigenous transmission of Japanese Encephalitis in urban areas of National Capital Territory of Delhi, India. *Tropical Medicine & International Health*, 18(6):743–749. 10.1111/tmi.12104

Le Flohic, G., Gonzalez, J.-P. (2011). When Japanese Encephalitis Virus Invaded Eastern Hemisphere—The History of the Spread of Virus Genotypes. In *Flavivirus Encephalitis*, Ruzek, D., Ed., InTech.

Le Flohic, G., Porphyre, V., Barbazan, P., Gonzalez, J.-P. (2013) Review of climate, landscape, and viral genetics as drivers of the Japanese encephalitis virus ecology. *PLoS Neglected Tropical Diseases*, 7:e2208. 10.1371/journal.pntd.0002208

Lee, DW, Choe, YJ, Kim, JH, et al. (2012). Epidemiology of Japanese encephalitis in South Korea, 2007–2010. *International Journal of Infectious Diseases*, 16(6), e448–e452.

Lian, W.-C., Liau, M.-Y., Mao, C.-L., Lian, W., Liu, M., Mao, C. (2002). Diagnosis and Genetic Analysis of Japanese Encephalitis Virus Infected in Horses. *Journal of Veterinary Medicine Series B*, 49, 361–365.

Lindenbach, BD, Rice, CM (2003). Molecular biology of flaviviruses. *Advances in Virus Research*, 59, 23–62.

Lvov, D. K.1978. The role of ixodid ticks in the preservation and transmission of arboviruses in the USSR. In Tick-Borne Disease and their Vectors, ed. T. K. H Wilde, pp. 482–86. Tonbridge, UK, Lewis Reprints. 573 pp.

Mansfielda, K. L., Hernández-Trianaa, L. M., Banyarda, Fooksa, A. R., Johnson, N. Japanese encephalitis virus infection, diagnosis and control in domestic animals. *Veterinary Microbiology*, 201 (2017) 85–92.

Marten, G.G., Astaiza, R., Suarez, M.F., Monje, C., Reid, J.W. (1989). Natural control of larval Anopheles albimanus (Diptera: Culicidae) by the Predator mesocyclops (Copepoda: Cyclopoida). *Journal of Medical Entomology*, 26, 624–627.

MichaelJ, Conway, Tonya M, Colpitts, Erol Fikring (2014). Role of the vector in Arbovirus transmission. *Annual Review of Virology*, 1(71), 88.

Miller, R.H., Masuoka, P., Klein, T.A., Heung-Chul, K., Todd, S., John, G., 2012. Ecological niche modeling to estimate the distribution of Japanese encephalitis virus in Asia. *PLoS Neglected Tropical Diseases*, 6, e1678.

MitamuraT, KitaokaM, MoriK, OkubaK (1938) Isolation of the virus of Japanese epidemic encephalitis from mosquitoes caught in nature. *Tokyo IjiShinshi*, 62:820–831.

Nah, J.-J., Yang, D.-K., Kim, H.-H., Song, J.-Y., (2015). The present and future of veterinary vaccines for Japanese encephalitis in Korea. Clinical and Experimental Vaccine Research, 4, 130–136.

Nga P.T., Parquet M.D.C., Cuong V.D., Ma S.-P., Hasebe F., Inoue S., Makino Y., Takagi M., Nam V.S., MoritaK. (2004) Shift in Japanese encephalitis virus (JEV) genotype circulating in northern Vietnam: Implications for frequent introductions of JEV from southeast Asia to East Asia. The Journal of General Virology, 85:1625–1631. 10.1099/vir.0.79797-0

NVBDCP (2006) Guidelines for surveillance of acute encephalitis syndrome (with special reference to Japanese encephalitis). Directorate of National Vector Borne Disease Control Programme, Government of India, 28 pp.

Okuno T, Okada T, Kondo A, Suzuki M, Kobayashi M, Oya A. Immunotyping of different strains of Japanese encephalitis virus by antibody-absorption, haemagglutination-inhibition and complement-fixation tests. Bulletin of the World Health Organization, 1968, 38(4):547.

Operational Guidelines, National Programme for Prevention and Control of Japanese Encephalitis/ Acute Encephalitis Syndrome. Government of India, Ministry of Health & Family Welfare, Directorate General of Health Services, National Vector Borne Disease Control Programme, 2014: 1–3.

Paterson PY, Ley HL, Wisseman CL, Pond WL, Smadel JE, Diercks FH, Hetherington DG, Sneath PHA, Witherington DH, Lancaster WE (1952) Japanese encephalitis in Malaya. I. Isolation of virus and serologic evidence of human and equine infection. *American Journal of Hygiene*, 56:320–330.

Philip Samuel. 2000. Japanese Encephalitis Virus Infection in Mosquitoes and its Epidemiological Implications. *ICMR Bulletin*, 30(4).

Phukan, A. C., Borah, P. K., Mahanta, J. (2004) Japanese encephalitis in Assam, northeast India. *The Southeast Asian Journal of Tropical Medicine and Public Health*, 35, 618–622.

Poland, JD, Cropp, CB, Craven, RB, Monath, TP. Evaluation of the potency and safety of inactivated Japanese encephalitis vaccine in US inhabitants. *The Journal of Infectious Diseases*, 1990, 161: 878–82.

Pond, WL, Smadel, JE (1954) neurotropic viral diseases in the Far East during the Korean War. *Army Med Serv Graduate Sch*4:219–233.

Rappleye, WC (1939) Epidemiology of Japanese B encephalitis. Epidemic Encephalitis: Third Report by the Matheson Commission Columbia University Press. New York.

Rawlins, S.C., Martinez, R., Wiltshire, S., Clarke, D., Prabhakar, P., Spinks, M. Evaluation of Caribbean strains of Macrocyclops and Mesocyclops (Cyclopoida: Cyclopidae) as biological control tools for the dengue vector Aedesaegypti. *Journal of the American Mosquito Control Association*, 1997, 13, 18–23.

Reuben, R., Gajanana, A. (1997) Japanese encephalitis in India. *Indian Journal of Pediatrics*, 64: 243–251.

Reuben, R., Tewari, S. C., Hiriyan, J., & Akiyama, J. (1994). Illustrated keys to species of Culex (Culex) associated with Japanese encephalitis in Southeast Asia (Diptera: Culicidae). *Mosquito Systematics*, 26(2), 75–96.

Saxena, V, Dhole, TN, Preventive strategies for frequent outbreaks of Japanese encephalitis in Northern India.*Journal of Biosciences*, 2008, 33(4): 505–514.

Scholte, E.J., Knols, B.G., Samson, R.A., Takken, W. (2004). Entomopathogenic fungi for mosquito control: A review. *Journal of Insect Science*.

Scholte, E.J., Takken, W., Knols, B.G. (2007). Infection of adult Aedesaegypti and Ae. albopictus mosquitoes with the entomopathogenic fungus Metarhiziumanisopliae. *Acta Tropica*, 102, 151–158.

Schuh, A.J., Li, L., Tesh, R.B., Innis, B.L., Barrett, A.D.T. (2010) Genetic characterization of early isolates of Japanese encephalitis virus: Genotype II has been circulating since at least 1951. *The Journal of General Virology*, 91, 95–102. 10.1099/vir.0.013631-0

See, E, Tan, HC, Wang, D, Ooi, EE, Lee, MA (2002). Presence of hemagglutination inhibition and neutralization antibodies to Japanese encephalitis virus in wild pigs on an offshore island in Singapore. *Acta Tropica*, 81(3), 233–236.

Shield, J, Hanna, IN, Phillips, DA (1996) Reappearance of the Japanese encephalitis virus in the Torres Strait. *Communicable Diseases Intelligence*, 1120, 191–194.

SpicklerA.R. (2016). *Japanese Encephalitis. Center for Food Security and Public Health Established at Iowa State*, Ames, IA, USA.

Subramaniam, J., Murugan, K., Panneerselvam, C., Kovendan, K., Madhiyazhagan, P., Dinesh, D., Kumar, P.M., Chandramohan, B., Suresh, U., Rajaganesh, R., et al. Multipurpose effectiveness of Couroupitaguianensis-synthesized gold nanoparticles: High antiplasmodial potential, field efficacy against malaria vectors and synergy with Aplocheiluslineatus predators. *Environmental Science and Pollution Research International*, 2016, 23, 7543–7558.

Swe, T, Thein, S, Myint, MS (1979). Pilot seroepidemiological survey on Japanese encephalitis in north-western Burma. *Biken Journal*, 22(4): 125–129.

Thongcharoen, P (1985). Japanese encephalitis in Thailand. *Journal of the Medical Association of Thailand* 68, 534–544.

Ting, SH, Tan, HC, Wong, WK, Ng, ML, Chan, SH, Ooi, EE (2004). Seroepidemiology of neutralizing antibodies to Japanese encephalitis virus in Singapore: continued transmission despite abolishment of pig farming? *Acta Tropica*, 92(3), 187–191.

Tiwari, Sarika, Singh, Rishi Kumar, Tiwari, Ruchi, Dhole, Tapan N. (2012) Japanese encephalitis: a review of the Indian perspective. *The Brazilian Journal of Infectious Diseases*, 16(6): 564–573.

Touch, S, Hills, S, Sokal, B, et al. Epidemiology and burden of disease from Japanese encephalitis in Cambodia: results from two years of sentinel surveillance. *Tropical Medicine & International Health* 2009, 14(11): 1365–1373).

Trosper, JH, Ksiazek, TG, Cross, JH Isolation of Japanese encephalitis virus from the Republic of the Philippines. *Transactions of the Royal Society of Tropical Medicine and Hygiene*, 1980, 74(3):292–295.

Tu, Y.The discovery of artemisinin (qinghaosu) and gifts from Chinese medicine. *Nature Medicine* 2011, 17, 1217–1220.

Turtle, L., Solomon, T. (2018). Japanese encephalitis - the prospects for new treatments. *Nature Reviews Neurology*, 14(5), 298–313. 10.1038/nrneurol.2018.30

Tyagi, B.K., R. Rajendran, I.P. Sunish, P. Philip Samuel, S.C. Tewari, D. Ramesh, A. Venkatesh (2011). *Establishment of a field station at Vridhachalam, South Arcot District, Tamil Nadu – "Two decadal results on Japanese encephalitis control trial studies" – established in 1987 at Vridhachalam, Tamil Nadu. ICMR Centenary Celebrations (1911–2011).* Centre for Research in Medical Entomology, 455 pp.

Tyagi, B.K., V. Thenmozhi and S. Karthigai Seli, (2014). Transmision dynamics of Japanese encephalitis, with emphasis on gaps in understanding and priority areas on Japanese encephalitis and other acute encephalitis syndromes in India. *The Journal of Communicable Diseases*, 1(1):1–10.

Uchil, PD, Satchidanandam, V (2001). Phylogenetic analysis of Japanese encephalitis virus: Envelope gene based analysis reveals a fifth genotype, geographic clustering, and multiple introductions of the virus into the Indian subcontinent. *The American Journal of Tropical Medicine and Hygiene*, 65, 242–51.

Upadhyay, R.K. Epidemiology, Disease Transmission and Pathogenesis Caused by JE Virus: Its Prevention and Control Ravi Kant Upadhyay. *American Journal of Infectious Diseases and Microbiology*, 2015a, 3(1), 38–64.

Upadhyay, R. K. (2015b). Epidemiology, disease transmission and pathogenesis caused by JE virus: its prevention and control. *American Journal of Infectious Diseases and Microbiology*, *3*, 38–64.

UT Wise Cases and Deaths due to Japanese Encephalitis, 2013–2017, 2018. Available from: http://www.cbhidghs.nic.in/WriteReadData/l892s/

Vajpayee, A, Mukherjee, MK, Chakraborty, AK, Chakraborty, MS. Investigation of an outbreak of Japanese encephalitis in Rourkela City (Orissa) during 1989. *The Journal of Communicable Diseases*, 1991, 23: 18–21.

van den Hurk, A.F., Ritchie, S.A., Mackenzie, J.S., 2009. Ecology and geographical expansion of Japanese encephalitis virus. *Annual Review of Entomology*, 54: 17–35.

Wang, H., Li, Y., Liang, X., Liang, G. (2009). Japanese encephalitis in mainland China. *Japanese Journal of Infectious Diseases*, 62(5):331–336.

Wang, H., Liang, G. (2015). Epidemiology of Japanese encephalitis: Past, present, and future prospects. *Therapeutics and Clinical Risk Management*, 11:435–448.

Wu. C. J., Wu. S. Y. 1957. Isolation of virus of B type encephalitis from *Lasioheleataiwana*Shiraki- a blood sucking midge. *Acta Microbiologica Sinica*, 5:22–26 (In Chinese).

Wuryadi, S. Suroso, T. (1989) Japanese encephalitis in Indonesia. *Southeast Asian Journal of Tropical Mrdicine and Public Health*, 20:575–580.

Yang, DK, Kim, BH, Kweon, CH, Kwon, JH, Lim, SI, Han, HR (2004). Biophysical characterization of Japanese encephalitis virus (KV1899) isolated from pigs in Korea. *Journal of Veterinary Science*, 5(2):125–130.

Yu, Y. (2010). Phenotypic and genotypic characteristics of Japanese encephalitis attenuated live vaccine virus SA14-14-2 and their stabilities. *Vaccine*, 28: 3635–41.

Zug, R., Hammerstein, P. (2012). Still a host of hosts for Wolbachia: Analysis of recent data suggests that 40% of terrestrial arthropod species are infected. *PLoS One*, 7:e38544.

7 West Nile Virus in India

R. Balasubramanian
ICMR-National Institute of Virology, Kerala unit, Alappuzha, India

Mangesh Gokhle
Entomology Group, ICMR-National Institute of Virology, Pune, India

A.B. Sudeep Balan
ICMR-National Institute of Virology, Microbial Containment Complex, Pashan, Pune, India

B.K. Tyagi
Chandigarh University, Mohali, India

7.1 INTRODUCTION

Researchers and health managers in India's various states are becoming increasingly concerned about newly emerging and reemerging mosquito-borne diseases. An arthropod-borne virus of public health significance is the West Nile virus (WNV). The Japanese encephalitis virus (JEV) antigenic complex, which is part of the family flaviviridae, includes WNV, which is a member of the flavivirus genus. Although vertical and horizontal transmission strategies also appear to be important, biological transmission between blood-feeding arthropods and susceptible vertebrate hosts is the major way that the WNV is kept in nature. (Hayes et al., 1982). The WNV coming under flavivirus spread via the bite of infected *Culex quinquefasciatus* and *Cx. tritaeniorhynchus* mosquitoes, which are widespread in the various areas where the virus is reported to be endemic. Between September and April is often when these mosquitoes are most active, with peak activity in the warm and humid months of January and March (Balasubramanian et al., 2021). Although the disease was initially reported in the state of Maharashtra, it was subsequently confirmed in a total of seven Indian states. The disease is steadily spreading to the states in the north, the north-east, and the south (Anukumar et al., 2014; Mammen et al., 2020; Khan et al., 2017). It was discovered that residents of villages and cities had both a history of high mosquito exposure in residential areas and a high seroprevalence of WNV. In humans, WNV is typically asymptomatic or manifests as a mild, self-limiting febrile illness in 20% of infected individuals, with neuroinvasive disease only infrequently occurring. The danger of encephalitis and death is highest in elderly patients and children, and little is known about the variety of illnesses and risk factors for neuroinvasive disease. There is no particular antiviral therapy, and the clinical management and patient supportive care are both insufficient. Despite the prevalence of mosquitoes and exposure to mosquito bites, many individuals are still unaware of the diseases caused by mosquitoes (Rajaiah et al., 2023).

The prevalence of WNV has been linked to a wide range of conditions. Villagers in low socioeconomic situations typically reside near the paddy fields, whereas the associated villagers frequently visit outdoor activities for livestock maintenance. This can result in a greater proliferation of mosquitoes and, as a result, a higher incidence and exposure to mosquito-borne diseases. Some

DOI: 10.1201/9781003327028-8

studies were unable to demonstrate a link between mosquito abundance and disease transmission. Numerous local conditions, such as vegetation cover, domestic animals, and contact between people and wild animals, particularly birds, might disrupt this process (Kanojia and Geevarghese, 2005; Reisen et al., 2014). Mosquito populations are more prevalent in areas with a high density of livestock since they are increasingly common in residential areas (Balasubramanian et al., 2021).

7.2 VECTOR ECOLOGY AND VIRUS DETECTION

Mosquitoes are principal vectors of WNV. The primary technique for tracking virus activity and measuring the rate of transmission is mosquito-based surveillance. In several regions of India, WNV has been found in samples from a variety of mosquito species. The most significant vectors are *Cx. quinquefasciatus, Cx. tritaeniorhynchus*, and *Cx. vishnui*, from which the greatest number of isolates has been, generated (Paramasivan et al. 2003). The expansion of rice-growing areas, irrigation systems, and coastal regions in Asia, including hill station areas, have all contributed to an increase in WNV vectors (Balasubramanian et al., 2021; Khan et al., 2011). In South India, mosquitoes of the Culex species increased in number between February and October, coinciding with the state's rice harvest. However, after harvest, when the fields are dry, their predominance decreased significantly. The incidence of JE/WNV cases in the states was discovered to be connected to a high Culex mosquito population from September through January. Similar to this, locations where double paddy crops were grown saw two maxima in JE/WNV vector activity (Kanojia et al., 2003; Arunachalam et al., 2009; Soman et al., 1976). Although an increase in vector abundance always precedes epidemics, other factors, such as the rate of mosquito infection, are also at stake (Gajanana et al., 1995). JEV/WNV-neutralizing antibodies were found in ardeid birds, primarily pond herons and cattle egrets, in southern India. This suggests that ardeid birds may be involved in the normal cycle of WNV in India. Domestic pigs in southern India have been proven to produce WNV antibodies; however, they are poor hosts.

7.3 NATURAL CYCLE

In nature, WNV is maintained by a cycle involving mosquitoes and animal hosts, with birds serving as the primary and preferred reservoir (Figure 7.1 shows a schematic of mosquito-mammal transmission).

Because the dominant viral vectors prefer to feed on heron and egrets, it is believed that these wild bird species are the primary hosts for the maintenance and transmission of WNV in India. The low amount of viremia in mammals is typically insufficient to transfer WNV to mosquitoes, putting a stop to the transmission cycle. As a result, humans are regarded as "dead-end" hosts for WNV (Colpitts et al., 2012). In nature, the WNV is kept alive and multiplied by enzootic transmission cycles that primarily include birds and mosquitoes of the Culex species, with humans and horses acting as incidental hosts. *Culex vishnui* group and *Culex quinquefasciatus* are two ornithophilic mosquito species that frequently act as amplifying hosts. *Culex bitaeniorhynchus* and *Cx. univittatus*, two mosquito species that frequently switch hosts during the transmission season, may be significant bridge vectors (Turell et al., 2005). In contrast to most arboviruses, it lacks the vector specificity that prevents it from using a particular type of mosquito for its enzootic cycles. More than 60 different mosquito species have been found to carry the WNV. However, based on the frequency of WNV discovery in these species and laboratory vector competence, it appeared that *Cx. quinquefasciatus* was the most significant WNV vector in India (Goddard et al., 2002; Bolling et al., 2012). Although this species is anthrophilic, as expected for an enzootic vector, some studies have found that a portion of their meals come from mammals, indicating that these mosquitoes may be able to transmit WNV to mammals in addition to continuing the avian enzootic transmission cycle. Breastfeeding, organ transplantation, and blood transfusions are among the recognized ways that humans might get infected.

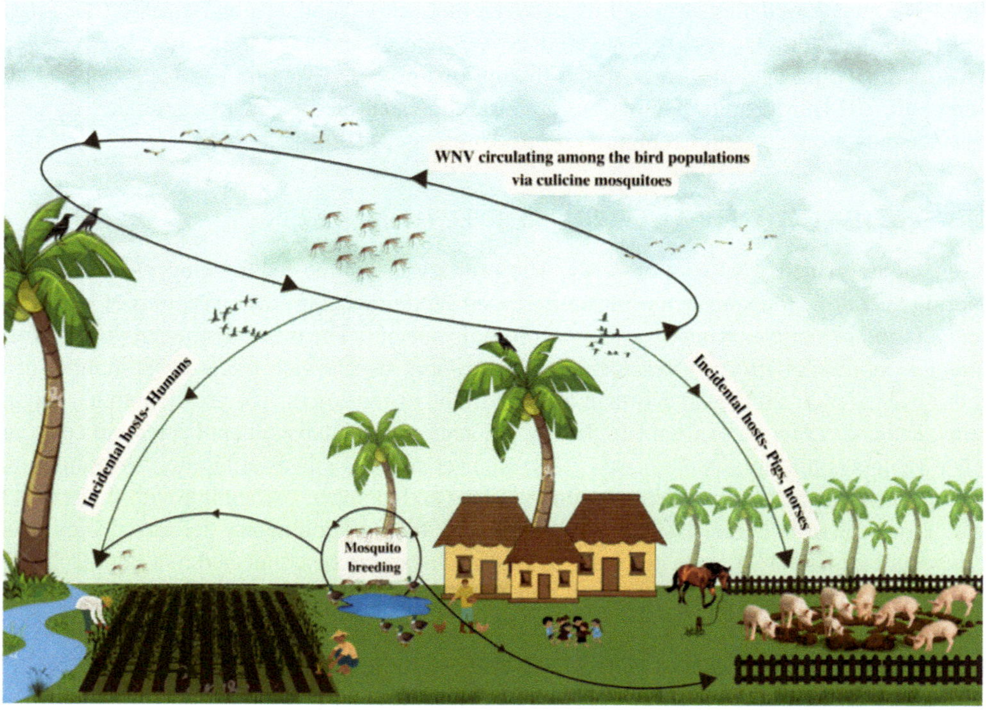

FIGURE 7.1 Transmission cycle of WNV, an example of mosquito-borne arboviruses with avian host reservoir hosts and other mammals as incidental host.

7.4 PATHOGEN PREVALENCE IN VECTORS

It is possible to find vector-borne pathogens in the vector, the receiver, and the donor. Pathogens detection in vectors is essential for virus surveillance research in order to provide early warning or examine the geographic patterns of virus occurrence. Because many wild species are protected, it often becomes impossible to detect pathogens in wildlife. Although companion animals like dogs and horses have been employed as sentinels for human infection, they are not suitable as means of pathogen detection since they can develop low-level infections in the blood of the animal. A direct insight into public health-relevant vector species is provided by pathogen identification in vectors collected from bird roosting sites. These findings may also be utilized to assess transmission hazards. An effective approach to assess the relative significance of hosts to the vector population as well as the distribution and prevalence of infections in feeding hosts may be blood-meal analysis in vectors combined with pathogen detection.

In temperate places, WNV outbreaks occurred at the same time as Culex mosquito populations grew during the summer. Worldwide, there are regional variations in WNV vectors. Two mosquito species, *Cx. restuans* and *Cx. salinarius*, have been attributed for spreading the virus in the USA. This virus has been mainly isolated from *Cx. univittatus* in Egypt and South Africa, *Cx. pipiens* and *Cx. perexiguus* in Israel, *Cx. molestus* in France, and *Cx. vishnui* complex in Pakistan. During the inter-enzootic periods, the role of non-Culicine arthropods in the maintenance of WNV has also been taken into consideration. *Aedes albopictus, Cx. pipiens, Ochlerotatus triseriatus*, and *Ochlerotatus japonicus japonicus* are among the other mosquito species that are known to spread pathogens, according to later investigations. WNV has been isolated from arthropods other than mosquitoes. These include chicken mites (Ornithonyssus sylviarum), soft ticks (*Ornithodoros maritimus* and *Argas hermanni*), and hard ticks (*Hyalomma marginatum* and *Rhipicephalus sanguineus*; Paul et al., 1970).

This virus has been found in mosquitoes, domestic pigs, and humans in India. Two WNV strains were discovered from *Cx. quinquefasciatus* in Maharashtra in a different investigation. Dandawate et al. isolated WNV from *Cx. vishnui* mosquito pools collected from Tamil Nadu. *Aedes albopictus, Cx. tritaeniorhynchus, Cx. vishnui, Cx. bitaeniorhynchus, Cx. univittatus*, and *Cx. quinquefasciatus* have all been identified as potential WNV vectors in experimental research (Sudeep et al., 2015). In Andhra Pradesh, the WNV isolate was found in *Cx. vishnui* mosquitoes. Experimental evidence of WNV transovarial transmission in the *Cx. vishnui* mosquito has been found. The hypothesis of this species serving as a "bridge vector" is supported by observations made in rural and urban areas of India that *Cx. quinquefasciatus* is mostly anthropophilic while *Cx. tritaeniorhynchus* outdoors feeds more commonly on mammals. However, samples from Kerala suggest that the *Cx. vishnui* complex of mosquitoes may easily feed on mammals, including humans (Arunachalam et al., 2009).

7.5 GEOGRAPHICAL DISTRIBUTION

In the West Nile district of Uganda, WNV was originally isolated in 1937 from a feverish woman (Smithburn et al., 1940). As a result of its global expansion, it is now present on all continents, with the exception of Antarctica, making it the most widely dispersed arbovirus worldwide (Kramer and Bernard, 2001). The world is currently being infected with WNV. There have been a number of epidemics reported in various nations, ranging in severity. The first known WNV epidemic in the northern United States was discovered in 1995, and as a result of the broad virus activity, WNV is currently becoming endemic in the north-eastern United States.

WNV is very common in India and typically affects people with a mild, non-fatal sickness similar to dengue. West Nile antibodies were first detected in humans in India from Bombay in 1952. WNV has been isolated from mosquitoes and sporadic encephalitis patients. In a post-encephalitis outbreak study conducted in 1982, Risbud et al. (1991) showed that neutralizing antibodies to WNV were more common than those to JEV, although Gajanana et al. discovered a low frequency of WNV in the same region in 1989 and 1990. Human sera from Tamil Nadu, Karnataka, Andhra Pradesh, Maharashtra, Gujarat, Madhya Pradesh, Orissa, and Rajasthan have been found to contain WNV-neutralizing antibodies (Paramasivan et al., 2003; Khan et al., 2017). In the Kuttanad region of Kerala, southern India, domestic ducks (Anas platyrhynchos var. Domesticus) have been found to exhibit WNV and JEV antibodies. In 2010, WNV antibodies were also found in horses from these five northern Indian states (Kalaiyarasu et al., 2016).

7.6 VIRUS BIOLOGY

7.6.1 Genomic Characteristics

WNV are small spherical, enveloped particles containing single-stranded positive-sense RNA genome. The length of the viral genome is roughly 11,000 nucleotides. Three structural (C, M, and E) and seven non-structural (NS1, NS2a, NS2b, NS3, NS4a, NS4b, and NS5) proteins are encoded by the viral genome. It has a single big open reading frame (ORF), which is flanked at both ends by non-coding regions (NCR). The flaviviruses spread by mosquitoes don't have a poly (A) tail at the 3' end. WNV is mostly split into lineage-1 and lineage-2 genetic lineages. The lineage-1 strains have been linked to cases of clinical encephalitis in humans. The lineage-2 members are still present in enzootic foci in Africa and have not been linked to human encephalitis. WNV's genome characterization has previously been discussed. A phylogenetic analysis of the isolates found in both Europe and Africa revealed that WNV was brought into Europe by migratory birds from sub-Saharan Africa. The lineage-1 and -5 strains reported from India from the cases of clinical encephalitis in humans (Chowdhury and Khan, 2021; Figure 7.2).

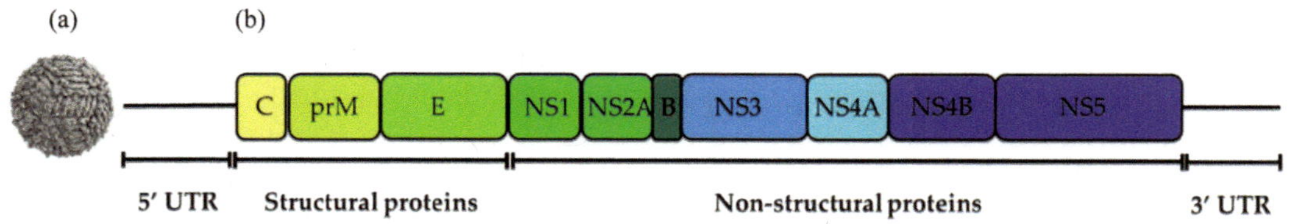

FIGURE 7.2 The structure of WNV virion (a) and 11 kb long viral genome represented with one ORF encoding three structural and seven non-structural proteins (b).

Source: From De Filette et al. (2012) and Habarugira et al. (2020).

7.6.2 Diagnosis

Different diagnostic techniques, including virus isolation, RT-PCR, serology, and histological inspection, are frequently used to identify WNV infection. IgM and IgG antibodies to WNV should be detected by ELISA in serum or cerebral fluid in individuals who have a suspicion of having the virus. Three to seven days after exposure, antibodies can still be seen in serum or cerebral fluid. However, it has been noted that there may be cross reactivity with other flaviviruses, particularly JEV (Tardei et al., 2000; Roehrig et al., 2003). The plaque reduction neutralization test is incredibly helpful in identifying and verifying viral isolates. The virus neutralizing test (VNT) for WNV is still the gold standard test because of its excellent specificity and ability to measure neutralizing titres as well as detect neutralizing antibodies to the virus (Paramasivan et al., 2003). A quick and effective diagnostic method has been a nucleic acid-based molecular approach. The diagnosis of WNV viral infections frequently uses it. Reverse transcription polymerase chain reaction (RT-PCR), quantitative RT-PCR (qRT-PCR), and in situ hybridization are the most frequently utilized molecular diagnostic methods (Shi and Kramer, 2003). Samples of serum or cerebrospinal fluid (CSF) should be used for the testing. Human CSF that tested negative using a traditional cell culture technique has been successfully tested for WNV using the TaqMan assay. The clinical and experimental study of WNV infections has benefited from the pathological diagnosis (Habarugira et al., 2020).

7.7 CONCLUSION

A spontaneous WNV illness outbreak occurred in India. This includes the southern and north-eastern regions of the states as two significant WNV epidemic hotspots. Since there is no specific WNV medication or vaccine, robust surveillance programmes, ongoing mosquito eradication efforts, and community outreach are the cornerstones of human illness prevention. Studies such as the vector index for abundance of WNV-infected mosquitoes indicating that the combined mosquito abundance and WNV infection rate produce a more complete estimate of entomological risk of exposure to WNV are rare. Considering the many ecogeographical regions of India, the WN virus epidemiology is still lacking. In India, the WNV is still a major public health concern. In India, the WNV is still a major public health concern, and programmes for surveillance, prevention, and control will be more effective if they are based on an understanding of the disease's intricate transmission cycle. For the successful execution of mosquito control measures and to inform the public about crucial times when use of personal protection measures, such as repellents, should be stressed, knowledge of seasonal patterns of activity for mosquito vectors and WNV is essential.

REFERENCES

Anukumar, B, Sapkal, GN, Tandale, BV, Balasubramanian, R, Gangale, D (2014). West Nile encephalitis outbreak in Kerala, India, 2011. *Journal of Clinical Virology*, *61*, 152–155.

Arunachalam, N, Narayanamurty, US, Narahari, D, Balasubramanian, A, Philip Samuel, P,Thenmozhi, V, Paramasivan, R, Rajendran, R, Tyagi, BK (2009). Longitudinal Studies of Japanese Encephalitis

Virus Infection in Vector Mosquitoes in Kurnool District, Andhra Pradesh, India. *Journal of Medical Entomology*, 46, 633–639.

Balasubramanian, R, Nadh, VA, Sahina, S (2021). Ecology of breeding habitats of mosquito population and screening for virus of Japanese encephalitis and West Nile in the coastal area of Kerala, India. *J Vector Borne Dis*, 58(3):232–239. doi: 10.4103/0972-9062.318307

Bolling, BG, Olea-Popelka, FJ, Eisen, L, Moore, CG, Blair CD (2012). Transmission dynamics of an insect-specific flavivirus in a naturally infected Culex pipiens laboratory colony and effects of co-infection on vector competence for West Nile virus. *Virology*, 427(2), 90–97.

Chowdhury, P, Khan, SA (2021). Global emergence of West Nile virus: Threat & preparedness in special perspective to India. *Indian Journal of Medical Research*, 154(1), 36–50. doi: 10.4103/ijmr.IJMR_642_19.

Colpitts, TM, Conway, MJ, Montgomery, RR, Fikrig, E (2012). West Nile virus: Biology, transmission, and human infection. *Clin Microbiol Rev.*, 25(4), 635–48. doi: 10.1128/CMR.00045-12

De Filette, M, Ulbert, S, Diamond, M, Sanders, N (2012). Recent progress in West Nile virus diagnosis and vaccination. *Vet. Res.*, 43: 16.

Gajanana, A, Thenmozhi, V, Samuel, PP, Reuben, R (1995). A community based study of subclinical flavivirus infections in children in an area of Tamil Nadu, India, where Japanese encephalitis is endemic. *Bull World Health Organ*, 73, 237–44.

Goddard, LB, Roth, AE, Reisen, WK, Scott, TW (2002);Vector competence of California mosquitoes for West Nile virus. *Emerg Infectious Diseases*, 8, 1385–91.

Habarugira, G, Suen, WW, Hobson-Peters, J, Hall, RA, Bielefeldt-Ohmann, H (2020). West Nile virus: An update on pathobiology, epidemiology, diagnostics, control and "One Health" implications. *Pathogens*, 9(7), 589. doi: 10.3390/pathogens9070589.

Hayes, CG, Baqar, S, Ahmed, T, Chowdhry, MA, Reisen, WK (1982). West Nile virus in Pakistan. Sero-epidemiological studies in Punjab Province. *Trans R Soc Trop Med Hyg*, 76: 431–6.

Kalaiyarasu, S, Mishra, N, Khetan, RK, Singh, VP (2016). Serological evidence of widespread West Nile virus and Japanese encephalitis virus infection in native domestic ducks (*Anas platyrhynchos* var domesticus) in Kuttanad region, Kerala, India. *Comparative Immunology, Microbiology & Infectious Diseases*, 48:61–8.

Kanojia, PC, Geevarghese, G. A new mosquito record of an area known for Japanese encephalitis hyperendemicity, Gorakhpur district, Uttar Pradesh, India. *J Am Mosq Cont Assoc*2005; 21:1–4.

Kanojia, PC, Shetty, PS, Geevarghese, G (2003). A long-term study on vector abundance and seasonal prevalence in relation to the occurrence of Japanese encephalitis in Gorakhpur district, Uttar Pradesh. *Indian J Med Res*, 117, 104–10.

Khan, SA, Chowdhury, P, Choudhury, P, Dutta, P (2017). Detection of West Nile virus in six mosquito species in synchrony with seroconversion among sentinel chickens in India. *Parasit Vectors*, 10: 13.

Khan, SA, Dutta, P, Khan, AM, Chowdhury, P, Borah, J, Doloi, P, et al. (2011). West Nile virus infection, Assam, India. *Emerg Infect Dis*, 17: 947–8.

Kramer, LD, Bernard, KA (2001). West Nile virus infection in birds and mammals. *Ann N Y Acad Sci*, 951:84–93.

Mammen, S, Nair, A, Kumar, S, Zonderveni, Z, Prabhakar, AT, Vijay Prakash, T, Aaron, S, Alexander, M, Zachariah, A, Abraham, AM (2020). Clinical Features of Four West Nile Virus Cases and Its Molecular Characterization from a South Indian Tertiary Care Hospital. *Case RepInfect Dis.*, 16, 1315041. doi: 10.1155/2020/1315041

Paramasivan, R,Mishra, A,Mourya, D (2003). West Nile virus: The Indian scenario. *Indian J. Med. Res.*, 118, 101–108.

Rajaiah, P, Mayilsamy, M, Kumar, A (2023). West Nile virus in India: An update on its genetic lineages. *Journal of Vector Borne Diseases* 60(3), 225–237. doi: 10.4103/0972-9062.374039

Reisen, WK, Fang, Y, Martinez, VM (2014). Effects of temperature on the transmission of West Nile virus by *Culex tarsalis (Diptera:Culicidae)*. *J Med Entomol*, 43, 309–17.

Risbud, AR, Sharma, V, Mohan Rao, CVR, Rodrigues, FM, Shaikh, BH, Pinto, BD, et al. (1991). Post-epidemic serological survey for JE virus antibodies in South Arcot district (Tamil Nadu). *Indian J Med Res*, 93, 1–5.

Roehrig, JT, Nash, D, Maldin, B, Labowitz, A, Martin, DA, Lanciotti, RS, Campbell, GL (2003). Persistence of virus-reactive serum immunoglobulin m antibody in confirmed west nile virus encephalitis cases. *Emerg. Infect. Dis.*, 9, 376–379.

Paul, SD, Rajagopalan, PK, Sreenivasan, MA (1970). Isolation of the west nile virus from the frugivorous bat, Rousettusles chenaulti. *Indian Journal of Medical Research*, 58(9), 1169–1171.

Smithburn K., Hughes T., Burke A. (1940). A neurotropic virus isolated from the blood of a native of Uganda. *American Journal of Tropical Medicine* 20, 471–492.

Shi, P-Y,Kramer, LDMolecular detection of West Nile virus RNA. *Expert Rev. Mol. Diagn.*2003; 3: 357–366.

Soman, RS, Kaul, HN, Guru, PY, Modi, GB, Guttikar, SN, Dhanda, V (1976). A report on the mosquitoes collected during an epidemic of encephalitis in Burdwan and Bankura districts, West Bengal. *Indian J Med Res*, 64: 808–13.

Sudeep, AB, Mandar, P, Ghodke, YK, George, RP, Gokhale, MD (2015). Vector competence of two Indian population of *Culex quinquefasciatus* (*Diptera:Culicidae*) mosquitoes to three West Nile virus strains. *J Vector Borne Dis* 52: 185–92.

Tardei, G, Ruta, S, Chitu, V, Rossi, C, Tsai, T, Cernescu, C (2000). Evaluation of immunoglobulin M (IgM) and IgG enzyme immunoassays in serologic diagnosis of West Nile virus infection. *J. Clin. Microbiol*, 38, 2232–2239.

Turell MH, et al. (2005). An update on the potential of north American mosquitoes (Diptera: Culicidae) to transmit West Nile Virus. *Journal of Medical Entomology*, 1(42), 57–62.

Section II

Diagnosis

8 Diagnosis of Malaria

Susanta K. Ghosh
Ex-ICMR-National Institute of Malaria Research, Bengaluru Field Station, India

Chaitali Ghosh
Tata Institute for Genetics and Society, Bengaluru, India

Rajeev K. Tyagi
Biomedical Parasitology and Translational-immunology Lab, CSIR-Institute of Microbial Technology, Chandigarh, India

B.K. Tyagi
Chandigarh University, Mohali, India

8.1 INTRODUCTION

Diagnosis of a pathogen causing infectious diseases is a primary step for treatment and case management (NVBDCP, https://nvbdcp.gov.in/WriteReadData/l892s/SOP-Quality-Assurance-Microscopy.pdf). Malaria is a protozoan parasitic infection transmitted by certain *Anopheles* mosquitoes. The malaria parasite was first described by French scientist Charles Louis Alphonse Laveran in 1880 in Algeria in unstained blood samples (Figure 8.1; CDC, 2015).

In 1879, Paul Ehrlich first used fuchsine (acidic) and methylene blue (basic) dyes to differentiate tissue cells. In 1891, Dmitri Leonidovich Ramanowsky, a Russian Physician, developed methods to stain blood smears using a mixture of eosin Y and methylene blue that produced a distinctive shade of purple. This was the beginning of staining to detect malaria parasites under microscope. Subsequently, in 1901–2, William Boog Leishman and James Homer Wright introduced methanol as a solvent. In 1904, Gustav Giemsa partially modified and improved this technique by adding glycerol to increase stability of the stain. Subsequently in 1944, the quick and rapid method of malaria blood staining was developed by Jaswant Singh and S. Bhattacharji, commonly referred to as JSB stain. The JSB stain is widely used throughout India but rarely used in other countries.

Five parasites, *Plasmodium vivax, P. falciparum, P. malariae, P. ovale*, and *P. knowlesi*, are responsible for causing human malaria. *P. ovale* consists of two genetically distinct subspecies *P. ovale curtisi* and *P. ovale wallikeri*. In India, the first two species are widely prevalent. *P. falciparum* is the dominant species in most tribal areas, whereas *P*. vivax is in the rest of the country. Very low level of *P. malariae* is found in high malaria-endemic tribal areas. *P. knowlesi* has been reported in the forested areas of Andaman and Car-Nicobar Islands. Sporadic cases of *P. ovale* including the two subspecies have also been reported.

DOI: 10.1201/9781003327028-10

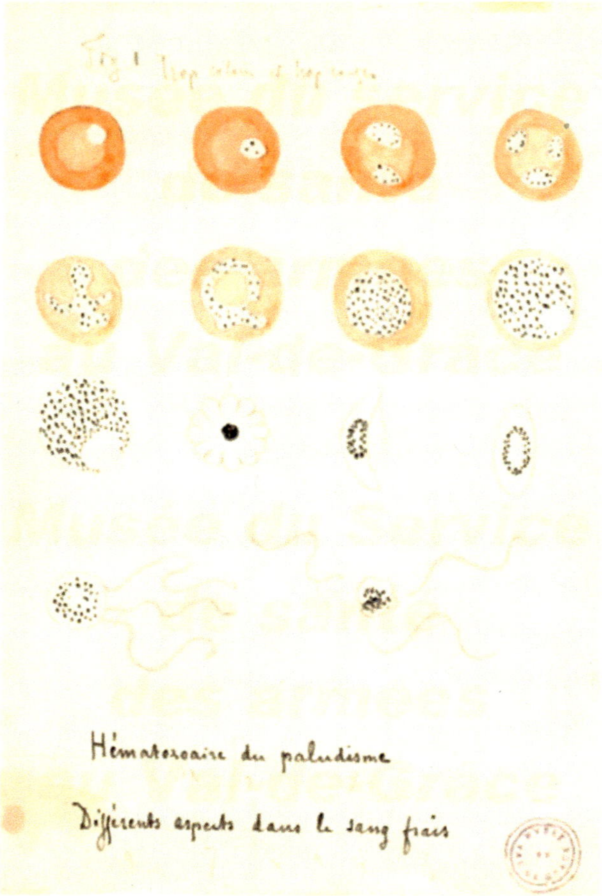

FIGURE 8.1 Illustration drawn by Laveran of various stages of malaria parasites (***Plasmodium falciparum***) from fresh blood. Dark pigments are present. Also, filiform elements show an exflagellating male gametocyte at the bottom.

Source: CDC, 2015.

8.2 METHODS OF MALARIA DIAGNOSTICS

8.2.1 Malaria Microscopy

Malaria parasites undergo a series of complex life cycle. The asexual cycle is exhibited in human host, and the sexual stages found in mosquitoes. Malaria microscopy detects the asexual stages of ring, trophozoites, schizonts and gametocytes. In *P. falciparum*, mainly rings and sausage-shaped gametocytes are observed. All the species possess specific identifying characteristics. A microscopist can identify each species of parasite based on these specific characteristics. Many standard operating procedures (SOPs) and handbooks are available from time to time (NCVBDC, n.d.; WHO 2010a, 2010b, 2015). Details have been described including training, resource mobilization, quality assurance and overall supervision.

8.2.2 Method of Staining

Microscopic micro-glass slides (75 × 25 × 1.5 mm) are used for making the smears from sterile finger-pricked blood. The slides should be clean and grease free; if not, these are cleaned with boiling water with detergent solution. The slides are washed thoroughly to remove all traces of detergent and grease.

8.2.3 Smear Preparation

The tip of the ring finger of the left hand or toe finger (in case of infant) is cleaned with a spirit swab and allowed to dry completely before pricking. Three drops of whole blood are collected for a thick smear and one drop for a thin smear is made in such a way that the edge of the smear is approximately 1 cm apart from the edge of the slide. The distance between the edge of the thick and thin smears should preferably be 0.75 cm. The thin smear is fixed with methanol. Proper care should be taken so that no traces of methanol (including its vapour) come in contact with the thick smear. The smears are stained after fully drying.

Stains: In India, JSB stain is widely used for malaria microscopy, while Giemsa stain is used for specific purposes.

JSB Stain: This stain consists of Solution I and Solution II.

Solution I: Thoroughly dissolve the following ingredients:

Methylene blue (preferably E. Merck brand)	0.5 gm
Sulphuric acid (H_2SO_4) 1%	3 mL
Potassium di-chromate ($K_2Cr_2O_7$)	– 0.5 gm
Di-sodium hydrogen phosphate (Na_2HPO_4)	– 3.5 gm
Distilled water	– 500 mL

The resulting mixture is heated to a slow boil in a refluxing condenser for one hour. The stain is allowed to cool at room temperature and kept for in a closed bottle.

Solution II: Prepared by dissolving 1 mg of yellowish aqua eosin powder (preferably E. Merck brand) in 500 mL of distilled water.

Both solutions should be kept for a couple of months to mature before use.

Buffer solution: Buffer solution is prepared by dissolving the following ingredients and adjusting the pH at 6.8.

Potassium hydrogen phosphate	– 0.752 gm
Di-sodium hydrogen phosphate (Na_2HPO_4)	– 3.5 gm
Distilled water	– 2 L

Before staining, the thick smear is dehaemoglobinized, and the thin smear is fixed with methanol. After the smears are dried, the fixed slides are dipped into the Solution II for two to three seconds (one to three dips) and washed thoroughly in the buffer water. The slides are then dipped in Solution I for 45 seconds. They are again washed gently and thoroughly in buffer water. After drying, the slides are examined under an oil immersion lens (100X). A slide should be declared negative after examining 200 thin or 100 thick fields. In certain situations, even 500 thick fields can be performed.

1. **Giemsa stain:** The stock solution is prepared from Giemsa powder (preferably E. Merck brand)

Giemsa powder	– 3.8 gm
Glycerol (pure)	– 250 mL
Methyl alcohol (pure)	– 250 mL

The stain is prepared by mixing all the ingredients in a porcelain mortar. The prepared stock solution is filtered and kept in a bottle.

Currently, readymade commercial brand Giemsa stains are available.

2. **Dilution of stock stain:** Stock solution must be diluted at a specific amount. Generally, 5% fresh stock solution is prepared for staining. Both thick and thin smears are dipped in the stain solution for 45 minutes after dehaemoglobinization; the smears are washed and dried for examination as mentioned earlier. A buffer solution of pH 7.2 is used for dilution as follows:

Potassium dihydrogen phosphate (KH_2PO_4)	– 0.7 gm
Di-sodium hydrogen phosphate (Na_2HPO_4)	– 1.0 gm
Distilled water	– 1 L

Currently, buffer tablets with specific pH values are available commercially and can be dissolved in distilled water.

8.3 THIN SMEAR EXAMINATION

Detailed morphology of the malaria parasites can be seen in properly made thin smears. Thin smear examination is the best method for species identification. With this method, loss of parasites is not possible since the smears are fixed with methanol.

8.3.1 *Plasmodium vivax*

All the erythrocytic stages of the *P. vivax* parasites are likely to be seen in the smear. The presence of schizonts is dependent on the time of schizogony. In certain cases, it is sometimes difficult to differentiate between *P. falciparum* and other sibling parasites like *P. vivax* based on the ring stage. In the initial stage of the rings, the cytoplasm is thin, and it becomes thick in the developmental stage when the erythrocytes also enlarge. In the trophozoite stage, the parasite is fairly amoeboid-shaped, and there is an appreciable increase in the size of the infected erythrocytes. The presence of Schüffner's dots in the erythrocyte cytoplasm is very characteristic of vivax malaria. Schüffner's dots are also present in schizont and gametocyte stages. In the schizont stage, the host cells are enlarged almost twice the size of the normal cells, have 12–24 merozoites, yellowish-brown, coalesced pigment. At the gametocyte stage, the vacuole of the parasite disappears, and the pigments are scattered, dark brown distributed in the cytoplasm.

8.3.2 *Plasmodium falciparum*

In the peripheral smears, only rings and gametocytes are seen. The other stages of trophozoites and schizonts are usually not found in peripheral smears. On some rare occasions, in hyperparasitaemic cases, these stages are seen. In the early ring, the cytoplasm is usually very thin and hair-like. They are present in different forms, such as marginal, tenue, and with double and multiple chromatins placed along the sides or lying in opposite poles.

Multiple infections are also seen in falciparum infections. The presence of Maurer's dots/clefts is characteristic of this parasite species. The dots present on the erythrocyte are coarse and usually few in numbers. The gametocytes have got characteristic sausage or crescent shapes. The stretched erythrocyte outlines can be seen in most of the parasites. Some pigments can be detected at this stage.

8.3.3 *Plasmodium malariae* (Quartan Malaria)

This parasite is very seldom found because of its low prevalence. It is found in most of the tribal areas, and its presence is an indicator of high malaria endemicity. This species can be identified very easily by its characteristic features. The rings are quite stocky and solid. The cytoplasm is thicker than vivax. The pigments can be seen in the cytoplasm even in the ring stage. Majority of the trophozoites are band shaped. Coarse blackish pigments can be seen in the cytoplasm. Following incorporation into erythrocytes, early trophozoites begin to accumulate haemozoin and the pink-staining Ziemann's dots. There is no enlargement in the erythrocyte containing the schizont. In fully mature schizont, 6–12 merozoites are seen. The pigments are concentrated in the centre, giving an appearance of rosette. The gametocytes are round in shape. Here the erythrocytes are not enlarged. They appear to be smaller in some parasites. Blackish pigment granules are found scattered in the cytoplasm. They can be identified easily, having a round shape and size.

8.3.4 *Plasmodium ovale*

Like *P. vivax, P. ovale* exhibits similar characters except the size and shape. Here the infected cells are oval-shaped and in some cases are with fimbricated margins (comet cell). The mature schizonts of *P. ovale* never have more than 12 nuclei within them, and this is the only reliable way of distinguishing character between *P. vivax* and *P. falciparum*. The genetically distinct subspecies *P. ovale curtisi* and *P. ovale wallikeri* can be detected through a molecular test.

8.3.5 *Plasmodium knowlesi*

Plasmodium knowlesi, a parasite of primate origin is the latest addition in human malaria. It is found throughout Southeast Asia and is the most common cause of human malaria in Malaysia. Microscopy is the diagnostic method of choice but has limitations for *P. knowlesi* as ring stages resemble *P. falciparum* while the trophozoites resemble *P. malariae*, resulting misidentification. Hence, molecular diagnosis is the best method of diagnosis.

8.4 THICK SMEAR EXAMINATION

Thick smear examination for malaria parasites was introduced by the national programme. In fact, thick smear is made of 10–12 layers of thin smears. In this method, it is possible to detect the parasites quickly with certain expertise. Since parasite morphology is distorted species of the parasites becomes difficult in certain cases. In this method, loss of parasites is common.

Like thin smear, the parasites are identified easily in thick smears. It is essential to associate the pink coloured chromatin with blue cytoplasm of each parasite especially the ring stage. It is very important to differentiate between the real parasites and the artefacts. Detection of the presence of mixed infection of two parasite species may be difficult to identify. For this, proper smear preparation and staining are required. The advanced stages of *P. vivax*, as well as *P. falciparum*, gametocytes could be identified easily in the thick smears.

8.5 DETERMINATION OF PARASITAEMIA

8.5.1 Thick Smear

Infected erythrocytes (RBCs) are counted in relation to a predetermined number of leukocytes (WBCs). Generally, an average number of 8,000/μL is considered as standard. Approximately 300 WBCs are counted in 100 micro fields, if >10 parasites are found, the following formula is applied for calculating the parasitaemia.

$$\frac{\text{Number of parasites}}{\text{Number of WBCs counted}} \times 8{,}000 = \text{Number of parasites per } \mu\text{L blood}$$

When parasites are < 10, about 500 WBCs should be counted

The '+' system is less precise as variation in the thickness of the thick results in false variation in parasite count.

+	= 1–10 parasites per 100 thick fields
++	= 11–100 parasites per 100 thick fields
+++	= 1–10 parasites per thick field
++++	= >10 parasites per thick field

8.5.2 Thin Smear

Determination of percent parasitaemia is essential for some specific purposes. The number of infected erythrocytes (not number of parasites) counting in 10,000 RBCs is calculated to percentage. Likewise, parasite per μL blood can also be calculated taking RBC per μL basis.

8.6 ADVANCEMENTS OF DIAGNOSTIC TOOLS AND THEIR APPLICATIONS

8.6.1 Rapid Diagnostic Tests

The recent advancement in rapid diagnostic tests (RDTs) has revolutionized the point-of-care diagnosis not only for malaria but other diseases also. In the last two decades, immunochromatographic malaria RDTs (mRDTs) with histidine rich protein 2 and 3 (HRP2/3) are being used extensively and has transformed the landscape of malaria diagnosis. In the initial stage, the dip-stick method was developed and subsequently handier with the card system. Principally, parasite antigens are detected to know the current events of infections, but antibodies can also be detected using this method. HRPs (histidine rich protein) are water soluble protein synthesized by the parasites and released from the infected erythrocytes. The system has an affinity purified monoclonal antibodies (earlier polyclonal antibodies) specific against immunodominant repeating sequence from PfHRP2 molecule. About 10 μL blood (either finger pricked or whole blood) is added on the fibrous pad at the base of the test card. Then four to five drops of buffer solution (supplied in the kit) are added on the pad that dissolve the blood. The blood and the gold conjugated antibody mixture thus released from the pad. The blood-buffer mixture moves along the nitrocellulose strip. After 10 to 15 minutes, the strip gets clear. In positive samples a pin-coloured band is noticed along with control band. In the negative samples only one band, i.e., control band is seen. If no bands are found the test is considered invalid (Figure 8.2). Generally, HRP2 also cross react with HRP3.

Recently parasites with HRP2 deletion have been reported and has been a matter of concern. First detected in Amazon region in 2010 and now in the East African countries. In India, around 25% cases are reported to have the deletion of this gene. Figure 8.3 shows the global pattern of HRP2/3 deletion. The most disturbing aspect is that this method of diagnosis is limited in sub-microscopic/

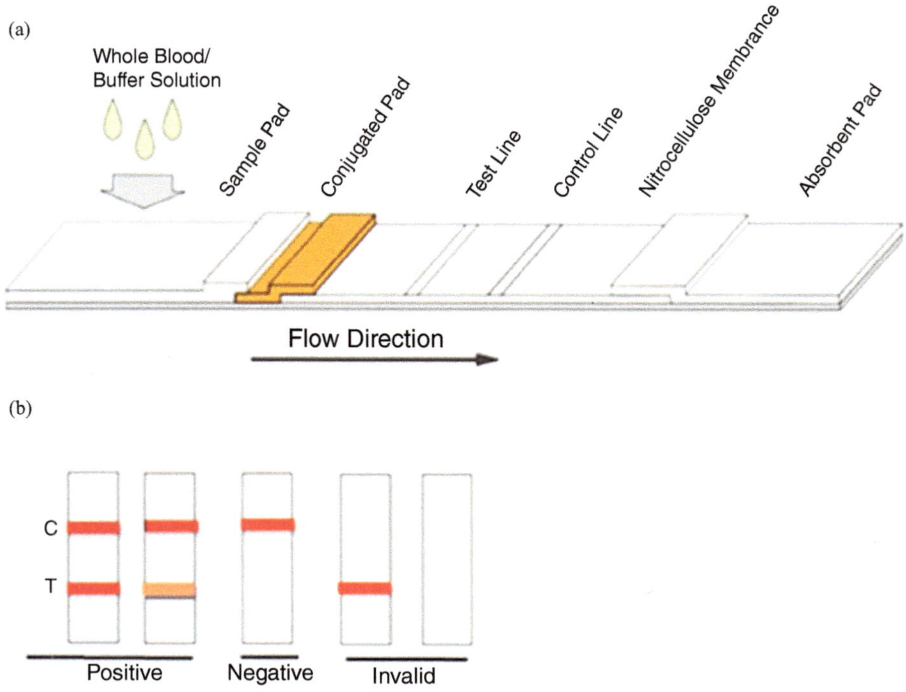

FIGURE 8.2 Schematic diagram of an RDT. (a) How a test is conducted. (b) Positive, negative and invalid readings.

Source: Free Wikipedia.

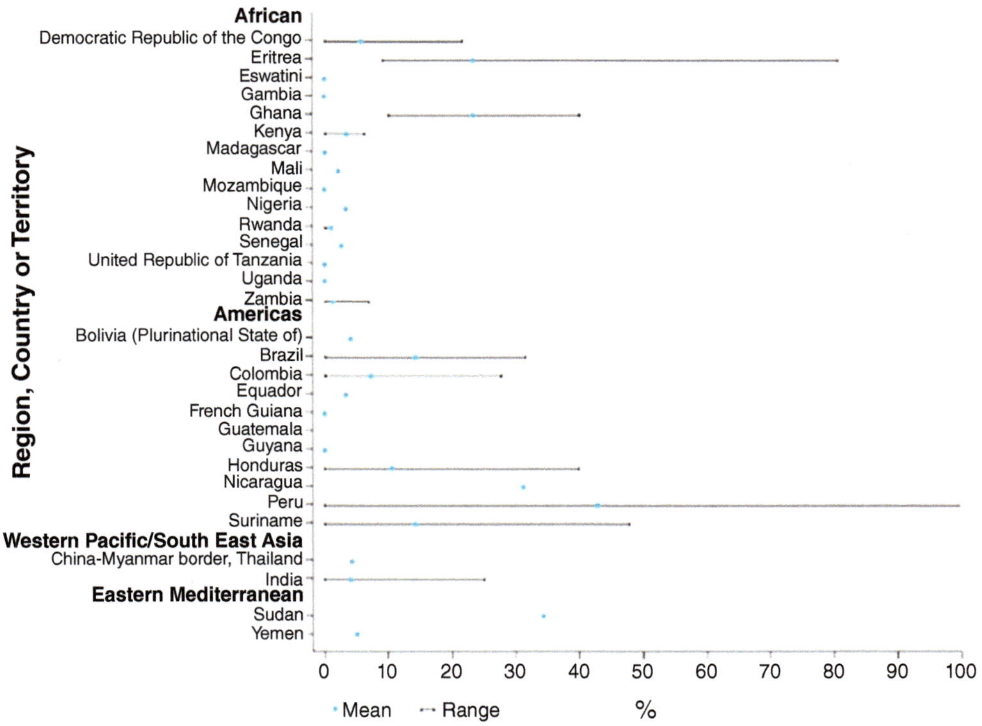

FIGURE 8.3 Weighted average estimates for *Pfhrp2* deletions in *P. falciparum* patients tested by country.

Source: Ghosh and Ghosh, 2019; WHO, 2020.

asymptomatic cases having <100 parasites/μL blood. Gene deletion has added more problem. Another issue of 'prozone phenomenon' means excess HRP causes exhaustion of antibodies showing false negative cases (Poti et al., 2020).

Now ultra-RDT with falciparum aldolase or parasite lactate dehydrogenase (pLDH) is now available to counter the problems with HRP2-based RDTs. These antigens are used for detection of other malaria parasites, such as vivax malaria.

8.6.2 Nucleic Acid Probe

8.6.2.1 DNA Probe

The eurokaryotic genome contains highly repetitive species-specific sequences. The ability to clone such deoxyribonucleic acid (DNA) region provides a means to produce a new array of diagnostic reagents based on the DNA–DNA hybridization procedure. The binding of a DNA probe to its specific target DNA is usually detected by tagging the probe with radiolabel (P^{32}) and exposing the hybridized DNA to an X-ray film.

Since most DNA probes contain repeats of the 21-base pair sequences, synthetic oligodeoxynucleotides which can be radio-labelled with P^{32} have also been used as a diagnostic probe.

8.6.2.2 RNA Probe

RNA probes can be constructed that are specific but are not very sensitive in detecting low grade parasitaemia. An rRNA would be a better proposition where there are 0.2–1.0 pg of RNA per parasite compared to 0.02 pg of DNA. This is a simplified procedure where the blood samples are applied directly onto the nitrocellulose membrane. The parasites are detected using a single 21-mer oligonucleotide probe complementary to stem-loop region of the parasite's small sub-unit RNA. One parasite/μL blood is detected after one hour of autoradiography.

8.6.2.3 Polymerase Chain Reaction-Based Diagnosis

Over three decades, one of the exciting developments in the field of recombinant DNA technology has been the polymerase chain reaction (PCR). It is possible to synthesize millions of copies *in vitro* of specific target DNA sequences. In malaria, PCR technology has opened the prospects of increasing sensitivity for detection as well as genes involved in drug resistance. Using this non-isotopically labelled probe following PCR amplification, it is possible to detect 1 parasite/μL blood.

Loop-mediated isothermal amplification (LAMP) method of malaria diagnosis is emerging as an alternative to routine PCR. In this method, whole blood samples with minimal processing with purified DNA or RNA can be employed giving high sensitivity as low as 0.5 to 0.05 parasites/μL compared to PCR and RDT (Puri et al., 2022).

Recently we have developed an indigenously developed chip-based microPCR-based point-of-care diagnosis Truenat® malaria device that can detect malaria parasites (*P. vivax* and *P. falciparum*) as low as 1.3 parasites/μL blood. Similar devices have been used for the diagnosis of Covid-19, Tuberculosis, and many other diseases also such as dengue, chikungunya, etc. (Nair et al., 2016).

8.6.2.4 Droplet Digital PCR Method

Efforts are being made to find out methods to detect malaria parasites from saliva or urine samples. Highly sensitive molecular techniques of droplet digital PCR (ddPCR) are used to detect *Plasmodium* DNA targeting multicopy sites from saliva samples either whole saliva or buccal swabs (Costa et al., 2022). Such a method needs more research for practical use. Now malaria is under the elimination phase and would achieve this goal by 2023. Since 1900, 130 countries have achieved malaria elimination. In the elimination phase, serological tests such as immunofluorescence, indirect haemagglutination (IHA), ELISA, and also radio-immuno assay (RIA) may be used to find out the antibody load in specific areas.

8.6.2.5 Quantitate Buffy Coat

The quantitative buffy coat (QBC) technique is a method of diagnosing malarial parasites based on micro-centrifugation, fluorescence, and density gradient of infected red blood cells. This method involves spinning of blood in special capillary tubes precoated with acridine orange (AO) in which parasite DNA is stained with AO. A small precision moulded plastic floats presses the parasitized erythrocytes against the wall of the tube where the stained parasites can be seen in an UV light microscope. This method can detect 5 parasites/µl blood (Kochareka et al., 2012). This is an expensive method and not recommended in the national programme.

8.6.2.6 Gazelle Device

A novel malaria diagnostic tool, Gazelle (Hemex Health, USA) detects haemozoin. This is a magneto-optical detector for identifying haemozoin crystals produced by all species of human malaria parasites. A beam of polarized light is passed through the lysed diluted blood sample under the influence of high (~.55 T) and low magnetic fields. The difference in light transmission through the sample between the high and low magnetic fields indicates presence of Hz, indicating possible malarial infection. This device is also used in the detection of *P. knowlesi* parasites (Kumar et al., 2022; Tan et al., 2023).

8.7 APPLICATION OF ARTIFICIAL INTELLIGENCE

With the development of artificial intelligence (AI), its application has been employed in disease diagnosis. Several applications have been developed in this direction. Expert or trained microscopists analyse images of parasites to diagnose malaria. Modern deep learning techniques may be used to do this analysis automatically. The automated convolutional neural network (CNN)-based model for the diagnosis of malaria from the microscopic blood smear images can be used for fast and accurate diagnosis of malaria. Modified YOLOV3 and YOLOV4 models is found to be promising even smart phones are used (Abdurahman et al., 2021; Fuhad et al., 2020).

8.8 NON-INVASIVE METHOD

Sensor-based diagnosis: A near-infrared *Plasmodium* absorption peaks acquired non-invasively through the skin using a miniaturized hand-held near-infrared spectrometer. Using spectra from the ear, these absorption peaks and machine learning techniques enabled non-invasive detection of malaria-infected human subjects with varying parasitaemia levels in less than ten seconds (Garcia et al., 2022).

8.9 DISCUSSION AND CONCLUSION

In the last four decades, several methods have been developed for malaria diagnosis. Among them, RDTs are most promising. Practical and cost-effective methods are need of the hour. Many methods cannot find their use because of this factor. Some methods such as radio-labelled technique are used for research purpose. RDTs using HRP2/3 antigens may not find its use due to the gene deletion and other issues mentioned earlier (Poti et al., 2020). Many other methods, such as haematology analyser XN31 (Komaki-Yasuda et al., 2022), AI-based devices, and Fourier transform infrared (FTIR) spectroscopy (Heraud et al., 2019), need more studies for their practical use. Many researchers are working on APTAMAR and CRISPAR-Cas9-based technologies for better diagnostic tools in the future (Ghosh and Ghosh, 2019). However, malaria microscopy will remain the mainstay in malaria diagnosis and should not be diluted in the face of other techniques since microscopy has many added advantages.

REFERENCES

Abdurahman, F., Fante, K. and Aliy, M. (2021). Malaria parasite detection in thick blood smear microscopic images using modified YOLOV3 and YOLOV4 models. *BMC Bioinformatics*, 22, 112. https://doi.org/10.1186/s12859-021-04036-4.

Centre for Disease Prevention and Control, USA (2015). History – Laveran and the discovery of malaria. Available from: https://www.cdc.gov/malaria/about/history/laveran.html

Costa, G.L., Alvarenga, D.A.M., Aguiar, A.C.C., Louzada, J., Pereira, D.B., de Oliveira, T.F., Fonseca Júnior, A.A., Carvalho, L.H., Ferreira, Alves de Brito. C. and Nóbrega de Sousa, T. (2022). Improving the Molecular Diagnosis of Malaria: Droplet Digital PCR-Based Method Using Saliva as a DNA Source. *Frontiers in Microbiology*, 13, 882530. doi: 10.3389/fmicb.2022.882530

Fuhad, K.M.F., Tuba, J.F., Sarker, M.R.A., Momen, S., Mohammed, N. and Rahman, T. (2020). Deep Learning Based Automatic Malaria Parasite Detection from Blood Smear and its Smartphone Based Application. Diagnostics (Basel). *Diagnostics*, 10(5), 329. doi: 10.3390/diagnostics10050329

Garcia, G.A., Kariyawasam, T.N., Lord, A.R., da Costa, C.F., Chaves, L.B., Lima-Junior, J.D.C., Maciel-de-Freitas, R. and Sikulu-Lord, M.T. (2022). Malaria absorption peaks acquired through the skin of patients with infrared light can detect patients with varying parasitemia. *PNAS Nexus*, 1(5), 272. doi: 10.1093/pnasnexus/pgac272.

Ghosh, S.K. and Ghosh, C. (2019). New ways to tackle malaria. In a book chapter 'Current Topics in the Epidemiology of Vector-Borne Diseases'. IntechOpen, UK, DOI: 10.5772/intechopen.89467. https://www.intechopen.com/online-first/new-ways-to-tackle-malaria

Heraud, P., Chatchawal, P., Wongwattanakul, M., Ippayawat P, Doerig C, Jearanaikoon P, Perez-Guaita D. and Wood BR. (2019). Infrared spectroscopy coupled to cloud-based data management as a tool to diagnose malaria: a pilot study in a malaria-endemic country. *Malaria Journal*, 18: 348. https://doi.org/10.1186/s12936-023-04780-3

Kochareka, M., Sarkar, S., Dasgupta, D. and Aigal, U. (2012). A preliminary comparative report of quantitative buffy coat and modified quantitative buffy coat with peripheral blood smear in malaria diagnosis. *Pathogens and Global Health*, 106, 335–339. doi: 10.1179/2047773212Y.0000000024

Komaki-Yasuda, K., Kutsuna, S., Kawaguchi, M. Kamei, M., Uchihashi, K., Nakamura, K., Nakamoto, T., Ohmagari, N. and Kano, S. (2022). Clinical performance testing of the automated haematology analyzer XN-31 prototype using whole blood samples from patients with imported malaria in Japan. *Malaria Journal*, 21, 229. https://doi.org/10.1186/s12936-022-04247-x

Kumar, R., Verma, A.K., Shrivas, S., Thota, P., Singh, M.P., Rajasubramaniam, S., Das, A. and Bharti, P.K. (2022). First successful field evaluation of new, one-minute haemozoin-based malaria diagnostic device. *eClinical Medicine*, 22, 100347. doi: 10.1016/j.eclinm.2020.100347

Nair, C.B., Jagannath, J., Pradeep, A.S., Prakash, B.N., Manoj, N.M., Malpani, S., Pullela, P.K., Subbarao, P.V., Ramamoorthy, S. and Ghosh, S.K. (2016). Differential Diagnosis of Malaria on Truelab Uno1, a Portable, Real-Time, MicroPCR Device for Point-Of-Care Applications. *PLoS One*, 11(1), e0146961. doi:10.1371/journal.pone.0146961

National Center for Vector Borne Diseases Control (NCVBDC) (n.d.). Laboratory Diagnosis of Malaria. Quality Assurance Microscopy. NVBDCP; Government of India. https://nvbdcp.gov.in/WriteReadData/l892s/SOP-Quality-Assurance-Microscopy.pdf

Poti, K.E., Sullivan, D.J., Dondorp, A.M. and Woodrow, C.J. (2020). HRP2: Transforming Malaria Diagnosis, but with Caveats. *Trends in Parasitology*, 36, 122–126. doi: 10.1016/j.pt.2019.12.004

Puri, M., Kaur Brar, H., Madan, E., Srinivasan, R., Rawat, K., Gorthi, S.S., Kumari, G., Sah, R., Ojha, S.B., Panigrahi, S., Dhangadamajhi, G., Muthuswami, R., Singh, S. and Madhubala R. (2022). Rapid diagnosis of *Plasmodium falciparum* malaria using a point-of-care loop-mediated isothermal amplification device. *Frontiers in Cellular and Infection Microbiology*, 12, 961832. doi: 10.3389/fcimb.2022.961832

Tan, A.F., Thota, P., Sakam, S.S.B., Lew, Y.L., Rajahram, G.S., William, T., Barber, B.E., Kho, S., Anstey, N.M., Bell, D. and Grigg, M.J. (2023). Evaluation of a point-of-care haemozoin assay (Gazelle device) for rapid detection of *Plasmodium knowlesi* malaria. *Scientific Reports*, 13, 4760. https://doi.org/10.1038/s41598-023-31839-7

WHO (2010a). *Basic Malaria Microscopy: Part I: Learner's Guide*. 2nd edition. World Health Organization, Geneva. https://www.who.int/publications/i/item/9241547820

WHO (2010b). *Basic Malaria Microscopy: Part II. Tutor's Guide*. 2nd edition. World Health Organization, Geneva.

WHO (2015). *Malaria Microscopy Quality Assurance Manual – Version 2*. World Health Organization, Geneva. https://www.who.int/publications/i/item/9789241549394

WHO (2020). *World Malaria Report 2020*. World Health Organization, Geneva. https://www.who.int/teams/global-malaria-programme/reports/world-malaria-report-2020.

9 Diagnosis of Lymphatic Filariasis

Susanta K. Ghosh
Ex-ICMR-National Institute of Malaria Research, Bengaluru Field Station, India

Chaitali Ghosh
Tata Institute for Genetics and Society, Bengaluru, India

B.K. Tyagi
Chandigarh University, Mohali, India

9.1 INTRODUCTION

Lymphatic filariasis (LF) or commonly known as elephantiasis (*Hathipaon* in Hindi) is a mosquito-borne neglected tropical disease. Eight parasites are responsible for filarial infections. Three parasites, *Wuchereria bancrofti* (Cobbold, 1877), *Brugia malayi* (Brug 1927), and *Brugia timori* (Partono et al., 1977), are responsible for LF that affects the lymphatic system, leading to severe organ deformities and social stigma. The most impacted body parts are legs and genitals: hydrocele in males and breasts in females. In India, the first two species *W. bancrofti* and *B. malayi* are present. *W. bancrofti* contributes 99.4% of the total burden. It is a roundworm nematode parasitic helminthiasis group of diseases under the Filarioidea type of infection. In India, LF is the second most vector-borne disease after malaria and globally ranks third after malaria and tuberculosis. The World Health Organization (WHO) estimated that LF is found in 47 tropical and subtropical countries, with 120 million infected cases and 1 billion people at risk; 863 million people are threatened, whereas 40 million people are disfigured by this infection (WHO, 2023; Ghosh and Srivastava, 2020).

Wuchereria bancrofti is the most predominant species, sharing 90% of the burden. India contributes around 40% of the global burden, with 272 LF endemic districts from 16 states and 5 UTs covering almost 670 million people are at risk (Tripathi et al., 2022; Figure 9.1).

The two pillars, viz., (i) annual mass drug administartion (MDA) and (ii) morbidity management, are the key components of LF elimination. The National Health Policy in 2002 set a goal to eliminate LF in India by 2015, which was later extended to 2017. The goal set by the Global Programme for Elimination of Lymphatic Filariasis (ELF) was to eliminate LF by 2020. Subsequently, the global goal was revised to eliminate LF by 2030 (Tripathi et al., 2022). To achieve the elimination goal and routine LF surveys diagnosis remains the main domain. Microscopy is the main method of LF diagnosis, even though many options are now available, such as rapid diagnostic tests (RDTs; Ghosh and Srivastava, 2020).

DOI: 10.1201/9781003327028-11

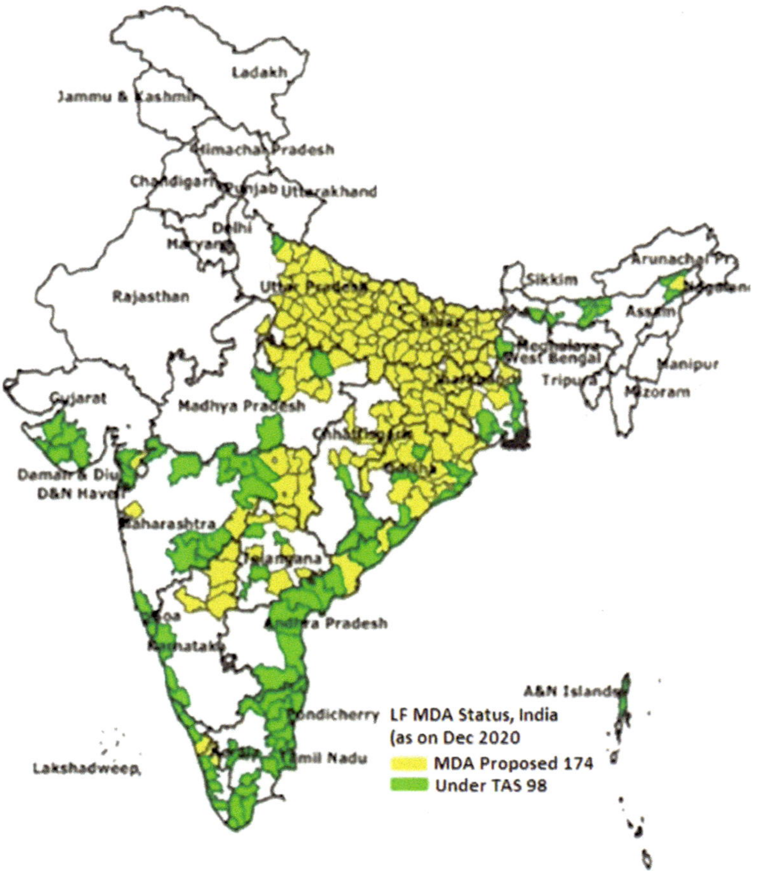

FIGURE 9.1 Distribution of lymphatic filariasis endemic districts in India, 2021 showing the progress of mass drug administration (MDA).

Source: Tripathi et al., 2022.

9.2 METHODS OF DIAGNOSIS

Microscopy remains the main method for the detection of microfilariae (mf) in blood smears. In recent years, several significant achievements have been made to diagnose LF. Rapid diagnostic tests (RDTs, PCR-based and other initiatives are under way.

9.2.1 Microscopic Method

Thick Blood Smear: Examination of Thick Blood Smear on 20 μL blood (TBS 20) is a routine method for detecting mf. It is important that the time of blood collection should be closer to the peak activity time of the mf based on species and strains. About 20 μL whole blood is drawn during night hours from sterile finger prick on a clean micro-slide, evenly spreaded, dried, carefully dehaemoglobinized, and stained. Generally, JSB (Jaswant Singh and Bhattacharya) and Giemsa stains are used to stain the *mf* (microfilarae), like in the case of malaria parasites. It has been observed that > 40 μL blood is more sensitive than TBS 20 (Vinkeles Melchers et al., 2020; WHO, 1987).

Detailed comparative diagnostic characteristics of *W. bancrofti* and *B. malayi* are shown in Tables 9.1 and 9.2, and Figure 9.2a–9.2d.

TABLE 9.1
Comparative Morphological Identifying Characteristics of Human Filarial Parasites (Microfilariae) in Blood

Characteristics	Wuchereria bancrofti	Brugia malayi
Mean length (µm)	260 (245–300)	220 (177–230)
Cephalic space (length:breath)	1:1	2:1
Sheath colour	Unstained	Pink
Nuclei	Regularly spaced and separately situated	Irregularly spaced and overlapping
Tail	Single row of nuclei does not reach the tail end	Single row of nuclei reaches the tail end
Terminal nuclei	None	Two nuclei closely spaced which clearly bulges the cuticle
Appearance in blood film	Smoothly curved	Kinky

TABLE 9.2
Distinguishing Characteristics of Two Human Filarial Parasites (Microfilariae) in Blood

Microfilaria	Head	Tail
Wuchereria bancrofti	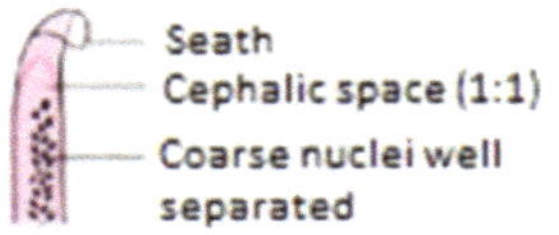	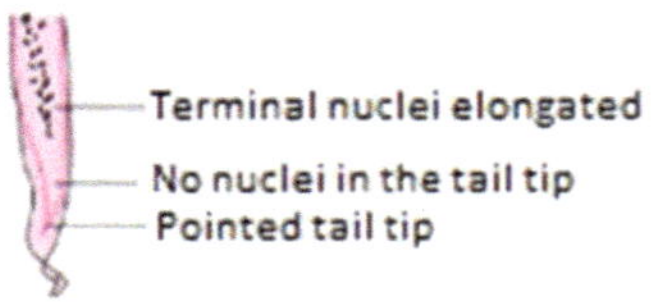
Brugia malayi	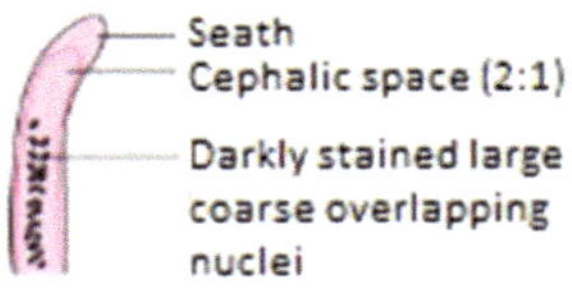	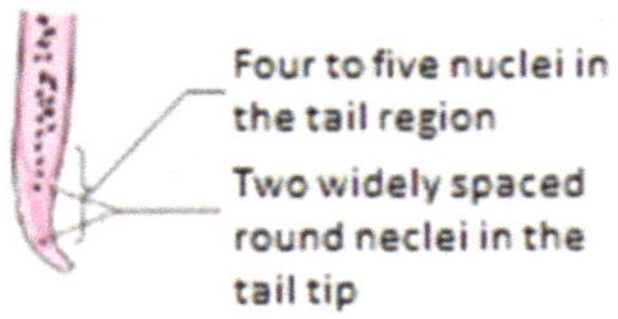

The mf or the larval stages of *W. bancrofti* are sheathed measuring from 245 to 300 µm. One end of the body is round, while the other end is pointed. Nuclei do not appear at the end of the tail, and this is the major difference from other mf of *B. malayi*. The morphology of *B. malayi* mf, like that of *W. bancrofti* has the main differentiating species character. The mf range from 177 to 230 µm. Here, the nuclei extend to the tip of the tail (Figure 9.2). Both species lack a digestive system because they absorb nutrients from their hosts.

9.2.2 Concentration Technique

9.2.2.1 Membrane Filtration Technique

This is a very sensitive method of mf detection. Anticoagulated venous blood samples of 1–10 mL are filtered through millipore or nitrocellulose membrane filters of 3–5 µm porosity. The blood is haemolyzed with Teepol. The filter is washed three times with 0.9% normal saline and then fixed with formal saline (1 part formalin + 9 parts normal saline). Wash twice with distilled water. Blow air through the filter. The dried filter paper is immersed in hot Harris' haematoxylin for five minutes. "Blue" briefly in gentle running water. Dry on a clean microscope slide. Cover with immersion oil and a coverslip. Examine under low power lens (10X). This is an expensive method and should

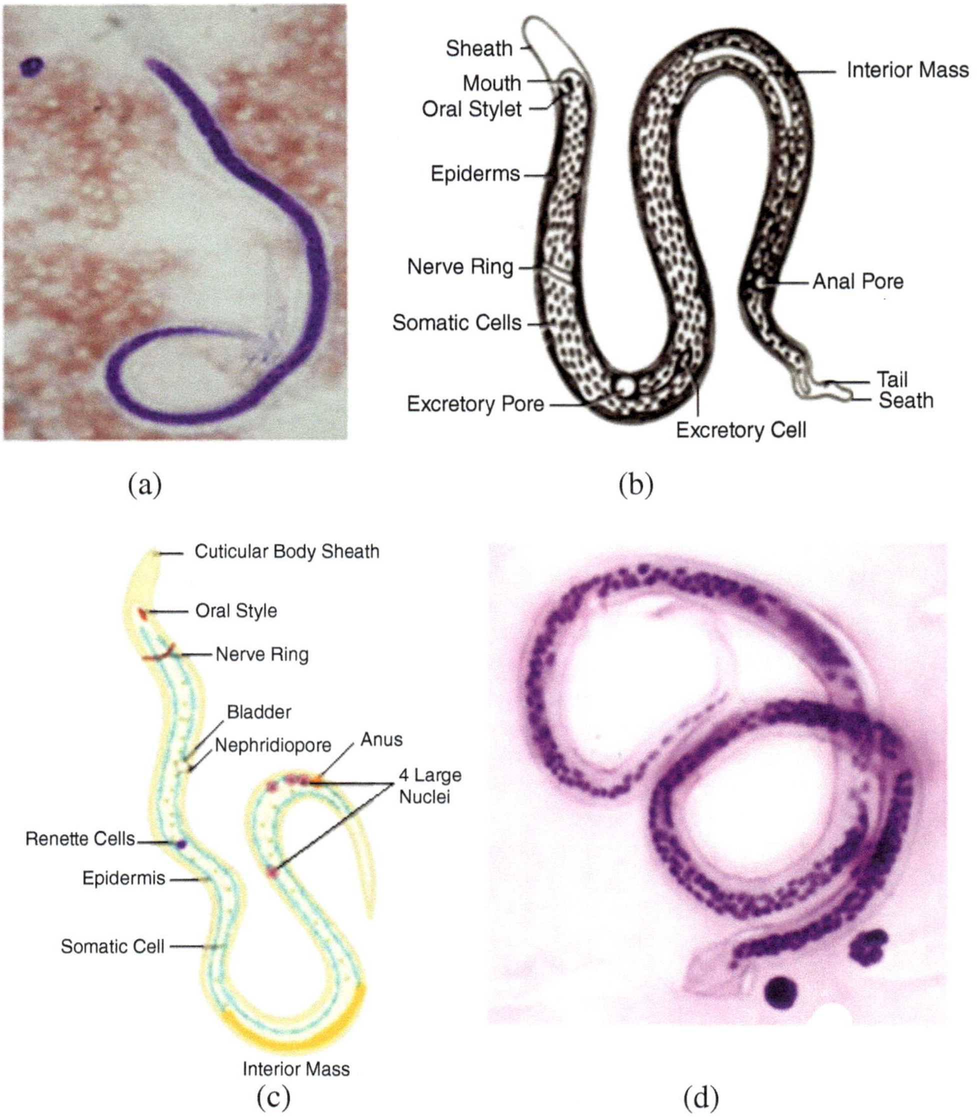

FIGURE 9.2 (a) Microfilariae of *Wuchereria bancrofti* (b & c). Detailed morphological characteristics of *W. bancrofti*. (d) Microfilariae of *Brugia malayi*. Note the extended nuclei on the tip of the tail.

Source: Free Wikipedia.

be done for special cases only, such as diagnosis of individual cases, evaluation of treatment, as a research tool, or in pilot control studies (WHO, 1987).

9.2.2.2 Knott's Concentration Technique

This is an alternative technique when membrane filters are not available. It has lower sensitivity than membrane filters, as mf are likely to be missed.

Mix 1 mL venous blood with 10 mL of dilute formalin (2 mL/L) in a conical centrifuge tube. Leave the tube for 12 hours or centrifuge for 5 minutes at 350g. Discard the supernatant. Examine the sediment for mf. A few drops of 1% methylene blue will help show the mf (WHO, 1987).

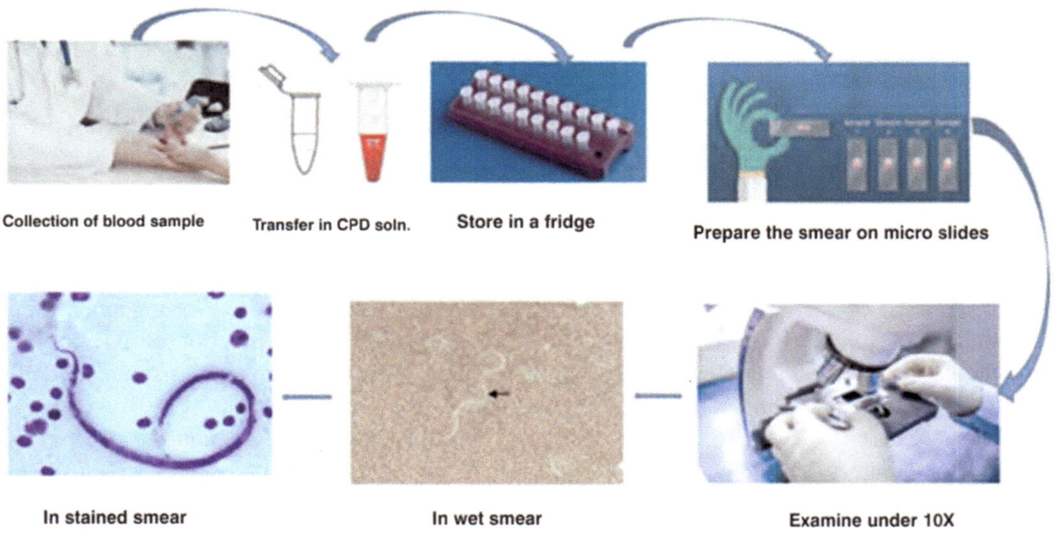

FIGURE 9.3 Schematic diagram of wet blood smear examination for detection of live mf.

Source: Several links from Wikipedia.

9.2.2.3 Concentration Technique for Live Parasites

Detection of live mf in wet smears has been developed by Ghosh and Srivastava (2020). This is a very simple concentration technique and can be done easily under the field condition. It has been observed that blood smear preparation on the micro-slides is a cumbersome process in the field during night hours. In most situations, it takes considerable time to dry the thick blood, resulting in uneven layers of the thick smears.

The finger-pricked whole blood (50 µL) is collected in a citrate-phosphate-dextrose (CPD) solution charged (25 µL) in 1.5 mL microfuge tubes. The tubes can be kept in a +4°C freezer and can be examined within 48 hours. In our experience, even one week period was good. CPD-mixed whole wet blood (20 µL) is drawn by micropipette and placed in a micro-slide. The blood is smeared on the micro-slides and examined under a 10X microscope when the blood is wet. In positive samples, live, wrangling, moving parasites can be seen easily. This is a very simple method and can be easily executed. If needed, the dry smears can be stained for future reference and species diagnosis. We have observed that there is no loss of parasites due to the staining process in this method. The detailed procedure is shown in Figure 9.3.

9.2.2.4 Counting Chamber Technique

The counting chamber technique is fast, quantitative, and cheap. An aliquot of 100 µL of finger-pricked blood is added to a tube containing 0.9 mL of 3% acetic acid. Microfilariae are counted in a counting chamber using a low-power objective of a compound microscope. This technique is suitable for both routine hospital diagnosis and field surveys in areas that are endemic for only one filarial species. This is because species identification is difficult with the counting chamber technique (WHO, 1987).

a. **In hydrocele fluid or urine:** Sometimes mf are absent in blood samples but present in hydrocele fluid or urine. The samples are processed and examined using the concentration technique (WHO, 1987; Chenthamarakshan et al., 1993).
b. **Provocative test:** This method is applied when a patient is suspected of filarial infection. In this test, 100 mg of diethylcarbamazine citrate (DEC) is administered orally (adult dose;

50 mg for children). The peripheral blood is collected after 30–45 minutes of DEC administration. DEC provokes the mf in the lung capillaries to invade the peripheral blood (WHO, 1987).

9.3 DISCUSSION AND CONCLUSION

Human lymphatic filariasis is under the process of elimination (ELF). ELF has to be achieved by 2030. LF has remained a neglected disease since investment in such programmes does not have commercial interest. Many diagnostic methods have been developed in recent years. The Filariasis Test Strip (FTS) of Alere (now Abbott Diagnostics) is an RDT recommended for mapping, monitoring, and transmission assessment surveys (TAS) for the qualitative detection of *W. bancrofti* antigen in human blood samples. Now, the FTS has replaced the Binax Now filariasis immunochromatographic test (ICT), which also detects the same antigen in blood samples. The Brugia Rapid Point-of-Care Cassette Test (BRT) manufactured by Reszon Diagnostics is recommended for use during TAS to detect the IgG4 antibody against *Brugia spp.* in human blood samples (Ghosh and Srivastava, 2020).

Other methods such as DNA-based (Nuchprayoon, 2009), LAMP-based (Nancy et al., 2021), RFLP (Thanomsub et al., 2000), recombinant antigens-based (Pastor et al., 2021), indirect fluorescent antibody, enzymelinked immunosorbent assays (Lim et al., 1984), and image analysis on peripheral blood smears (Mallawaarachchi et al., 2013; Kumar and Babulal, 2024) have been developed. Methods like QBC, ultrasonography, lymphoscintigraphy (LSG), X-ray diagnosis, and haematology have been developed, but all these have very limited use. A systematic review and meta-analysis showed that the counting chamber technique is more sensitive than other methods (Vinkeles Melchers et al., 2020). Our method of concentration techniqueusing wet blood that detects live mf is very useful and should be recommended in the programme for achieving the ELF (Ghosh and Srivastava, 2020).

A fast colour change mf detection method 'FILACHEK' was developed by the Department of Biochemistry, Mahatma Gandhi Institute of Medical Sciences, Sevagram, Wardha, Maharashtra, India in 1990. The test detected the excretory–secretary (ES) antigen of *W. bancrofti* stained with iodine solution in the laboratory trials. Pink colour lasted for 30 minutes in the whole blood samples indicated positive for mf. However, this method was discontinued since the sensitivity and specificipy were lower than the standard levels in the field trials conducted in Rourkela (Odisha) and Shahjahanpur (Uttar Pradesh).

Recently, the BIRAC (Biotechnology Industry Research Assistance Council), a public sector enterprise set up by the DBT (Department of Biotechnology), Government of India has initiated projects for the diagnosis of LF. Also, the ICMR (Indian Council of Medical Research) has invited applications to develop filarial vaccines. The ICMR-Vector Control Research Centre has identified 12 antigens. Such antigens may be used for indigenously developed diagnostic techniques (ICMR, 2023).

ACKNOWLEDGEMENT

The authors thankfully acknowledge Dr Bhupendra Tripathi, the Bill and Melinda Gates Foundation, India, for allowing us to reproduce Figure 9.1.

REFERENCES

Chenthamarakshan, V., Padigel, U.M., Ramaprasad, P. Reddy, M.V.V. and Harinath, B.C. (1993). Diagnostic utility of fractionated urinary filarial antigen. *Journal of Bioscience*, 18: 319–326. https://www.bing.com/search?FORM=U523DF&PC=U523&q=Diagnostic+utility+of+fractionated+urinary+filarial+antigen&PC=U316&FORM=CHROMN

Ghosh, S.K., and Srivastava, P.K. (2020). A new outlook in lymphatic filariasis elimination in India. In: *Public Health in Developing Countries – Challenges and Opportunities*. IntechOpen, UK. https://www.intechopen.com/books/p.arasitology-and-microbiology-research/a-new-outlook-in-lymp.atic-filariasis-elimination-in-india

Indian Council of Medical Research (2023). Expression of Interest (EoI) for Joint collaboration for development of vaccine against Lymphatic filariasis (LF). https://www.icmr.gov.in

Kumar, P., and Babulal, K. (2024). Automated detection of microfilariae parasite in blood smear using OCR-NURBS image segmentation. *Multimedia Tools Applications*, 83: 63571–63591. https://link.springer.com/article/10.1007/s11042-023-17958-3, https://doi.org/10.1007/s11042-023-17958-3

Lim, P.K.C., Mak, J.W., Yong, H.S., and Dhaliwal, S.S. (1984). Immunodiagnosis of Brugian filariasis: Comparative efficiency of the indirect fluorescent antibody test and the enzymelinked immunosorbent assay. *Tropical Biomedine*, 1: 125–131. https://www.researchgate.net/publication/234002342_Lim_PKC_Mak_JW_Yong_HS_Dhaliwal_SS_1984_Immunodiagnosis_of_Brugian_filariasis_Comparative_efficiency_of_the_indirect_fluorescent_test_IFAT_and_the_enzyme-linked_immunosorbent_assay_ELISA_Trop_Biomed_1

Mallawaarachchi, S., Premalal, G. V. A., Wimalana, K. W. S. S., Liyanage, A. S., Samarasinghe, S., and Nanayakkara, N. D. (2013). Detection of microfilariae in peripheral blood smears using image analysis. *2013 IEEE 8th International Conference on Industrial and Information Systems*, Peradeniya, Sri Lanka (pp. 300–303). doi: 10.1109/ICIInfS.2013.6731999

Nancy, K., Lillian, W, Wilkinson, M., Claire, M., Luna, K., Dorcas, W., Rosemary, G., Japheth, L., Christine, I., Emily, W. and Jim, K. (2021). Optimization of a Loop-Mediated Isothermal Amplification Assay as a Point-of-Care Tool for the Detection of *Wuchereria bancrofti* in Human Blood in Tana River Delta, Kenya. *Journal of Parasitological Research*, 21:6650870. doi: 10.1155/2021/6650870

Nuchprayoon, S. (2009). DNA-based diagnosis of lymphatic filariasis. *Southeast Asian Journal of Tropical Medicine & Public Health*, 40(5), 904–913. https://pubmed.ncbi.nlm.nih.gov/19842372/

Thanomsub, B.W., Chansiri, K., Sarataphan, N, and Phantana, S. (2000). Differential diagnosis of human lymphatic filariasis using PCR-RFLP. *Molecular and Cellular Probes*, 14(1), 41–46. doi: 10.1006/mcpr.1999.0283

Tripathi, B., Roy, N., and Dhingra, N. (2022). Introduction of triple-drug therapy for accelerating lymphatic filariasis elimination in India: Lessons learned. *American Journal of Tropical Medicine & Hygiene*, 106(5), 29–38.

Pastor, A.F., Silva, M.R., Dos Santos, W.J.T., Rego, T., Brandão, E., de-Melo-Neto, O.P. and Rocha, A. (2021). Recombinant antigens used as diagnostic tools for lymphatic filariasis. *Parasites & Vectors*, 14(1),474. doi: 10.1186/s13071-021-04980-3

Vinkeles Melchers, N.V.S., Coffeng, L.E., de Vlas, S.J. and Stolk W.A. (2020). Standardization of lymphatic filariasis microfilaraemia prevalence estimates based on different diagnostic methods: A systematic review and meta-analysis. *Parasites & Vectors*, 13(1), 302. doi: 10.1186/s13071-020-04144-9.

WHO (1987). *Control of lymphatic filariasis: A manual for health personnel*. World Health Organization, Geneva. https://iris.who.int/handle/10665/39356

WHO (2023). *Lymphatic Filariasis*. World Health Organization, Geneva. https://www.who.int/news-room/fact-sheets/detail/lymphatic-filariasis

10 Arboviruses

R. Balasubramanian
ICMR-National Institute of Virology-Kerala Unit, Alappuzha, India

A.B. Sudeep Balan
Ex-ICMR-National Institute of Virology, Microbial Containment Complex, Pashan, Pune, India

Mangesh Gokhale
Ex-Entomology Group, ICMR-National Institute of Virology, Pune, India

B.K. Tyagi
Chandigarh University, Mohali, India

10.1 INTRODUCTION

According to the World Health Organization (WHO) report, vector-borne pathogens account for more than 17% of all infectious diseases. Vector-borne diseases (VBD) are widespread in our country. These diseases are transmitted by the bite of infected haematophagous arthropods, such as mosquitoes, ticks, sand flies, and mites. Arboviruses that are native to India have widened their geographic range and increased the frequency of human outbreaks over the past few decades. As the most common virus transmitted by mosquitoes in India, dengue is now endemic over the whole nation. In addition, with fresh importation from endemic areas, there occurs the autochthonous spread of additional arboviruses, including chikungunya and Zika viruses. Dengue, a mosquito-borne viral infection is regarded as a major public health problem globally. *Aedes aegypti* and *Aedes albopictus* are the primary vectors of the flavivirus dengue virus (DENV), which has multiple serotypes (DENV 1–4). Dengue fever, dengue haemorrhagic fever, and dengue shock syndrome can all result from DENV infection. Dengue illness poses a risk to over 2.5 billion people worldwide, with 390 million cases being reported annually in most subtropical and tropical areas. There are currently no treatments for dengue infection, and even the most cutting-edge vaccination has little use. DENVs are, however, amongst the most difficult arboviruses to be isolated and propagated in vitro. Every year, around 100 million people become sick, requiring medical attention and around 22,000 people die due to the DENV infection globally (CDC. About Dengue: What You Need to Know. 2020. Available online: https://www.cdc.gov/dengue/about/index.html).

Each year, large numbers of dengue cases among children occur in many places coinciding with monsoon season and period of high vector prevalence. The sudden onset of dengue epidemics can be greatly mitigated using rapid and reliable diagnostic methods for the control and management of this disease. A definite diagnosis, which depends on the isolation of the virus, can be made only in the laboratory by detecting viral antigen. Until recently, detection of the virus relied solely on viral isolations, however, current procedures can detect DENVRNA and specific virus antigens in the different tissues. Surveillance of mosquitoes infected with DENVs can help monitor the infection rates within vector mosquito population harbouring specific serotype and provides an early warning sign to predict an impending epidemic (Lorono-Pino et al., 1999). Some workers have recently advocated use

DOI: 10.1201/9781003327028-12

of virus infection rates in vector species, as well as the density of vectors in developing early warning tool (Tewari et al., 2004). This fact underscores the importance of vector mosquitoes in being employed to yield quick and reproducible results without any ethical implications.

An accurate laboratory diagnosis in mosquito is very helpful in controlling this disease. An early diagnosis in mosquitoes before disease coming to human allows epidemiologist to track variation of disease transmission risk spatially and temporally. It is crucial for the implementation of appropriate vector control activities which help to prevent the spread of the disease infection. Additionally, early detection in mosquitoes provides key data on the epidemiology risk which is very useful for accurate public health surveillance. Currently, various types of sensitive, specific and rapid diagnostic tools are available, each having its own advantages and disadvantages, and can be effectively used to find out the infected mosquitoes under a given set of laboratory conditions, albeit extensive cross reactions among the flaviviruses obfuscating identification of the particular serotype. The best method for diagnosis is considered the one that is rapid, specific and inexpensive. Obviously, there is no one 'best' method for all arboviruses; therefore, a mélange of approaches fulfils the requirement. The present review brings together the available information on diagnostic methods employed for this purpose on vector mosquitoes and the relative significance of each technique to yield dependable results. This will help to get a reliable estimate of the dengue infection in the mosquito vectors that can be included in the dengue surveillance programmes in the future.

10.2 VIRUS ISOLATION

Virus isolation is a basic, reliable, and widely used method for diagnosing many viral diseases and remains the gold standard for viral diagnosis. Arboviruses can be cultured and isolated from many specimens, including mosquito specimens and various tissue types. No single isolation system is adequate for all arboviruses. Identification of the infecting DENV serotype depends upon the isolation of the virus in a sensitive host system followed by serotype identification. A successful isolation of the virus followed by serotyping usually takes more than two weeks which procrastinates the inevitable vector control under epidemic situations. DENVs do not grow well in vitro normally. However, at present, some more sensitive isolation systems comprising certain insect species such as non-bloodsucking *Toxorhynchites* mosquitoes are available wherein the inoculation and isolation of DENVs are successfully accomplished. Several types of techniques are available to isolate arboviruses in multiple host systems, including those described for flaviviruses. *Aedes albopictus, Ae. aegypti*, and Toxorhynchites are mosquito habitats. Subsequently, an Ae- derived C6/36 cell line was developed. Vero cell lines derived from humans and African green monkey kidneys were commonly available cell cultures for virus isolation. Different cell lines have different success rates for virus isolation.

10.2.1 Suckling Mice

All the four DENVs have been successfully isolated in BS-C-1 cells (African green monkey kidney cells) or 1- to 3-day old baby mice using a homogenate prepared from Ae. aegypti (Singhraj et al., 1966; Carey et al., 1964). Baby mice are very insensitive before inculcating evidence of infection (Lorono-Pino et al., 1999). In spite of this, suckling mice are important, as it is generally not possible to detect the virus in other animal host body (e.g., mosquitoes) when in low quantity. Mice are inoculated intracranially with classified suspensions of clinical specimens or macerated arthropod pools or animal tissues. Since the suckling mice are readily available in all laboratories and have certain practical advantages over others, the supernatant of the mosquito homogenate after centrifugation is inoculated intracerebrally into suckling mice for virus isolation.

Dengue serotypes 1 and 4 were isolated from Ae. aegypti in 1961 from Vellore, in Tamil Nadu State, by inoculating infant mice (Carey et al., 1966). In Singapore from 1960 to 1961, six strains of DEN-2 serotype were isolated, five from pools of Ae. aegypti and one from a pool of Ae. albopictus.

The DENV infection rates based on minimum virus isolation rates per 1000 were 18.6 for Ae. aegypti, and 0.8 for Ae. albopictus. Majorities of the virus isolates were adapted to infant mice with difficulty and required many serial passages before illness appeared (Rudnick, 1966). From 1962 to 1963, nearly 200 isolates of DENVs were recovered from human and arthropod materials in Thailand by passage in suckling mice and/or tissue culture (Udomsakdi et al., 1966). All the five pools of Ae. aegypti mosquitoes from Rajasthan were positive in mice (Ghosh et al., 1974). Using mouse inoculation test, many investigators were unsuccessful in isolating virus from human cases during epidemics in India (Chakraborty et al., 1980; Mohan Rao et al., 1983; Mathur et al., 1982). However, using the same system, virus isolations were made from about 24% (Rashmi et al., 1990) and 30% of patients (Mathur et al., 1990). Prasad Rao et al. (1982) reported only 3% isolation rate during the 1978 outbreak (Prasad Rao et al., 1982). One advantage of using baby mice is that other arboviruses that cause dengue like illness may be isolated with this system (Gubler, 1998). Virus isolation using suckling mice is time-consuming, slow, and expensive. Moreover, because of the low sensitivity of this method many viruses cannot be isolated with baby mice. Those that are isolated frequently require numerous passages to the adapt the virus growth in mice. This method is no longer recommended for isolation of DENVs (Gubler et al., 1984).

10.2.2 Mosquito Cell Cultures

The choice depends on the availability of a host-cell cultures or mice that serve an indicator of virus infection, i.e., cytopathic effects in cell cultures, sign of illness or death in mice. This is the most common method for virus isolation. Grace (1966) reported the establishment of the first mosquito cell line in the world from Ae. aegypti mosquitoes. Singh (1967) established two cell lines at the National Institute of Virology (NIV) Pune, from the larval tissues of *Ae. aegypti* and *Ae. albopictus*. The first Culex cell line in the country was established from the embryonic tissue of *Cx. bitaeniorhynchus* mosquitoes (Pant and Dhanda, 1980). Cell lines from other two species of Culicine mosquitoes viz., *Cx. infula* and *Cx. ambiguour*, which are members of *Cx. bitaeniorhynchus* complex, were also established subsequently (Pant and Dhanda, 1985). These cell lines also supported the multiplication of other arboviruses of public health importance in India, viz., chikungunya, dengue and sindbis viruses (Pant et al., 1982). A new cell line from the forest mosquito *Ae. krombeini* was established in the early 1990s, which proved highly sensitive to several arboviruses (Pant et al., 1992). It was extremely sensitive to Japanese encephalitis (JE) and DENVs, and a very low titre of these viruses could be detected using the indirect immunofluorescent technique (IFA) while *Cx. bitaeniorhynchus* and C6/36 cell lines could not detect these viruses when infected with the same dilution (Sudeep and Pant, 1999). Recently, a new cell line from the embryonic tissue of *Cx. tritaeniorhynchus* mosquitoes was established. Immunofluorescence antibody (IFA) technique is now routinely being used in the detection of virus antigen in infected cell lines and mosquitoes (Sudeep et al., 1997; Ilkal et al., 1984). Mosquito cell cultures particularly C6/36 (Ae. albopictus), AP61 (*Ae. pseudoscutellaris*), *Ae. krombeini*, TR-248 (*Toxorhynchites amboinensis*), and other established mammalian cell culture lines (LLC-MK2 cells), are common methods for virus isolation (Pant et al., 1992; Igarashi, 1978; Race et al., 1979; Kuno et al., 1985; Maneekarn et al., 1993; Guzman and Kouri, 1996). These are not only the most widely used but also provide a relatively simple and economical method for DENV isolation and assay (Tesh, 1979). The development of direct plaque assay using LLC-MK2 cells provided a faster and more sensitive method but still had the disadvantage of requiring the adoption of the virus to the cell culture (Yull et al., 1968). Several continuous mosquito cell lines have been shown to be highly susceptible to DENV infection. The C6/36 clone of Ae. albopictus cells was chosen for virus isolation because they demonstrated high sensitivity to DENV infection and ease of handling (Igarashi, 1978; Tesh, 1979). These cell lines are highly stable and have optimal growth at lower temperatures than mammalian cells. The virus titre can be determined in all groups of mosquitoes using BHK-21 cells (Morens et al., 1985). In Peninsular Malaysia, virus isolation was carried out using cell culture (C6/36 clone) of Ae. albopictus and detection of DENV by peroxidase staining. All positive isolations were further confirmed by the reverse transcriptasepolymerase chain reaction (RT-PCR) (Ahmad et al., 1997).

For DENV isolation mosquito cell cultures have proved to be more sensitive than mice or mammalian cell culture systems. Some prefer to use cytopathic effect to detect infection especially with AP-61 cells. This has the disadvantage of not producing cytopathic effects (rounding, refraction of light, detachment from the substrate) and requires a secondary step for recognizing the presence of the virus in the culture. Intrathoracic inoculation of *Toxorhynchites* mosquitoes (which do not take blood meals) or Aedes mosquitoes has also been used. In routine diagnosis, the C6/36 cell lines have become the most widely used. Following a period of 7–10 days post-inoculation in cell lines or 14 days post-inoculation in mosquitoes, virus identification is done by immunofluorescence assay with serotype-specific monoclonal antibodies (Guzman and Kouri, 1996; Henchal et al., 1983; Vorndam and Kuno, 1977). Viral isolation rates were obtained up to 36%with C6/36 cell lines or up to 80% by direct mosquito inoculation (Vorndam and Kuno, 1977; Kuno et al., 1985). A rapid centrifugation assay for DENV improved the isolation rate, even in tissue samples (Rodriguez et al., 2000). IFA has become an important tool in the detection of virus antigen with its simplicity and specificity (Gubler and Trent, 1994). Several virus isolations were made in mosquito cell lines from field-collected mosquitoes at NIV, Pune (Dhanda et al., 1989; Mourya et al., 1989; Naik et al., 1990). The *Cx. bitaeniorhynchus* cell line for JE virus and Ae. albopictus cell line or C6/36 cells31 for DEN viruses are mainly being used in the isolation of these viruses from mosquitoes collected during epidemic periods. Ravi et al. (1989) reported isolation rates of 3.7%using *Ae. albopictus* (Ravi et al., 1989) and 7.5%using *Ae. pseudoscutellaris* cell cultures (Burke et al., 1985).

A new cell line from *Cx. tritaeniorhynchus* supported the multiplication of JE and West Nile viruses but not of any of the dengue serotypes (Athawale et al., 2002). Due to the presence of certain cytoplasmic inclusions in the cells (Sreenivasan et al., 1994), this cell line was not used for further research especially for the isolation of viruses from field-collected mosquitoes (Sudeep et al., 2005). However, this system was less sensitive than mosquito inoculation (Gubler and Sather, 1988). The sensitivity of the mosquito cell lines may also vary with the strain of the virus. In samples from an epidemic in Mozambique, more than twice as many DEN-3 viruses were isolated by mosquito inoculation than by using mosquito cells (Gubler et al., 1986). None of the standard laboratory animals or mammalian cell cultures are sufficiently susceptible to DENV infection to use for routine DENV isolation (Gubler, 1987). Although the use of this method continues in some laboratories, it is not recommended (Vorndam and Kuno, 1977; Gubler and Sather, 1988).

10.2.3 Mosquito Inoculation

Mosquito inoculation techniques are reported for detection and viral amplification (Dhanda et al., 1989; Mourya et al., 1989; Kuberski and Rosen, 1977; Gubler et al., 1985; Gubler et al., 1979; Thet-Win, 1982; Lam et al., 1986). This technique provided for the first time a sensitive method for isolation and assay of DENVs. *Tx. splendens*, a non-haematophagous mosquito, was evaluated as a bioassay host for the detection and propagation of DENVs. All DENV serotypes and strains attained titres in *Tx. splendens* comparable to those observed for two strains of *Ae. aegypti*. Peak virus titres occurred in *Tx. splendens* approximately 6 days post-inoculation; however, specific fluorescence for all viruses was not observed in 100%of mosquito heads until 12 days post-inoculation. A 100% correlation was noted between specific fluorescence in *Tx. splendens* heads and the recovery of the virus from the corresponding thorax, and abdomens. The volume of inoculum tolerated by *Tx. spendens* was approximately five times greater than that injected into *Ae. aegypti*. The overall survival rate for *Tx.splendens* following intrathoracic inoculation with DENVs was higher compared to *Ae. aegypti*. These findings imply that *Tx. splendens* would be more efficient than *Ae. aegypti* as a laboratory assay host for detecting DENVs in the blood of infected patients and for use in experimental investigations (Rosen and Shroyer, 1985).

The mosquito inoculation technique was used for alltitrations (Rosen and Gubler, 1974). After incubation for ten days at 32°C, individual mosquitoes were examined for the presence or absence of viral antigen in the salivary glands and brain tissue by the direct IFA. Mourya (1990) demonstrated fluorescence 24 h after intracerebral inoculation of *Tx. splendens* larvae. This is a simple technique to determine DENV infection in the mosquito (Kuberski and Rosen, 1977). Variations of the mosquito inoculation technique include inoculation of adult (Rosen and Gubler, 1974; Gubler and Rosen, 1976) and larval Toxorhynchites mosquitoes (Gajanana et al., 1995; Padbidri et al., 1996). Isolation of viruses by intracerebral inoculation of the fourth instars of Tx. splendens larvae is routinely followed for dengue diagnosis (Mourya et al., 1989; Thet-Win, 1982; Lam et al., 1986; Thu et al., 1998; Mitchel and Miller, 1990; Rosen et al., 1985; Freier and Grimstad, 1983). The viruses are detected by indirect immunofluorescence (IIFA) using a type-specific dengue monoclonal antibody (Mourya, 1990; Thenmozhi et al., 2000). Mosquito inoculation techniques are sensitive for isolation of flaviviruses (Gajanana et al., 1995). Inoculation of Tx. splendens larvae is relatively simple and safe and have been employed for isolation of DENV (Lam et al., 1986) and JE virus from field-caught mosquitoes (Gajanana et al., 1997).

Both the inoculation techniques have similar sensitivity to intrathoracic inoculation of adult mosquitoes but are more difficult and labour-intensive. The main advantage is that viruses can be isolated in a few days. Inoculation of samples directly into mosquitoes, especially adult or larval inoculation in *Tx. amboinensis* and *Tx. splendens* is the best isolation system in terms of sensitivity (Gubler and Rosen, 1976; Rosen et al., 1985). NIV, Pune and Centre for Research in Medical Entomology(CRME), Madurai have developed necessary expertise in this discipline (Tewari et al., 2004; Gajanana et al., 1995; Gajanana et al., 1997; Mahadev et al., 1993; Mahadev et al., 1997; Philip Samuel et al., 1998a, 1998b; 2000; Thenmozhi et al., 2001; Arunachalam et al., 2002; Arunachalam et al., 2004; Das et al., 2004; Thenmozhi et al., 2005; Dhanda et al., 1997). Detection of virus antigen is a promising tool for surveillance. Distinct advantage is that it can be performed under field conditions (Burke et al., 1985). With the availability of a fluorescence microscope objective attachment to a standard laboratory microscope for field use (Polsuwan et al., 1992; Makler, 1992), it should now be possible to perform the virus antigen detection test even in peripheral laboratories. Insect bioassays are cumbersome and special facilities are required (Scott et al., 1987). It has the disadvantage of being labour-intensive and requires rearing and maintenance of mosquitoes (Gajanana et al., 1995), the expertise for which is not available in most virology laboratories. The mosquito inoculation technique has the disadvantages of being labour-intensive and requiring an insectary to produce large numbers of mosquitoes for inoculation. Also, unless strict safety precautions are maintained, the chance of laboratory infections increase, although this risk can be eliminated by using male Aedes mosquitoes or non-biting *Toxorhynchites* species for inoculation (Rosen and Gubler, 1974). However, the possibility of laboratory acquired infections through the bite of mosquitoes should be borne in mind and, therefore, rigid safety precautions must be followed in utilizing this technique. The use of non-biting mosquitoes (Watt et al., 1982) can eliminate the risk of laboratory infection through infective bite, but the slow breeding and protracted life history of this mosquito make it rather difficult to get them insufficiently large numbers for experimental purposes. Unfortunately, this method is not available in most endemic countries.

10.3 DIRECT VISUALIZATION (DETECTION) OF VIRAL ANTIGENS IN FIELD-COLLECTED MOSQUITO VECTORS OF DENGUE

10.3.1 Detection of Antigen by Enzyme Linked Immunosorbant Assay (ELISA)

ELISA is the most widely used in routine practice (Kuno et al., 1991; Miagostovich et al., 1999; Nawa et al., 2001). DENVs are difficult to isolate and propagate, as they do not grow well in laboratory animals or cultures of mammalian cells. Although there are very sensitive methods to isolate DENVs in mosquito cell lines (Tesh, 1979) and mosquitoes (Rosen and Gubler, 1974), they are time-consuming, labour-intensive and expensive (Kuno et al., 1985; Gubler and Rosen, 1976). Isolation and recovery of arboviruses from field-collected mosquitoes can be most efficiently accomplished by pooling adult specimens prior to testing in various in vivo and invitro assay systems (Broom et al., 1998). Such pooling of larvae has also become a routine practice for evaluating the role of transovarial transmission in the ecology of arboviruses (Thenmozhi et al., 2001; Arunachalam et al., 2002; Arunachalam et al., 2004). It is sometimes important to detect virus in individual mosquitoes, although this is more labour-intensive than detection in pooled specimens. Use of individual specimens greatly reduces the chance of mistakes caused by introducing a misidentified specimen into a large pool. More importantly, the evaluation of individual mosquitoes can generate a precise estimate of vector infection rate. This parameter, combined with the survival rate and human biting rate, can be used to make a meaningful estimation of local transmission risk (Sithiprasna et al., 1994).

Direct detection of a virus can provide a rapid diagnosis. In the epidemic situations, the early detection will help to devise appropriate public health measures. Therefore the more rapid, sensitive and specific diagnostic methods would be better in situations for which time is a critical element. In order to develop dengue-specific ELISA, dengue serotype-specific monoclonal antibody (MABs) that captures only dengue DEN1-DEN4 was used. By this method studies undertaken in the rural dengue surveillance in Vellore district and dengue epidemic in Chennai, south India, were investigated. In Vellore, pools of mosquitoes of adult *Ae. aegypti* and *Ae. albopictus* were tested by ELISA and found positive for flavivirus antigen. By employing this ELISA and/or Toxo-IFA system the natural vertical transmission of DENVs in Ae. aegypti in Vellore district was also investigated and DEN-2virus was identified in Vellore (Tewari et al., 2004; Thenmozhi et al., 2000). ELISA can be used as an inexpensive way to screen large number of mosquito specimens with relatively little effort (Tewari et al., 2004; Gajanana et al., 1995; Gajanana et al., 1997; Philip Samuel et al., 1998a, 1998b; 2000; Arunachalam et al., 2002; Arunachalam et al., 2004; Das et al., 2004; Thenmozhi et al., 2005; Dhanda et al., 1997; Tsai et al., 1987; Scott et al., 1987; Murthy et al., 2000). Detection of virus infection in wild-caught Aedes species should form an essential component of a surveillance system for dengue. ELISA technique is shown to be a sensitive alternative to insect bioassay for monitoring arboviruses in wild mosquito population. This technique will be useful in DENV surveillance for monitoring the DENV activity in endemic areas and also to develop an early warning to plan control strategies (Philip Samuel et al., 2000). Thus ELISA-based methods using specific MABs can also lead to definite diagnosis (Kuno et al., 1985; Broom et al., 1998; Tsai et al., 1987; Hunderkar et al., 2002; Knox et al., 2003). The antigen capture ELISA for virus detection is the most useful procedure currently available and it is widely recommended for virological surveillance. During routine surveillance in our study, initially some 600 pools were tested by Toxo-IFA and those found positives were confirmed by ELISA. This procedure took nearly 1 yr. After adopting the revised strategy, in about 6 months, over 4,000 pools were first screened by ELISA and more than 70 per cent of about 300 pools were examined by Toxo-IFA also (Gajanana et al., 1995).

10.3.2 Detection of Antigens by Immunofluorescence

Identification of DENV isolates by using monoclonal antibodies in an indirect immunofluorescence assay was developed (Henchal et al., 1983; Henchal et al., 1982). The method of choice for DENV identification is IFA with serotype-specific monoclonal antibodies produced in tissue culture or mouse ascitic fluids and an anti-mouse immunoglobulin G-fluoresce in isothiocyanate conjugate. This test can be easily performed with infected cell cultures, mosquito brain or tissue squashes, mouse brain squashes, or even on formalin-fixed tissues embedded in paraffin and sectioned for histopathological testing. It is a simple and reliable method which that allows the detection of multiple viruses in patients with concurrent infections with more than one serotype.

In those laboratories that do not have immunofluorescence capability, DENVs can also be identified using monoclonal antibodies in an antigen capture ELISA (Kuno et al., 1985; Gubler and Sather, 1988).

These methods usually take two or more weeks, delaying after the laboratory diagnosis for initiating the control operation.

10.3.3 Immune Electron Microscopy

Its advantages include rapidity and lack of requirement for specific reagents. High viral density in liquids is required. Only small tissue areas can be examined by this sectioning; however, techniques have been described for enhancing chances for virus detection. The most difficult task in diagnosing viruses by electron microscopy is to determine whether unusual structure are indeed viruses rather than spherical structures and membrane debris in negative stains and normal cellular organelles in thin sections. Electron microscopy can be an important adjunct to other methods for virus identification but possibly cannot be relied on its own. Recently structural identity of the Chandipura virus was confirmed by electron microscopy from a large encephalitis outbreak in children in many districts of Andhra Pradesh (Rao et al., 2004).

Polymerase chain reaction (PCR) has been applied to dengue diagnosis with sera, tissue from fatal cases, mosquito pools, infected cell cultures, and mosquito larvae, mosquito surveillance and genetic strain characterization. Several PCR protocols for dengue detection have been described that vary in the RNA extraction methods, genomic location of primers, specificity, and sensitivity, and the methods to detect PCR products and to determine the serotype105-108. Reverse transcriptase (RT)-PCR has provided one of the most important steps in the molecular diagnosis of DENV (Maneekarn et al., 1993). A rapid assay has been developed followed by a second PCR with specific primers allowing serotype identification (DNA products) of different sizes according to the dengue serotype obtained (Lanciotti et al., 1992).

10.3.3.1 Polymerase Chain Reaction

Nucleic acid detection tests for dengue diagnosis were first developed over 20 years ago and are widely used in research worldwide. Diagnosis of DENV was made from mosquitoes during dengue epidemics when viremia levels are high, and viral RNA can be detected by cell culture or mouse inoculation methods. Recently, several RT-PCR assays, including the increasingly popular real-time polymerase chain reaction, have been developed for DENV detection and serotyping with high sensitivity and specificity in mosquitoes. As an automated RNA extraction protocol, these tests have enabled more rapid and accurate diagnosis of mosquito DENV. However, as RT-PCR assays became more sensitive, faster and more affordable, it became easier to diagnose dengue from single mosquito sample collected during epidemics.

RT-PCR has been developed for a number of RNA viruses, including DENVs (Sudiro et al., 1997). Unlike most other techniques which require screening of pools of mosquitoes to detect viruses, RT-PCR carries out the job with solitary specimen. The technique allows for the multi-fold biological amplification of viral nucleic acid and has been used to rapidly diagnose viral diseases (Maneekarn et al., 1993; Saiki et al., 1988; Henchal et al., 1991; Morita et al., 1991). The primary advantage of this molecular tool lies in the speed at which specimens can be screened for the presence of DENVs and also by its highly sensitive and specific detection. It is able to monitor the infection rate in mosquitoes, both adults and larvae, with a high degree of precision (Chan et al., 1994; Rohani et al., 1997). RT-PCR can also detect small quantities of virus (Fonseca and Fonseca, 2002). This method has also been employed for detecting and typing DENV RNA in the field-caught Aedes mosquitoes, besides determining the infection rate in local Aedes mosquitoes (Pinheir et al., 2005). Pinheir et al. (2005) have developed a single RT-PCR followed by a semi-nested PCR using an upstream consensus primer and four type-specific primers within the non-structural protein gene (NS3) of DENVs to type DENVs in field populations of female Aedes mosquitoes. Alternatively, RT-PCR offers the potential for the rapid, highly sensitive and specific detection of DENVs up to the serotype level.

In the dengue-sensitive areas in Singapore, a rapid and sensitive, semi-nested, RT-PCR assay using non-structural protein three gene primers for the type-specific detection of DENVs in artificially infected and in field-collected adult Aedes mosquitoes were employed (Chow et al., 1998). In laboratory experiments, the assay was sensitive enough to detect one virus-infected mosquito head in pools of up to 59 uninfected heads. Use of RT-PCR in epidemiological investigations has been intensified recently as geographic locations of the virus-infected Aedes mosquitoes, detected as early as six weeks before the start of the dengue outbreaks, were traced to have a correlation with the residence or workplaces of patients. Virologic surveillance using RT-PCR for detecting dengue virus-infected Aedes mosquitoes in the field may serve as an additional early warning monitoring system for predicting dengue outbreaks (Chow et al., 1998).

Since a change in serotype is particularly important to bring about a surge in DF/DHF cases, RT-PCR is further advantageous in detecting the specific serotype circulating in natural populations of mosquitoes (Armstrong and Rico-Hesse, 2003). The PCR can be designed to be broadly cross-reacting or extremely specific. The extraction techniques used to isolate RNA destroy the infectivity of the virus, therefore, the risk of exposure to bio-hazardous materials is eliminated once the RNA is extracted from the sample. Despite its advantages, the PCR is not routinely employed in arbovirus diagnostic laboratories because it is expensive. Improvements in automated handling of PCR as well as detection of product are currently being developed but are not available to most laboratories performing arbovirus diagnosis. PCR is notoriously prone to contamination. Every batch of assays of mosquitoes should accompany a no-DNA (negative control) sample, and stringent precautions must be taken to avoid carryover contamination (Kow et al., 2001). PCR is used for quick detection, and it requires a small quantity of the sample.

Although RT-PCR has a similar sensitivity to virus isolation systems that use C6/36 cell cultures, poor handling, storage, and time of sample collection and quantum of virus present in the samples greatly affect the results. A number of methods involving primers from different locations in the genome and different approaches to detect the RT-PCR products have been developed over the past several years (Deubel, 1997; Guzman and Kouri, 1996; Vorndam and Kuno, 1977). Since RT-PCR is highly sensitive to amplicon contamination, without proper controls, false-positive results may occur.

10.3.3.2 Hybridization Probes for Detection of Viral Nucleic Acid

Recent advances in molecular technology such as RT-PCR and hybridization, nucleic acid probes are made possible for the detection of viral genomic materials in mosquitoes. The hybridization techniques are used for the detection of many important arboviruses. Hybridization probe method detects viral nucleic acids with cloned hybridization probes (Henchal et al., 1987). Probes with variable specificity ranging from dengue complex to serotype-specific can be constructed depending on the genome sequences used. A PCR-amplified serotype-specific complementary DNA (cDNA) was cloned and used as a non-radioactive nucleic acid hybridization probe in subsequent assays (Deubel et al., 1990). Hybridization techniques are rapid, specific and relatively simple especially good for the diagnosis of viruses that are difficult to cultivate. Using degenerate primers, RT-PCR forms a sensitive and specific method for the detection of DENVs in clinical specimens. The use of radio labeled serotype probes allows for an additional level of specificity (96–100% specific), unavailable in assays using RT/PCR alone (89%–100%; Henchal et al., 1991). Despite several advantages, hybridization techniques have certain drawbacks like lack of sensitivity. Moreover, a labeled probe specific for each virus to be detected is required.

Preliminary data suggest (Gubler et al., 1981) that this method is less sensitive than RT-PCR, but like PCR the outcome of the test is not influenced by the presence of inhibitory substances. Difficulties encountered during working with RNA and the technical expertise required to obtaining reproducible results make this method more suitable as a research tool than as a routine diagnostic test (Gubler, 1998). Moreover, in this technique, as a pre-requisite, viraemia levels should also correlate with mosquito infection rates.

One of the top concerns for preventative techniques and tracking viral circulation to manage endemic conditions is routine surveillance of possible vector populations. However, in endemic locations, especially those with significant transmission, arbovirus testing or surveillance in field-caught mosquitoes is still not common practice. Several investigations (Grard et al., 2014, Guerbois et al., 2016, Diaz-Quinonez et al., 2016, Faye et al., 2013, Ayllon et al., 2017, Smartt et al., 2017) reported the discovery of arboviral RNA in mosquitoes using conventional RT-PCR or real-time RT-PCR. The prompt application of insecticides after cases were reported, difficulty in approaching homes with patients, and the delay between diagnosing illness in people and conducting field investigations have all been suggested as explanations for the lack of arbovirus detection from wild mosquitoes (Diallo and Diallo, 2016). Additionally, it can be challenging to store and handle RNA in endemic areas since it is more fragile and prone to degradation than DNA. Arboviruses have RNA genomes, so future monitoring of arbovirus in wild mosquitoes may call for skilled personnel, RNA storage devices, and RNA handling equipment.

A newly identified virus-derived DNA (vDNA) in the life cycle of an arthropod-infecting RNA virus was first discovered in Drosophila melanogaster infected with flock house virus (FHV; Goicetal., 2013). Evidence for the emergence of vDNA in other arbovirus infections in mosquitoes has also been reported from studies using conventional PCR (Goicetal, 2016; Nagand Kramer, 2017). Since the presence of vDNA may indicate the presence of RNA viruses, vDNA may be a good target candidate for molecular surveillance due to its superior stability.

A loop-mediated isothermal amplification (LAMP) method was developed to detect vDNA instead of viral RNA in arbovirus-infected mosquitoes. LAMP is a simple and rapid DNA detection method using BstDNA polymer as it allows DNA amplification with high sensitivity and specificity under other malconditions (Notomietal, 2000). It has been used to detect malaria parasites in anthropogenically infected mosquitoes and filariae in both anthropogenically infected and wild-borne mosquitoes (Aonumaetal, 2008, 2009). LAMP recognized her ZIKV vDNA from both cultured cells and vector mosquitoes, leading to the development to fvDNA-targeted LAMP as a potential method for field surveillance of arbovirus-carrying mosquitoes.

This vDNA-LAMP does not require RNA storage or manipulation and can be easily applied to field studies of viral infection in wild mosquitoes. Further experiments with wild mosquito populations are needed to confirm whether vDNA-LAMP is a viable alternative to traditional surveillance methods. The introduction of vDNA-LAMP in endemic areas may be a promising strategy for monitoring vector-mediated viral circulation (Aonumaetal, 2022).

10.4 EMERGING NOVEL TECHNOLOGIES FOR POINT-OF-CARE VIRUS DETECTION

The increased incidence of vector-borne arboviral diseases has prompted further efforts to develop rapid and reliable diagnostic techniques for use in the field for the detection of pathogens in vectors. In the recent past a rapid emergence of novel technologies has been witnessed. These methods can reduce test consumables and duration alike. Currently, there is a need for point-of-care (POC) technologies that can be cheaper and reliable in various health care sectors. The RT-LAMP assay, first developed by Notomi in 2000 consists of four to six primers, this method has been developed as a promising pathogen detection technique that uses only water bath or a simple heating block to amplify the viral target at a constant temperature. Usually, the RT-LAMP starts transcripting the target to DNA, followed by creating a loop structure that creates copies of the target. The amplified target can be visualized by the naked eye or under UV light (Kim et al., 2018). The RT-LAMP assay they designed showed a 100% and 98.9% success rate for diagnosing clinical patients. Targeting the C-prM gene of all DENV serotypes, Li et al. have reported a single tube reaction method utilizing the RT-LAMP primers within 30 min. Li et al. (2011) also published a single tube RT-LAMP method focusing

30-NCR gene of all DENV serotypes and detected the DENV serum by analysing the change in turbidity, with LOD of 10 RNA copies per reaction (Lau et al., 2015). RT-LAMP has many advantages, such as it is rapid, cost-effective, isothermal, highly sensitive and specific. However, in the case of multiplexing and viral quantification applications, PCR is superior to LAMP-based assays (Wong et al., 2018).

LAMP is a technique of nucleic acid amplification (Notomi et al., 2000) based on *Bst* DNA polymerase-mediated strand displacement. DNA synthesis occurs at isothermal conditions and relatively low temperatures (60°C–65°C) so that there is no need for a thermal cycler. This method has been applied recently for the diagnosis of several human pathogenic viruses in addition to the rapid identification of microbial infections. Furthermore, LAMP is being used increasingly by various investigators for rapid detection and typing of emerging arthropod-borne diseases such as dengue, West Nile, ZIKA, chikungunya, JE (Kurosaki et al., 2017; Aonuma et al., 2022; Silva et al., 2024; Rafferty et al., 2024; Kim et al., 2024; Lamb et al., 2018). The LAMP reaction can be achieved using six primers thus providing a higher specificity to the reaction than conventional PCR methods. One of the main drawbacks of conventional PCR methods is the inhibition of the reaction by mosquito debris and other biological components (Aonuma et al., 2022); however, it has been reported that LAMP reactions are resistant to such inhibitory materials (Perera et al., 2009) consistent with the observation of successful RT-LAMP detection of FHV from a mixture of mosquito debris and DNA. Another advantage of using LAMP is based on the fact that the amplification from stem-loop structures leads to the accumulation of large amounts of products of various lengths, ultimately making the detection of amplified DNA much easier including the by-product of the reaction, magnesium pyrophosphate, a white-coloured precipitate seen easily by the naked eye (Silva et al., 2024). Thus, these features of the LAMP reaction would be of value especially in emergency cases such as the management of an outbreak of viral diseases.

10.5 DISCUSSION AND CONCLUSION

Reliable estimation of natural mosquito infection with arboviruses forms a key element in any surveillance system and is essential for vector incrimination and monitoring control measures (Gajanana et al., 1995). A laboratory diagnosis of dengue infection can be accomplished by detecting either the virus or anti-dengue antibodies. Methods selected for virus isolation depend much upon the laboratory facilities available. In the case of virus isolation, the baby mice virus isolation method is very time-consuming, slow and expensive. Moreover, because of the low sensitivity of the method, many wild-type viruses cannot be isolated. Those that are isolated frequently require numerous passages to adapt the viruses to grow in mice (Gubler, 1988). This method is no longer recommended for the isolation of DENVs.

Mosquito cell cultures are recent addition to DENV isolation methodologies. Three cell lines of comparable sensitivity are most frequently used. The C6/36 clone of Ae. albopictus cells is less sensitive than the mosquito inoculation method. The use of these cell lines has provided a rapid, sensitive and economical method for DENV isolation. The sensitivity of mosquito cell lines may vary with the strain of the virus. Even though cell cultures are less sensitive than mosquito inoculation, large number of samples can be processed in a relatively short time. Mammalian cell cultures have many of the same disadvantages as baby mice for isolation of DENVs, such as being expensive, slow, and intensive.

Mosquito inoculation is a method sensitive enough for routine successful virologic confirmation of dengue. Moreover, there are many endemic DENV strains that can be recovered only by this method. Four mosquito species have been used for virus isolation, viz., *Ae. aegypti, Ae. albopictus, Tx. amboinensis*, and *Tx. splendens*. Male and female mosquitoes are equally susceptible. A recent variation on this method involves intracerebral inoculation of larval and intrathoracic inoculation of adult *Toxorhynchities* mosquitoes. However, intracerebral inoculation did not provide any advantage over intrathoracic inoculation method since it was rapid. The mosquito inoculation technique has the disadvantage of being labour-intensive and requires an insectary to produce large numbers of mosquitoes for inoculation. The possibility of laboratory acquired infections through the bite is possible, and therefore, safety precautions must be followed in utilizing this technique (Ilkal et al., 1984). Unless rigid safety precautions are maintained, the chance of laboratory infections increases, although this risk can be eliminated by using male Aedes mosquitoes or non-biting *Toxorhynchites* species for inoculation (Watt et al., 1982). Conventional methods, i.e., culture of virus in suckling mice and the use of micro inoculation of *Toxorhynchites* adults or larvae and subsequent detection with fluorescent antibody staining, for detecting virus in the vector, are laborious and time-consuming (Ahmad et al., 1997). Virus detection in the mosquito, regardless of the species, is generally performed by the IIFA test on mosquito tissues, usually brain or salivary glands (Henchal et al., 1983; Kuberski et al., 1977; Rosen et al., 1985; Waterman and Monath, 1982). Another recent advance is the development of specific monoclonal antibodies for the identification of DENVs (Tesh, 1979; Gentry et al., 1982). For large-scale surveillance on wild-caught mosquito vectors of dengue to estimate true positives, the ELISA technique can be used as a convenient and rapid system for screening a large number of samples (Peiris et al., 1992). For the diagnosis up to serotype level, if the quantity of the sample is sufficient, serotype-specific antibodies can be used for ELISA.

In comparison to the foregoing description of various techniques employed in detection/isolation of DENVs, the RT-PCR method is rapid, highly sensitive, simple, reliable, and reproducible for the specific detection of DENVs. It can be effectively used to detect viral RNA in mosquitoes (Chan et al., 1994).The improvements in RT-PCR technology are being continuously improvised, which make it even more useful in its application for DENV detection. When the sample quantity is very less or insufficient all the positives can be propagated by mosquito inoculation technique/tissue culture/RT-PCR for further virus isolation studies.

DENV does not grow well in any of the laboratory animals or mammalian cell cultures normally used in virology laboratories. DENVs were first isolated by inoculation of baby mice. This system is very insensitive, and several blind passages are usually required to allow adaptation of the virus. The development of the mosquito inoculation technique provided, for the first time a sensitive and relatively rapid method for the isolation and assay of DENVs. The IFA test provides a simple technique to determine DENV infection in the mosquitoes, but it requires expertise to maintain a mosquito colony, which is not available in many laboratories. This can be done using mosquito cell lines that are less labour-intensive. Even though mosquito cell lines are highly susceptible to DENV infection, the techniques are less sensitive than the mosquito inoculation technique for routine isolation of DENVs. *Toxorhynchites* inoculation technique followed by IFA is very sensitive for the detection of infection in wild-caught vector mosquitoes and gives an estimate of true positives. However, it is time-consuming and cumbersome for testing a large number of samples. The ELISA test on the other hand is a convenient and rapid system and can be used for large-scale screening of vectors. Further ELISA positives should be again screened by the Toxo-IFA system. By this new strategy, a large number of pools can be screened within a short time. The cost and labour involved are greatly reduced. Thus utilization of these two techniques will give a better estimate of mosquito infection rates and help to forecast the associated risk of a dengue epidemic. Other methods can be employed for propagation of these viruses and can be used for virus isolation later. According to the laboratory facility available, certain technical improvements can be made on the diagnostic procedures which can also be standardized and selected for the routine surveillance use. This system allows the DENVs being transmitted in an area to be monitored with a minimal amount of effort and provides the early warning capability necessary to predict epidemic dengue. On-site mosquito monitoring in endemic areas is likely prevented by these regulations, so the development of a quicker and simpler approach will be necessary for successful surveillance.

REFERENCES

Ahmad, R., Ismail, A., Saat, Z., Lim, L.H. (1997). Detection of dengue virus from field *Aedes aegypti* and *Aedes albopictus* adults and larvae. *Southeast Asian Journal of Tropical Medicine and Public Health*, 28, 138–42.

Aonuma, H., Iizuka, I., LiJ.-C., Ote, M., Tajima, S., Saijo, M., et al. (2022). LAMP detection of virus-derived dna of Zika virus in vector mosquito. *Frontiers in Tropical Disease*, 3, 759375.

Aonuma, H., Suzuki, M., Iseki, H., Perera, N., Nelson, B., Igarashi, I., et al. (2008). Rapid identification of plasmodium-carrying mosquitoes using loop-mediated isothermal amplification. *Biochemical and Biophysical Research Communications*, 376, 671–6.

Aonuma, H., Yoshimura, A., Perera, N., Shinzawa, N., Bando, H., Oshiro, S., et al. (2009). Loop-mediated isothermal amplification applied to filarial parasites detection in the mosquito vectors: *Dirofilaria immitis* as a study model. *Parasites and Vectors*, 2, 15.

Armstrong, P.M., Rico-Hesse, R. (2003). Efficacy of dengue serotype 2 virus strains to infect and disseminate in *Aedes aegypti. American Journal of Tropical Medicine and Hygeine*, 68, 539–44.

Arunachalam, N., Murthy, S., Kabilan, L., Balasubramanian, A., Thenmozhi, V., Narahari, D., et al. (2004). Studies on dengue in rural area of Kurnool district in Andhra Pradesh. *Journal of American Mosquito Control Association*, 20, 87–90.

Arunachalam, N., Philip Samuel, P., Hiriyan, J., Thenmozhi, V., Balasubramanian, A., Gajanana, A., et al. (2002). Vertical transmission of Japanese encephalitis virus in *Mansonia* species, in an epidemic-prone area of southern India. *Annals of Tropical Medicine and Parasitology*, *96*, 419–20.

Athawale, S.S., Sudeep, A.B., Barde, P.V., Jadi, R.S., Pant, U., Mishra, A.C., et al. (2002). A new cell line from the embryonic tissues of *Culex tritaeniorhynchus* and its susceptibility to certain flaviviruses. *Acta Virologica*, 46, 237–40.

Ayllon, T., Campos, R.M., Brasil, P., Morone, F.C., Câmara, D.C.P., Meira, G.L.S., et al. (2017). Early evidence for Zika virus circulation among *Aedes aegypti* mosquitoes, Rio De Janeiro, Brazil. *Emerging Infectious Diseases*, 23, 1411–2.

Broom, A.K., Hall, R.A., Johansen, C.A., Oliveria, N., Howard, M.A., Lindsay, M.D., et al. (1998). Identification of Australian arboviruses in inoculated cell cultures using monoclonal antibodies I ELISA. *Pathology*, 30, 286–8.

Burke, D.S., Lorsomrudee, W., Leake, C.J. (1985). Fatal outcome in Japanese encephalitis. *American Journal of Tropical Medicine and Hygiene*, 85(34), 1203–10.

Carey, D.E., Myers, R.M., Reuben, R., Rodrigues, F.M. (1966). Studies on dengue in Vellore South India. *Bulletin of the World Health* Organization, 35, 61.

Carey, D.E., Myers, R.M., Reuben, R. (1964). Dengue types 1 and 4 viruses in wild caught mosquitoes of south India. *Science*, 143, 131–2.

Chakraborty, M.S., Chakraborty, S.K., Mukherjee, K.K., Mitra, A.C., Mitra, K.K., Gupta, B., et al. (1980). Recurrent Japanese encephalitis epidemics in West Bengal. *Indian Journal of Medical Research*, 72, 1–6.

Chan, S.Y., Kautner, I.M., Lam, S.K. (1994). Detection and serotyping of dengue viruses by PCR: a simple, rapid method for the isolation of viral RNA from infected mosquito larvae. *Southeast Asian Journal of Tropical Medicine and Public Health*, 25, 258–61.

Chow, V.T., Chan, Y.C., Yong, R., Lee, K.M., Lim, L.K., Chung, Y.K., et al. (1998). Monitoring of dengue viruses in field-caught *Aedes aegypti* and *Aedes albopictus* mosquitoes by type specific polymerase chain reaction and cycle sequencing. *American Journal of Tropical Medicine and Hygeine*, 58, 578–86.

Das, B.P., Kabilan, L., Sharma, S.N., Lal, S., Regu, K., Saxena, V.K. (2004). Detection of dengue virus in wild caught *Aedes albopictus* (Skuse) around Kozhikode Airport, Malappuram district, Kerala, India. *Dengue Bulletin*, 28, 210–2.

Deubel, V., Laille, M., Hugnot, J.P., Chungue, E., Guesdon, J.L., Drouet, M.T., et al.Identification of dengue sequences by genomic amplification: rapid diagnosis of dengue virus serotypes in peripheral blood. *Journal of Virological Methods*, 1990, 30, 41–54.

Deubel, V. (1997). The contribution of molecular techniques to the diagnosis of dengue infection. In: Gubler DJ, KunoG, editors. *Dengue and dengue hemorrhagic fever*. London, United Kingdom: CAB International, 335–6.

Dhanda, V., Mourya, D.T., Mishra, A.C., Ilkal, M.A., Pant, U., Jacob, P.G., et al. (1989). Japanese encephalitis virus infection in mosquitoes reared from field collected immatures and in wild caught males. *American Journal of Tropical Medicine and Hygiene*, 41, 732–6.

Dhanda, V., Thenmozhi, V., Kumar, N.P., Hiriyan, J., Arunachalam, N., Balasubramanian, A., et al. (1997). Virus isolation from wild-caught mosquitoes during a Japanese encephalitis outbreak in Kerala in 1996. *Indian Journal of Medical Research*, 106, 4–6.

Diallo, D., Diallo, M. (2016). Why is Zika virus so rarely detected during outbreaks and how can detection be improved? *BMC Research Notes*, 10, 524.

Diaz-Quinonez, J.A., López-Martínez, I., Torres-Longoria, B., Vázquez-Pichardo, M., Cruz-Ramírez, E., Ramírez-González, J.E., et al. (2016). Evidence of the presence of the Zika virus in Mexico since early 2015. *Virus Genes*, 52, 855–7.

Faye, O., Faye, O., Diallo, D., Diallo, M., Weidmann, M., Sall, A.A. (2013). Quantitative real-time PCR detection of Zika virus and evaluation with field-caught mosquitoes. *Virology Journal*, 10, 311.

Freier, J.E., Grimstad, R.P. (1983). Transmission of dengue virus by orally infected *Aedes triseriatus*. *American Journal of Tropical Medicine and Hygiene*, *32*, 1429–34.

Gajanana, A., Rajendran, R., Philip Samuel, P., Thenmozhi, V.,Tsai, T.F., Kuroda, J.K., et al. (1997). Japanese encephalitis in SouthArcot District, Tamil Nadu, India: A three-year longitudinal study of vector abundance and infection frequency. *Journal of Medical Entomology*, 34, 651–9.

Gajanana, A., Rajendran, R., Thenmozhi, V., Philip Samuel, P., Tsai, T.F., Reuben, R. (1995). Comparative evaluation of bioassay and ELISA for detection of Japanese encephalitis virus in field collected mosquitoes. *Southeast Asian Journal of Tropical Medicine and Public Health*, 26, 91–7.

Gentry, M.K., Henchal, E.A., McCown, J.M., Brandt, W.E., Dalrymple, J.M. (1982). Identification of distinct determinants on dengue 2 virus using monoclonal antibodies. *American Journal of Tropical Medicine and* Hygiene, 31, 548–55.

Ghosh, S.N., Pavri, K.M., Singh, K.R.P., Sheikh, B.H., D'Lima, L.V., Mahadev, P.V., et al. (1974). Investigations on the outbreak of dengue fever in Ajmer city, Rajasthan state in1969. Part I. Epidemiological, clinical and virological study of the epidemic. *Indian Journal of Medical Research*, 62, 511–22.

Goic, B., Stapleford, K.A., Frangeul, L., Doucet, A.J., Gausson, V., Blanc, H., et al. (2016). Virus-derived DNA drives mosquito vector tolerance to arboviral infection. *National Communications*, 7, 12410.

Goic, B., Vodovar, N., Mondotte, J.A., Monot, C., Frangeul, L., Blanc, H., et al. (2013). RNA mediated interference and reverse transcription control the persistence of RNA viruses in the insect model drosophila. *National Immunology*, 14, 396– 403.

Grace, T.D.C. (1966). Establishment of a line of mosquito (*Aedesaegypti*, L) cells grown *in vitro*. *Nature*, 211, 366–7.

Grard, G., Caron, M., Mombo, I.M., Nkoghe, D., Mboui Ondo, S., Jiolle, D., et al. (2014). Zika virus in Gabon (Central Africa) - 2007: A new threat from *Aedes albopictus*?*PloS Neglected Tropical Diseases*, 8, e2681.

Gubler, D., Trent, D.W. (1994). Emergence of epidemic dengue/dengue hemorrhagic fever as a public health problem in the Americas. *Infectious Agents and Diseases*, 2, 383–93.

Gubler, D.J., Kuno, G., Sather, G.E., Valez, M., Oliver, A. (1984). Mosquito cell cultures and specific monoclonal antibodies in surveillance for dengue viruses. *American Journal of Tropical Medicine and Hygiene*, 33, 158–65.

Gubler, D.J., Reed, D., Rosen, L., Hitchcock, J.C.Jr. (1985). Epidemiologic, clinical and virologic observations on dengue in the Kingdom of Tonga. *American Journal of Tropical Medicine and Hygiene*, 27, 581–9.

Gubler, D.J., Rosen, L.A. (1976). Simple technique for demonstrating transmission of dengue virus by mosquitoes without the use of vertebrate hosts. *American Journal of Tropical Medicine and Hygiene*, 25, 146–50.

Gubler, D.J., Sather, G.E., Kuno, G., Cabral, J.R. (1986). Dengue 3 virus transmission in Africa. *American Journal of Tropical Medicine and Hygiene*, 35, 1280–4.

Gubler, D.J., Sather, G.E. (1988). Laboratory diagnosis of dengueand dengue hemorrhagic fever. In: HommaA, CunhaJF,editors. *Proceedings of the international symposium on yellow fever and dengue*. Riode Janeiro, Brazil, May 15–19, 201–302.

Gubler, D.J., Suharyono, W., Lubis, I., Eram, S., Saroso, J.S. (1979). Epidemic dengue hemorrhagic fever in rural Indonesia I.Virological and epidemiological studies. *American Journal of Tropical Medicine and Hygiene*, 28, 701–10.

Gubler, D.J., Suharyono, W., Tan, R., Abidin, M., Sie, A. (1981). Viraemia in patients with naturally acquired dengue infections. *Bulletin of World Health Organization*, 59, 623–30.

Gubler, D.J. (1987). Current research on dengue. In: Kerry F. Haris, editor. *Current topics in vector research.* New York:Springer-Verlag, 337–56.

Gubler, D.J. (1998). Dengue and dengue hemorrhagic fever. *Clinical Microbiology Reviews*, 11, 480–96.

Gubler, D.J. (1988). Dengue. In: MonathTP, editor. *Epidemiology of arthropod-borne viral diseases*. Boca Raton, Fla: CRC Press, 223–60.

Guerbois, M., Fernandez-Salas, I., Azar, S.R., Danis-Lozano, R., Alpuche-Aranda, C.M., Leal, G., et al. (2016). Outbreak of Zika virus infection, Chiapas State, Mexico, 2015, and first confirmed transmission by *Aedes Aegypti* mosquitoes in the Americas. *Journal of Infectious Diseases*, 214, 1349–56.

Guzman, M.G., Kouri, G. (1996). Advances in dengue diagnosis. *Clinical and Diagnostic Laboratory Immunology*, 3, 621–7.

Henchal, E.A., Gentry, M.K., Mc Cown, J.M., Brand, W.E. (1982). Dengue virus specific and flavivirus group determinants identified with monoclonal antibodies by indirect immunofluorescence. *American Journal of Tropical Medicine and Hygiene*, 31, 830–6.

Henchal, E.A., Mc Cown, J.M., Seguin, M.C., Gentry, M.K., Brandt, W.E. (1983). Rapid identification of dengue virus isolates by using monoclonal antibodies in an indirect immunofluorescence assay. *American Journal of Tropical Medicine and Hygiene*, 32, 164–9.

Henchal, E.A., Narupiti, S., Feighny, R., Padmanabhan, R., Vakharia, V. (1987). Detection of dengue virus RNA using nucleic acid hybridization. *Journal of Virological Methods*, 15, 187–200.

Henchal, E.A., Polo, S.L., Yaemsiri, V.V.C., Hoke, C.H. (1991). Sensitivity and specificity of a universal primer set for the rapid diagnosis of dengue virus infections by polymerase chain reaction and nucleic acid hybridization. *American Journal of Tropical Medicine and Hygiene*, 45, 418–28.

Hunderkar, S.L., Thakare, J.P., Gokhle, M.D., Barde, P.V., Argade, S.V., Mourya, D.T. (2002). Development of monoclonal antibody based antigen capture ELISA to detect chikungunya virus antigen in mosquitoes. *Indian Journal of Medical Research*, 115, 144–8.

Igarashi, A. (1978). Isolation of a Singh's *Aedes albopictus* cellclone sensitive to dengue and chikungunya viruses. *Journal of General* Virology, 40, 531–44.

Ilkal, M.A., Dhanda, V., Rodrigues, J.J., Mohan Rao, C.V.R., Mourya, D.T. (1984). Xenodiagnosis of laboratory acquired infectionby mosquito inoculation and immune fluorescence. *Indian Journal of Medical Research*, 79, 587–90.

Kim, D., DeBriere, T.J., Eastmond, B.H., Alomar, A.A., Yaren, O., McCarter, J., Bradley, K.M., Benner, S.A., Alto, B.W., Burkett-Cadena, N.D. (2024). Rapid detection of West Nile and Dengue viruses from mosquito saliva by loop-mediated isothermal amplification and displaced probes. *PLoS One*, 19(2), e0298805. https://doi.org/10.1371/journal.pone.0298805

Kim, J.G., Baek, S.H., Kim, S., Kim, H.I., Lee, S.W., Phan, L.M.T., Kailasa, S.K., Park, T.J. (2018). Rapid discriminative detection of dengue viruses via loop mediated isothermal amplification. *Talanta*, 190, 391–96.

Knox, T.B., Kay, B.H., Hall, R.A., Ryan, P.A. (2003). Enhanced vector competence of *Aedes aegypti* (Diptera: Culicidae) from the Torres strait compared with mainland Australia for Dengue 2 and 4 viruses. *Journal of Medical Entomology*, 40, 950–6.

Kow, C.Y., Koon, L.L., Yin, P.F. (2001). Detection of dengue viruses in field caught male *Aedes aegypti* and *Aedes albopictus* (Diptera: Culicidae) in Singapore by type-specific PCR. *Journal of Medical Entomology*, 38, 475–9.

Kuberski, T.T., Rosen, L. (1977). A simple technique for the detection of dengue antigen in mosquitoes by immune fluoresence. *American Journal of Tropical Medicine and Hygiene*, 26, 533–7.

Kuno, G., Gomez, I., Gubler, D.J. (1991). An ELISA procedure for the diagnosis of dengue infections. *Journal of Virological Methods*, 33, 101–3.

Kuno, G., Gubler, D.J., Santiago de Well, N. (1985). Antigen capture ELISA for the identification of dengue viruses. *Journal of Virological Methods*, 12, 93–103.

Kurosaki, Y., Martins, D.B.G., Kimura, M., Catena, A.D.S., Borba, M.A.C.S., Mattos, S.D.S., Abe, H., Yoshikawa, R., de Lima Filho, J.L., Yasuda, J. (2017). Development and evaluation of a rapid molecular diagnostic test for Zika virus infection by reverse transcription loop-mediated isothermal amplification. *Scientific Reports*, 7, 13503. https://doi.org/10.1038/s41598-017-13836-9

Lam, S.K., Chew, C.B., Poon, G.K., Ramalingam, S., Seow, S.C., Pang, T. (1986). Isolation of dengue viruses by intracerebral inoculation of mosquito larvae. *Journal of Virological Methods*, 14,133–40.

Lamb, L.E., Bartolone, S.N., Tree, M.O., Conway, M.J., Rossignol, J., Smith, C.P., Chancellor, M.B. (2018). Rapid detection of Zika virus in urine samples and infected mosquitos by reverse transcription-loop-mediated isothermal amplification. *Scientific Reports*, 8, 3803.

Lanciotti, R.S., Calisher, C.H., Gubler, D.J., Chang, G.J., Vorndam, A.V. (1992). Rapid detection and typing of dengue viruses from clinical samples using reverse transcriptase-polymerase chain reaction. *Journal of Clinical Microbiology*, 30, 545–51.

Lau, Y.L., Lai, M.Y., Teoh, B.T., Abd-Jamil, J., Johari, J., Sam, S.S., Tan, K.K., AbuBakar, S. (2015). Colorimetric Detection of Dengue by Single Tube Reverse-Transcription-Loop-Mediated Isothermal Amplification. *PLoS ONE*, 10, e0138694.

Li, S., Fang, M., Zhou, B., Ni, H., Shen, Q., Zhang, H., Han, Y., Yin, J., Chang, W., Xu, G., et al. (2011). Simultaneous detection and differentiation of dengue virus serotypes 1–4, Japanese encephalitis virus, and West Nile virus by a combined reverse-transcription loop-mediated isothermal amplification assay. *Virology Journal*, 8, 360.

Lorono-Pino, M.A., Cropp, C.B., Farfan, J.A., Vorndam, A.V., Rodriguez-Angulo, E.M., Rosado-Paredes, E.P., et al. (1999). Common occurrence of concurrent infections by multipledengue virus serotypes. *American Journal of Tropical Medicine and Hygiene*, 61, 725–30.

Mahadev, P.V.M., Prasad, S.R., Ilkal, M.A., Mavale, M.S., Bedekar, S.S., Banerjee, K. (1997). Activity of Dengue -2 virus and prevalenceof *Aedes aegypti* in the Chirimiri colliery area, Madhya Pradesh, India. *Southeast Asian Journal of Tropical Medicine and Public Health*, 28, 126–37.

Makler, M.T. (1992). Fluorescent microscope objective. *Transactions of the Royal Society of Tropical Medicine and Hygiene*, 86, 108.

Maneekarn, N., Morita, K., Tanaka, M., Ugarashi, A., Usawattanakul, W., Srisanthana, V., et al. (1993). Application of polymerase chain reaction for identification of dengue virus isolated from patient sera. *Microbiology and* Immunology, 37, 41–7.

Mathur, A., Chaturvedi, U.C., Tandon, H.O., Agarwal, A.K., Mathur, G.P., Nag, D., et al. (1982). Japanese encephalitis epidemic inUttar Pradesh, India during 1978. *Indian Journal of Medicine and Research*, 75, 161–9.

Mathur, A., Rashmi, K., Sandhya, S. (1990). Rapid diagnosis of Japanese encephalitis by immune fluorescent examination of cerebrospinal fluid. *Indian Journal of Medical Research*, 19, 1–4.

Mitchel, C.J., Miller, B.R. (1990). Vertical transmission of dengue viruses by strains of *Aedes albopictus* recently introduced into Brazil. *Journal of the American Mosquito Control Association*, 6, 251–3.

Mohan Rao, C.V.R., Prasad, S.R., Rodrigues, J.J., Sharma, N.G.K., Shaikh, B.H., Pavri, K.M., et al. (1983). The first laboratory provenoutbreak of Japanese encephalitis in Goa. *Indian Journal of Medical Research*, 78, 745–50.

Morens, M.D., Halstead, S.B., Repik, M.P. (1985). Simplified plaque reduction neutralization assay for dengue viruses by semi-micro methods in BHK-21 cells. Comparison of BHK suspension tests with standard plaque reduction neutralization. *Journal of Clinical Microbiology*, 22, 205–14.

Morita, K., Tanka, M., Igarashi, A. (1991). Rapid identification of dengue virus serotype by polymerase chain reaction. *Journal of Clinical Microbiology*, 29, 2107–10.

Mourya, D.T., Ilkal, M.A., Mishra, A.C., George, J.P., Pant, U., Ramanujam, S., et al. (1989). Isolation of Japanese encephalitis virusfrom mosquitoes collected in Karnataka state, India, from1985 to 1987. *Transactions of the Royal Society of Tropical Medicine and Hygiene*, 83, 550–2.

Mourya, D.T. (1990). Rapid detection of Japanese encephalitis virus by immunofluorescence after intra cerebral inoculation of mosquito larvae. *Transactions of the Royal Society of Tropical Medicine and Hygiene*, 84, 580.

Murthy, U.S., Singh, T.G., Arunachalam, A., Philip Samuel, P. (2000). Epidemiology of Japanese encephalitis in Andhra Pradesh, India- a brief overview. *Tropical Biomedicine I*, 17, 97–102.

Nag, D.K., Kramer, L.D. (2017). Patchy DNA Forms of the Zika Virus RNA Genome Are Generated Following Infection in Mosquito Cell Cultures and in Mosquitoes. *Journal of General Virology*, 98, 2731–7.

Notomi, T., Okayama, H., Masubuchi, H., Yonekawa, T., Watanabe, K., Amino, N., et al. (2000). Loop-Mediated Isothermal Amplification of DNA. *Nucleic Acid Research*, 28, E63.

Padbidri, V.S., Mahadev, P.V.M., Thakare, J.P., Pant, U., Ilkal, M.A., Geevarghese, G., et al. (1996). Virological and entomological investigations of an outbreak of dengue fever in Dhule district, Maharashtra. *Indian Journal of Medical Microbiology*, 14, 5–32.

Pant, U., Banerjee, K., Athawale, S.A., Dhanda, V. (1982). Susceptibilityof *Culex bitaeniorhynchus* cell line to some arboviruses. *Indian Journal of Medical Research, 76*, 789–94.

Pant, U., Dhanda, V. (1985). Arthropod tissue culture and its uses inarboviral research. *ICMR Bulletin*, 16, 33–8.

Pant, U., Dhanda, V. (1980). Establishment of a cell line from *Culex bitaeniorhynchus. Journal of Tissue Culture Methods*, 6, 61–3.

Pant, U., Sudeep, A.B., Dhanda, V., Mourya, D.T., Banerjee, K. (1992). New embryonic cell line from *Aedes krombeini* (H) (Diptera: Culicidae). *In Vitro Cellular and Developmental Biology (Animal)*, 28, 567–8.

Peiris, J.S.M., Amerasinghe, F.P., Amerasinghe, P.H., Ratnayake, C.B., Karunaratne, S.H.P., Tsai, T.F. (1992). Japanese encephalitis in Sri Lanka- a study of an epidemic: vector incrimination, porcine infection an human disease. *American Journal of Tropical Medicine and Hygiene*, 86, 307–13.

Perera, N., Aonuma, H., Yoshimura, A., Teramoto, T., Iseki, H., Nelson, B., Igarashi, I., Yagi, T., Fukumoto, S., Kanuka, H. (2009). Rapid identification of virus-carrying mosquitoes using reverse transcription-loop-mediated isothermal amplification. *Journal of Virological Methods*, 156(1–2), 32–6.

Philip Samuel, P., Hiriyan, J., Gajanana, A. (2000). Japanese encephalitis virus infection in mosquitoes and itsepidemiological implications. *ICMR Bulletin*, 30, 37–43.
Philip Samuel, P., Hiriyan, J., Thenmozhi, V., Balasubramanian, A.A. (1998a). Note on first isolation of Japanese encephalitis virus from *Culex infula* Theobald (Diptera:Culicidae). *Journal of Communicable Diseases*, 30, 199–200.
Philip Samuel, P., Hiriyan, J., Thenmozhi, V., Balasubramanian, A. (1998b). A system for studying vector competence of mosquitoesfor Japanese encephalitis virus. *Indian Journal of Malariology*, 35, 146–50.
Pinheir, V.C., Tadei, W.P., Barros, P.M., Vascocelos, P.F., Cruz, A.C. (2005). Detection of dengue virus serotype 3 by reverse transcription-polymerase chain reaction in *Aedes aegypti* Diptera-Culicidae captured in Manaus, Amazonas. *Memorias do Instituto Oswaldo Cruz*, 100, 833–9.
Polsuwan, C., Lumlertdaccha, B., Tepsumethanon, W., Wilde, H. (1992). Using the UV paralens adapter on a standard laboratory microscope for fluorescent rabies antibody detection. *Transactions of the Royal Society of Tropical Medicine and Hygiene*, 86, 107.
Prasad Rao, G.L.N., Rodrigues, F.M., Nambiapan, M., Nagarajan, M., Ghalsasi, G.R., Rodrigues, J.J., et al. (1982). Aetiology of the 1978 outbreak of encephalitis in Tirunelveli and other district of Tamil Nadu. *Indian Journal of Medical Research*, 76, 36–46.
Race, M.W., Williams, M.C., Agostini, C.F.M. (1979). Dengue in the Caribbean: virus isolation in a mosquito (*Aedes pseudoscutellaris*) cell line. *Transactions of the Royal Society of Tropical Medicine and Hygiene*, 73, 18–22.
Rafferty, C., Raise, G., Scaife, J., et al. (2024). Loop-mediated isothermal amplification assay to detect invasive malaria vector *Anopheles stephensi* mosquitoes. *Emerging Infectious Diseases*, 30(9), 1779–89. https://doi.org/10.3201/eid3009.240444
Rao, B.L., Basu, A., Wairagkar, N.S., Gore, M.M., Arankalle, V.A., Thakare, J.P., et al. (2004, September 4). A large outbreak of acute encephalitis with high fatality rate in children in Andhara Pradesh, India, in 2003, associated with Chandipura virus. *The Lancet*, 364, 869–74.
Rashmi, K., Asha, M., Arvind, K. (1990). Clinical features andprognostic indicators of Japanese encephalitis in childrenin Lucknow (India). *Indian Journal of Medical Research, 91*, 321–7.
Ravi, V., Vanajakshi, S., Gowda, A., Chandramuki, A. (1989). Laboratory diagnosis of Japanese encephalitis usingmonoclonal antibodies and correlation of findings with the outcome. *Journal of Medical Virology*, 29, 221–3.
Rodriguez, R., Alvarez, M., Guzman, M.G., Morier, L., Kouri, G. (2000). Isolation of dengue 2 in C6–36/HT cells by rapid centrifugation/shell vial assay. Comparison with conventional virus isolation method. *Journal of Clinical Microbiology*, 38, 3508–10.
Rohani, A., Ismail, A., Saat, Z., Lee, H.L. (1997). Detection of dengue virus from field *Aedes aegypti* and *Aedes albopictus* adults and larvae. *Southeast Asian Journal of Tropical Medicine and Public Health*, 28, 138–42.
Rosen, L., Gubler, D.J. (1974). The use of mosquitoes to detect and propagate dengue viruses. *American Journal of Tropical Medicine and Hygiene*, 23, 1153–60.
Rosen, L., Roseboom, L.E., Gubler, D.J., Lien, J.C., Chaniotis, B. (1985). Comparative susceptibility of mosquito species and strains to oral and parenteral infection with dengue and Japanese Samuel & Tyagi: detection & isolation of dengue virus in mosquitoes. *The American Journal of Tropical Medicine and Hygiene*, 34, 627.
Rosen, L., Shroyer, D. A. (1985). Comparative susceptibility of five species of Toxorhynchites mosquitoes to parenteral infection with dengue and other flaviviruses. *The American Journal of Tropical Medicine and Hygiene*, *34*(4), 805–809.
Rudnick, A. (1966). Dengue viruses isolated from mosquitoes in Singapore, 1960–61. *Bulletin of World Health Organization*, 35, 63.
Saiki, R.K., Gelfand, D.H., Stoffel, S., Scharf, S.J., Higuchi, R., Horn, G.T., et al. (1988). Primer directed enzymatic amplification of DNA with a thermostable DNA polymerase. *Science*, 239, 487–91.
Scott, T.W., Olson, J.G., Lewis, T.E., Carpenter, J.W., Lorenz, L.H., Lembeck, L.A., et al. (1987). A prospective field evaluation of an enzyme immunoassay. Detection of eastern equine encephalomyelitis virus antigen in pools of *Culiseta melanura. Journal of American Mosquito Control Association*, 3, 412–7.
Silva, S.J.R.D., Magalhães, J.J.F., Matthews, Q., Divarzak, A.L.L., Mendes, R.P.G., Santos, B.N.R., Cabral, D.G.A., Silva, J.B.D., Kohl, A., Pardee, K., Pena, L. (2024). Development and field validation of a reverse transcription loop-mediated isothermal amplification assay (RT-LAMP) for the rapid detection of chikungunya virus in patient and mosquito samples. *Clinical & Microbiological Infections*, 30, 810–15. https://doi.org/10.1016/j.cmi.2024.03.004

Singh, K.R.P. (1967). Cell cultures derived from larvae of *Aedes albopictus* (Skuse) and *Aedes aegypti* (L). *Current Science*, 36, 506–8.

Singhraj, P., Simasathien, P., Sukhavachana, P., Halstead, S.B., Scanlon, J.E. (1966). Recovery of dengue and other viruses inmice and tissue culture from Thai mosquitoes. *Bulletin of World Health Organization*, 35, 67–8.

Sithiprasna, R., Strickman, D., Innis, B.L., Linthicum, K.J. (1994). ELISA for detecting dengue and Japanese encephalitis viral antigen in mosquitoes. *Annals of Tropical Medicine and Parasitology*, 88, 397–404.

Smartt, C.T., Stenn, T.M.S., Chen, T.Y., Teixeira, M.G., Queiroz, E.P., Souza Dos Santos, L., et al. (2017). Evidence of Zika Virus RNA Fragments in Aedes Albopictus (Diptera: Culicidae) Field-Collected Eggs From Camaçari, Bahia, Brazil. *Journal of Medical Entomology*, 54, 1085–7.

Sreenivasan, M.A., Pant, U., Athawale, S., Arankalle, V. (1994). Detection by electron microscopy of endogenous viruses in *Aedes krombeini* (H) line. *In Vitro Cellular and Developmental Biology (Animal)*, 30, 142–4.

Sudeep, A.B., Mourya, D.T., Mishra, A.C. (2005). Insect cell culture in research: Indian scenario. *Indian Journal of Medical Research*, 121, 725–38.

Sudeep, A.B., Pant, U. (1999). *Aedes krombeini* (H) cell line- A new approach to arboviral research. Paper presented at the 4thinternational symposium on vectors and vector borne diseases. *Proceedings of the National Academy of Vector Borne Diseases, Bhubaneswar and Defence Research and Development Establishment, Gwalior*, p. 45.

Sudiro, T.M., Ishiko, H., Green, S., Vaughn, D.W., Nisalak, A., Kalayanarooj, S., et al. (1997). Rapid diagnosis of dengue viremia by reverse transcriptase- polymerase chain reaction using 3′-noncoding region universal primers. *American Journal of Tropical Medicine and Hygiene*, 56, 424–9.

Tesh, R.B. (1979). A method for the isolation and identification of dengue viruses using mosquito cell cultures. *American Journal of Tropical Medicine and Hygiene*, 28, 1053–9.

Tewari, S.C., Thenmozhi, V., Katholi, C.R., Manavalan, R., Munirathinam, A., Gajanana, A. (2004). Dengue vector prevalenceand virus infection in a rural area in South India. *Tropical Medicine and International Health*, 4, 499–507.

Thenmozhi, V., Kabilan, L., Philip Samuel, P., Dash, A.P. (2005). Detection of dengue virus antigens in desiccated mosquitoes: an improved surveillance system. *Tropical Medicine and International Health*, 10, 187–9.

Thenmozhi, V., Rajendran, R., Philip Samual, P., Hiriyan, J., Ayanar, K., Balasubramanian, A., et al. (2001). Natural vertical transmission of JE virus in south Indian mosquitoes. *Tropical Biomedicine*, 18, 19–27.

Thenmozhi, V., Tewari, S.C., Manavalan, R., Balasubramanian, A., Gajanana, A. (2000). Natural vertical transmission of dengue viruses in *Aedes aegypti* in southern India. *Transactions of the Royal Society of Tropical Medicine and Hygiene*, 94, 507.

Thet-Win (1982). Detection of dengue virus by immunofluorescence after intracerebral inoculation of mosquitoes. *Lancet*, i, 53–4.

Thu, H.M., Aye, K.M., Thein, S. (1998). The effect of temperature and humidity on dengue virus propagation in *Aedes aegypti* mosquitoes. *Southeast Asian Journal of Tropical Medicine*, 29, 280–4.

Tsai, T.F., Bolin, R.A., Montoya, M., Bailey, R.E., Francy, D.B., Jozen, M., et al. (1987). Detection of St. Louis encephalitis virus antigen in mosquitoes by capture enzyme immunoassay. *Journal of Clincal Microbiology*, 25, 370–6.

Udomsakdi, S., Nisalak, A., Halsted, S.B. (1966). Identification of dengue and chikungunya viruses. *Bulletin of World Health Organization*, 35, 68.

Vorndam, V., Kuno, G. (1977). Laboratory diagnosis of dengue virusinfections. In: GublerDJ, KunoG, editors. *Dengue and dengue haemorrhagic fever*. New York: CAB International, 313–33.

Waterman, S.H., Monath, T.P. (1982). Fluorescent antibody techniques applied to the identification of dengue virus in infected tissues. *Acta Virology*, 26, 367–81.

Watt, D.M., Harrison, B.A., Nisalak, A., Scott, R., Mc, N., Burke, D.S. (1982). Evaluation of *Toxorhynchites splendens* (Diptera: Culicidae) as a bioassay host for dengue virus. *Journal of Medical Entomology*, 19, 54–9.

Wong, Y.P., Othman, S., Lau, Y.L., Radu, S., Chee, H.Y. (2018). Loop-mediated isothermal amplification (LAMP): A versatile technique for detection of micro-organisms. *Journal of Applied Microbiology*, 124, 626–43.

Yull, T.M., Sukhavachana, P., Nisalak, A., Russell, P.K. (1968). Dengue virus recovery by direct and delayed plaques in LLC-MK2cells. *American Journal of Tropical Medicine and Hygiene*, 17, 441.

Section III

Control

11 Mosquito Control

Part I: Chemicals Including Repellents, Non-chemical, and Biological

Appadurai Daniel Reegan
National Centre for Disease Control, Karnataka, India
ICMR-Vector Control Research Centre, Puducherry, India

Prabhakargouda B. Patil
Maharashtra Hybrid Seeds Company Pvt. Ltd., Jalna, India

E. Pushpalatha
University of Calicut, Malappuram, India

T. Mariappan
Ex-Centre for Research in Medical Entomology (now Vector Control Research Centre), Madurai, India

S. K. Dasgupta
Ex-Tata Institute of Genetics and Society, Bengaluru, India

Arti Mathur
MLS University, Udaipur, India

Devinder Kaur Kocher
Punjab Agricultural University, Ludhiana, India

Minu Bharti
S. B. S. Government College, Dakshin Dinajpur, India

Prabhakar Mishra
REVA University, Bangalore, India

DOI: 10.1201/9781003327028-14

N. Chandrasekaran
Vellore Institute of Technology, Vellore, India

D. S. Suman
Zoological Survey of India, (Ministry of Environment, Forest & Climate Change, GoI), Kolkata, India

Vibha Ahuja
Biotech Consortium India Limited, New Delhi, India

S. Sridharan
Department of Public Health and Preventive Medicine, Chennai, India

Manas Sarkar
Reckitt Benckiser Ltd., Gurugram, India

C. Kandasamy
C. Kandasamy, Consultant – Agrochemicals, Chennai, India

S. K. Ghosh
Ex-ICMR-National Institute of Malaria Research, Bengaluru, India

M. J. Mendki
Naval Materials Research Laboratory (DRDO), Ambarnath, India

R. S. Sharma
Ex-National Centre for Disease Control, Delhi, India

Vijay Veer
Ex-DRDO-Defence Research Laboratory, Tezpur, India

Myra Mistica
University of the Philippines Manila, SEAMEO TROPMED Regional Centre for Public Health, Hospital Administration, Environmental and Occupational Health, Manila, Philippines

Lilian de las Llagas
University of the Philippines Manila, Philippines

Dhiraj Saha
University of North Bengal, Darjeeling, India

B.K. Tyagi
Chandigarh University, Mohali, India

11.1 INTRODUCTION

Mosquito control is an important component in the control of vector-borne diseases (VBD) and effective preventive measures to reduce or interrupt disease transmission. There are several different methods available for larval and adult control under integrated mosquito management (IMM). This is based on applying multiple methods like chemical, biological, physical, and environmental to control mosquito-borne diseases and to improve public health. James and Harwood (Herm's Textbook of Medical Entomology, 1969) stated that control programmes must have their objectives, the reduction of a vector population below a level that can maintain a disease. The term 'eradication' is always mentioned when control of VBD is being discussed. An example is malaria eradication versus malaria control. Eradication is total elimination of the vector in every breeding site of the vector. Control is a strategy based on preventing transmission and reducing outbreaks or epidemics of VBD. The objectives, methods and strategies are different for eradication and control programmes. Further definitions of vector control:

i. Vector control is the effort to reduce populations of the vectors or to reduce human contact with the vectors
ii. Vector control utilizes various control strategies that are available to reduce the population rise of vectors
iii. Any action that is deliberately initiated, maintained or intensified by humans to prevent the build-up of the vector population
iv. Vector control is suppression of target population by reducing its reproductive capacity, the gross reproductive rate (GRR) and the net reproductive rate (NRR)
v. Vector control is the application of measures that will protect humans from the infective bites or contacts with the vectors.
vi. Vector control is the effort to reduce populations of the vectors or to reduce human contact with the vectors.

The aims of vector control are primarily (1) to utilize the various techniques in regulating the reproductive potential of the vector to reduce the population of the vector below a level that can maintain a disease Critical Density Transmission Index (CTDI) and (2) to prevent an effective contact between the vector and the human host. There are major categories of techniques or measures or/interventions that can be used to respond to the two aims of vector control:

i. To reduce the GRR, which is a function of fecundity and longevity through (a) chemical application, use of adulticides to kill the females and males to shorten their lives and eventually all the mosquito requirements for reproduction and (b) genetic manipulation of the vectors or the use of genetically modified mosquitoes (GMM).

ii. To reduce the NRR, which are measures that prevent/control insect survival from egg stage to adult emergence. This is to regulate the vectors in breeding and establish their populations in their breeding habitats.

Since no single methodology for pest and vector control is 100% effective, foolproof, economical and, above all, environment friendly, a more comprehensive integrated vector management (IVM) strategy is exercised instead (Figure 11.1).

IVM is defined as a rational decision-making process to optimize the use of all available resources for vector control. It promotes various interventions, alone or in combination, selected on the basis of local knowledge about the vectors, diseases and disease determinants. The IVM approach, therefore, facilitates tackling the area under influence of single or multiple vectors simultaneously. The IVM is encouraged globally by the World Health Organization (WHO) as it ensures benefits to community through judicious and evidence-based use of chemicals as vector control measures. While promoting the combination of novel tools, IVM also advocates elimination of persistent organic pollutants (POPs), which have long run impacts on the environment. DDT is one of the POPs, and it's phasing out for indoor residual spray (IRS) under malaria control programmes and development of alternatives as per commitments under the Stockholm Convention is crucial for the successful implementation of IVM strategy.

The IVM comprises the following five key elements:

- Advocacy, social mobilization and legislation
- Collaboration within the health and non-health sectors
- Integrated approach to disease control
- Evidence-based decision-making
- Capacity building

As far as India is concerned, the IVM with respect to mosquito pest and/or vector control is brought about following a well-defined hierarchy of actions (Figure 11.2).

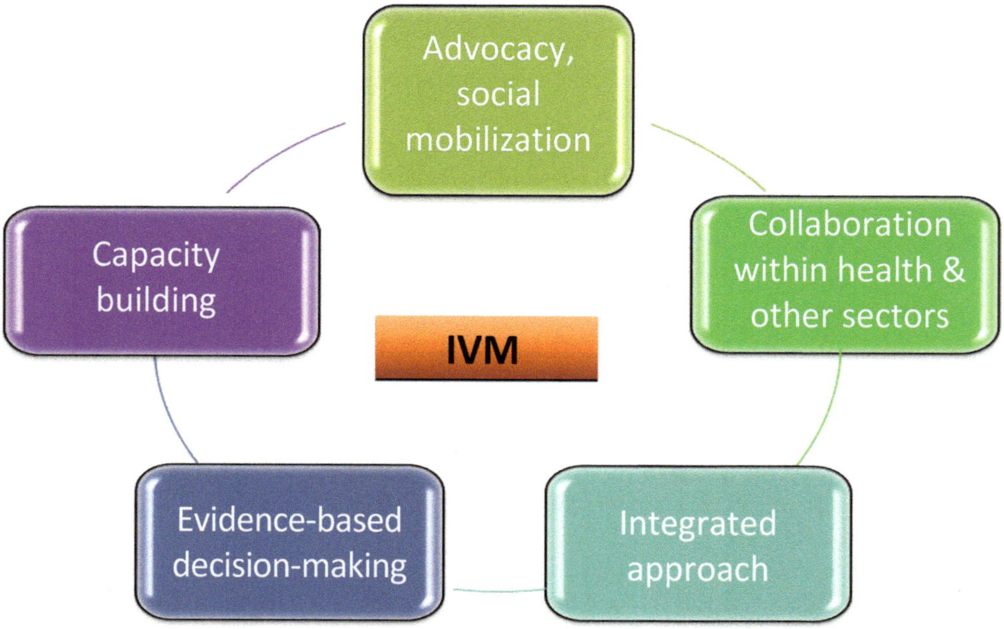

FIGURE 11.1 Graphic representation of the principal components of IVM.

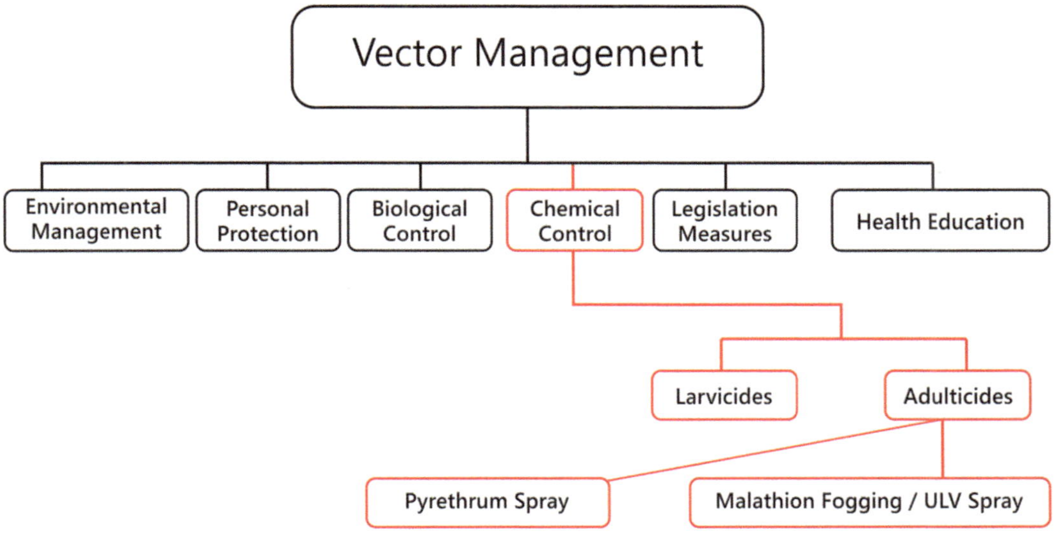

FIGURE 11.2 Hierarchical progression in the vector management under the IVM.

The aims of vector control are primarily (1) to utilize the various techniques in regulating the reproductive potential of the vector to reduce the population of the vector below a level that can maintain a disease CTDI and (2) to prevent an effective contact between the vector and the human host.

There are major categories of techniques or measures or/interventions that can be used to respond to the two aims of vector control:

i. To reduce the GRR, which is a function of fecundity and longevity through (a) chemical application, use of adulticides to kill the females and males to shorten their lives and eventually all the mosquito requirements for reproduction, and (b) genetic manipulation of the vectors or the use of GMM.
ii. To reduce the NRR, which are measures that prevent/control insect survival from egg stage to adult emergence. This is to regulate the vectors in breeding and establish their populations in their breeding habitats.

11.2 CHEMICAL CONTROL

When millions of humans are killed or disabled annually from insect-borne diseases, and the world toll of insects, VBD, weeds, and rodents is estimated at $100 billion annually, it becomes apparent that the control of various harmful organisms is essential for the future development of human health, agriculture and industry. In the process of accommodating these vital human requirements, pesticides have consequently become an indispensable part of the global process. In the 1940s, the chemical era was opened with the discovery of synthetic organic insecticides, and a totally new concept of insect control began.

11.2.1 Classification of Chemical Insecticides

Insecticides used in mosquito control belong to four major chemical groups, chlorinated hydrocarbons, organophosphates, carbamates, and pyrethroids, and a special class, the so-called insect growth regulators (Table 11.1).

TABLE 11.1
Commonly Available Insecticides with Their Properties

Insecticide	Insecticide Class	Toxicity Class	Use	Type of Poison
DDT	OC	II	Adulticide	Sodium channel modulators (neurotoxic)
Methoxychlor	OC	III N	Adulticide	Sodium channel modulators (neurotoxic)
Temephos	OP	III N	Larvicide	Acetylcholinesterase (AChE) inhibitors (neurotoxic)
Chlorpyrifos	OP	II	Larvicide and Adulticide	Acetylcholinesterase (AChE) inhibitors (neurotoxic)
Chlorpyrifos-methyl	OP	II	Larvicide	Acetylcholinesterase (AChE) inhibitors (neurotoxic)
Diazinon	OP	II	Larvicide and Adulticide	Acetylcholinesterase (AChE) inhibitors (neurotoxic)
Dichlorvos/DDVP	OP	I B	Larvicide and Adulticide	Acetylcholinesterase (AChE) inhibitors (neurotoxic)
Fenitrothion	OP	II	Larvicide and Adulticide	Acetylcholinesterase (AChE) inhibitors (neurotoxic)
Fenthion	OP	I B	Larvicide and Adulticide	Acetylcholinesterase (AChE) inhibitors (neurotoxic)
Jodfenphos	OP	III N	Larvicide and Adulticide	Acetylcholinesterase (AChE) inhibitors (neurotoxic)
Malathion	OP	III	Larvicide and Adulticide	Acetylcholinesterase (AChE) inhibitors (neurotoxic)
Naled	OP	II	Larvicide and Adulticide	Acetylcholinesterase (AChE) inhibitors (neurotoxic)
Pirimiphos-methyl	OP	III	Larvicide and Adulticide	Acetylcholinesterase (AChE) inhibitors (neurotoxic)
Alpha cypermethrin	PY	II	Adulticide	Sodium channel modulators (neurotoxic)
Bioresmethrin	PY	III	Adulticide	Sodium channel modulators (neurotoxic)
Cypermethrin	PY	III N	Adulticide	Sodium channel modulators (neurotoxic)
Cyfluthrin	PY	II	Adulticide	Sodium channel modulators (neurotoxic)
Deltamethrin	PY	II	Larvicide and Adulticide	Sodium channel modulators (neurotoxic)
Etofenprox	PY	III N	Larvicide	Sodium channel modulators (neurotoxic)
Lambdacyhalothrin	PY	II	Adulticide	Sodium channel modulators (neurotoxic)
Permethrin	PY	II	Larvicide and Adulticide	Sodium channel modulators (neurotoxic)
Resmethrin	PY	III	Adulticide	Sodium channel modulators (neurotoxic)
Bendiocarb	C	II	Adulticide	Acetylcholinesterase (AChE) inhibitors (neurotoxic)
Propoxur	C	II	Adulticide	Acetylcholinesterase (AChE) inhibitors (neurotoxic)

(Continued)

TABLE 11.1
(Continued)

Insecticide	Insecticide Class	Toxicity Class	Use	Type of Poison
Imidacloprid	NN	II	Larvicide	Nicotinic acetylcholine receptor (nAchR) (neurotoxic)
Thiacloprid	NN	I	Larvicide	Nicotinic acetylcholine receptor (nAchR) (neurotoxic)
Thiamethoxam	NN	III	Larvicide	Nicotinic acetylcholine receptor (nAchR) (neurotoxic)
Diflubenzuron	IGR	III N	Larvicide	Inhibitors of chitin biosynthesis (Growth regulation)
Pyripoxiphen	IGR	III	Larvicide	Inhibitors of chitin biosynthesis (Growth regulation)
Methoprene	IGR	III N	Larvicide	Juvenile hormone mimics (Growth regulation)
Triflumuron	IGR	III	Larvicide	Inhibitors of chitin biosynthesis (Growth regulation)
Spinosad	SP	III	Larvicide	Nicotinic acetylcholine receptor (nAchR) allosteric activator (neurotoxic)
Chlorfenapyr	PR	III	Adulticide	Oxidative phosphorylation via disruption of the proton gradient (Respiration target)
Pyrethrins	NO	II	Adulticide	Sodium channel modulators (neurotoxic)
Surface film	SF	-	Larvicide	Tracheal system (disruption of surface forces)

OC: Organochlorines; OP: Organophosphates; PY: Pyrethroids; C: Carbamates; NN: Neonicotinoids; IGR: Insect Growth Regulators; SP: Spinosyns; PR: Pyrroles; NO: Natural Organic; SF: Surface Film

11.2.2 Chemical Insecticides as Larvicides and Adulticides

These insecticides are used directly to kill larval and adult mosquitoes. Larvicides are applied in the waterbodies to kill the immature stages (larvae and pupae) of the mosquitoes. Adulticides are used to kill the adult mosquitoes, and usually they are applied in the form of fogging and indoor space spray. The following are the insecticides used in the mosquito control programme against the larvae and adult stages of the vector mosquitoes (Table 11.2).

Insecticides with residual effects are used to spray indoors to control the adult female mosquitoes that transmit malaria since they tend to rest on the walls of the houses after their blood meal. Therefore, the IRS provides effective control of malarial mosquitoes. The formulation, doses, preparation, and application of these insecticides for IRS are given in Table 11.3.

Synthetic pyrethroid (PY) insecticides are used for impregnation of bed nets, which are used in the malaria control programme. The insecticide formulation and dosage for impregnation of mosquito nets have been given in Table 11.4.

Space spray is a technique used for killing indoor resting adult mosquitoes, which is carried out using a fine pesticide sprayer. Also, fogging can be carried out for controlling outdoor resting adults using a fine pesticide spray directed by a blower. Insecticides used for indoor space spray and outdoor fogging are given in Tables 11.5 and 11.6.

TABLE 11.2
Dosage and Formulations of Different Larvicides

Sl. No.	Name of Larvicide	Class of insecticide	Commercial Formulation	Preparation of Ready to Spray Formulation	Dosage of Suspension			Frequency of Application	Equipment required
					1 sq. m.	50 Linear m.	Hectare		
1	MLO	—	100% petroleum product	As it is	20 cc	1 lit.	200 lit.	Weekly	Mop and bucket
2	Temephos (EC)		50% EC	2.5 cc in 10 lit. of potable water	20 cc	1 lit.	200 lit.	Weekly	Knapsack/Hand compression sprayer
3	Diflubenzuron 25% WP	IGR	25% Wettable powder	100 g (25 g a.i.) in 100 lit/ of water (10 g in 10 lit.)	—	—	100 lit.	Weekly	Knapsack/Hand compression sprayer
				200 g (50 gm a.i.) in100 lit. of water (20 g in 10 lit.)			100 lit.	Weekly	Knapsack/Hand compression sprayer
4	Pyriproxyfen GR	Insect Growth Regulator	0.5% Granular	Ready-to-use	—	—	2 kg 4 kg	3 Weekly	Manual
5	*Bacillus thuringiensis* var. *israelensis* aqueous suspension (12AS)	Bio-larvicide	Aqueous suspension	1 lit. in 200 lit. of water (50 cc in 10 lit. clean water) 2 lit. in 200 lit. of water (or 100 cc in 10 lit. of polluted water)	—	—	200 lit.	Weekly	Knapsack/Hand compression sprayer

Source: NCVBDC.

TABLE 11.3
Formulation, Doses, Preparation and Application of Synthetic Insecticides for IRS

Sl. No.	Name of Insecticide	Class of Insecticide	Amount of Insecticide to Prepare 10 Litres of Suspension	Dosage per sq.m of Active Ingredient	Residual Effect in Weeks	Area (in sq.m) to be Covered by 10 Litres of Suspension	Requirement of Insecticide per Million Population (in MT)
1	DDT 50% WP	Organochlorine	1.000 kg	1 gm	10–12	500	150.00
2	Malathion 25% WP	Organophosphate	2.000 kg	2 gm	6–8	500	900.00
3	Deltamethrin 2.5% WP	Synthetic pyrethroid	0.400 kg	20 mg	10–12	500	60.00
4	Cyfluthrin 10% WP	Synthetic pyrethroid	0.125 kg	25 mg	10–12	500	18.75
5	Lambdacyhalothrin 10% WP	Synthetic pyrethroid	0.125 kg	25 mg	10–12	500	18.75
6	Alphacypermethrin 5% WP	Synthetic pyrethroid	0.250 kg	25 mg	10–12	500	37.50
7	Bifenthrin 10% WP	Synthetic pyrethroid	0.125 kg	25 mg	10–12	500	18.75

Source: NCVBDC.

TABLE 11.4
Insecticides Formulation and Dosage for Impregnation of Mosquito Nets

Sl. No.	Insecticide	Formulation	Dosage	Frequency of Reimpregnation
1	Deltamethrin	2.5%	25 mg per sq. m	6 months
2	Lambdacyhalothrin	10%	25 mg per sq. m	6 months
3	Cyfluthrin	10%	50 mg per sq. m	5 months

Source: NCVBDC.

TABLE 11.5
Indoor Space Spray and Formulation

Sl. No.	Name of Insecticide	Class of Insecticide	Commercial Formulation	Preparation of Formulation
1	Pyrethrum Extract	Plant extract	2.0% extract	1:19, i.e.,1 part of 2% pyrethrum extract in 19 parts of kerosene (50 ml in 1 litre kerosene oil)
2	Cyphenothrin	Synthetic pyrethroid	5% EC	0.5 mg, a.i, per sq. m (5 ml in 200 ml kerosene oil for 500 sq. m (flow rate @ 20 ml per minute)

Source: NCVBDC.

TABLE 11.6
Outdoor Fogging and Formulation

Sl. No.	Name of Insecticide	Class of Insecticide	Commercial Formulation	Preparation of Formulation	Equipment Required
1	Malathion	Organophosphate	Technical malathion (95%)	1:19, i.e., 1 part of malathion in 19 parts of diesel (50 mL in 950 mL diesel)	Shoulder- mounted fogging machine or vehicle- mounted thermal fogging machine
2	Cyphenothrin	Synthetic pyrethroid	5% EC	3.5 gm *a.i.* per hectare (70 mL in 500 ml diesel)	

Source: NCVBDC.

11.2.3 Insect Growth Regulators

IGRs are insecticides that mimic hormones, interfering with growth and development, which leads to mortality before the insect reaches adulthood. IGRs disrupt the hormone pathway of the targeted insect and do not interfere with other co-existing organisms. Juvenile hormones (JHs) and ecdysteroids are important hormones in insects that help in moulting and metamorphosis. These hormones are accurately regulated by soluble esterases and microsomal epoxide hydrolases during the biogenesis process. The basic pathway of juvenile hormone metabolism has been documented (Dekort and Granger, 1996). Amalraj et al. (1988a, 1988b) have studied the insect growth regulatory efficacy of a few compounds. Inhibition of juvenile hormone esterase (JHE) leads to delayed metamorphosis

TABLE 11.7
Some Repellents in Public Health Programmes

Repellent Product	Active Ingredient	Insecticide Class	Concentration Range (%)
Mosquito coil	d-allethrin	PY	0.1–0.3
	d-trans allethrin	PY	0.05–0.3
	transfluthrin	PY	0.02–0.05
	prallethrin	PY	0.03–0.08
Incense stick	d-allethrin	PY	0.1–0.3
	d-trans allethrin	PY	0.05–0.3
	transfluthrin	PY	0.02–0.05
	prallethrin	PY	0.03–0.08
Repellent cream/ lotion	DEET (N,N-diethylmethyl -3-methyl benzamide)	Monoamide	12–15% W/W
Repellent spray	DEET	Monoamide	12–30% W/W
Liquid vaporizer	d-allethrin	PY	3–6
	d-trans allethrin	PY	0.8–1.5
	transfluthrin	PY	1.5–6.0
	prallethrin	PY	0.6–1.5
	S-bioallethrin	PY	1.2–2.4
Vaporizer mat (electric)	d-allethrin	PY	25–60 mg/mat
	d-trans allethrin	PY	15–30 mg/mat
	transfluthrin	PY	6–15 mg/mat
	prallethrin	PY	6–15 mg/mat
	S-bioallethrin	PY	15–25 mg/mat

PY: Pyrethroids.

in insects. There are various JHs like JH 0, JH I, JH II, and JH III have been identified in the Lepidopteran insects. Among them, JH III is the most important JH in the Dipteran insects. IGRs are likely to provide another tool for mosquito control, supplementing organophosphorus, pyrethroid, and microbial larvicides. Methoprene is one such IGR that does not kill the mosquito larvae; instead, it interferes with the growth of the mosquito larvae, which leads to the death of the larvae (Nishiura et al., 2003). The U.S. Environmental Protection Agency (EPA) (1999, 2000) approved methoprene in the year of 1975, and it is available in the market as pellets, granules and liquids.

11.2.4 Chemical Repellents

Repellents are insecticides used to repel blood-sucking arthropods in order to prevent arthropod-borne diseases. It is usually applied on the skin, clothes, curtains, etc., to repel or deter the arthropods. Currently, formulations are made for controlled release, and fabrics are impregnated with pyrethroids. Repellents are available in a variety of formulations (liquids, lotions, solid waxes, creams, foams, impregnated wipe-on towelettes, soaps) and can be dispensed from tubes, pressurized cans, roll-ons, etc. For general protection against mosquitoes, all exposed parts of the body, such as arms, legs, and face (except the eye area) should be treated. Impregnating clothing with a repellent provides extra protection, and temporary protection can be achieved with a spray-on application. For longer lasting effects, clothes should be dipped in a solution of a repellent and a pyrethroid (Table 11.7).

Insect repellents, primarily N,N-diethyl meta toluamide (DEET), are used to prevent nuisance bites from mosquitoes (as well as ticks, biting flies, and mites) and may aid in lowering disease transmission from these pests. However, they should not be relied upon to prevent disease transmission, particularly where Lyme disease or encephalitis are endemic or malaria, yellow fever, or other VBDs are prevalent. Repellents, mosquito coils, and permethrin clothing treatment products are subject to EPA pesticide registration performance requirements. Information on safe use of repellents is located at the EPA Office of Pesticide Programs website. Citronella and its oil for mosquitoes and 30 other active ingredients are exempted from EPA pesticide registration. However, some of these products may not be efficacious.

11.2.5 Chemical Attractants

Several attempts have been made to use attractive lures to kill arthropod vectors (Akhoundi et al., 2018; Roiz et al., 2016). Moreover, improving the efficacy of trapping methods has several scientific and practical purposes. It would support the local survey of mosquito species and their population abundance, surveillance of invasive mosquito species, monitoring of vector-borne pathogens, and prediction of disease epidemics (Akhoundi et al., 2018; Englbrecht et al., 2015). Mosquito attractants may thus play a key role in future programmes of IVM for the control and surveillance of vector mosquitoes. The use of attractants may help reduce the use of chemical insecticides in coming years (Mafra-Neto and Dekker 2019; Mweresa et al., 2020). The efficacy of trapping methods depends upon both the trapping device and the use of effective mosquito attractants. Many trapping devices (types of traps) are now available and have been already evaluated in numerous comparative studies for different mosquito species (Bazin and Williams 2018; Kline 2006; Thornton et al., 2016). In particular, visual stimuli play a role in the convergence of male and female *Anopheles* mosquitoes to a swarming site (Charlwood et al., 2002; Diabate et al., 2011). Consequently, recent research has demonstrated the possibility of developing 'kill zones' to disrupt or enhance mosquito swarms using artificial swarm markers or landmarks (Benelli and Mehlhorn, 2016; Benelli et al., 2016).

11.2.6 Insecticide Application Procedure

i. ***Larvicides*:** Detection of large numbers of immature mosquitoes in areas where source reduction or biological control is not feasible may require larvicide treatment to prevent the emergence of adult mosquitoes. Use of larvicides is less controversial than use of adulticides, although use of larvicides may lead to public concern about their effects on untargeted beneficial aquatic arthropods and vertebrates.

ii. ***Adulticides*:** Effective sustainable IMM programmes strive to prevent large flights or swarms of mosquitoes through all the measures described earlier, but heavy precipitation, flooding, high tides, environmental constraints, inaccessible larval habitats, missed breeding sites, and human disease outbreaks, as well as budget shortfalls, absent employees, or equipment failures, may necessitate use of adulticides.

iii. ***Surfactants*:** Surfactants (surface active agents) are non-ionic, biologically degradable chemicals that decrease the surface tension between a liquid and a gas. Surfactants may act as detergents, emulsifying agents, foaming agents, wetting and spreading agents, and sticking agents, and are reported to enhance the effectiveness of insecticides (Sukkanon et al., 2016). Surfactants form a monomolecular film on the water surface form barrier on the surface, and the larvae/pupae of the target species die off due to anoxia (Sukkanon et al., 2016). The films also lower the surface tension of the treated water body (25–29 cm/dynes), causing suffocation in the tracheal system of larvae/pupae and disallowing inflation of the wings of the emerging adults (Sukkanon et al., 2016). Surfactants are considered an alternative measure to managing resistant mosquitoes. Petroleum oil has been used as a surfactant to form a layer on the breeding grounds (water bodies) of mosquitoes to kill the larval and pupal stages (Walker and Lynch, 2007). Later, plant-derived oil surfactants were used as an alternative to petroleum oils to minimize environmental issues (Nayar and

Ali, 2003, Walker and Lynch, 2007). Mosquito larvicidal oil (MLO) is very effective, non-toxic, and widely used in India against mosquito larvae. MLO acts as a mechanical barrier between mosquito larval and atmospheric oxygen, leading to suffocation and death of the larvae. MLO may even enter the tracheal tube of the mosquito larvae and block the respiration entirely and kills the larvae. Recently, silicone-based surfactants has been developed and used as a larvicide in several studies (Bukhari and Knols, 2009, Webb and Russell, 2009, Webb and Russell, 2012, Bukhari et al., 2011, Wang et al., 2013, Mbare et al., 2014).

11.2.7 Insecticide-Impregnated Paint, Fabric, and Bed Nets

The impregnated paint technology was innovated in mid-20th century primarily to disinfect places like hospitals, aerodromes, movie halls, convention theatres, hotels, cinema theatres, ships, etc., and make them pest- and/vector-free on a long-term basis. This technology, involving embedding of long-lasting microencapsulated insecticides, or active ingredients (AIs), in the paint matrix with minute openings channelling a gradual release on the surface of the dried paint, which help to prevent its fall-out from the surface, became a sure way to safeguard such infrastructures as stated earlier in Europe and the United States where they were promoted against nuisance pests that dwell on walls and ceilings. In India, however, the technology has never been deployed on a large scale and the first ever evaluation of a commercially available insecticidal pint was carried out by Das et al. (1986a, 1986b) who have demonstrated the utility of Vernacide against both the filariasis vector mosquito *Culex quinquefasciatus* and the household pest *Periplanata americana* in urban conditions with a 100% effect for over one year. A major advantage of this technology is that the paint gained its effectiveness after a cold-water wash to the painted surface, although the insecticidal paints are available in both water-base and oil-base formulations. Earlier than insecticidal paints, insecticide-impregnated fabrics were also developed in various different shapes and forms – for example, hoods, face nettings, gowns, long sleeves, trousers, gloves, and socks, primarily to protect from the bites of mosquitoes and other flies. In the malaria endemic tribal villages of the Koraput district of Orissa (now Odisha), the effect of bed nets impregnated with permethrin (0.5 gm (ai)/sq. m.) on the reduction of the vector density of *An. culicifacies* was evaluated for more than three months. Similar investigations to enhance usefulness of fabrics, ropes and/or bed nets impregnated with cyfluthrin, deltamethrin, and lambdacyhalothin were conducted in various parts of the country by Sharma and Yadav (1995); Sharma and Mehrotra (1986), Sharma et al. (1986), Jambulingam et al. (1989), and Sahu et al. (2003).

11.2.8 Controlled Slow-Release Formulations

The spraying of insecticides over vast stretches of polluted lakes and marshy breeding habitats (e.g., fenthion or Baytex) is generally nonproductive and inhibitive to achieve desired suppression of the vector population. To overcome this serious operational constraint, slow-release formulations of varied types have been developed and made commercially available in different forms. A slow-release formulation has an active ingredient material and a carrier material which is neutral and biodegradable. As a result, a larvicide with a relatively high concentration dose is imbibed in the base biodegrades carrier material which by a process of larvicide dilution spontaneously releases a desired concentration on a long-term basis in the breeding habitat for an eventual abatement of the targeted vector species. Similarly, several slow-release DEET-based repellent formulations have been reported (Mapossa et al., 2021). As a rule, such slow-release formulations as briquettes are sprayed on a weekly basis to achieve substantial vector control, and such methodologies have been found quite cost-effective and time saving in the long run in controlling serious lymphatic filariasis and Japanese encephalitis–transmitting vector species such as *Culex quinquefasciatus* and *Cx. tritaniorhynchus*.

11.2.9 Mode of Action of Insecticides

The mode of action of each insecticide class is different, and it is important to prevent insecticide resistance management (IRM). The mode of action against mosquitoes differs from insecticide to insecticide. General modes of action are stomach poison, contact poison, fumigant poison, etc. Organochlorines (OC), such as DDT, known for producing tremors and hyperexcitability in vector mosquitoes, lead to loss of insect movement (ataxia), subsequently paralyzing the mosquitoes. Organophosphates (OP) act at the synaptic junction of the nervous system and inhibit cholinesterase (ChE) enzymes like acetylcholinesterase (AChE), resulting in the accumulation of acetylcholine (ACh) at the synaptic junction, which leads to continuous twitching of muscles and death of the treated mosquitoes (Corbett, 1974; Maddrell, 1980). The carbamates (C) mode of action is similar to organophosphates. They affect the activity of acetylcholinesterase (AChE) enzyme and hydrolyse acethylcholine (ACh) at the synaptic junction in the nervous system of the mosquitoes. Following this, hyperexcitation of the nervous system leads to convulsions and the death of the treated mosquitoes (Eldefrawi, 1985). Pyrethroids (PY) produced neurotoxicity in mosquitoes. The action of pyrethroids on the nerve membrane modulates the sodium channels by changing protein arrangements at the lipid-protein interface. Pyrethroids usually act on both peripheral and central neurons, while the knockdown effect is most probably produced by peripheral intoxication (Jerome et al., 2011). IGRs are classified into two major groups: (i) chitin synthesis inhibitors and (ii) JH analogues:

i. The chitin synthesis inhibitors interfere with new cuticle formation, resulting in moult disruption. Example: diflubenzuron, cyromazine, and triflumuron.
ii. The JH analogues interfere with the metamorphic processes affecting development to the adult stage. Example: Methoprene. Neonicotinoids (NN) binds to the acetylcholine site on nicotinic acetylcholine receptors (nAChR) and modulates nAChRs, causing a range of symptoms from hyperexcitation to lethargy and paralysis. Example: Neonicotinoids insecticides, viz., Clothianidin, Imidacloprid, Butenolides, and Flupyradifurone. Spinosyns (SP) interferes with the neural mechanism of the mosquitoes.

The spinosyns and spinosoids have a novel mode of action, primarily targeting binding sites on the nicotinic acetylcholine receptors (nAChRs) of the insect's nervous system (Kirst, 2010). Pyrroles (PR) are broad-spectrum insecticides that show contact and stomach toxicity (Gunning and Moores, 2002; N'Guessan et al., 2007). Pyrroles act at the cellular level and disrupt respiratory pathways and proton gradients through the uncoupling of oxidative phosphorylation in mitochondria (Black et al., 1994; Oxborough et al., 2015). Pyrethrins are a natural organic (NO) mixture of six lipophilic, and the mode of action is related to a high rate of photodegradation and a short "knockdown" (rapid paralysis) effect in insects, which are often accompanied by a lack of coordination and tremors (Jerome et al., 2011). Surface films (SF) act on the mosquito larvae by flooding the tracheal system with oil, by disrupting the surface forces that allow larvae to rest at the surface, by toxicity, or by eliciting chemosensory responses. Larvae that experienced tracheal flooding applied their mouthparts to the siphon (tail nibbling), often losing contact with the surface while doing so and falling to the bottom and dying (Corbet et al., 2000).

11.2.10 Insecticide Classes Based on the Mode of Action

i. **Class 1**. These insecticides are Acetylcholinesterase (AChE) inhibitors that inhibit AChE, causing hyperexcitation. AChE is the enzyme that terminates the action of the excitatory neurotransmitter acetylcholine at nerve synapses. Example: Carbamates insecticides, viz., Bendiocarb, Organophosphates, Malathion, Pirimiphos-methyl.

ii. **Class 3**. These insecticides are Sodium channel modulators that keep sodium channels open, causing hyperexcitation and, in some cases, nerve block. Sodium channels are involved in the propagation of action potentials along nerve axons. Example: Pyrethroids insecticides, viz., Alpha-cypermethrin, Bifenthrin, Deltamethrin, trans-cyphenothrin, Etofenprox, Lambdacyhalothrin, Permethrin, Prallethrin, S-bioallethrin, Transfluthrin.
iii. **Class 4**. These insecticides are Nicotinic acetylcholine receptor (nAChR) competitive modulators that bind to the acetylcholine site on nAChRs, causing a range of symptoms from hyperexcitation to lethargy and paralysis. Acetylcholine is the major excitatory neurotransmitter in the insect central nervous system. Example: Neonicotinoids insecticides, viz., Clothianidin, Imidacloprid, Butenolides, Flupyradifurone.
iv. **Class 7**. Juvenile hormone mimics like pre-metamorphic in larval instar that disrupt and prevent metamorphosis. It reduces fecundity and fertility in female adults. Example: Pyriproxyfen.
v. **Class 13**. These insecticides are uncouplers of oxidative phosphorylation via disruption of the proton gradient that protonophores that short-circuit the mitochondrial proton gradient so that adenosine triphosphate (ATP) cannot be synthesized. Example: Pyrroles like Chlorfenapyr.

The toxicity and specificity of compounds from different plants showed appropriate biochemical modes of action at the cellular or physiological levels in various insect pests as well as mosquitoes. Rattan (2010) described the toxicity of plants affecting directly or indirectly the insect pest proteins, receptors, signalling molecules, ion-channels, structural proteins, nucleic acids, bio-membranes, cellular components, regularity in the nervous system, neurotransmitter synthesis, storage, release, binding, and reuptake, receptor activation and function and enzymes involved in signal transduction pathway. Living beings are susceptible to neurotoxic insecticides and are generally acting as agonists and antagonists. Blockage of the neural system by toxic molecules causes brain death and leads to paralysis, convulsions, hyperactivity, and spasms in all vertebrates, as well as the invertebrate animals having developed neuronal systems. Mostly, neurotoxic insecticides affect the normal function of acetylcholinesterase (AChE), g-aminobutyric acid receptors (GABA-R), octapomine receptors, sodium channels, and nicotine acetylcholinesterase receptors (nACh-R), which are most important in the neural development and brain function in various living beings and also affect the respiratory mechanism in insects (Matsuda et al., 2001; Casida and Quistad, 2004; Sanchez-Bayo and Goka, 2006; Tomlin, 2009; Hertlein et al., 2010). Most of the plant compounds act as stomach poison; some plant compounds act as neurotoxins against mosquito larvae; for example, asaricin 1 and isoasarone 2 against *Ae. aegypti*, *Ae. albopictus* and *Cx. quinquefasciatus* (Hematpoor et al., 2016), stigmasterol and hexacosanol against *Ae. aegypti* and *Cx. quinquefasciatus* (Srinivasan et al., 2015), thujone against *Ae. aegypti* (Gade et al., 2016), lawsone against *Cx. quinquefasciatus* (Sandhanam et al., 2018), and plumbagin against *An. stephensi* (Pradeepa et al., 2016). Some plant compounds act as growth inhibitors and regulators (IGR) that disrupt the development of metamorphosis in insects. Insecticides including benzoylureas, tebufenozide, and azadirachtin from plants prompt the moulting and premature moulting in larval stages, and pupae to immature moulting into adults. Some insecticides such as ascryolite, sulcofuron sodium, and cry toxins act as stomach poison (Dhadialla et al., 1998; Dinan et al., 1999; Brock et al., 2009). Certain compounds act as IGRs – for example, pellitorine (Perumalsamy et al., 2013), benzopyrans HP1_HP3 (da Silva et al., 2013), and ecbolin A and ecbolin B against *Ae. aegypti* (Reegan et al., 2016), b-sitosterol against *Cx. quinquefasciatus* (Angajala and Subashini, 2018), and proanthocyanidins against *An. gambiae* (Muema et al., 2016). Plant compounds used to control the larvae of mosquitoes primarily affect the midgut region, malpighian tubules, gastric caeca, and mostly act as stomach poison. Larvae of mosquito treated with plant compounds exhibited severe damages in the midgut region as observed in the histology studies (David et al., 2000). Niloticin-, tiliamosine-, and (4R, 5S) 4-acetoxy-7-tigloyloxy carvotanacetone-, etc., treated mosquito larvae exhibited (1) serious damage to the midgut epithelial columnar cells (CC), (2) total rupture of peritrophic membrane (PM), and (3) oozing out of midgut content (MC) into ectoperitrophic space. These results suggested that molecules from plants had strong larvicidal mode of action in mosquitoes (Reegan et al., 2014; Sandhanam et al., 2019; Sivanandhan et al., 2021).

Third-generation insecticides act as IGRs that strongly disrupt the endocrine system of the insects and thus interfere with the normal process of development, reproduction, or metamorphosis of various insect pests (Su and Scheffrahn, 1990). Thus, several mechanisms of the insect's body are vulnerable to plant metabolites. Some essential oils inhibit acetylcholinesterase, morphogenesis, behaviour, and memory of the cholinergic system. Thymol disrupted GABA-gated chloride channel in larvae of *Cx. pipiens* (Youssefi et al., 2019) and *An. arabiensis* (Damtie and Mekonnen, 2021); pyrethrin caused ion-exchange disruption of sodium in Cx. quinquefasciatus (Li et al., 2012) and potassium in *Ae. aegypti* larvae (Rusconi et al., 2018); rotenone inhibited cellular respiration in *An. stephensi* (Giulivi et al., 2008); ryanodine blocked the calcium channels; sabadilla acted on the nerve cell membrane; thymol acted on octopamine receptors in larvae of *Cx. pipiens* (Youseffi et al., 2019); azadirachtin disrupted hormone and poisons mitosis (Senthilnathan et al., 2008; Rattan, 2010).

11.2.11 Resistance to Chemical Insecticide and Its Management

Resistance to certain larvicides and adulticides has occurred periodically. Failure of mosquito control indicating resistance must be verified by laboratory analysis or use of test kits because other factors (improper equipment calibration, dilution, timing and other application errors, off-specification products, and climatic factors) can prevent insecticides from providing satisfactory control in the field. Resistance may occur between insecticides within a class or could be passed from immature to adult stages subject to the same insecticidal mode of action. Additionally, different species of mosquitoes may inherently vary in susceptibility to different larvicides and adulticides (Chareonviriyaphap et al., 2013). Insecticides with different modes of action can be alternated to prevent resistance. Even though source reduction and the use of predators such as larvivorous fish are also used for sustainable IMM, only two chemical classes of adulticides (organophosphates and pyrethroids) with different modes of action are available. Biological controls (including birds and bats) may be present, but often not in sufficient numbers to provide satisfactory alternative control, particularly in coastal areas where salt-marsh mosquitoes are abundant or when human disease outbreaks occur. Therefore, sustained IMM requires alternative use of different classes of insecticides, in conjunction with resistance monitoring, source reduction, biological control, and public education. Further, the following methods help avoid insecticide resistance:

i. ***Plan*:** IRM should be considered an integral part of any vector control programme and included during the planning phase.
ii. ***Monitor*:** The susceptibility status of the target mosquito populations should be monitored regularly to guide the design of the IRM programme, and choice of intervention.
iii. ***Rotation*:** Guided by susceptibility monitoring data, plan to rotate insecticides by Mode of Action (MoA) class, either temporally or spatially. In the absence of susceptibility data, the rotation of products between Issue, Rule, Application, Conclusion – Mode of Action (IRAC-MoA) classes will reduce selection pressure for resistance development.
iv. ***Mixtures*:** The use of mixtures of insecticides will have the greatest IRM benefit if the insecticides in the mixture are from different classes, and the target mosquito population is fully susceptible to both. Include mosquito larvicides, or other interventions, with alternative MoA in the vector control programme where appropriate.

11.3 NON-CHEMICAL CONTROL

11.3.1 Expanded Polystyrene Beads

In certain specific breeding habitats such as, for example, unused wells, cess pools and cess-pits, quite a few non-chemical or physical measures or products like expanded polystyrene (EPS) beads

– a synthetic product of a very light thermocol-like material that are inert, nontoxic, non-wettable and resistant (do not interact) to sea water, salt solutions, soap and wetting agents, and of spherical shape or form measuring approximately 0.2–0.8 mm in diameter – have been found highly effective to cease the breeding mosquitoes like *Cx. quinquefasciatus* and *An. stephensi*. The EPS is widely available commercially at a price of Rs. 80/kg in major metropolises. The major advantage using the EPS is that they are non-biodegradable and can remain effective for months in the treated water bodies such as unused or discarded wells and latrine pits. Upon application, the EPS beads are superlatively light and keep buoyant irrespective of the level of the breeding habitat's water at any time. However, they need to be protected from direct exposure to sunlight in the field since they can turn yellowish and brittle. The EPS beads, when applied at 500 g to 1 kg/m^2 in different habitats, control serious nuisance and vector mosquitoes, like the primary filariasis vector *Cx. quinquefasciatus*, a great deal in unused wells with turbid and/or polluted water and latrine pits, and the urban malaria vector *Anopheles stephensi* in live wells with good drinking water (Curtis and Minjas 1985; Dua et al., 1989). In addition to EPS beads, Soltani et al. (2012) applied Shredded Waste Polystyrene (SWAP) beads for control of mosquitoes.

11.3.2 Pneumatic Technique

Pneumatic pest control means the suction of vector insects using a vacuum device. Limited research has been done on this technique against pest insects including mosquitoes. Research is being carried out to improve this technique in terms of technical and physical parameters (Vincent and Boiteau, 2001).

11.4 BIOLOGICAL CONTROL

Biological control or biocontrol is a method of controlling disease-transmitting arthropods, such as mosquitoes, fleas, ticks, mites, sand flies, etc., using predatory organisms. It relies on predatory potential and parasitism but also involves an active human management role. Biological control is an environmentally sound and effective means of reducing disease-transmitting arthropods through the use of natural enemies. In recent years, the interest in biological control strategies to control vector mosquitoes has increased, and it is reported to be possible by the use of different predators, including insects (Quiroz-Martínez and Rodríguez-Castro, 2007; Mogi, 2007; Shaalan and Canyon 2009). Though they occur under natural conditions, they seldom reduce vector populations too low to maintain a disease. For this reason, the biocontrol agents must be manipulated to be able to effectively control the vector populations. There are two types of manipulating biocontrol agents: agents found naturally occurring with the host are reared and introduced in a larger quantity to be able to effectively lower the host population, and the other is rearing a known effective agent outside its natural habitat and then introducing the laboratory-reared biocontrol agents to the habitat and making the habitat conducive to the compatibility of the agent and the vector host (De Las Llagas and Portus 2016; Figure 11.3).

11.4.1 Invertebrate Predators

11.4.1.1 Insect Predators

The importance of insect predators of mosquitoes has been known for a long time. The nymphs and adults of dragonflies and damselflies of several families (Order: Odonata) predate mosquitoes (Saha et al., 2012; Choo et al., 2021). In feeding experiments, nymphs of dragonflies and damselflies proved to be extremely voracious. According to Choo et al. (2021), urban odonates (e.g., *Crocothemis servilia, Ischnura senegalensis, Orthetrum sabina*) can consume substantial numbers of mosquito larvae (up to 44 per day) under experimental conditions. Nymphs of *Aeshna cyanea* were reported to consume up to 100 mosquito larvae each day with an average of 30 larvae per day

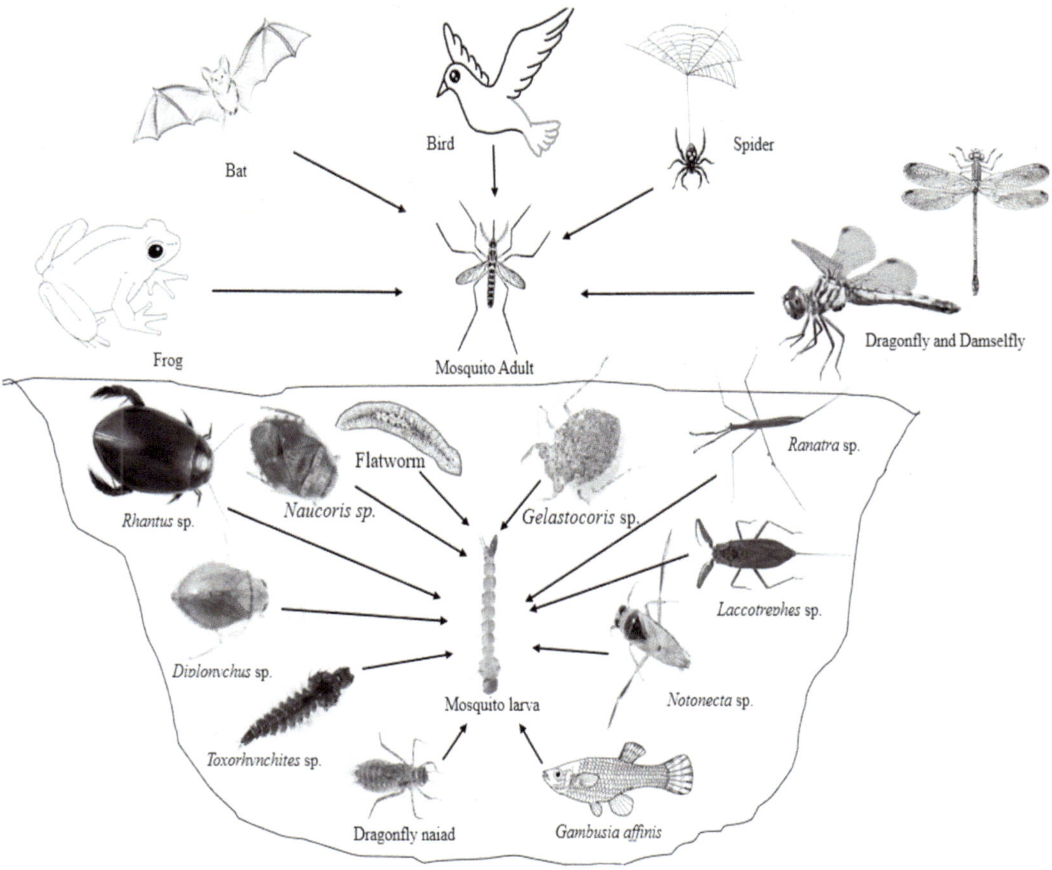

FIGURE 11.3 Biocontrol agents in nature many of which have been found useful in reducing vector mosquito population.

Source: This work.

(Kögel, 1984). In another study, a dragonfly 12th instar larva reported consuming approximately 133 ± 2 all stages of larvae of *Aedes aegypti* per 24 hours. In the case of dragonfly *Bradinopyga geminata* controlling dengue/chikungunya vector *Ae. aegypti*, in the laboratory, the feeding rate of eighth instar manifested maximum predation on 1st instar mosquito larvae (86%), followed by second, third, and fourth instars (72%, 66%, and 48%), respectively (Venkatesh and Tyagi 2015). Sebastian et al. (1990) inundatively deployed dragonfly *Crocothemis servilia* to successfully control yellow fever/dengue vector breeding in bamboo stumps, ornamental vases, and other water holding pots in Myanmar. Morales et al. (2003) confirmed that dragonfly nymphs (Libellulidae) feed well on *Anopheles* larvae, and that mosquito DNA, although rapidly digested, can be successfully recovered and detected from within nymphal digestive tract. Odonate larvae and mosquito larvae share many breeding habitats (e.g., cement cisterns, over-head tanks, ponds, tree holes) where they co-occur with larvae of *Aedes* and *Anopheles* mosquitoes. In a natural cycle, the odonate nymph predate mosquito larvae and minimize its populations. Further, odonates have a lengthy nymphal period and take months to develop into adults. Thus, odonate nymphs spend a longer duration in the aquatic breeding habitats and predate mosquito larvae continuously, providing a significant level of biological control.

Aquatic and semiaquatic bugs of several families under the Hemipteran order, including Gelastocoridae, Dytiscidae, Naucoridae, Nepidae, Belostomatidae, and Notonectidae are voracious consumers of different mosquito larvae and are important biological control agents. Notonectids or backswimmers (Heteroptera: Notonectidae) have been considered the most promising (Mogi, 2007, Shaalan and Canyon, 2009). According to Saha et al. (2020), water bugs like *Ranatra elongata, Ranatra filiformis*, and *Laccotrephes griseus* (Heteroptera: Nepidae) showed a preference for mosquito larvae under multi-prey conditions. In another study, the backswimmer was reported to be the most aggressive predator of *Anopheles* mosquito larvae (95% of the exposed larvae consumed), with daily mean predation of 71.5 larvae (Eba et al., 2021). Similarly, *Diplonychus rusticus* (Heteroptera: Belostomatidae), a relatively bigger sized tropical water insect, serves as the main predator of mosquito larvae (Reegan et al., 2020). According to Saha et al. (2007), a single adult *A. bouvieri* could consume 28–34 fourth instar mosquito larvae per day. Similarly, *D. rusticus* could consume 11–87 fourth instar mosquito larvae per day, and *D. annulatus* could consume 33–122 fourth instar mosquito larvae per day, depending upon the prey and predator densities (Saha et al., 2007). In a survey, the Hemipteran predators were more abundant (54.2%) compared to other insect predators (Odonata: 22.9% and Coleoptera: 22.9%; Dida et al., 2015). In a study, Carlson et al. (2004) recorded that insect predator colonization of aquatic habitats was strongly influenced by habitat type and not associated with location characteristics. Similarly, some species of Dytiscidae were reported predating mosquitoes. For example, *Rhantus signatus* was reported as a predator of mosquito larvae in the temporary pools in Argentina (Campos et al., 2004; Fischer et al., 2000). Also, *Rhantus consputus* was reported to predate the larvae of *Aedes vexans* mosquitoes in central and eastern European countries (Kögel, 1984, 1987; Aditya and Saha, 2006). In another study, *Colymbetes paykulli* Erichson was reported to predate *Culex* mosquito larvae more often (Lundkvist et al., 2003). Similarly, *Naucoris congrex* (Naucoridae) was reported to predate final instar mosquito larvae with a type II functional response at all temperatures tested from 5°C–25°C (McCoull et al., 1998). Gelastocoridae and Ochteridae also participate in suppressing populations of these blood-sucking dipterans (Papáček, 2001). Further, *Toxorhynchites splendens* (a non-biting mosquito) have been extensively experimented with for their potential to control vector larval population under specific natural conditions (Tyagi, 2020; Tyagi et al., 2020; Corbet, 1986). Single *Tx. splendens* fourth instar can consume 114 larvae of the vector species in laboratory conditions.

11.4.1.2 Crustacean Predators

Crustaceans of the order Copepoda are natural predators of mosquito larvae. The scientific research on copepods being used as biological control agents of mosquito larvae started during the 1980s in French Polynesia, where *Mesocyclops aspericornis* recorded a 99.3% mortality in the *Aedes* population (Lazaro et al., 2015). To date, the predatory efficiency of about 50 copepod species belonging to 15 genera has been tested mostly in the laboratory but also in the field, mainly against *Aedes* and other species of mosquitoes (Kocher and Jamwal, 2016). They predate mainly first instar larvae of mosquitoes. A single copepod has the capacity to kill more than 40 *Aedes* larvae/day and reduce *Aedes* production by 99%–100% (Marten and Reid, 2007), whereas it is reported to kill fewer *Anopheles* and *Culex* larvae (Marten and Reid, 2007). Species of *Mesocyclops* in small habitats are capable of killing up to 50 *Aedes* larvae per day (Kocher et al., 2018; Thakur and Kocher, 2018). Vu et al. (2005) and Nobuko et al. (2020) deployed *Mesocyclops* spp. and others to predate and control mosquito larvae in several villages in Vietnam. Kay et al. (2002) also conducted many studies with *Mesocyclops* against *Ae. aegyti* and *Ae. albopictus* and achieved complete control. Therefore, the use of copepods remains a vital step in IVM programmes in recent years, especially against *Ae. aegypti*.

11.4.1.3 Arachnid Predators

Spiders are general predators, and they are known to predate adult mosquitoes (Service, 1973). Some species under the families *Pisauridae* and *Lycosidae* predate aquatic organisms moving on the water surface, and they also capture emerging adult mosquitoes and even larvae or pupae (Bishop and Hart, 1931). *Argyroneta aquatica* is one such species; it builds webs under the water to catch aquatic organisms, including mosquito larvae. In laboratory experiments, *A. aquatica* consumed 29 fourth instar mosquito larvae per day (Kögel, 1984). In natural water bodies, these spiders play a major role in controlling mosquitoes. Water mites (Hydrachnellae) can also attack mosquito larvae and suppress mosquito populations. Kögel (1984) reported that the small water mite *Piona nodata* attacked up to 18 mosquito larvae per day in laboratory experiments. Kirkhoff et al. (2013) conducted a survey of water mite larvae parasitizing adult mosquitoes and reported that 21 species of mosquitoes were parasitized by mites, viz., *Parathyas* sp., *Arrenurus* spp., and *Arrenurus morphotypes*. Simmons and Hutchinson, (2016) mentioned in their review that most of the mosquito parasitizing mites were *Arrenurus* Sp. or *Parathyas barbigera*. In another study, 38 mosquito individuals were found to be parasitized by 93 different water mite larvae (Werblow et al., 2015).

11.4.1.4 Flatworm Predators

Turbellarians (Platyhelminthes, Tricladida, Paludicola), often referred to as planarians, are free-living flatworms that live in freshwater (Tranchida et al., 2009). Flatworms are excellent predators of invertebrates, including mosquitoes (Dambach, 2020). Predation of mosquito larvae by free-living flatworms was first recorded by Lischetti (1919), which led to the extensive investigation of predatory potential of different species of flatworms on mosquito larvae. In the field, *Mesostoma* sp. showed a considerable reduction of *Cx. tarsalis* and *An. freeborni* mosquito populations in Californian rice fields (Case and Washino, 1979). In the laboratory, *Girardia anceps* (Tricladida, Paludicola, Dugesiidae) killed 52% of *Ae. aegypti* or *Cx. pipiens* mosquito larvae (Tranchida et al., 2009). Similarly, *Mesostoma ehrenbergii* and *Bothromesostoma evelinae* (Rhabdocoela, Typhloplanoida, Typhloplanidae) killed 100% of *Ae. aegypti* or *Cx. pipiens* mosquito larvae (Tranchida et al., 2009). Melo and Andrade (2001) reported that flatworm *Girardia tigrina* (Girard) killed more *Culex quinquefasciatus* L. larvae than *Aedes albopictus* (Skuse) larvae in automobile tires. In another study, Suprakash and Aditya (2003) reported a lower predation on first instar mosquito larvae due to their faster movements (Suprakash and Aditya, 2003). However, the predatory capacity of each species of flatworms depends on their body size. Adults of *B. evelinae* flatworm are reported to be smaller than the other flatworm species (Noreña-Janssen and Faubel, 1992). *M. ehrenbergii* and *G. anceps* were capable of preying on all larval instars because of their larger size (Tranchida et al., 2009).

11.4.1.5 Freshwater Snails (Gastropods)

Freshwater snails are molluscs, live in fresh water and consume plants and detritus (Lombardo and Cooke, 2002). *Lymnaea (Galba) palustris* was reported to be predacious on mosquito larvae both in the laboratory and in the field (Becker et al., 2010).

11.4.1.6 Leeches

Leeches (Hirudinea) are not aggressive predators of mosquito larvae. Hence, they are of no practical importance. However, a leech *Herpobdella octoculata* is reported to be a predator of mosquitoes (Becker et al., 2010).

11.4.2 Vertebrate Predators

11.4.2.1 Larvivorous Fish

Fish, especially indigenous species, can be integrated with other vector control approaches. Larvivorous fish feed on the larvae of mosquitoes (WHO, 2003). Larvivorous fish are surface feeding in nature, hardy and adaptable to local conditions, with high reproduction capacity. Approximately, 315 species belonging to 32 genera under seven families are recorded. Family *Cyprinodontidae* contributes maximum number of genera (15) and species (300) followed by *Cyprinidae, Hemirhamphidae, Chilidae, Goodeidae, Poeciliidae*, and *Anabantidae* (Ghosh and Dash, 2007).

The renowned entomologist, Leland Ossian Howard, was first to advocate the use of *Poeciliids* for the control of mosquito larvae. Smith stated that no fish native to New Jersey (United States) could control mosquito larvae as efficiently as the mosquito fish. Nearly 10,000 *Gambusia* were released in the vicinity of Camden in 1905. This was the first attempt to introduce *Gambusia* into northern waters as a mosquito control measure. The introduction of *Gambusia* into Europe was due to the efforts of Dr Massimo Sella, who with the aid of the Red Cross, arranged to have shipments of *Gambusia affinis holbrooki* sent from Augusta, Georgia, to Spain and Italy in 1921. Presently, both the subspecies of *Gambusia* have been widely distributed in Europe, Asia, and Africa and nearby islands. Beginning in 1908, another larvivorous fish, *Poecilia reticulata* (guppy), a native of South America, was introduced for malaria control into British India and many other countries. The use of larvivorous fish as a vector control for aquatic stages of the mosquitoes has been documented as early as 1952 by Ejercito. It was only in 1987 that the use of tilapia fish in streams has been described in detail by the WHO (VBC 187). The predation by tilapia on aquatic disease vectors such as mosquito larvae occurs most likely because of the tilapia's opportunistic feeding habits and because the larvae are found closely associated with the usual vegetable/detrital food of tilapia. A pilot project using tilapia fish (De las Llagas and Ventura, 1991) as a biological control agent was conducted in Sapang Palay, Bulacan. It has been shown that the fish feed on anopheline larvae under laboratory and field situations. Though larvivorous fish has been found effective in controlling the aquatic stages of anophelines, this vector control measure needs a community-based fish farming technology to be sustainable as an intervention. (Veneracion, 1994, ibid.). Some of the most successful larvivorous fish species introduced in different countries are the top minnow or mosquito fish (*Gambusia affinis*) and the guppy (*Poecilia reticulata*), *Aplocheilus blochii*, and *Macropodus* (Poopathi and Tyagi, 2006; Tyagi, 1994). *Gambusia affinis* has been widely used to control immature stages of mosquito vectors. It is efficient in clean water, but it cannot survive in temperatures below 10°C (WHO, 2003). *Poecilia spp*. have been found effective in rice fields in warm areas since they can tolerate high temperature and organically polluted water. Other fish species used in breeding sites are *Tilapia spp*., *Cynolebias, Nothobranchius*, and *Aphyosemion* (WHO, 2003). Two freshwater fishes – namely, rainbow fish, *Melanotaenia splendida*, and fly-specked hardyhead *Craterocephalus stercusmuscarum* – preyed on immature *Ae. aegypti* in north Queensland while the larvivorous fish *Oreochromis spilurus* was found effective against malaria vectors in Somalia (as cited by Poopathi and Tyagi, 2006). The use of larvivorous fish in the rice fields must take into consideration the irrigation practices and presence of native fish populations that might be displaced by the introduced fish (Wada, 1987). Thus, the compatibility of the chosen fish with local fauna and flora is of high importance (as cited by Poopathi and Tyagi, 2006).

In India, successful introduction of larvivorous fish was possible during the bioenvironmental control of malaria. It was documented that *P. reticulata* are most suited for well ecosystems with sub-column feeding zones and the capacity to survive in polluted water bodies. On the contrary, *G. affinis* is a surface feeder and considered ideal for large water body ecosystems like ponds and lakes. *Gambusia* have also been used successfully in the mixed edible fish culture related to income generating schemes for benefit of communities. For instance, grass carp fish (*Ctenopharyngodon idella*) can be added in ponds to clear all the aquatic weeds and thus favouring *Gambusia affinis* to devour on the mosquito larvae making the pond unfit for mosquito breeding. Fish-based intervention for vector control was becoming increasingly popular, making rounds in the country. A feasibility study to control other mosquito-borne diseases, a special fish-based intervention programme, was launched by the Indian Council of Medical Research (New Delhi) for control of vectors of Japanese encephalitis (JE) in Gorakhpur district of Uttar Pradesh (UP) reporting most case fatalities annually. A block-level task force model for fish release has been very effective (Ghosh and Rashi, 2019). Currently, the larvivorous fish has been used in mosquito control programmes worldwide. But its impact has been sceptical for lack of sufficient evidence and proper study designs. The meta-analysis also suggested that there is no concrete evidence of adverse effect of these fishes on the local environment/ecosystem. Nevertheless, routine monitoring and proper care are some of the important aspects for fish survival and sustained supply which must be incorporated (Ghosh and Rashi, 2019).

11.4.2.2 Frogs

Frogs, toads, and their tadpoles also serve as predators of mosquito larvae and help to control the mosquito population (Becker et al., 2010; Poopathi and Tyagi, 2006; Raghavendra et al., 2008).

11.4.2.3 Birds

In general, birds are not considered to be important predators of mosquitoes, although mosquitoes are a source of food for some bird species (Blotzheim and Bauer, 1985). For example, the duck *Anas platyrhynchos* was repeatedly observed filtering out larvae of *Ae. vexans* from mass breeding sites. Flycatcher is an insectivorous bird which primarily feeds on small insects and flies and plays an important role in biological control of insects including mosquitoes. The birds are highly motivated, efficient and cost-effective pest controllers; the feasibility of use of insectivorous flycatcher birds in mosquito control has been reviewed (Singh and Mohan, 2013).

11.4.2.4 Bats

Bats are active during the night. Hence the activity pattern of most mosquitoes and bats are similar. Most bats are insectivorous but exhibit a large variety of hunting strategies (Becker et al., 2010). Female mosquitoes are active for blood meal during dusk hours, and male mosquitoes form copulation swarms. This enables bats to capture adult mosquitoes during dusk hours and helps to minimize the mosquito population. Hence bats are considered potential predators of mosquitoes.

11.4.3 Microbials as Insecticides

At present, there are only two biological/biocontrol agents – namely, *Bacillus thuringensis* H-14 (Bti) and *B. sphaericus* – which are in operational use for mosquito control. Due to its high specificity, Bti has been found more effective against mosquitoes which breed in relatively unpolluted water, including those of *Aedes* species. Bti has low mammalian toxicity; hence, its proper use in drinking water should not cause any problems (WHO, 2009).

11.4.4 Parasites as Control Agents

Certain helminths belonging to the family *Mermethidae* have been documented for attacking and destroying mosquito larvae, especially in paddy fields breeding vector mosquitoes of JE. Two species such as, for example, *Romanomermis culicivorax* and *R. iyengari*, whose infective preparasite stage (nema) actually penetrates the mosquito larvae, have been bioprospected in Indian laboratories and extensively explored for their antilarval kill *Cx. vishnui* sub-group species larval populations breeding in paddy fields near Bangalore (now Bengaluru), southern India. After six to eight days when the parasite emerges out as a postparasite, this invariably results in the death of the immature mosquito larvae. Initial evidence on *R. iyengari* possibly came from observations made by Ross (1906) who reported the presence of mermithid nematodes in *Culex* larvae from India. Iyengar (1927) found juvenile mermithid nematode parasites in the larvae of seven different species of anopheles. In the laboratory, mermithid nematode *Romanomermis iyengari* killed 100% L1 *An. gambiae* larvae within 24 hours post-treatment and killed 100% of both L2 and L3 larvae within 7 days of treatment (Abagli et al., 2019). In field experiments, the monthly applications of 3,500 to 5,000 pre-parasitic nematodes per m^2 eliminated *Anopheles* larvae development (Abagli et al., 2019). Operational potential of these mermethids faces a unique challenge, as the infective preparasite stage, nema, can tolerate only very low levels of salinity, alkalinity and pollution, restricted for application only in fairly freshwater bodies such as the stagnated fresh water in rice fields. *Microsporidia* are one of the biggest and most diversified genera of parasitic fungi associated with mosquito species in the natural world. Indeed, all mosquitoes are likely to be hosts for one or more *microsporidia* parasites. They are parasitic eukaryotes that have developed a unique and highly specialized mechanism for invading host cells by infectious spores (Andreadis, 2007).

11.4.5 Pathogens as Control Agents

Many entomopathogenic microbes are reported to be potential against various insect pests, which produces specialized spores and kill the hosts (Evans et al., 2018). *Lagenidium giganteum* is one such fungal larvicide reported to be effective in both laboratory and field conditions (Evans et al., 2018). Further, fungi such as *Coelomomyces, Lagenidium,* and *Culicinomyces* are effective against larval or adult stages of mosquitoes and have been extensively investigated (Scholte et al., 2004). According to Darbro et al. (2011), 93 isolates of entomopathogenic fungi, such as *B. bassiana, M. anisopliae, Isaria fumosorosea, Isaria farinosa, Isaria flavovirescens, and Lecanicillium spp.*, could be used as biological control agents for *Ae. aegypti* mosquito.

11.5 ZERO RESISTANCE DEVELOPMENT

Mosquito-borne diseases are now increasing due to the development of insecticide resistance in mosquito vectors. increased metabolic detoxification of insecticides is mainly responsible for insecticide resistance (Liu, 2015). Resistance monitoring is an important aspect of a mosquito control programme. Further, IRM, as an integral component, helps to prevent the formation of resistance in vector mosquitoes and helps to maintain the effectiveness of insecticides. To maintain zero resistance, the following four strategies under IRM are recommended:

i. **Insecticide Rotations:** Changing insecticides with different modes of action at a particular area (may be an endemic pocket)
ii. **Insecticide Varieties:** Application of insecticides of different modes of action in adjacent areas
iii. **Insecticide Mixtures:** Application of insecticide mixtures prepared with two or more insecticides with different modes of action
iv. **Insecticide Combinations:** Application of insecticide combination with two classes of insecticides with different modes of action in a particular endemic place by IRS and fogging for adult control

11.6 OTHER NON-CHEMICAL MOSQUITO CONTROL METHODS

11.6.1 Source Reduction

Source reduction is an activity included in a mosquito control programme to remove the breeding sources of mosquitoes, and sometime permanently destruct the breeding sources. The main aim of the source reduction is to prevent the development of immature mosquito stages (larvae and pupae) by reducing the breeding sources. Source reduction activities are environment friendly method followed mainly to control container breeding mosquitoes under the genus *Aedes*. Source reduction activities include eliminating/removing of breeding containers like drums, barrels, bottles, plastic tubs, coconut shells, tyres, other solid waste. Further, cleaning and drying of water storing containers on a weekly basis is recommended (Forsyth et al., 2020).

11.6.2 Modification of Breeding Habitats (Environmental Manipulation)

A form of environmental management consisting of any planned recurrent activity aimed at producing temporary conditions unfavourable to the breeding of vectors in their habitats. Water salinity changes, flushing, regulation of the water level in the reservoirs, altering irrigation settings, vegetation removal, shading and exposure to sunlight, and other activities can prevent the vector from laying eggs, thus preventing vector population development (Hawaria et al., 2020). Manipulation techniques seek to capitalize on the natural factors unfavourable to the vector and to extend and intensify them. The two most commonly used measures are water management and vegetation control.

11.6.2.1 Water Management

Where the body of water cannot be eliminated, the habitat may be manipulated to maintain a biological imbalance unfavourable to the vector. Examples are water-level fluctuation in large impoundments and reservoirs, and the intermittent irrigation of paddy fields. In China, the wet irrigation method has been successfully used for control of the malaria vector *Anopheles sinensis* breeding in rice fields. Instead of flooding the rice fields, water just enough to wet the rice fields is let in, and this is repeated weekly. No water was available for the mosquito to breed, and the species was controlled. This method also reduced the water usage by half and increased rice yield by 10% (Lu, 1999).

11.6.2.2 Vegetation Control

Clearing vegetation to remove the water habitat is an effective method for vector control, specifically mosquito control. Some species of mosquito vectors are sunlight loving; planting of trees will discourage breeding. Some authors call these environmental management manipulations as ecological measures, defined as "the manipulation of the existing environmental factors to effect change on the vector or the pest population." For a long time, environmental management operations have been considered as basic vector control measures in disease prevention and control programmes (Nazareth et al., 2020). They were first used in operations against malaria vectors in the early twentieth century. The advantages are the following:

i. They are effective. Once a body of water is drained or filled and dried or manipulated in any other way to discourage the propagation of disease vectors, the vector population declines and with it the prevalence of the disease transmitted.
ii. Once implemented, their long-term effects continue for years, needing little or no maintenance. Any maintenance required is moderate in cost and simple in operation.
iii. Their overall costs are lower. Environmental modification works are often operations that are performed once and do not have to be repeated. The experiences of Malaysia, India, and the Philippines have proved their worth in their initial campaigns.
iv. Their side benefits are considerable. Most environmental management works contribute to increased crop yields and better-quality agricultural produce.
v. The environmental impact is considerably less harmful, since their operations are limited to environmental alteration, either of a temporary or sometimes of a permanent nature, which is in no way harmful to the quality of the environment. No environmental contamination will result, which might not be the case with other methods of vector control.
vi. No special safety measures are required. The residents of the area do not have to be protected from the other hazards.

11.6.3 Electric Bats

The electric bats/electric mosquito killer rocket is a household device used widely to kill adult mosquitoes. It consists of a lengthy handle with on/off button and a racket-shaped head covering electrical screen which is free from chemicals/insecticides. When the electric screen touches the adult mosquito, an electric flashlight is generated, and the mosquitoes are incinerated.

11.6.4 Sound-based Control

Sound-based mosquito control or electronic mosquito repellents (EMRs) are designed to repel adult female mosquitoes by producing high-pitched sounds against the mosquitoes, and it is inaudible to the human ear. It can be used both indoors and outdoors, and is claimed to repel adult mosquitoes up to 2.5 metres (Kutz, 1974; Helson and Wright, 1977). Commercially available EMRs are claimed to be effective in repelling adult mosquitoes and preventing mosquito-borne diseases

(Enayati et al., 2007). In a few studies, the frequency of the ultrasound used to repel adult mosquitoes were mentioned from 125 Hz to 74,600 Hz (Belton, 1981; Sylla et al., 2000). However, many of the commercially available EMRs were found to be ineffective at the laboratory testing (Foster and Lutes, 1985; Jensen et al., 2000; Andrade and Bueno, 2001; Cabrini and Andrade, 2006; Coro and Suarez, 2000).

11.6.5 Light-Based Attractants

The attract and kill techniques against insects were evaluated since from the 1990s (Crans, 1995). Due to the behavioural differences between the mosquito species, a variety of traps with lights as attractant have been developed and they were further improved with different types of lights and reported to be very effective (Barr et al., 1963, Burkett and Butler, 2005, Hoel, 2005). In a study, 15 mosquito species of 7 genera were collected with diode-equipped sticky cards (DESCs) attached to LED lights, and the number of mosquitoes captured was found to be high with blue LEDs, followed by green, red, and infrared LED lights (Bentley et al., 2009). The wavelength preference of host-seeking mosquitoes are recorded to be high in the blue-green range (400–600 nm), and the preference/attraction decreased as the wavelengths increase (> 600 nm) (Ali et al., 1989; Burkett et al., 1998).

11.7 ILL EFFECTS OF CHEMICAL INSECTICIDES

Urbanization and other man-made activities, including usage of insecticides, led to the decline of many organisms, including natural predators in the ecosystems. Also, the continuous application of insecticides against mosquito vectors affects beneficial organisms and affects human health. Insecticide fogging against adult mosquitoes has become the most common activity in mosquito control programmes to prevent the spread of mosquito-borne diseases (Lee et al., 2020a, 2020b). The following are some of the harmful effects of various mosquito control insecticides.

11.7.1 Chemical Insecticides on Human Health

Chemical insecticides create both acute and chronic health issues in humans. The acute health issues in humans include headaches, giddiness, skin irritation, eye irritation, vomiting, weakness, mental confusion and perplexity, fits, difficulty in breathing, and unconsciousness; they may even cause death. Chronic health problems such as nerve weakness, reproductive issues, immune-system disorders, cancer, hyperexcitability, and gastrointestinal problems may appear after many years of exposure to these chemical insecticides (Ray and Forshaw, 2000). Pregnant women, older people, and children below 5 years of age are the more vulnerable groups to these chemical insecticides (Landrigan, 2001). Organochlorines and organophosphates cause both acute and chronic health problems in human beings. In particular, exposure to temephos causes skin irritation and other complications in children. The continuous exposure to pyrethroids, such as fenvalerate, sumithrin, *d-trans* allethrin, and permethrin are reported to cause reproductive issues like imbalance in reproductive hormone secretions (Garey and Wolff, 1998). Similarly, DEET is reported to produce problems in the respiratory system of humans, and it an allergen for children and pregnant woman (Corbel et al., 2009). Malathion produce headaches, vomiting, respiratory illness, blurred vision, suffocation, variation in heartbeat, and leads to coma and it affects our central nervous system (CNS). Prolonged exposure to malathion causes chronic health issues in human reproductive system, and produce hereditary disabilities (Brenner, 1992). Similarly, synthetic pyrethroids and carbamates are also reported to affect human health (Lee et al., 2020a, b).

11.7.2 Chemical Insecticides on Other Organisms

Chemical insecticides not only kill mosquito stages but also affect other associated organisms and disturb the ecosystem (Stratton, 1986). Particularly, organochlorines and organophosphates affect beneficial insects present in the water. In a study, Rico et al. (2011) reported that malathion is highly toxic and produced acute toxicity to the tested insects – namely, *Hydrophilus* sp., *Buenoa unguis, Palustra laboulbeni*, and fishes – namely, *Colossoma macropomum, Hyphessobrycon erythrostigma, Paracheirodon axelrodi, Nannostomus unifasciatus, Otocinclus affinis*, crustacean *Macrobrachium ferreirai*, and freshwater snail *Pomacea dilioides*. Further, malathion also affects several invertebrates (Key et al., 1998; Van Wijngaarden et al., 2005; Saler and Saglam, 2005).

11.7.3 Chemical Insecticides and Environmental Pollution

The uncontrolled and nonstop use of chemical insecticides against the vector mosquitoes leads to environmental pollution. Particularly, large quantities of organochlorines and organophosphates are used in the mosquito control programme in India that create environmental pollution (Kellogg et al., 2000). Insecticides such as temephos, malathion, aldrin, and dieldrin are reported to pollute small waterbodies and also pollute groundwater systems (Howard et al., 1991). Similarly, organochlorine insecticides such as DDT are also responsible for environmental pollution (Howard et al., 1991; Bingham, 2007).

11.8 PRACTICAL IMPLICATIONS

Vector-borne disease like malaria, filaria and Leishmaniasis are under elimination stage in India. In this context, greatest challenges will be to make optimal use of limited resources including insecticides to achieve mosquito control and improve public health. In recent years, many issues have arisen in the mosquito control programme, especially the use of insecticides. Hence, IMM is suggested for the programme.

11.9 MOSQUITO CONTROL IN THE CURRENT SCENARIO

Currently, mosquito larvae and adults are targeted in the mosquito control programme with insecticides, IGRs, and microbial agents. Mosquito breeding source reduction/elimination is employed in dengue vector control. Further, indoor residual spraying is employed in malaria control, and insecticide-treated bed nets are also issued under the mosquito control programme.

11.10 FUTURE PERSPECTIVES

The WHO Vector Control Advisory Group issued new guidance on the collection of field trial data with epidemiological aspects and these data will be used by WHO to recommend new interventions in future (WHO, 2007). Further, several research studies demonstrated improved guidance on appropriate trial design for vector control, especially for dengue mosquito control (Wolbers et al., 2012; Wilson et al., 2015). Furthermore, new insecticides are warranted, and appropriate application methods continue to be developed. Mosquito control by novel and advanced techniques like attract and kill technology, toxic sugar bait technology, sterile insect technique (SIT), genetic manipulation, and gene drive by CRISPR-Cas9 technology need to be studied and improved. However, the vector control approaches should be effective in reducing vector density and give long-term (or even permanent) results with respect to the control of disease transmission.

11.11 CONCLUSION

In conclusion, mosquito control is a comprehensive approach to control vector mosquitoes to prevent transmission. However, this may only be possible provided several specific criteria are met. The IMM, offering a desired degree of control at a lower cost or with a greater long-term benefit, invariably emphasizes local needs and problems, and, therefore, the operational strategy changes from area to area accordingly. The success of an IMM strategy rests largely on the extent of health education, community participation, and supportive legislation or regulations. In essence, an IMM strategy, including new interventions, would be effective and specific followed in an integrative approach, implying that each tool or control measure or intervention technology is complementarily integrated into the rest of the methodology.

REFERENCES

Abagli, A.Z., Thiery, B.C., Alavo, Perez-Pacheco R. and Platzer, E.G. (2019). Efficacy of the mermithid nematode, *Romanomermis iyengari*, for the biocontrol of *Anopheles gambiae*, the major malaria vector in sub-Saharan Africa. *Parasites & Vectors*, 12(1), 253.

Aditya, G. and Saha, G.K. (2006). Predation of the beetle *Rhantus sikkimensis* (Coleoptera: Dytiscidae) on the larvae of *Chironomus* Meigen (Diptera: Chironomidae) of the Darjeeling Himalayas of India. *Limnologica*, 36(4), 251–257.

Ali, A., Nayar, J.K., Knight, J.W. and Stanley, B.H. (1989). Attraction of Florida mosquitoes (Diptera: Culicidae) to artificial light in the field. *Proceedings of Californian Mosquito Vector Control Association*, 57, 82–88.

Amalraj, D., Vasuki, V., Kalyanasundaram, M. and Tyagi, B.K. (1988a). Laboratory and field evaluation of three insect growth regulators against mosquito vectors. *Indian Journal of Medical Research*, 87, 24–31.

Amalraj, D., Vasuki, V., Sadanandane, C., Kalyanasundaram, M., Tyagi, B.K. and Das, P.K. (1988b). Evaluation of two new juvenile hormone compounds against mosquito vectors. *Indian Journal of Medical Research*, 87, 19–23.

Andrade, C.F.S., Bueno, V.S. (2001). Evaluation of electronic mosquito repelling devices using *Aedes albopictus* (Skuse) (Diptera: Culicidae). *Neotropical Entomology* 30(3), 497–499.

Andreadis, T.G. (2007). Microsporidian parasites of mosquitoes. *Journal of the American Mosquito Control Association*, 23, 3–30.

Angajala, G., Subashini, R. (2018). Evaluation of larvicidal potential of β-sitosterol isolated from indigenous *Aegle marmelos* Correa crude leaf extracts against blood feeding parasites and its binding affinity studies towards sterol carrier protein. *Biocatalysis and Agricultural Biotechnology*, 16, 586–593.

Barr, A.R., Smith, T.A., Boreham, M., White, K.E. (1963). Evaluation of some factors affecting the efficiency oflight traps on collecting mosquitoes. *Journal of Economic Entomolology* 56, 123–127.

Becker, N., Petrić, D., Zgomba, M., Boase, C., Madon, M., Dahl, C., Kaiser, A. (2010). Chapter 11: Biological Control. Mosquitoes and Their Control, pp 405–431.

Belton, P. (1981). An acoustic evaluation of electronic mosquito repellers. *Mosquito News*, 41, 751–755

Benelli, G., Jeffries, C.L., Walker, T. (2016). Biological control of mosquito vectors: Past, present, and future. *Insects*, 7(4), 52.

Benelli, G., Mehlhorn, H. (2016). Declining malaria, rising of dengue and Zika virus: Insights for mosquito vector control. *Parasitology Research*, 115, 1747–1754.

Bentley, M.T., Kaufman, P.E., Kline, D., Hogsette, J. (2009). Response Of Adult Mosquitoes To Light-Emitting Diodes Placed in Resting Boxes and in the Field. *Journal of the American Mosquito Control Association*, 25(3), 285–291.

Bingham, S. (2007). *Pesticides in rivers and groundwater*. Environment Agency, UK.

Bishop. S.C., Hart, R.C. (1931). Notes on some natural enemies of the mosquito in Colorado. *Journal of the New York Entomological Society*, 39(2), 151–157.

Black, B.C., Hollingworth, R.M., Ahmmadsahib, K.I., Kukel, C.D. (1994). Donovan S. Insecticidal action and mitochondrial uncoupling activity of AC-303,630 and related halogenated pyrroles. *Pesticide Biochemistry and Physiology*, 50, 115–128.

Blotzheim, G.V.U.N., Bauer, K.M. (1985). Handbuch der Vögel Mitteleuropas, Band 10/1. AULA-Verlag, Wiesbaden.

Brenner, L. (1992). Malathion. *Journal of Pesticide Reform*, 12, 29.

Brock, T.C.M., Roessink, I., Belgers, J.D.M., Bransen, F., Maund, S.J. (2009). Impact of a benzoyl urea insecticide on aquatic macro-invertebrates in ditch mesocosms with and without non-sprayed sections. *Environmental Toxicology and Chemistry*, 28, 2191–2205.

Bukhari, T., Knols, B.G. (2009). Efficacy of aquatain, a monomolecular surface film, against the malaria vectors *Anopheles stephensi* and *An. gambiae s.s.* in the laboratory. *American Journal of Tropical Medicine and Hygine*, 80, 758–763.

Bukhari, T., Takken, W., Githeko, A.K., Koenraadt, C.J. (2011). Efficacy of aquatain, a nomomolecular film, for the control of malaria vectors in rice paddies. *Plos One*, 6(6), e21713.

Burkett, D.A., Butler, J.F. (2005). Laboratory evaluation of colored light as an attractant for female *Aedes aegypti, Aedes albopictus, Anopheles quadrimaculatus* and *Culex nigripalpus*. *Florida Entomologist*, 88, 383–389.

Burkett, D.A., Butler, J.F., Kline, D.L. (1998). Field evaluation of colored light-emitting diodes as attractants for woodland mosquitoes and other Diptera in north central Florida. *Journal of the American Mosquito Control Association*, 14, 186–195.

Cabrini, I., Andrade, C.F.S. (2006). Evaluation of seven new electronic mosquito repellers. *Entomologia Experimentalis et Applicata*, 121(2), 185–188.

Campos, R.E., Fernandez, L.A., Sy, V.E. (2004). Study of the insects associated with the floodwater mosquito *Ochlerotatus albifasciatus* (Diptera: Culicidae) and their possible predators in Buenos Aires province. *Hydrobiologia*, 524(1), 92–102.

Carlson, J., Keating, J., Mbogo, C.M., Kahindi, S., Beier, J.C. (2004). Ecological limitations on aquatic mosquito predator colonization in the urban environment. *Journal of Vector Ecology*, 29(2), 331–339.

Case, T.J., Washino, R.K. (1979). Flatworm control of mosquito larvae in rice fields. *Science*, 206, 1412–1414.

Casida, J.E., Quistad, G.B. (2004). Organophosphate toxicology: safety aspects of nonacetylcholinesterase secondary targets. *Chemical Research in Toxicology*, 17(8), 983–998.

Chareonviriyaphap, T., Bangs, M.J., Suwonkerd, W., Kongmee, M., Corbel, V., Ngoen-Klan, R. (2013). Review of insecticide resistance and behavioral avoidance of vectors of human diseases in Thailand. *Parasites & Vectors*, 6, 280.

Charlwood, J.D., Pinto, J., Sousa, C.A., Ferreira, C., do Rosario, V.E. (2002). Male size does not affect mating success (of *Anopheles gambiae* in Sao Tome). *Medical and Veterinary Entomology*, 16, 109–111.

Choo, M.Z.J., Low, B.W., Ngiam, R.W.J., Yeo, D.C.J. (2021). Predation of mosquitos by odonates in a tropical urban environment: insights from functional response and field mesocosm experiments. *Biological Control*, 161, 104702.

Corbel, V., Stankiewicz, M., Pennetier, C., Fournier, D., Stojan, J., Girard, E., Dimitrov, M., Molgó, J., Hougard, J.M., Lapied, B. (2009). Evidence for inhibition of cholinesterases in insect and mammalian nervous systems by the insect repellent DEET. *BMC Biology*, 7, 1–11.

Corbet, P.S. (1986). Using dragonflies to suppress mosquitoes in domestic water-storage containers. *Waterlines*, 4, 10–11.

Corbet, S.A., Tiley, C., Moorhouse, T., Giam, C., Pursglove, S., Raby, J., Rich, M. (2000). Surface films as mosquito larvicides: partitioning the mode of action. *Entomologia Experimentalis et Applicata*, 94(3), 295–307.

Coro, F., Suarez, S. (2000). Review and history of electronic mosquito repellers. *Wing Beats*, 11(2), 6–7.

Crans, W.J. (1995). Resting boxes as mosquito surveillance tools. *Proceedings of the New Jersey Mosquito Control Association*, 82, 53–57.

Curtis, C.F., Minjas, J. (1985). Expanded polystyrene for mosquito control. *Parasitology Today*, 1(1), 36.

da Silva, O.S., da Silva, F.C., de Barros, F.M.C., da Silva, J.L.R., de Loreto Bordignon, S.A., Eifler-Lima, V.L., Prophiro, J.S. (2013). Larvicidal and growth-inhibiting activities of extract and benzopyrans from *Hypericum polyanthemum* (Guttiferae) against *Aedes aegypti* (Diptera: Culicidae). *Industrial Crops and Products*, 45, 236–239.

Dambach, P. (2020). The use of aquatic predators for larval control of mosquito disease vectors: opportunities and limitations. *Biological Control*, 150, e104357.

Damtie, D., Mekonnen, Y. (2021). Toxicity and oviposition deterrent activities of thyme essential oils against Anopheles arabiensis. *Psyche*, 7. https://doi.org/10.1155/2021/6684156

Darbro, J.M., Graham, R.I., Kay, B.H., Ryan, P.A., Thomas, M.B. (2011). Evaluation of entomopathogenic fungi as potential biocontrol agents of the dengue mosquito, *Aedes aegypti* (Diptera: Culicidae). *Biocontrol Science and Technology*, 21, 1027–1047.

Das, P.K., Tyagi, B.K., Kalyasundaram, M. (1986b). Vernacide, a new insecticide impregnated paint for controlling mosquito vector, *Culex quinquefasciatus* and cockroach, *Periplanata americana*. *Indian Journal of Medical Research*, 83, 268–270.

Das, P.K., Tyagi, B.K., Somachari, N., Vasuki, V. (1986a). Efficacy of Arosurf, a molecular surface film in controlling *Culex quinquefasciatus* Say, *Anopheles stephensi* Liston and *Aedes aegypti* (L.). *Indian Journal of Medical Research*, 83, 271–276.

David, J.P., Rey, D., Pautou, M.P., Meyran, J.C. (2000). Differential toxicity of leaf litter to dipteran larvae of mosquito developmental sites. *Journal of Invertebrate Pathology*, 75, 9–18.

De Las Llagas, L.A., Portus, L.M. (2016). Promoting Health Communication and Empowerment: Intervention towards Malaria Mosquito Control. *International Journal of Health Wellness & Society*, 6(3), 61–75.

DeKort, C.A.D., Granger, N.A. (1996). Regulation of JH titers: the relevance of degradative enzymes and binding proteins. *Archives of Insect Biochemistry and Physiology*, 33, 1–26.

Dhadialla, T.S., Carlson, G.R., Le, D.P. (1998). New insecticides with ecdysteroidal and juvenile hormone activity. *Annual Review of Entomology*, 43, 545–569.

Diabate, A., Yaro, A.S., Dao, A., Diallo, M., Huestis, D.L., Lehmann, T. (2011). Spatial distribution and male mating success of *Anopheles gambiae* swarms. *BMC Evolutionary Biology*, 11, 184.

Dida, G.O., Gelder, F.B.,. Anyona, D.N., Abuom, P.O., Onyuka, J.O., Matano, A.S., Adoka, S.O., Kanangire, C.K., Owuor, P.O., Ouma, C., Ofulla, A.V.O. (2015). Presence and distribution of mosquito larvae predators and factors influencing their abundance along the Mara River, Kenya and Tanzania. *Springerplus*, 4, 136.

Dinan, L., Savchenko, T., Whiting, P., Sarker, S.D., (1999). Plant natural products as insect steroid receptor agonists and antagonists. *Pesticide Science*, 55, 331–335.

Dua, V.K., Sharma, S.K., Sharma, V.P. (1989). Use of expanded polystyrene beads for the control of mosquitoes in an industrial. complex at Hardwar, India. *Journal of the American Mosquito Control Association*, 5(4), 614–615.

Eba, K., Duchateau, L., Olkeba, B.K., Boets, P., Bedada, D., Goethals, P.L.M., Mereta, S.T., Yewhalaw, D. (2021). Bio-Control of *Anopheles* Mosquito Larvae Using Invertebrate Predators to Support Human Health Programs in Ethiopia. *International Journal of Environmental Research and Public Health*, 18, 1810.

Enayati, A., Hemingway, J., Garner, P. (2007). Electronic mosquito repellents for preventing mosquito bites and malaria infection. *Cochrane Database of Systematic Reviews*, 2, CD005434.

Evans, H.C., Elliot, S.L., Barreto, R.W. (2018). Entomopathogenic fungi and their potential for the management of *Aedes aegypti* (Diptera: Culicidae) in the Americas. *Memórias do Instituto Oswaldo Cruz*, 113, 206–214.

Fischer, S., Marinone, M.C., Fontanarrosa, M.S., Nieves, M., Schweigmann, N. (2000). Urban rain pools: Seasonal dynamics and entomofauna in a park of Buenos Aires. *Hydrobiologia*, 441(1), 45–63.

Forsyth, J.E., Mutuku, F.M., Kibe, L., Mwashee, L., Bongo, J., Egemba, C., Ardoin, N.M., LaBeaud, A.D. (2020). Source reduction with a purpose: Mosquito ecology and community perspectives offer insights for improving household mosquito management in coastal Kenya. *PLoS Neglected Tropical Diseases*, 14(5), e0008239.

Foster, W.A., Lutes, K.I. (1985). Tests of ultrasonic emissions on mosquito attraction to hosts in a flight chamber. *Journal of the American Mosquito Control Association*, 1(2), 199–202.

Garey, J., Wolff, M.S., (1998). Estrogenic and antiprogestagenic activities of pyrethroid insecticides. *Biochemical and Biophysical Research Communications*, 251(3), 855–859.

Ghosh, S.K., Dash, A.P. (2007). Larvivorous fish against malaria vectors. *Transactions of the Royal Society of Tropical Medicine Hygiene*, 103(2), 210–211.

Ghosh, S.K., Rashi, M. (2019). Malaria elimination in India-The way forward. *Journal of Vector Borne Diseases*, 56(1), 32–40.

Giulivi, C., Ross-Inta, C., Horton, A.A., Luckhart, S. (2008). Metabolic pathways in *Anopheles stephensi* mitochondria. *Biochemical Journal*, 415(2), 309–316.

Gunning, R.V., Moores, G.D. (2002). Chlorfenapyr resistance in Helicoverpa armigera in Australia. In: Pests and diseases proceedings of the British crop protection council conference. London, UK: *British Crop Protection Council*, 793–798.

Hawaria, D., Demissew, A., Kibret, S., Lee, M.C., Yewhalaw, D., Yan, G. (2020). Effects of environmental modification on the diversity and positivity of anopheline mosquito aquatic habitats at Arjo-Dedessa irrigation development site, Southwest Ethiopia. *Infectious Diseases of Poverty*, 9, 9.

Helson, B.V., Wright, R.E. (1977). Field evaluation of electronic mosquito repellers in Ontario. *Proceedings of the Entomological Society of Ontario*, 108, 59–61.

Hematpoor, A., Liew, S.Y., Chong, W.L., Azirun, M.S., Lee, V.S., Awang, K. (2016). Inhibition and Larvicidal activity of phenylpropanoids from Piper sarmentosum on acetylcholinesterase against mosquito vectors and their binding mode of interaction. *PLoS One*, 11(5), e0155265.

Hertlein, M.B., Mavrotas, C., Jousseaume, C., Lysandrou, M., Thompson, G.D., Jany, W., Ritchie, S.A. (2010). A Review of Spinosad as a Natural Product for Larval Mosquito Control. *Journal of the American Mosquito Control Association*, 26(1), 67–87.

Hoel, D. (2005). Response of *Aedes albopictus* (Diptera: Culicidae) to traps, attractants, and adulticides in north central Florida. [Ph.D. dissertation]. University of Florida, Gainesville, FL.

Howard, P.H., Boethling, R.S., Jarvis, W.F., Meylan, W.M., Michalenko, E.M. (1991). *Handbook of environmental degradation rates*. Ed. Heather Taub Printup. Lewis Publishers. Chelsea, MI.

Iyengar, M.O.T. (1927). Parasitic nematodes of *Anopheles* in Bengal. In: *Transactions of the Far Eastern Association of Tropical Medicine*, 7th Congress, 3, 128–135.
Jambulingam, P.K., Gunasekharan, K., Sahoo, S.S., Hota, P.K., Tyagi, B.K., Kalyanasundaram, M. (1989). Effect of permethrin-impregnated bednets in reducing population of malaria vector, *Anopheles culicifacies* in a tribal village of Orissa State (India). *Indian Journal of Medical Research*, 89, 48–51.
Jensen, T., Lampman, R., Slamecka, M.C., Novak, R.J. (2000). Field efficacy of commercial antimosquito products in Illinois. *Journal of the American Mosquito Control Association*, 16(2), 148–152.
Jerome, J., Schleier, I.I., Peterson, R.K.D. (2011). Chapter 3: Pyrethrins and Pyrethroid Insecticides. *Green Trends in Insect Control.* Edited by Oscar Lopez and Jose G. Fernfmdez-Bolafios. Royal Society of Chemistry. www.rsc.org
Kay, B.H., Vu, S.N., Tien, T.V., Yen, N.T., Phong, T.V., Diep, V.T.B., Ninh, T.U., Bektas, A., Aaskov, J.G. (2002). Control of *Aedes* vectors of dengue in three provinces of Vietnam by use of *Mesocyclops* (Copepoda) and community-based methods validated by entomologic, clinical, and serological surveillance. *American Journal of Tropical Medicine and Hygiene*, 66(1), 40–48.
Kellogg, R.L., Nehring, R., Grube, A., Goss, D.W., Plotkin, S. (2000). Environmental indicators of pesticide leaching and runoff from farm fields. The Wayback Machine. *United States Department of Agriculture Natural Resources Conservation Service*, 213–256.
Key, P.B., Fulton, M.H., Scott, G.I., Layman, S.L., Wirth, E.F. (1998). Lethal and sublethal effects of malathion on three life stages of the grass shrimp, *Palaemontes pugio. Aquatic Toxicology*, 40(4), 311–322.
Kirkhoff, C.J., Simmons, T.W., Hutchinson, M. (2013). Adult mosquitoes parasitized by larval water mites in Pennsylvania. *Journal of Parasitology*, 99(1), 31–39.
Kirst, H.A. (2010). The spinosyn family of insecticides: realizing the potential of natural products research. *The Journal of Antibiotics*, 63, 101–111.
Kocher, D.K., Jamwal, S. (2016). Predatory potential of cyclopoid copepods: A biological approach for control of mosquito larvae. *Journal of Insect Science*, 29(1), 216–223.
Kocher, D.K., Jamwal, S., Bakshi, D.K., Kumar, A. (2018). Mosquito larvae specific predation by native cyclopoid copepod species, *Mesocyclops aspericornis* (Daday, 1906). *Journal of Entomology and Zoology Studies*, 6(3), 983–987.
Kögel, F. (1984). Die Prädatoren der Stechmückenlarven im Ökosystem der Rheinauen. Ph. D. thesis, University of Heidelberg, pp 347.
Kögel, F. (1987). Zur Biologie und O˙kologie von *Rhantus consputus* Strm. *Entomologische Arbeiten aus dem Museum*, 35(36), 5–19.
Kutz, F.W. (1974). Evaluation of an electronic mosquito repelling device. *Mosquito News*, 34, 369–375.
Landrigan, P.J. (2001). Pesticides and PCBs: Does the evidence show that they threaten children's health? *Contemporary Pediatrics*, 18(2), 110.
Lazaro, A., Han, W.W., Manrique-Saide, P., George, L., Velayudhan, R., Toledo, J., Ranzinger, S.R., Horstick, O. (2015). Community effectiveness of copepods for dengue vectorcontrol: systematic review. *Tropical Medicine and International Health*, 20(6), 685–706.
Lee, N.S.M., Clements, G.R., Ting, A.S.Y., Wong, Z.H., Yek, S.H. (2020a). Persistent mosquito fogging can be detrimental to non-target invertebrates in an urban tropical forest. *PeerJ*, 8, e10033.
Lee, W.S., Lim, Y.H., Kim B.N., Shin, C.H., Lee, Y.A., Kim, J.I., Hong, Y.C., Kim, K.N. (2020b). Residential pyrethroid insecticide use, urinary 3-phenoxybenzoic acid levels, and attention-deficit/hyperactivity disorder-like symptoms in preschool-age children: the environment and development of children study. *Environmental Research*, 188, 109739.
Li, T., Reid, W.R., Xu, Q., Dong, K., Liu, N. (2012). Multiple mutations and mutation combinations in the sodium channel of permethrin resistant mosquitoes, Culex quinquefasciatus. *Scientific Reports*, 2, 781.
Lischetti, A.B., (1919). Un verme del género Planaria, enemigo natural de las larvas del mosquito. *Physis (Buenos Aires)*, 4, 591–595.
Liu, N. (2015). Insecticide Resistance in Mosquitoes: Impact, Mechanisms, and Research Directions. *Annual Review of Entomology*, 60(1), 537–559.
Lombardo, P., Cooke, G.D. (2002). Consumption and preference of selected food types by two freshwater gastropod species. *Archiv fur Hydrobiologie*, 155(4), 667–685.
Lu, B.L. (1999). *Integrated management of mosquito*, 2nd edn. Science Press, Beijing.
Lundkvist, E., Landin, J., Jackson, M., Svensson, C. (2003). Diving beetles (Dytiscidae) as predators of mosquito larvae (Culicidae) in field experiments and in laboratory tests of prey preference. *Bulletin of Entomological Research*, 93, 219.
Mapossa, A.B., Focke, W.W., Tewo, R.K., Androsch, R., Kruger, T. (2021). Mosquito-repellent controlled-release formulations for fighting infectious diseases. *Malaria Journal*, 20, 165.

Marten, G.G., Reid, J.W. (2007). Cyclopoid copepods. *Journal of American Mosquito Control Association*, 23(2 Suppl), 65–92.
Matsuda, K., Buckingham, S.D., Kleier, D., Rauh, J.J., Grauso, M., Sattelle, D.B. (2001). Neonicotinoids: insecticides acting on insect nicotinic acetylcholine receptors. *Trends in Pharmacological Sciences*, 22(11), 573–580.
Mbare, O., Lindsay, S.W., Fillinger, U. (2014). Aquatain® Mosquito Formulation (AMF) for the control of immature *Anopheles gambiae sensu stricto* and *Anopheles arabiensis*: dose-responses, persistence and sub-lethal effects. *Parasite & Vectors*, 7, 438.
McCoull, C.J., Swain, R., Barnes, R.W. (1998). Effect of temperature on the functional response and components of attack rate in *Naucoris congrex* Stål (Hemiptera: Naucoridae). *Australian Journal of Entomology*, 37(4), 323–327.
Melo, A.S., Andrade, C.F. (2001). Differential predation of the planarian Dugesia tigrine on two mosquito species under laboratory conditions. *Journal of American Mosquito Control Association*, 17, 81–83.
Mogi, M. (2007). Insects and other invertebrate predators. In: T.G. Floore (ed.). Biorational Control of Mosquitoes. *Journal of American Mosquito Control Association*, 23(Suppl. 2), 93–109.
Morales, M.E., Wesson, D.M., Sutherland, I.W., Impoinvil, D.E., Mbogo, C.M., Githure, J.I., Beier, J.C. (2003). Determination of *Anopheles gambiae* larval DNA in the gut of insectivorous dragonfly (Libellulidae) nymphs by polymerase chain reaction. *Journal of American Mosquito Control Association*, 19, 163–165.
Muema, J.M., Bargul, J.L., Nyanjom, S.G., Mutunga, J.M., Njeru, S.N. (2016). Potential of Camellia sinensis proanthocyanidins-rich fraction for controlling malaria mosquito populations through disruption of larval development. *Parasites & Vectors*, 9(1), 1–10.
N'Guessan, R., Boko, P., Odjo, A., Akogbeto, M., Yates, A., Rowland, M. (2007). Chlorfenapyr: a pyrrole insecticide for the control of pyrethroid or DDT resistant *Anopheles gambiae* (Diptera: Culicidae) mosquitoes. *Acta Tropica*, 102, 69–78.
Nayar, J.K., Ali, A. (2003). A review of monomolecular surface films as larvicides and pupicides of mosquitoes. *Journal of Vector Ecology*, 28, 190–199.
Nazareth, T., Craveiro, I., Moutinho, A., Seixas, G., Gonçalves, C., Gonçalves, L., Teodósio, R., Sousa, C.A. (2020). What happens when we modify mosquitoes for disease prevention? A systematic review. *Emerging Microbes & Infections*, 9(1), 348–365.
Nobuko, T., Vu, P.T., Masahiro, T. (2020). Climate Change May Restrict the Predation Efficiency of *Mesocyclops aspericornis* (Copepoda: Cyclopidae) on *Aedes aegypti* (Diptera: Culicidae) Larvae. *Insects*, 11(5), 307.
Noreña-Janssen, C., Faubel, A. (1992). Revision of the Subfamily Mesostominae, Rhabdocoela, Platyhelminthes. *Mitteilungen aus dem Hamburgischen Zoologischen Museum und Institut*, 89, 7–47.
Oxborough, R.M., N'Guessan, R., Jones, R., Kitau, J., Ngufor, C., Malone, D., Mosha, F.W., Rowland, M.W. (2015). The activity of the pyrrole insecticide chlorfenapyr in mosquito bioassay: towards a more rational testing and screening of non-neurotoxic insecticides for malaria vector control. *Malaria Journal*, 14, 124.
Papáček, M. (2001). Small aquatic and ripicolous bugs (Heteroptera: Nepomorpha) as predators and prey: The question of economic importance. *European Journal of Entomology* 98, 1–12.
Perumalsamy, H., Kim, J.R., Oh, S.M., Jung, J.W., Ahn, Y.J., Kwon, H.W. (2013). Novel histopathological and molecular effects of natural compound pellitorine on larval midgut epithelium and anal gills of *Aedes aegypti*. *PLoS One*, 8(11): e80226.
Poopathi, S., Tyagi, B.K. (2006). The Challenge of Mosquito Control Strategies: FROM Primordial to Molecular Approaches. *Biotechnology and Molecular Biology Reviews*, 1(2), 51–65.
Pradeepa, V., Senthilnathan, S., Narayanan, S.S., Rani, S.S., Srinivasan, P.V., Thanigaivel, A., Ponsankar, A., Edwin, E.S., Bagavathy, M.S., Kalaivani, K., Murugan, K., Duraipandiyan, V., Al-Dhabi, N.A. (2016). Potential mode of action of a novel plumbagin as a mosquito repellent against the malarial vector *Anopheles stephensi*, (Culicidae: Diptera). *Pesticide Biochemistry and Physiology*, 134, 84–93.
Quiroz-Martínez, H., Rodríguez-Castro, A. (2007). Aquatic insects as predators of mosquito larvae. In: T.G. Floore (ed.). Biorational Control of Mosquitoes. *Journal of American Mosquito Control Association*, 23(Suppl. 2), 110–117.
Raghavendra, K., Sharma, P., Dash, A.P. (2008). Biological control of mosquito populations through frogs: opportunities & constrains. *Indian Journal of Medical Research*, 128(1), 22–25.
Rattan, R.S. (2010). Mechanism of action of insecticidal secondary metabolites of plant origin. *Crop Protection*, 29(9), 913–920.
Ray, D.E., Forshaw, P.J. (2000). Pyrethroid insecticides: poisoning syndromes, synergies, and therapy. *Journal of Toxicology: Clinical Toxicology*, 38(2), 95–101.

Reegan, A.D., Bentrock, B.B., Asharaja, A.C., Tennyson, S., Raveen, R. (2020). Toxicity and sub-lethal effects of temephos, lambda-cyhalothrin and cypermethrin on predatory insect *Diplonychus rusticus* Fabricius (Hemiptera: Belostomatidae). *International Journal of Tropical Insect Science*, 41(9), 1–8.
Reegan, A.D., Gandhi, M.R., Paulraj, M.G., Balakrishna, K., Ignacimuthu, S. (2014). Effect of niloticin, a protolimonoid isolated from *Limonia acidissima* L. (Rutaceae) on the immature stages of dengue vector *Aedes aegypti* L. (Diptera: Culicidae). *Acta Tropica*, 11(139C), 67–76.
Reegan, A.D., Gandhi, M.R., Sivaraman, G., Cecilia, K.F., Ravindhran, R., Balakrishna, K., Ignacimuthu, S. (2016). Bioefficacy of Ecbolin A and Ecbolin B isolated from *Ecbolium viride* (Forssk.) Alston on dengue vector *Aedes aegypti* L. (Diptera: Culicidae). *Parasite Epidemiology and Control*, 1, 78–84.
Rico, A., Waichman, A.V., Geber-Corrêa, R., Van den Brink, P.J. (2011). Effects of malathion and carbendazim on Amazonian freshwater organisms: comparison of tropical and temperate species sensitivity distributions. *Ecotoxicology*, 20(4), 625–634.
Ross, R. (1906). Notes on the parasites of mosquitoes found in India between 1895 and 1899. *Journal of Hygiene*, 6, 101–108.
Rusconi, T.R., Hopkins, C.R., Denton, J.S., Piermarini, P.M. (2018). Pharmacological inhibition of inward rectifier potassium channels induces lethality in larval *Aedes aegypti*. *Insects*, 9(4), 163.
Saha, N., Aditya, G., Bal, A., Saha, G.K. (2007). A comparative study of predation of three aquatic heteropteran bugs on *Culex quinquefasciatus* larvae. *Limnology*, 8, 73–80.
Saha, N., Aditya, G., Banerjee, S., Saha, G.K. (2012). Predation potential of odonates on mosquito larvae: Implications for biological control. *Biological Control*, 63(1), 1–8.
Saha, N., Kundu, M., Saha, G.K., Aditya, G. (2020). Alternative prey influence the predation of mosquito larvae by three water bug species (Heteroptera: Nepidae). *Limnological Review*, 20(4), 173–184.
Sahu, S., Jambulingam, P., Vijayakumar, T., Kalyanasundaram, M. (2003). Impact of alphacypermethrin treated bed nets on malaria in villages of Malkangiri district, Orissa, India. *Acta Tropica*, 89(1), 55–66.
Saler, S., Saglam, N. (2005). Acute toxicity of malathion on *Daphnia magna* Straus, 1820. *Journal of Biological Sciences*, 5(3), 297–299.
Sanchez-Bayo, F., Goka, K. (2006). Ecological effects of the insecticide imidacloprid and a pollutant from antidandruff shampoo in experimental rice fields. *Environmental Toxicology and Chemistry*, 25(6), 1677–1687.
Sandhanam, S.D., Ganesan, P., Stalin, A., Michael, G.P., Balakrishna, K., Pandikumar, P., Ignacimuthu, S., Naif, A.A. (2018). Effect of compound isolated from *Lawsonia inermis* (L.) (Myrtales: Lythraceae) on the immature stages of filarial vector Culex quinquefasciatus Say (Diptera: Culicidae) and its docking analysis with acetylcholinesterase (AChE1), *Biocatalysis and Agricultural Biotechnology*, 15, 210–218.
Sandhanam, S.D., Pathalam, G., Antony, S., Samuel, R., Michael, G.P., Kedike, B., Pandikumar, P., Ignacimuthu, S., Al-Dhabi, N.A. (2019). Effect of tiliamosine, a bis, benzylisoquinoline alkaloid isolated from Tiliacora acuminata (Lam.) Hook. f. & Thom on the immature stages of filarial mosquito Culex quinquefasciatus Say (Diptera: Culicidae). *Experimental Parasitology*, 204, 107719.
Scholte, E.J., Knols, B.G., Samson, R.A., Takken, W. (2004). Entomopathogenic fungi for mosquito control: A review. *Journal of Insect Science*, 4, 19.
Sebastian, S., Sein, M.M., Thu, M.M., Corbet, P.S. (1990). Suppression of *Aedes aegypti* (Diptera: Culicidae) using augmentative release of dragon fly larvae (Odonata: Libellulidae) with community participation in Yangon, Mynmar. Bull. *Entolnological Reasearch*, 80, 223–232.
Senthilnathan, S., Hisham, A., Jayakumar, G. (2008). Larvicidal and growth inhibition of the malaria vector Anopheles stephensi by triterpenes from Dysoxylum malabaricum and Dysoxylum beddomei. *Fitoterapia*, 79(2), 106–111.
Shaalan, E.A., Canyon, D.V. (2009). Aquatic insect predators and mosquito control. *Tropical Biomedicine*, 26, 223–261.
Sharma, V.P., Mehrotra, K.N. (1986). Malaria resurgence in India: A critical study. *Social Science & Medicine*, 22, 835–845.
Sharma, V.P., Sharma, R.C., Gautam, A.S. (1986). Bioenvironmental control of malaria in Nadiad, Kheda district, Gujarat. *Indian Journal of Malariology*, 23, 95–117.
Sharma, V.P., Yadav, R.S. (1995). Impregnating mosquito nets with cyfluthrin. Study in the mining settlements of Orissa, India to control malaria. *Public Health*, 12, 8–17.
Simmons, T.W., Hutchinson, M.L. (2016). A Critical Review of All Known Published Records for Water Mite (Acari: Hydrachnidiae) and Mosquito (Diptera: Culicidae) Parasitic Associations from 1975 to present. *Journal of Medical Entomology*, 53(4), 737–752.
Singh, S.P., Mohan, L. (2013). Biological control of mosquitoes by insectivorous flycatcher birds. *Journal of Entomological Research*, 37(4), 359–364.

Soltani, A., Vatandoost, H., Jabbari, H., Mesdaghinia, A.R., Mahvi, A.H., Younesian, M., Hanafi-Bojd, A.A., Bozorgzadeh, S. (2012). Field efficacy of expanded polystyrene and shredded waste polystyrene beads for mosquito control in artificial pools and field trials, Islamic Republic of Iran. *Eastern Mediterranean Health Journal*, 18(10), 1042–1048.

Srinivasan, R., Huang, B.S., Venugopal. S., Johnston, A.D., Chai, H., Zeng, H., Golshani, P., Khakh, B.S. (2015). Ca(2+) signaling in astrocytes from Ip3r2(-/-) mice in brain slices and during startle responses in vivo. *Nature Neuroscience*, 18(5), 708–717.

Stratton, G.E. (1986). The inheritance of courtship behavior and its role as a reproductive isolating mechanism in two species of Schizocosa wolf spiders (Araneae; Lycosidae). *Evolution*, 40, 129–141.

Su, N.Y., Scheffrahn, R.H. (1990). Potential of insect growth regulators as termiticides: A review. *Sociobiology* 17(2), 313–328.

Sukkanon, C., Yaicharoen, R., Ngrenngarmlert, W. (2016). Comparative effectiveness of monomolecular surface film on *Aedes aegypti* (L.) and *Anopheles minimus* (Theobald) (Diptera: Culicidae). *Agriculture and Natural Resources*, 50, 465e469.

Suprakash, K., Aditya, A.K. (2003). Biological control of mosquitoes by aquatic planaria. *Tiscia*, 34, 15–18.

Sylla, H. K., Lell, B., Kremsner, P.G. (2000). A blinded, controlled trial of an ultrasound device as mosquito repellent. *Wiener Klinische Wochenschri*, 112(10), 448–450.

Thakur, A., Kocher, D.K. (2018). Evaluation of predatory and competitive potential of cyclopoid copepod, *Mesocyclops aspericornis* (Daday, 1906) against *Aedes aegypti* L. mosquito larvae. *Journal of Insect Science*, 31(1–2), 57–62.

Tomlin, C.D.S. (2009). The e-pesticide manual. 12 ed. City. *British Crop Protection Council*, 49, 647–648.

Tranchida, M.C., Maciá, A., Brusa, F., Micieli, M.V., Ga, J.J. (2009). Predation potential of three flatworm species (Platyhelminthes: Turbellaria) on mosquitoes (Diptera: Culicidae). *Biological Control*, 49, 270–276.

Tyagi, B.K. (1994). Distribution of arthropod vector-borne communicable diseases and control of their vectors in India. *Indian Review of Life Sciences*, 14, 223–243.

Tyagi, B.K. (2020). *Mosquito Hunters: A history of hostilities against man's deadliest foe – the mosquito – since 1881*. Scientific Publisher (India), Jodhpur. 474.

Tyagi, B.K., Naik R.B., Bhattacharya, S. (2020). *Dr Ronald Ross: mosquito, malaria, India and the Nobel Prize – an untold story of the First Indian Nobel Laureate*. Scientific Publishers (India), 258 pp.

U.S. Environmental Protection Agency. (1999). *Product Performance Test Guidelines, OPPTS 810.3700, Insect Repellents for Human Skin and Outdoor Premises*. Washington: U.S. Government Printing Office. Available from: http://www.epa.gov/OPPTS_Harmonized

U.S. Environmental Protection Agency. Notice to Manufacturers, Formulators, Producers and Registrants of Pesticide Products on Minimum Risk Pesticides Exempted under FIFRA Section 25(b) Pesticide Registration (PR). Notice 2000-6, 2000. Labeling Unit, Registration Division (7505C). Washington: Office of Pesticide Programs, EPA; 2000. Available from: URL: http://www.epa.gov/opppmsd1/PR_Notices/#2000

U.S. Environmental Protection Agency. (n.d.) Using insect repellents safely. EPA Office of Pesticide Programs. Available from: http://www.epa.gov/pesticides/citizens/insectrp.htm

Van Wijngaarden, R.P.A., Brock, T.C.M., Van den Brink, P.J. (2005). Threshold levels for effects of insecticides in freshwater ecosystems: a review. *Ecotoxicology*, 14, 355–380.

Venkatesh, A., Tyagi, B.K. (2015). Predatory potential of *Bradinopyga geminata* and *Ceriagrion coromandelianum* larvae on dengue vector *Aedes aegypti* under controlled conditions (Anisoptera: Libellulidae; Zygoptera: Coenagrionidae; Diptera: Culicidae). *Odonatologica*, 42(2), 139–149.

Vincent, C., Boiteau, G. (2001). Pneumatic Control of Agricultural Insect Pests. In: Vincent, C., Panneton, B., Fleurat-Lessard, F. (eds) *Physical Control Methods in Plant Protection*. Springer, Berlin, Heidelberg. https://doi.org/10.1007/978-3-662-04584-8_19

Vu, S.N., Nguyen, T.Y., Tran, V.P., Truong, U.N., Le, Q.M., Le, V.L., Le, T.N., Bektas, A., Briscombe, A., Aaskov, J.G., Ryan, P.A., Kay, B.H. (2005). Elimination of dengue by community programs using *Mesocyclops* (Copepoda) against *Aedes aegypti* in central Vietnam. *American Journal of Tropical Medicine and Hygiene*, 72(1), 67–73.

Wada, Y. (1987). Strategies for control of Japanese encephalitis in rice production systems in developing countries. In: *Vector-borne disease control in humans through rice agro ecosystems management: proceedings of the Workshop on Research and Training Needs in the Field of Integrated Vector-Borne Disease Control in Riceland Agroecosystems of Developing Countries*, 1987 Mar 9–14. International Rice Research Institute in collaboration with the WHO/FAO/UNEP Panel of Experts on Environmental Management for Vector Control.

Walker, K., Lynch, M. (2007). Contributions of *Anopheles* larval control to malaria suppression in tropical Africa: review of achievement and potential. *Medical and Vetetinary Entomol*ogy, 21, 2–21.

Wang, C.Y., Teng, H.J., Lee, S.J., Lin, C., Wu, J.W., Wu, H.S. (2013). Efficacy of various larvicides against *Aedes aegypti* immatures in the laboratory. *Japanese Journal of Infectious Diseases*, 66, 341–344.

Webb, C.E., Russell, R.C. (2012). Does the monomolecular film aquatain mosquito formula provide effective control of container-breeding mosquitoes in Australia? *Journal of American Mosquito Control Association*, 28, 53–58.

Webb, C.E., Russell, R.C.A. (2009). Laboratory investigation of the mosquito control potential of the monomolecular film Aquatain mosquito formula against immature stages of *Aedes aegypti* and *Culex quinquefasciatus*. *Journal of American Mosquito Control Association*, 25, 106–109.

Werblow, A., Martin, P., Dörge, D.D., Koch, L.K., Mehlhorn, H., Melaun, C., Klimpel, S. (2015). Hyperparasitism of mosquitoes by water mite larvae. *Parasitology Research*, 114(7), 2757–2765.

WHO, (2003). Malaria entomology and vector control: Learner's Guide. pp 67.

WHO (2007). Department of Control of Neglected Tropical Diseases. How to design vector control efficacy trials. Guidance on phase III vector control field trial design. Report No. WHO/HTM/NTD/VEM/2017.03 http://www.who.int/neglected_diseases/vector_ecology/resources/WHO_HTM_NTD_VEM_2017.03/en/

WHO (2009). Dengue, Guidelines for Diagnosis, Treatment, Prevention and Control. 147 p. http://www.who/2009/87:887

Wilson, A.L., Boelaert, M., Kleinschmidt, I., Pinder, M., Scott, T.W., Tusting, L.S., Lindsay, S.W. (2015). Evidence-based vector control? Improving the quality of vector control trials. *Trends in Parasitology*, 31, 380–390.

Wolbers, M., Kleinschmidt, I., Simmons, C.P., Donnelly, C.A. (2012). Considerations in the design of clinical trials to test novel entomological approaches to dengue control. *PLoS Neglected Tropical Diseases*, 6, e1937.

Youssefi, M.R., Tabari, M.A., Esfandiari, A., Kazemi, S., Moghadamnia, A.A., Sut, S., Dall'Acqua, S., Benelli, G., Maggi, F. (2019). Efficacy of two monoterpenoids, carvacrol and thymol, and their combinations against eggs and larvae of the West Nile vector Culex pipiens. *Molecules*, 24(10), 1867.

12 Mosquito Control
Part II: Plant-Based Products or Phytochemicals

E. Pushpalatha
University of Calicut, Malappuram, India

Devinder Kaur Kocher
Punjab Agricultural University, Ludhiana, India

Arti Prasad
MLS University, Udaipur, India

Ipsita Bhowmik
ICMR-Regional Medical Research Centre (NE), Dibrugarh (Assam), India

E. R. Anirudh
Nicolae Testemiţanu State University of Medicine and Pharmacy, Chişinău, Moldova

A.N. Anoopkumar and Embalil Mathachan Aneesh
University of Calicut, Malappuram, India

B.K. Tyagi
Chandigarh University, Mohali, India

12.1 INTRODUCTION

Mosquitoes act as a vector for many dreadful diseases like malaria, yellow fever, dengue fever, chikungunya, Zika virus disease, filariasis, Japanese encephalitis, West Nile virus, etc. (Table 12.1).

In the past, continuous and injudicious application of synthetic insecticides have caused (i) development of resistance in almost all vector species, (ii) biological magnification of toxic substances through the food chain, (iii) adverse effects on environmental quality and non-target organisms, and, most importantly, (iv) an array of human health issues including detection of these pesticides in undesirable quantities in humans, cattle milk, and tissue. The use of synthetic insecticides and development of resistance as a consequent emphasis on bioinsecticides that could replace conventional synthetic insecticide compounds with their potential applications to control mosquitoes has been

DOI: 10.1201/9781003327028-15

TABLE 12.1
Important Vector Mosquito Species and the Major Infections Transmitted by Them (WHO, 2014)

Sl. No.	Vector Species	Main Human Infections
1	*Aedes aegypti, Aedes albopictus*	Dengue, yellow fever, chikungunya, Zika virus
2	*Culex quinquefasciatus, Mansonia* spp.	Lymphatic filariasis
3	*Culex vishnui* Group species	Japanese encephalitis
4	*Anopheles* spp.	Malaria

reviewed by Sengül Demirak and Canpolat (2022). Under the integrated vector management (IVM) programmes, emphasis is being placed on the application of alternative strategies in mosquito control. The application of active toxic agents from plant extracts at suitable doses is highly effective and can certainly provide an alternative mosquito control strategy.

The plants have evolved for over 400 million years that have equipped them with plenty of chemical defence in the form of secondary metabolites to defend themselves from outside attack such as that by insects (Arivoli et al., 2007). More than 2,000 plant species have been known to be introduced in the biological pest control programmes, and among these, the secondary metabolites/products of some 344 species have been reported with significant activity against mosquitoes (Remia and Logaswamy, 2009). The secondary metabolites, usually referred to a phytochemicals, are noted as non-nutritive chemical elements formed through several chemical pathways in plants. The plant families Asteraceae, Lamiaceae, Solanaceae, Oocystaceae, Cladophoraceae, Rutaceae, Meliaceae, Myrtaceae, Verbenaceae, Poaceae, and Apocynaceae are well known for their larvicidal, adulticidal, and repellent activities against different species of mosquitoes (Shaalan et al., 2005).

12.1.1 Insecticides

The insecticidal effects of plant extracts mainly depend upon the plant species, geographical varieties, plant parts used, mosquito species, extraction methodology adopted, and the polarity of solvents used for extraction process (Tehri and Singh, 2015). Among these, the plant parts and plant species used significantly influence the efficacy of botanical derivatives. The plant parts such as roots, stems, leaves, flowers, fruits, fruit peel, seeds, bark and rhizomes have been reported to be used as source of botanicals in mosquito control. In addition, the phytocompounds from the plants are also reported with the significant contribution towards various fields focusing on the disease prevention, along with health promotion (González-Vallinas et al., 2013). *Chrysanthemum cinerariifolium* is noted as one of the major sources for natural pyrethrins which dominates over all the plant-derived chemicals used to date in mosquito control regimes around the world (Lybrand et al., 2020).

12.1.2 Repellents

Repellents are chemical products with an offensive smell or taste to mosquitoes. Plant-derived compounds with repellent properties are most likely chemicals that are produced in defence against insects that pose a threat to the plant itself. These chemicals can be grouped into different categories based on the functional groups present. They include nitrogen-containing compounds, terpenoids, phenolics, proteinase inhibitors and growth regulators. These compounds are generally produced to fight off a broad spectrum of insects, including mosquitoes. Plants with better repellent properties fall into distinct families, with the Poaceae family (citronella based especially *Cymbopogon spp.*) being the pre-eminent one. Species of Lamiaceae, Fabaceae and Asteraceae also show promising results.

Prior to the extensive use of synthetically produced repellents, aromatic/essential oils (EOs) were commonly used. Numerous essential oil-producing plants from the Lamiaceae, Poaceae, Rutaceae and Myrtaceae families have very wellknown repellent activity.

Plants offer a great promise, providing potential molecules that can be exploited to develop effective antimosquito products (Tyagi and Shahi, 2001; Tyagi, 2002a, 2002b; Tyagi and Shahi, 2002; Tyagi 2003a, b; Shahi et al., 2000; Rahana et al., 2016; Ahalya and Pushpalatha, 2020; Anju Viswan et al., 2020; Rahana and Pushpalatha, 2021; Athira and Pushpalatha, 2021). Most plants contain compounds, falling under several categories, including repellents, feeding deterrents, toxins and growth regulators, which they use in preventing attack from phytophagous (plant-eating) insects. These compounds may be categorized under (i) nitrogen compounds (primarily alkaloids), (ii) terpenoids, (iii) phenolics, (iv) proteinase inhibitors, and (v) growth regulators. Those volatile components released as a consequence of herbivory are now best known for being effective against mosquitoes and other biting insects (Pichersky and Gershenzon, 2002). Insects detect odours when volatile odours bind to odorant receptor (OR) proteins on ciliated dendrites of specialized odour receptor neurons (ORNs) often on the antennae and maxillary palps of the insect. Some ORNs, such as OR83b, which are important in olfaction and blocked by the gold standard synthetic repellent DEET (N, N-diethyl-3-methylbenzamide), are highly conserved across insect species. Plants commonly produce volatile 'green leaf volatiles' when leaves are damaged in order to deter herbivores. However, it is most likely that many plant volatiles are deterrent or repellent because they have high vapour toxicity to insects. In India, a large number of examples are there where the plants, whole or part of it (in dried or fresh and in different forms: leaf, fruit, seed, root, etc.), have been exploited for thousands of years, most simply by hanging bruised plants in houses or by burning in fire after preparing a formulation with animal dung (neem leaves are the best example of natural repellent). All over the world, there is now a common knowledge of using these plant-based 'natural' smelling repellents because plants are perceived as a safe and trusted means of mosquito bite prevention (Dam et al., 2000a, 2000b).

12.1.3 Some of the Best-Known Plants for Mosquito Repellency

12.1.3.1 Eucalyptus

Corymbia citriodora (Myrtaceae), also known as the lemon eucalyptus plant, has been known for many centuries to have insect repellency properties. Lemon eucalyptus essential oil, comprising 85% citronellal, is far more effective at repelling mosquitoes for several minutes. However, one of its ingredients, para-menthane-3,8-diol, which has a lower vapour pressure than volatile monoterpenes found in most plant oils, provides very high protection from a broad range of insect vectors over several hours. Nanotechnology has of course recently opened new vistas in using effectively the eucalyptus extracts (Sugumar et al., 2014).

12.1.3.2 Citronella

Best known as lemongrass, the EOs and extracts belonging to plants in the citronella genus (Family: Poaceae) are commonly used as ingredients of plant-based mosquito repellents mainly *Cymbopogon citratus, C. nardus, C. schoenanthus, C. winterianus*, and *C. jwarancusa*. Citronella, which mainly contains citronellal, citronellol, geraniol, citral, α pinene, and limonene, was used by the Indian Army to repel mosquitoes at the beginning of the twentieth century. Today, citronella is one of the most widely used natural repellents on the market, used at concentrations of 5%–10%. Citronella-based repellents only protect from host-seeking mosquitoes for about two hours, although formulation of the repellent is very important. Recently, the use of nanotechnology, particularly allowing vanillin (5%) with the essential oil, has allowed for a higher protection rate with slower release rates of oils. Encapsulated citronella oil nanoemulsion is prepared by high pressure homogenization of 2.5% surfactant and 100% glycerol to create stable droplets that increase the retention of the oil and slow down release.

12.1.3.3 Neem

Neem is widely appreciated globally for its manifold applications, one of which being the repellency against pests of various kinds, including hematophagous insects such as mosquitoes.

Several field studies from India have shown very high efficacy of neem-based preparations. It is just unfortunate that it could not be yet exploited fully for its unmatched and wonderful repellency characteristics, albeit considered as a natural alternative to DEET.

12.2 COMMONLY USED PHYTOCOMPOUNDS IN MOSQUITO CONTROL

The potential of various volatile organic compounds from the plants inducing toxic effects have been explored for destructing the life cycle of many of the arthropod-borne disease vectors, especially against the mosquito vectors (Schiestl and Ayasse, 2001). O-cresol, thujone, verbenone, ethyl acetate, limonene, farnesol, hexanal, 1-hexenol, (Z)-3-hexen-1-ol, ethyl propanoate, α-pinene, citral, nerol, and methyl propanoate, as well as ethyl butyrate, are found to produce significant toxic potential against the mosquito vectors (Nyasembe and Torto, 2014). Other compounds like phenyl ethyl alcohol, benzaldehyde, lilac aldehydes, (Z)-3-hexenyl acetate,) linalool oxide, phenyl acetaldehyde, linalool, (E)-β-farnesene, lilac alcohol, acetophenone, methyl salicyilate, (E)-2-nonenal, β-pinene, (Z)-, and (E)-β-ocimene also produced prominent range of toxicity against the mosquito population (Nyasembe et al., 2012), justifying the significance of the phytochemicals.

12.3 BIOLOGICAL ACTIVITY

Compounds like phenolics, terpenoids, and alkaloids present in plants elicit activity as repellents, antifeedants, oviposition deterrents, growth inhibitors, moulting hormones, antimoulting hormones, and juvenile hormone mimics, as well as attractants, which are responsible for interfering with the biological activity of target pests. While considering the biological properties of various chemical compounds from plants, it is evident that the antioxidant capacity, antimicrobial potential, chemopreventive potential, and inhibition towards α-Glucosidase and α-Amylase have been witnessed as the major biological activities, and they have definite links with the phytochemical composition of the bioactive component (Koss-Mikołajczyk et al., 2019). The efficacy of plant-derived compounds and the biological responses have been tested against various mosquito vectors, such as *Anopheles culicifacies, An. fluviatilis, An. baimaii, An. minimus, An. sundaicus, An. stephensi, Aedes aegypti, Ae. albopictus, Culex tritaeniorhynchus, Cx. vishnui, Cx. pseudovishnui*, and *Culex quinquefasciatus* and in general, the investigations have explored mostly the toxic effects of phytochemical elements on the vectors.

12.4 MAJOR SOURCES AND TYPES OF PHYTOCOMPOUNDS FOR MOSQUITO CONTROL

Plants are the principal source of phytocompounds and the degree of toxicity of phytochemicals towards mosquito vectors may vary prominently depending on the plant parts used, the species, the age of the parts used, and the solvent used for extraction purposes. Sukumar et al. (1991) enlisted the toxic properties of various medicinal plants against mosquito vectors such as, for example, *Citrus* sp., *Ocimum sanctum, Momordica charantia, Piper* sp., *Azadirachta indica, Curcuma domestica, Withania somnifera, Jatropha curcas, Piper retrofractum, Cestrum diurnum, Citrullus vulgaris, Tridax procumbens, Phyllanthus* sp., *Curcuma* sp., *Solanum* sp., *Solanum xanthocarpum, Azadirachta indica, Solanum villosum, Annona squamosa, Withania somnifera*, and *Melia azedarach.*

The secondary metabolites are extracted from the leaves, flowers, fruits, bark, roots, and seeds of plants and the common bioactive components constitute phenolics, trepenoids, alkaloids, phytosterols, and various other compounds which act as antifeedants, repellents, attractants, oviposition deterrents, moulting hormones, antimoulting hormones, growth inhibitors, and insect sterilants (Jacobson, 1975). There are more than 10,000 secondary metabolites known for their insecticidal properties, which comprise flavonoids, coumarines, phenols, tannins, quinones, phenolpropanoid derivatives, isoprene derivatives, diterpenes, triterpenes, saponins, steroids, kinolenes, alkaloids, sesquiterpenes, indole, glycosides, ethereal oils, and gums.

12.4.1 Phenolics

The phenolic compounds or polyphenols are recognized as vital compounds of plant origin and substances with more than 8,000 phenolic structures are widely distributed among the plant kingdom (Harborne, 1993). They form a heterogeneous group of chemical compounds which are highly unstable, easily oxidized, and susceptible to degradation. They classified into several types such as phenolic acids flavonoids xanthones, anthocyanins, coumarines, lignan polymers, and tannins (Carvalho et al., 2007). Among the phenolic compounds flavonoids and isoflavones occur as esters, ethers, or glycoside derivatives. The role of the various compounds belonging to this category are allied with cell wall thickening, osmoregulation, pigmentation, antimicrobial activity, fruit flavouring, hormone production, fruit protection, pigmentation, defence against stressors, potent toxicity, and anti-herbivory roles. Precocenes derived from certain plants like *Ageratum* sps. have been noted for its activity as anti-juvenile hormone and interfere with growth by transgressing developmental stages. Among these properties, the toxic potentials, capacity to defend the external stressors and interfering growth and metamorphosis have been utilized by the investigators to explore the applicability of phenolic compounds in mosquito control research

12.4.2 Terpenoids

The terpenes constitute one among the major group of secondary metabolites with more than 40,000 different molecules (Garcia and Carril, 2009). Among the various group of phytocompounds found in nature, the compounds belonging to the terpenoids are prominently known to exhibit significant range of toxicity towards the insect population (Isman, 2006). Terpenes generally act as a defence against herbivores and possess the same molecular structure of moulting hormones of the insects interrupting the process of moulting and make complex steroid precursors of animal hormones (Taiz and Zeiger, 2002). Several monoterpenes such as α pinene, β pinene, 3-carene, limonene, myrcene, a-terpenene, and camphene have been isolated and evaluated for their potential toxicity to various insect species, including mosquitoes. The insecticidal activities of these compounds are mainly accomplished due to the inhibition of acetylcholineresterase which might also cause certain retardation of biological activities such as appetite suppression, growth inhibition and reduced reproductive capacity (Marangoni et al., 2012). The toxic potentials of (R)-pulegone-1,8-cineole, 1,8-cineole-citronellal, (R)-pulegone-citronellal, (R)-pulegone-linalool, 1,8-cineole-linalool, and linalool-citronellal against *Aedes aegypti* have been well established (Sarma et al., 2019).

12.4.3 Alkaloids and Other Nitrogen-Containing Compounds

Recent years have witnessed the use of environment-friendly alkaloids against insects especially in mosquito control. The mode of action of alkaloids varies depending on the type of alkaloids and interferes with major cellular and physiological functions by affecting AChE receptors in the nervous system, regulating the hormonal activity, and causing toxicity. The inhibitory effect of natural alkaloids on AChE activity in *Aedes aegypti* was determined using molecular docking studies, and alpha-solanine has been found to fit into the AChE1 binding pocket so as to contribute as the best inhibitor of AChE1(Balachandran et al., 2021). The nitrogenous compounds like alkaloids have toxic and medicinal effects of on humans (Taiz and Zeiger, 2013), and compounds like nicotine and nornicotine are the most important alkaloids used to control a variety of pest species. A bioassay-guided chromatographic fractionation yielded about three kinds of following alkaloids with potent toxicity against the mosquito *Aedes albopictus* mosquito vectors; wuchuyuamide, rutaecarpine, and evodiamine (Liu et al., 2012). The neurotoxic effects of these compounds are primarily based on the chemical constitution of these compounds. The structures of the most prominent compounds could be elucidated by them using nuclear magnetic resonance spectroscopy and high-resolution electron impact mass spectrometry.

The natural compounds of plant origin with strong toxic potential towards the insect population has been reported in many of the studies, and among them, nitrogen-containing compounds are witnessed with great significance in the mosquito vector-control research. Lukevits et al. (1974) reported the strong toxic potential of nitrogen-containing organic silicon compounds which could be effectively used in insect control programmes. According to Tunón et al. (1994), of the 35 compounds elucidated, the nitrogen-containing compound stachydrine was found to exhibit extensive range of toxicity against the dengue fever vector *Aedes aegypti*.

12.5 TRENDS IN VECTOR-CONTROL RESEARCH

Recent years have witnessed the fact that mosquito-borne diseases cause an extensive range of morbidity and burden over the globe with prominent disproportionate impacts on the population. The vector-borne diseases, including filariasis, malaria, and dengue have contributed extensively to many proportions of severe problems in developing countries. Frequent and prolonged use of synthetic insecticides for insect control, especially mosquito control has, disturbed the natural biological systems, resulting in resurgences of the mosquito populations. The resultant effects of such practices are the development of insecticide resistance followed by ill effects on non-target organisms and the environmental as well as human health concerns. For these reasons, a lot of impetus is given to foster advanced research and developmental efforts to combat the aforementioned impacts inter alia the consideration of growing insecticide resistance in these dreadful vectors to broad-spectrum conventional chemical insecticides and the ill effects on nature and life. The best practice for mosquito management is a comprehensive mosquito control strategy that utilizes all available mosquito control methods to exploit the known vulnerabilities of vector control while maintaining a quality environment. The successful eradication/ prevention of vector-borne diseases relies on the use of suitable mosquitocides, education and organizing effective community awareness programmes (Gopalan et al., 2021), and proper vaccination. Besides all these conventional practices, features like more selective, less harmful and more compatible with biological control methods and incorporation of insect growth regulators (IGRs) become attractive alternatives to broad-spectrum insecticides which pose harmful effects on nature and humans.

To date, not less than 2,400 plant species have been reported to possess insecticidal properties; approximately 350 insecticidal compounds, more than 800 insect anti-feeding (deterrents) and quite a good number of IGRs and inhibitors have been reported yet, only a few have achieved the commercial status. Currently, marketed biopesticides in the world include pyrethrins-, nicotine-, rotenone-, sabadilla-, and neem-based products. During the last few years, certain plant EOs, monoterpenoids, and sesquiterpenoids are developed as ‘green pesticides,’ as these oils are well known insect toxins, repellents and deterrents. These EOs are marketed as Cinnamate, Valero, Eco Pco, and Bioganic.

However, reports have shown that there are still barriers existing for the commercialization of phytochemicals such as scarcity of the botanical resources for large scale production, standardization of extracts and quality control based on the active ingredients of phytochemicals. To isolate, characterize and synthesize/formulate the compounds of interest in vector/pest control becomes a constant challenge. The innovations in finding more effective phytocompounds like discovery of artemisinin from natural resources will give us more hope to overcome many of the complications met with developing effective strategies (Su and Miller, 2015). Due to the increasing demands for diverse phytochemical compounds, screening programmes for active compounds seeking proper analytical methodologies, which include extraction, isolation, characterization, and formulation of active ingredients from the natural products have become a thrust area of vector-control research.

12.6 SIGNIFICANCE IN THE CURRENT SCENARIO

To tackle the global burden of mosquito-borne diseases (MBDs), researchers all over the world are working to find the appropriate and efficient solutions. Presently, a vaccine for yellow fever has been discovered (Touret et al., 2018), but there are currently no commercially available drugs or vaccines for Zika or chikungunya. Moreover, a dengue vaccine (Dengvaxia), has been developed by Sanofi-Pasteur, which has been approved in several countries but this too has safety concerns for mass administration (Dayrit et al., 2020; Deng et al., 2020). Further, the emergence of new viruses and transformations of existing strains which could be transmitted by these vectors cannot be neglected. Like the co-occurrence of SARS-CoV-2 with MBDs, such as malaria and dengue in resource-limited settings represents an additional challenge to health-care systems. The overlapping incidence of COVID-19, malaria, dengue, Zika, and chikungunya could hinder the patient's timely diagnosis, treatment, and disease prevention, in part due to some similarity of clinical symptoms (Spinicci et al., 2020, Harapan et al., 2021; Masyeni et al., 2021). During the pandemic, vector-control measures might have decreased in the community, and dengue prevention programmes might have been paused in some countries. Therefore, countries that are experiencing or have high risk for dengue outbreak should maintain the vector-control measures during the pandemic (Wilder-Smith, 2020). One of the possible strategies is to improve the situation by encouraging community members to reduce mosquito breeding sites in and around their homes. On one hand, the lack of proper medication and vaccination for treating MBDs and on the other hand, the economic and ecological challenges being faced due to the human and environmental hazards caused by a majority of the synthetic insecticide chemicals being used for the control of mosquitoes (FAO, 1992), along with the emergence of insecticidal resistance among mosquitoes (WHO, 1992, Mittal et al., 2004), are creating havoc worldwide. After facing several problems due to injudicious and over application of synthetic insecticides in nature, the top priority is to refocus on phytochemicals that are easily biodegradable and have no ill effects on non-target organisms. Thus new bioactive compounds from the plant kingdom and an effort to determine their structure and commercial production are in high demand.

In the present scenario the use of herbal products is the best alternative for mosquito control. The plant-based insecticides are generally biodegradable, ecofriendly, target-specific, and relatively harmless to non-target organisms, including man (Maggi and Benelli, 2018; Ayinde et al., 2020). Another important fact is that the diversity of plants having insecticidal properties is enormous, which has resulted in significant interest in botanicals as a potential source of natural mosquito control material in recent years. Some phytochemicals act as general toxicants to all life stages of mosquitoes, whereas others interfere with growth and reproduction or act on the olfactory receptors, eliciting responses of attraction or repellency (Sukumar et al., 1991). A wide selection of plants from herbs, shrubs, and large trees have been used for extraction of products from various parts of plants like fruits, leaves, stems, bark, roots, peels, etc. (Ghosh et al., 2012, Krishnappa and Elumalai, 2014, Osanloo et al., 2019). Furthermore, unlike conventional insecticides which are based on a single active ingredient, plant-derived insecticides comprise an array of chemical compounds which act concertedly on both behavioural and physiological processes of the insects. Thus the chances of developing resistance to such bioinsecticides are unlikely (Adeyemi, 2010). The detection of active compounds in medicinal plants plays a strategic role in phytochemical investigation of crude plant extracts and is very important with regard to their potential pharmacological effects (Majaw and Moirangthem, 2009).

Plant-based products like EOs and extracts, due to their amazing insecticidal potential, have been used as traditional methods to solve the problem of various types of MBDs; thus, they have achieved a special interest in the society. On comparison with synthetic insecticides, the plant-based products are considered as beneficial in terms of cost, as well as environment safety (Maggi and Benelli, 2018). One of the most appealing features of phytochemicals in comparison to conventional synthetic insecticides is their low persistence in the environment and the absence of accumulation along the food chains (Isman, 2006). Since EOs/plant extracts are the mixtures of major as well as minor constituents, they act synergistically in the target individuals in contrast to chemical insecticides. Moreover, these plant-based products have been found to be potent against those species of insects which are resistant to synthetic products (Asadollahi et al., 2019). Undoubtedly, plant-based products are a more secured and trusted mode of prevention of mosquitoes and these bioinsecticides of plant origin provide an alternative to synthetic insecticides being effective, easily available, safe for humans and non-target species, biodegradable, inexpensive, and environment friendly (Ayinde et al., 2020). Cash cropping of plants used as larvicides/repellents provide a vital source of income for small-scale farmers in developing countries and can have beneficial environmental impacts when planted in intercropping systems to prevent soil erosions. The field of plant-based products is moving forward as consumers demand means of protection from mosquito bites that are safe, pleasant to use and environmentally sustainable (Senthi-Nathan, 2020). The positive results of initial studies regarding the potential of botanicals as insecticides encourage further investigation of bioactive compounds. Identifying botanical insecticides that are effective as well as appropriate and adaptive to overcome ecological hazards, are biodegradable and have broad-spectrum larvicidal properties will work as a new defence in the arsenal of insecticides and they may act as appropriate alternative products in the fight against MBDs (Meryem et al., 2022).

12.7 LIMITATION AND FUTURE PERSPECTIVES

12.7.1 Limitations

Plant products act as the reservoir pool of bioactive toxic agents against mosquito larvae. But only a few have been commercially produced and extensively used in vector-control programmes. At present, botanical insecticides make up < 1% of the world's pesticide market (Sola et al., 2014). The main reasons behind the failure in the laboratory to land movements of bioactive toxic phytochemicals are poor characterization and inefficiency in determining the structure of the active toxic ingredients responsible for insecticidal activity along with their lower residual efficacy and protection time as compared to synthetic products (Rehman et al., 2014). Reduction in efficacy of phytochemicals is attributed to their high volatility, thermal decomposition, biodegradability, and poor physiochemical stability, which make them unstable. Some other potential limitations to the efficacy of bioinsecticides, such as environmental conditions, mosquito fitness, and mosquito resistance, as well as the parts of the plants used, solvents used in extraction steps, insecticide dose, and exposure time (Tehri and Singh, 2015, Namias et al., 2020), should be considered for the successful assessment of novel bioinsecticides in mosquito control. While efficacy tests provide promising information on possible mosquitocidal effects, new compounds from plant origins, the identification of actual active ingredients for efficacy, and their modes of action are still waiting to be resolved. One of the reasons for their limited usage in the field is the formulation problem of overcoming their phytotoxic effects. The chemical composition of each compound should be formulated in such a way that it is bioactive to target insects and non-toxic to non-target organisms. In addition, the formulation of plant-based bioinsecticides should ensure that it can be produced in large quantities through biomass production of plants and administered in recommended dosages to minimize toxic effects and that biological activity can be maintained for a longer shelf-life.

Overly enthusiastic statements have often been made in a number of articles describing the potential utility of EOs as mosquito larvicides. In most cases, the efficacy of mixtures of intact EOs or of some purified components has been overrated, while any health and environmental hazards have been dismissed without experimental evidence. As demonstrated by the dearth of EOs that have been commercialized so far, a big disconnect exists between the laboratory trials reporting activity on mosquito larvae and the practice of mosquito larval control. In industrialized countries, unless plant extracts or oils comparable in bioactivity to conventional larvicides can be found, the chance to go beyond very small market niches seems unlikely. However, even in the developing countries, the use of EOs on wide spatial scales to control mosquito larvae breeding in natural areas could be hindered by the high LC_{50} values displayed by most of the compounds tested so far. On the contrary, EOs may be relevant to control mosquito larvae breeding in water containers in households such as against *Ae. aegypti* and *Ae. albopictus* (Masetti, 2016). In general, botanicals interfering with the function of insect nervous systems have very low toxic effects on dogs, cats, and humans, but are toxic to most fish, birds, reptiles and amphibians. Varma and Dubey (1998) stated that IGRs are generally selective in their toxicity and safe to both the environment and the natural enemies of insect pests. However, some antagonists (Anon, 2003) have stated that people falsely believe that these natural products, i.e., plant secondary compounds are safer than other pesticides because few phytochemicals have undergone toxicological testing, and many have striking adverse effects on mammals. This is a good point since many green phytochemicals considered to be natural or organic are more toxic than some synthetic insecticides (Ciccia et al., 2000, Yang et al., 2002, Kimball, 2003). However, research gaps concerning mammalian toxicity of many EOs suggested for this purpose remain to be filled.

Unfortunately, only a limited number of researchers have followed WHO protocols for evaluation by going to the extent of determining the effective concentration of botanical extracts that inhibit 50% of the treated larval population from emerging as adults (EC_{50}). The reasons for this are not clear but may be due to the protocol being difficult to source or restrictive. Insecticide efficacy methods are also not uniform and vary considerably despite the provision of WHO standard protocols (WHO, 1981) over several decades for the determination of lethal and effective concentrations. It is unfortunate that a primary screening method for potential larvicides is not universally used before time, funds, and energy are expended in detailed examinations of phytochemicals that are useless to vector control from a practical point of view. Further studies are needed with the implementation of new methods for the development of effective bioinsecticides from other plant-based bioactive compounds. Problems associated with their formulation and commercialization, non-standardization in evaluation of their bioactivities, and their persistence for longer durations should be resolved for development of effective and sustainable methods for their usage.

12.7.2 Future Perspectives

The evaluation of phytochemicals is yet in its infancy and much research remains to further characterize promising agents and discover the new ones. Several issues remain to be researched and improved upon, including residual efficacy (degradation and persistence), preservation of non-target organisms, development of resistance and standardization of testing protocols. The paucity in field trials suggests that most phytochemicals do not pass initial screening. The onus is thus placed upon the researcher to carry out field trials for promising phytochemicals that are not picked up initially by insecticide manufacturers. It should be mentioned, however, that the high degree of biodegradation exhibited by most phytochemicals is what makes them ecofriendly and attractive as replacements of synthetic chemicals in the first place.

The research has now moved on from the isolation of bioactive compounds with anti-vector potentials to formulate novel application methods. Apart from the direct application of plant metabolites in vector control, nanoparticles (NPs) synthesized from plants using green technology are emerging as a new trend. Nanotechnology is presently "revolutionizing" the manufacture of commercial pesticides. Although the activity of phytochemicals is generally attributed to some specific compounds, there is increasing evidence that the combination of botanicals and biopesticides will result in an increased bioactivity compared to single phytochemicals. Since phytochemicals have multiple modes of action and exert their effects on multiple target sites in insects, their efficacy can be enhanced when used as a blend (e.g., mixture of oils) against mosquitoes. In addition, insects are more likely to develop resistance to a single chemical compound rather than a mixture of compounds. Thus, a combinatorial usage of phytochemicals would limit development of resistance in mosquitoes (Senthil-Nathan and Kalaivani, 2006). Phytochemicals have a short residual half-life, which could be advantageous when synergistically used together with other biological control agents (Isman et al., 2011). It is encouraging that these features of phytochemicals make them alternative natural solutions for the development of suitable products to interfere with the mosquito-host interaction and reduce disease transmission. In recent years, new technologies, such as microencapsulation and nanoemulsion, have been used to overcome the problem of high volatility by enhancing the duration and efficacy of EOs (Gnankiné and Bassolé, 2017). Production of green NPs and nanoencapsulation compounds upsurges the permanence of EOs through the "slow release" phenomenon delivering sustained fortification against mosquito bites. Since insect-treated mosquito nets (ITNs) are one of the major intervention methods to control mosquitoes, the incorporation of plant-based insect repellents into fabrics seems a prompt and alternative way to provide safer protection against mosquito bites (Specos et al., 2010).

A new formulation in the form of tablets containing a lectin preparation showed mosquitocidal activity against different developmental stages of *Ae. aegypti* mosquitoes, and this formulation method is proposed as a new control strategy for *Ae. aegypti* populations (Silva et al., 2021). An upcoming approach in mosquito control is using the plant-based products with IGR effects, so identification of phytochemicals with growth inhibition properties combined with a considerable capacity to reduce adult emergence is the desired endpoint of botanical insecticide research nowadays. There is no doubt that there are more bioactive compounds that require exploring, and future research should focus on searching for plant-based products with the ultimate goal of deploying them as a reliable remedy to control mosquito populations and MBDs. The recent rapid progress in the field of biotechnology might ensure the development of bioinsecticides with improved efficacy, sustainability, and enhanced safety. In particular, current formulation technology and nanotechnology might accelerate the emergence of novel and effective plant-based products with long lasting results. Moreover, toxicological studies associated with multiomics can also be applied to find novel bioinsecticides, monitor target exposure to these products, track cellular responds to different doses, assess mechanism of action, and predict health risks associated with existing and newly developed phytochemicals.

12.8 CONCLUSION

The morbidity and mortality allied with the MBDs are found to be increasing day by day despite all the advancements in mosquito control research. The increasing prevalence of resistant mosquitoes, toxic effects on non-target organisms, and environmental issues have in turn warrants the discovery of safer, target-specific, biodegradable, and economically viable alternatives to replace the conventional mosquito control strategies. From this point of view, the phytochemicals gain growing importance to overcome problems faced with mosquito control programmes as being considered natural, environmentally safe, easily degradable, less toxic, cost-effective, and, more importantly, less prone to induction of mosquito resistance. However, the practical implementation of such perspectives faced certain limitations. As of now, there are a variety of phytocompounds showing a significant range of toxicity against the mosquito population, yet their efficacy in the field conditions and standardization of formulations for commercialization often meet with challenges. The advanced developments in the vector-control research have recently implemented the various approaches in mosquito management strategies, including computer-aided technologies that provide scope for the search of newer components of plant origins. The development of such approaches for mosquito control would help the scientific community to develop sustainable strategies for utilizing phytochemicals to combat threats by mosquitoes and various other dreadful arthropod-borne disease vectors.

REFERENCES

Adeyemi, M. M. (2010). The potential of secondary metabolites in plant material as deterrents against insect pests. *African Journal of Review & Pure Applied Chemistry*, 4, 243–246.

Ahalya, S.R. and Pushpalatha, E. (2020). IGR Activity of Three Indigenous Plants belonging to Asteraceae Family and its Potential Role on Control of Filarial Vector, *Culex quinquefasciatus* Say (Diptera: Culicidae). *Journal of Communicable Diseases - Special Issue on COVID-19*, 52(2), 57–62.

Anju Viswan, K. E., Pushpalatha, S. Arivudainambi and Selvamuthukumaran, T. (2020). Fractional extraction, isolation and identification of biologically active compounds from *Anamirta cocculus* (L.) Wight & Arn to control *Culex quinquefasciatus* Say and *Aedes albopictus* Skuse. *Journal of Communicable Diseases* 52(3), 53–57.

Anon. (2003). Botanicals. Golf and Plant Science Dept., Suny Dehli University. http://www.delhi.edu/page/plantsci/PlantScienceDelhi/Hortchem/

Arivoli, S., Tennyson, S. and Martin, J.J. (2007). Larvicidal efficacy of *Veronia cinerea* (L) (Asteracea) leaf extracts against the filarial vector *Culex quinquefasciatus* Say (Diptera : Culicidae). *Journal of Biopesticides*, 4 (1), 27–35.

Asadollahi, A., Khoobdel, M., Zahraei-Ramazani, A., Azarmi, S. and Mosawi, S.H. (2019). Effectiveness of plant-based repellents against different *Anopheles* species: a systematic review. *Malaria Journal*, 18(1), 1–20.

Athira, A., and Pushpalatha E. (2021). Essential oils of a few aromatic plants and their potential as knockdown, repellent and adulticidal agent to the filarial vector, *Culex quinquefasciatus* Say (Diptera: Culicidae). *International Journal of Mosquito Research*, 8(1), 35–40

Ayinde, A. A., Morakinyo, O. M., Sridhar, M. K. C. (2020). Repellency and larvicidal activities of *Azadirachta indica* seed oil on *Anopheles gambiae* in Nigeria. *Heliyon* 86(5), e03920.

Balachandran, C., Anbalagan, S., Kandeepan, C., Nagendran, N.A., Jayakumar, M., Abd Allah, E.F., Alqarawi, A.A., Hashem, A., Baskar, K. (2021). Molecular docking studies of natural alkaloids as acetylcholinesterase (ache1) inhibitors in *Aedes aegypti*. *Journal of Asia Pacific Entomology* 24, 645–652.

Carvalho, J.C.T., Gosmann, G., and Schenkel, E.P. (2007). Compostos fenolicos simplesheterosidicos C.M.O. Simoes, E.P. Schenkel, G. Gosmann, JCP. Mello, L.A. Mentz, PR Petrovick (Eds.), Farmacognosia: da planta ao medicamento, UFRGS, Port Alegre, pp. 519-535 Chromatography. *Proceedings of the National Academy of Sciences, USA* 61, 636–643.

Ciccia G., Coussio J. and E Mongelli (2000). Insecticidal activity against *Aedes aegypti* larvae of some medicinal South American plants. *Journal of Ethnopharmacology* 7(1–2), 145–151.

Dam, P.K., Ramnath, T., Yadav, S.P. et al. (2000a) Use of ethnomedicinal practices in the Thar desert, north western India. *Journal of Economic and Taxonomic Botany* 24, 303–318.

Dam, P.K., Yadav, S., Tyagi, B.K. (2000b). Thar Marukschetra mein aushadhiukta vanaspatiyan: ek purvalokana (in Hindi) [The medicinal plants of Thar Desert region: a review]. *ICMR Patrika* 14(8), 61–65.

Dayrit, M. M., Mendoza, R. U. and Valenzuela, S. A. (2020). The importance of effective risk communication and transparency: lessons from the dengue vaccine controversy in the Philippines. *Journal of Public Health Policy* 41, 252–267.

Deng, S. Q., Yang, X., Wei, Y., Chen, J. T., Wang, X. J. and Peng, H. J. (2020). A review on dengue vaccine development. *Vaccines* 8, 63.

FAO (1992). *Pesticide residues in food*. pp. 116–146. Report. Rome, Italy.

Garcia, A.A. and Carril, E. P.U. (2009). Metabolismo secundario de plantas Reduca (Biologia). *Serie Fisiologia Vegetal* 2(3), 119–145.

Ghosh, A., Chowdhury, N. and Chandra, G. (2012). Plant extracts as potential mosquito larvicides. *Indian Journal of Medical Research* 135, 581–598.

Gnankiné, O., and Bassolé. I. H. N. (2017). Essential oils as an alternative to pyrethroids' resistance gainst Anopheles species complex giles (Diptera: Culicidae). *Molecules* 22, 1321.

González-Vallinas, M., González-Castejón, M., Rodríguez-Casado, A., Ramírez de Molina, A., (2013). Dietary phytochemicals in cancer prevention and therapy: a complementary approach with promising perspectives. *Nutrition Reviews* 71, 585–599.

Gopalan, R.B., Babu, B.V., Sugunan, A.P., Murali, A., Ma, M.S., Balasubramanian, R., Philip, S. (2021). Community engagement to control dengue and other vector-borne diseases in Alappuzha municipality, Kerala, India. *Pathogens and Global Health* 115, 258–266.

Harapan, H., Ryan, M., Yohan, B., Abidin, R.S., Nainu, F. and Rakib, A. (2021). Covid-19 and Dengue: Double Punches for Dengue-Endemic Countries in Asia. *Reviews in Medical Virology* 31(2), e2161.

Harborne, J.B. (1993). *The flavonoids: Advances in Research since 1986*. Chapman and Hall, London.

Isman, M. (2006). Botanical insecticides, deterrents and repellents in modern agriculture and an increasingly regulated world. *Annual Review of Entomology* 51, 45–66.

Isman, M. B., Miresmailli, S., and Machial, C. (2011). Commercial opportunities for pesticides based on plant essential oils in agriculture, industry and consumer products. *Phytochemistry Reviews* 10, 197–204.

Jacobson, M. (1975). Insecticides from Plants a review of Literature 1954–1971. *Agriculture Hanbook* 461, USDA, Washington, DC, 138p.

Kimball, J. W. (2003). Kimball's Biology Pages. Online Biology Textbook http://users.rcn.com/jkimball.ma.ultranet/BiologyPages/

Koss-Mikołajczyk, I., Kusznierewicz, B., Bartoszek, A., (2019). The relationship between phytochemical composition and biological activities of differently pigmented varieties of berry fruits; comparison between embedded in food matrix and isolated anthocyanins. *Foods*, 8, 646.

Krishnappa, K., and Elumalai, K. (2014). Mosquitocidal activity of indigenous plants of Western Ghats, *Achras sapota* Linn. (Sapotaceae) and *Cassia auriculata* L. (Fabaceae) against a common malarial vector, *Anopheles stephensi* Liston (Culicidae: Diptera). *Journal of Coastal Life Medicine* 2, 402–410.

Liu, Z.L., Liu, Q.Z., Du, S.S., Deng, Z.W. (2012). Mosquito larvicidal activity of alkaloids and limonoids derived from Evodia rutaecarpa unripe fruits against Aedes albopictus (Diptera: Culicidae). *Parasitology Research* 111, 991–996.

Lukevits, E., Dremova, V., Simchenko, L., Voronkov, M. (1974). Nitrogen-containing organisilicon compounds XLIX. Sysnthesis and insect repellent properties of Si-substituted (amenomethyl) silanes (CH3) n (RO) 3– nSiCH2NR′ R″. *Pharmaceutical Chemistry Journal*, 8, 611–613.

Lybrand, D.B., Xu, H., Last, R.L., Pichersky, E. (2020). How Plants Synthesize Pyrethrins: Safe and Biodegradable Insecticides. *Trends in Plant Science*, 25(12), 1240–1251. doi: 10.1016/j.tplants.2020.06.012

Maggi, F. and Benelli, G. (2018). Essential oils from aromatic and medicinal plants as effective weapons against mosquito vectors of public health importance. *In*: *Mosquito-borne Diseases*, pp. 69–129. Springer, Cham.

Majaw, S., and Moirangthem, E., (2009). Qualitative and quantitative analysis of *Clerodendron colebrookiarium* Walp. leaves and *Zingibe cassumunar* Roxb. rhizomes. *Ethnobotanical Leaflets* 13, 578–589.

Marangoni, C., Moura, N.F., and Garcia F.R.M. (2012). Utilizacao de oleos essenciais e extratos de platas no controle de insectos. *Revista de Ciencias Ambientais*, 6(2), 95–112.

Masetti, A. (2016). The potential use of essential oils against mosquito larvae: a short review. *Bulletin of Insectology* 69(2), 307–310.

Masyeni, S., Santoso, M. S., Widyaningsih, P. D., Asmara, D. W., Nainu, F., Harapan, H. and Sasmono, R. T. (2021). Serological cross-reaction and coinfection of dengue and COVID-19 in Asia: Experience from Indonesia. *International Journal of Infectious Diseases* 102, 152–154.

Meryem, S., Sengül, D. and Emel, C. (2022) Plant-based bioinsecticides for mosquito control: Impact on insecticide resistance and disease transmission. *Insects* 13(162), 1–24.

Mittal, P. K., Wijeyaratne, P. and Pandey, S. (2004) *Status of Insecticide Resistance of Malaria, kala-Azar and Japanese Encephalitis Vectors in Bangladesh, Bhutan, India and Nepal (BBIN).* Activity Report 129. Environmental Health Project, Washington, DC.

Namias, A., Jobe, N. B., Paaijmans, K. P. and Huijben, S. (2020). The need for practical insecticide-resistance guidelines to effectively inform mosquito-borne disease control programs. *Life* 10, e65655.

Nyasembe, V.O., Teal, P.E., Mukabana, W.R., Tumlinson, J.H., Torto, B., 2012. Behavioural response of the malaria vector Anopheles gambiae to host plant volatiles and synthetic blends. *Parasites & Vectors* 5, 1–11.

Nyasembe, V.O., Torto, B., 2014. Volatile phytochemicals as mosquito semiochemicals. *Phytochemistry Letters* 8, 196–201.

Osanloo, M., Sedaghat, M. M., Sanei-Dehkordi, A. and Amani, A. (2019). Plant-derived essential oils; their larvicidal properties and potential application for control of Mosquito-Borne diseases. *Galen Medical Journal* 8, 1532.

Pichersky, E. and Gershenzon, J. (2002). The formation and function of plant volatiles: perfumes for pollinator attraction and defense. *Current Opinion in Plant Biology*, 5, 237–243. doi:10.1016/S1369-5266(02)00251-0

Rahana, V.K., Anju Viswan, K., and Pushpalatha, E. (2016). Response on growth regulatory activity of three indigenous plant extracts against dengue vector *Aedes albopictus* (Skuse). *International Journal of Mosquito Research*, 3(4), 1–5.

Rahana, V.K., and Pushpalatha, E. (2021). *Sterculia guttata* Roxb. As insect growth regulatory agent and its influence on developmental indices of filarial vector, *Culex quinquefasciatus* Say. *International Journal of Mosquito Research* 8(1), 19–24.

Rehman, J.U., Ali, A. and Khan, I.A., (2014). Plant based products: Use and development as repellents against mosquitoes: A review. *Fitoterapia* 95, 65–74.
Remia, K.M. and Logaswamy, S. (2009). Larvicidal ewfficacy of leaf extracts of two botanicals against the mosquito vector Aedes aegypti (Diptera : Culicidae). *Indian Journal of Natural Products and Resources*, 1(2), 208–212.
Sarma, R., Adhikari, K., Mahanta, S. and Khanikor, B., (2019). Combinations of plant essential oil based terpene compounds as larvicidal and adulticidal agent against *Aedes aegypti* (Diptera: Culicidae). *Scientific Reports* 9, 1–12.
Schiestl, F.P. and Ayasse, M. (2001). Post-pollination emission of a repellent compound in a sexually deceptive orchid: a new mechanism for maximising reproductive success? *Oecologia* 126, 531–534.
Sengül Demirak, M. S. and Canpolat, E. (2022). Plant-Based Bioinsecticides for Mosquito Control: Impact on Insecticide Resistance and Disease Transmission. *Insects* 13, 162. https://doi.org/10.3390/insects13020162s
Senthil-Nathan, S. and Kalaivani, K. (2006). Combined effects of azadirachtin and nucleopolyhedrovirus (SpltNPV) on *Spodoptera litura* Fabricius (Lepidoptera: Noctuidae) larvae. *Biological Control* 39, 96–104.
Senthi-Nathan, S.S. (2020). A review of resistance mechanisms and botanicals, phytochemicals and essential oils as alternative larvicidal agents against mosquitoes. *Frontiers in Physics.* https://doi.org/10.3389/fphys.2019.01591
Shaalan, E.A.S.D., Canyon, M.W.F., Younes, W.A.H. and Mansoura, A.H. (2005). A review of botanical phytochemicals with mosquitocidal potential. *Environment International*, 3, 1149–1166.
Shahi, A.K., Tyagi, B.K., Chaudhary, R.C. and Kaul, B.K. (2000). *Cymbopogon schoenanthus* (L.) Spreng. (Poaceae) – a versatile aromatic and medicinal plant of Indian Thar Desert. *FAFAI Fragrance & Flavour Association of India*, 1, 31–33
Silva, L. L. S., Silva, S. C. C., de Oliveira, A. P. S., Nascimento J. D. S., Silva, E. O., Coelho, L. C. B. B., Neto, P. J.R., Navarro, D. M. D. A. F., Napoleão, T. H. and Paiva, P. M. G. (2021). Effects of a solid formulation containing lectin-rich fraction of *Moringa oleifera* seeds on egg hatching and development of *Aedes aegypti* larvae. *Acta Tropica* 214, 105789.
Sola, P., Mvumi, B. M., Ogendo, J. O., Mponda, O., Kamanula, J. F., Nyirenda, S., Belmain, S. R. and Stevenson, P. C. (2014). Botanical pesticide production, trade and regulatory mechanisms in sub-Saharan Africa: Making a case for plant-based pesticidal products. *Food Security* 6, 369–384.
Specos, M. M., García, J. J., Tornesello, J., Marino, P., Vecchia, M. D., Tesoriero, M. D. and Hermida, L. G. (2010). Microencapsulated citronella oil for mosquito repellent finishing of cotton textiles. *Transactions of Royal Society of Tropical Medicine & Hygiene* 104, 653–658.
Spinicci, M., Bartoloni, A., Mantella, A., Zammarchi, L., Rossolini, G. M. and Antonelli, A. (2020). Low risk of serological cross-reactivity between dengue and COVID-19. *Memórias do Instituto Oswaldo Cruz - Fiocruz* 115, 55–67.
Su, X.-Z. and Miller, L.H., (2015). *The discovery of artemisinin and the Nobel Prize in physiology or medicine.* Springer.
Sugumar, S., Clarke, S.K. and Nirmala, M.J. (2014). Nanoemulsion of eucalyptus oil and its larvicidal activity against *Culex quinquefasciatus. Bulletin of Entomological Research* 104(3):393–402. doi:10.1017/S0007485313000710
Sukumar, K., Perich, M. J. and Boombar, L. R. (1991). Botanical derivatives in mosquito control: a review. *Journal of the American Mosquito Control Association* 7, 210–237.
Taiz, L. and Zeiger, E. (2002). *Plant physiology*. 3rd ed. Sunderland: Sinauer Associates. 690 pp.
Taiz, L. and Zeiger, E. (2013). *Fisiologia Vegetal*, 5th Ed. Editora Artmed, 954 pp.
Tehri, K. and Singh, N. (2015). The role of botanicals as green pesticides in integrated mosquito management—A review. *International Journal of Mosquito Research* 2, 18–23.
Touret, F., Gilles, M., Klitting, R., Aubry, F., de Lamballerie, X. and Nougairède, A. (2018). Live Zika virus chimeric vaccine candidate based on a yellow fever 17-D attenuated backbone. *Emerging Microbes & Infections* 7, 161.
Tunón, H., Thorsell, W., Bohlin, L., (1994). Mosquito repelling activity of compounds occurring in *Achillea millefolium* L.(asteraceae). *Economic Botany*, 48, 111–120.
Tyagi, B.K. (2002a). Future of phytochemicals against vector mosquitoes. *Pap. Abstr. 6th AZRA Congress on Advances in Applied Zoological Researches for Food Production and Environmental Safety, Cuttack*, 19–21 December 2002.
Tyagi, B.K. (2002b). Future of phytochemicals against vector mosquitoes. *Journal of Applied Zoological Research*, 14, 101–106.

Tyagi, B.K. (2003a). A review of plant-based materials active against vector mosquitoes with particular emphasis on repellent potential of Cymbopogon(Poaceae). *Journal of Economic and Taxonomic Botany* 28, 489–493.

Tyagi, B.K. (2003b). Cymbopogon essential oils: their chemistry and mosquito repellent activities. *Pap. Abstr. National Symposium on biochemical science health &environmental aspects*, Agra, pp. 140–142.

Tyagi, B.K. and Shahi, A.K. (2001). Evaluation of repellent potential of two cymbopogon species (Poaceae), endemic to the Thar Desert (India), against major vector mosquitoes. *International conference globalization of Indian essential oils, Agra*, 21–23 Sep 2001.

Tyagi, B.K. and Shahi, AK. (2002). Genus Cymbopogon (Poaceae) – a potent group of mosquito repellent bearing characteristics. *Pap. Abstr. XVI National Congress of Parasitology, Bareilly*, 31 Oct, 2 Nov 2002, p. s.

Varma, J. and Dubey, N.K. (1998). Prospectives of botanical and microbial products as pesticides of tomorrow. *Current Science* (Online) 75: http://www.iisc.ernet.in/currsci/jan25/articles22.htm

WHO (1981). Instructions for determining the susceptibility or resistance of mosquito larvae to insecticides. WHO/VBC/81.807.

WHO (1992). Vector resistance to pesticides. Fifteenth Report of the WHO Expert Committee on Vector Biology and Control. *World Health Organ Technical Report Series* 818, 1–62.

WHO (2014). *A global brief on vector-borne diseases*. World Health Organisation, Geneva.

Wilder-Smith, A. (2020). Dengue vaccine development: status and future. *Bundesgesundheitsblatt – Gesundheitsschutz*, 63, 40–44.

Yang, Y., Lee, S., Lee, H., Kim, M., Lee, S. and Lee, H. (2002). A piperidine amide extracted from *Piper longum* L fruit shows activity against *Aedes aegypti* mosquito larvae. *Journal of Agricultural and Food Chemistry*, 50, 3765–3767.

13 Mosquito Control Strategies Based on Genetic Manipulation

Ranil S. Dassanayake
University of Colombo, Colombo, Sri Lanka

Y. I. Nilmini Silva Gunawardene and Menaka Hapugoda
University of Kelaniya, Kelaniya, Sri Lanka

Tharaka Ranathunga
Eastern University, Vantharumoola, Sri Lanka

P. V. Desha Dilani
University of Colombo, Colombo, Sri Lanka

H.P.B.K.D. Ramyasoma
North Carolina State University, Raleigh, USA

P.D.S.U. Wickramasinghe
University of Colombo, Colombo, Sri Lanka

Prabhakargouda B. Patil
Maharashtra Hybrid Seeds Company Pvt. Ltd., Jalna, India

S. K. Dasgupta
Ex-Tata Institute for Genetics and Society (TIGS), in Stem New Building, NCBS (TIFR), Bangalore, India

B. K. Tyagi
Chandigarh University, Mohali, India

 DOI: 10.1201/9781003327028-16

13.1 INTRODUCTION

Mosquito-borne diseases are becoming a major public health problem worldwide. Common types of mosquito-borne diseases include malaria, dengue, West Nile virus, chikungunya, yellow fever, and Zika (Lee et al., 2018). Half of the population lives in areas at risk of mosquito-borne diseases, where malaria and dengue are the most prevalent diseases which impose severe economic burden on many developing tropical countries (Garg et al., 2008; Wijewickrama et al., 2018; Ranjha and Sharma, 2021). Malaria is responsible for approximately 200–400 million incidences and results in more than 400,000 deaths yearly (Bhatt et al., 2013; WHO, 2021). Dengue causes severe outbreaks in many urban areas, with an estimated incidence of 390 million annually. Moreover, outbreaks of yellow fever, chikungunya, have increased in size and frequency since 2014 (Wang et al., 2021). Given the health and economic burden attributable to mosquito-borne diseases, sustainable control measures are needed to better control and possibly eliminate these diseases.

In the absence of effective vaccines, therapeutics, and reliable diagnostics mechanisms, particularly for dengue and malaria, disease control primarily relies on mosquito vector control (Dilani et al., 2021; Gaudinski et al., 2021; Ariyaratne et al., 2022). The use of conventional mosquito control strategies through environmental management and chemical insecticides is only moderately effective, partly due to increased ecological consequences and the development of insecticide-resistant strains. Chemical interventions also have unintended effects on non-target organisms (Wang et al., 2021). As a consequence, it is now more important than ever to further optimize current strategies or develop alternative and innovative strategies for controlling the mosquito population. Recent advancements in genetics provide the underlying motivation to develop new genetics-based approaches (Wilke and Marrelli, 2012; Alphey et al., 2013). This chapter reviews various mosquito strategies based on genetic manipulation and discusses their future applications for sustainable mosquito-borne disease control.

13.2 SIT FOR MOSQUITO VECTOR CONTROL

13.2.1 What is SIT?

Sterile insect technology (SIT), a biological control method to control insect pests was developed by Knipling (1940) against screwworm (*Cochliomyia hominivorax*; Knipling, 1979). This is an effective and sustainable genetic approach to control populations of selected pest insects in area-wide integrated pest management (IPM) programmes (Robinson et al., 2009). The technique is based on induction of genetic sterility in the wild female insect.

This technique involves a few steps: mass rearing of target insect species, separation of male insects at the pupal stage, exposing male pupae to radiation to produce sterile male pupae, releasing large numbers of irradiated sterile male adult insects into the environment to mate with wild female insects. Ionizing radiation (gamma ray or X-ray) which blocks target insects' reproductive cycles is used for irradiation of insects in the SIT. When male insects are exposed to gamma irradiation, it causes large-scale random damage to the insects' chromosomes and/or dominant lethal mutations in the sperm (Knipling, 1979; Bakri et al., 2005). When these sterile male insects mate with wild females in the environment, viable offspring are rarely produced after mating. Sterile males cannot become established in the environment; therefore, the continuous release of artificially reared sterile males is necessary to suppress or eliminate the population of vector or pest insects (Bakri et al., 2005).

13.2.2 Importance of SIT

This technology is a species-specific, non-polluting, and eco-friendly method that has been used since 1977 to control insect pests (Klassen and Curtis, 2005). SIT has a strong record of success against a number of agricultural insect pests. The new world screw worm (*Cochliomya omnivorax*), a serious veterinary pest, has been successfully eliminated from all of North America, Central America, and

Panama (Knipling, 1985). SIT is used in a number of countries to control fruit fly pests, particularly the Mediterranean fruit fly (*Ceratitis capitata*) and the Mexican fruit fly (*Anastrepha ludens*; Franz, 2005) and lepidopteran species (Vreysen et al., 2016). Then SIT was developed to control medically and veterinary important insects in different parts of the world, such as the tsetse fly (Vreysen et al., 2000), medfly (Dantas et al., 2004) and melon fly (Koyama et al., 2004). Since the early 2000s, there has been a renewed interest in applying SIT against mosquito vectors of diseases (Lees et al., 2021).

13.2.3 SIT for Suppression of Mosquito Vectors

Nuclear technology, gamma or X-rays (electromagnetic waves) is used for the production of sterile male mosquitoes. In the current SIT-based suppression approach to control mosquito-borne diseases, laboratory-bred sterile male insects are released into a selected environment, where eventually the male mosquitoes will mate with wild female mosquitoes. Though the female mosquitoes lay eggs, they will be infertile eggs and, therefore, the next generation of mosquitoes will not be produced (Klassen, 2009). With the periodic release of sterile male insects, the population of targeted insects reduces to low levels or is completely eliminated. This process would result in the suppression of the mosquito target vector population and the reduction of the infected patients in the selected area. Therefore, there will be no risk in the replacement of the mosquito population, as only sterile male mosquitoes will be released into the environment. The SIT is a suitable technology for suppressing mosquitoes because (1) they can be mass-reared in a laboratory, (2) natural sexual dimorphism in many species aids sex separation, and (3) females become refractory after mating (Lees et al., 2021).

The successful application of SIT requires knowledge of the target population's ecology, including estimates of the absolute density of the adult population and how the density changes over time (Knipling, 1979; Itô and Yamamura, 2005). The number of required sterile males and the frequency of release depends on the population size, the extent of the area, and the required ratio (Hendrichs et al., 2005).

SIT has been researched extensively against a number of mosquito vectors. Recent progress on the development of the SIT package against mosquitoes allows for envisaging its deployment in several countries (Bouyer et al., 2020b, Vreysen et al., 2021). Simply, a SIT package includes mass rearing, sex separation, irradiation, and quality control, as well as small-scale field trials and large-scale operational population suppression programmes (IAEA, 2020a). In many instances, the optimal use of SIT would be within an integrated vector management (IVM) programme that used several approaches simultaneously (Alphey et al., 2010). At present, SIT has been utilized successfully to control mosquito-borne diseases caused by *Aedes*, *Anopheles*, *Culex*, and others in different parts of the world.

13.2.3.1 SIT for *Aedes* Control

Aedes aegypti (Linnaeus) and *Ae. albopictus* (Skuse) are vectors that transmit dengue fever, chikungunya, and Zika. Therefore, controlling the breeding of these two *Aedes* species in small artificial or natural water collecting in containers is important. Many of the currently employed vector control methods are not operationally feasible due to difficulty in reaching vector breeding sites. On such occasions, SIT has its specific advantages in bringing down the vector populations, which will in turn reduce the transmission of dengue in the high-risk areas. More importantly, SIT can be incorporated into the existing IVM for *Aedes* control programmes.

Most of countries follow a guidance framework, including phase-conditional approach for testing SIT against *Aedes* vectors (Bouyer et al., 2020b; WHO, 2020). The first phase in the phase-conditional approach is basic laboratory evaluation. The second phase is semi-field studies and small-scale field efficacy trials with entomological end points. The next phase would be to conduct a large-scale, strategically planned pre-operational trial with entomological and epidemiological end points. Further, IAEA has developed guidelines for production and releasing of sterile male *Aedes* mosquitoes and quality control of SIT procedure (IAEA, 2020a, 2020b). Oliva et al. (2021) have described a roadmap and good practice framework for designing, implementing, and evaluating SIT pilot field trials with *Aedes* mosquitoes. A communication plan targeting stakeholders and community should be put in place before releasing mosquitoes to avoid public resistance.

Many successful applications of SIT for *Aedes* have been reported around the globe: *Ae. aegypti* in Cuba (Gato et al., 2021); *Ae. albopictus* in Italy (Bellini et al., 2013, 2021), Spain (Tur et al., 2021), Germany (Becker et al., 2022), and Mauritius (Iyaloo et al., 2020); and both *Ae. aegypti* and *Ae. albopictus* in Mexico (Rivera et al., 2012). There are 39 globally distributed sterile males releasing projects including 34 in Phases I and II, four in Phase III, and one in Phase IV by 2024 (Bouyer, 2024).

The development of technology has improved. Recently, Bouyer et al. (2020a) reported a fully automated adult mosquito release system operated from an uncrewed aerial vehicle or drone, as it is crucial to release sterile mosquitoes by air to ensure homogeneous coverage, especially in large areas. They developed and tested the system in Brazil, and this will lead to improvements in areal coverage and savings in operational costs due to the requirement of fewer release sites and field staff.

13.2.3.2 SIT for *Anopheles* Control

A release of large numbers of sterile males of *Anopheles albimanus* was undertaken in El Salvador in the 1970s and an extensive trial was carried out from 1977 to 1979 (Helinski et al., 2006). The success stories for other pests have given impetus to the use of this technique for suppression or elimination of malaria vectors in some areas of Africa, including Sudan (El Sayed et al., 2009).

13.2.3.3 SIT for *Culex* Control

The first effective application of SIT to mosquitoes was reported in Florida in the 1960s with a pilot trial against *Culex quinquefasciatus* (Patterson et al., 1970).

13.2.4 SIT/IIT for Mosquito-Borne Diseases Control

Later, SIT was combined with the insect incompatible technique (IIT) using an intracellular bacteria (*Wolbachia*) in some countries and used for successful population suppression of the *Ae. polynesiensis* in French Polynesia (O'Connor et al., 2012), *Ae. albopictus* in China (Zheng et al., 2019), and *Ae. aegypti* in the United States (Mains et al., 2016), Thailand (Kittayapong et al., 2018), and Malaysia (Nazni et al., 2021). The SIT IIT programme in Guangzhou, China, is currently the largest release programme (Zheng et al., 2019).

13.3 GENETICALLY MODIFIED MOSQUITOES FOR VECTOR CONTROL

The advent of modern biotechnology has provided an unprecedented opportunity to genetically manipulate the mosquito genome to be used in mosquito population control. The invention of genetic tools has gradually revolutionized the capacity to functionally study and edit the mosquito genome from the generation of the first transgenic mosquitoes (Coates et al., 1998; Jasinskiene et al., 1998; Catteruccia et al., 2000; Grossman et al., 2001) to the creation of the first gene knock-outs (Aryan et al., 2013; DeGennaro et al., 2013; Smidler et al., 2013). When editing the mosquito genome, genetic engineering technologies enable the expression of heterologous gene expression or modify the endogenous gene in such a way that the expression will be manipulated as desired, and its success always depends on the selection of suitable gene transfer technology and the transformation marker system and regulatory elements.

13.3.1 Gene Transfer Technology

Genetic transformation permits exogenous DNA to be introduced into the mosquito genome through a vector system. This gene transfer may be either transient, expressing the gene temporally for a short period in the episomal state or stable following the integration of DNA into a host genome which is sustained for a long term in several generations. Stable transformation in mosquitoes targets the germ line cells where exogenous DNA is introduced into the developing embryo prior to pole cell formation. The introduced plasmid will be incorporated into developing pole cells and later integrate into the chromosomal DNA (Siriyasatien and Thavara, 2006). This type of integration is the basis for creating stably transformed lines in which the genetic modification will be transmitted through subsequent generations, eliminating the repeated release of mosquitoes. Different methods for delivering DNA have been used, including microinjection, biolistic, and electroporation. Nevertheless, microinjection remains the most common and best option for penetrating insect chorion and delivering DNA to germline cells (Atkinson et al., 2001, Sampath and Puttaraju, 2012).

The integration of the transgene into the mosquito genome is generally facilitated by an appropriate transformation system. Indeed, the interest in creating transgenic mosquitoes stems from the first insect transformation system based on the *Drosophila* P transposable element (Rubin and Spradling, 1982). In general, transposable elements are structurally diverse sets of genomic elements that are classified as either Class I or Class II. Only class II has been used to create insect gene vectors. Class II transposable elements are made up of DNA sequences that are less than 10 kb-long small inverted terminal repeats (ITR) of less than 100 base pairs bordering an internal transposase-encoding sequence (McDonald, 1993). Transposase catalyzes the specific excision and integration of the element. In the presence of transposase, any DNA sequence flanked by the ITR sequences will excise and integrate like a transposable element, and therefore, the transposable element functions as a vector for transgenes. A binary vector-helper system for transformation was later developed to facilitate the transformation using the transposase's ability to act in trans (Rubin and Spradling, 1982). The vector is a transposon that has a transgene between the ITR instead of transposase, and the helper plasmid is a transposon that has the transposase-encoding sequence but has been rendered non-autonomous by the deletion of the 3' ITR (Karess and Rubin, 1984). The co-injection of the vector and helper constructs will result in the virtual random transposition of the transgene into the host DNA. To aid successful transformation, mRNA or protein can now be co-injected instead of the helper plasmid. Unfortunately, the P transposable element has not been successful in the routine transformation of non-drosophilid insects since species-specific host cell factors are required for P-element transposition (Shirk and Bossin, 1991), and the technique has only had limited success in mosquito investigations. Miller et al. (1987) generated a single line of transgenic *Anopheles gambiae* that expressed an antibiotic-resistant gene, while McGrane et al. (1988) and Morris et al. (1989) reported transgenic *Aedes triseriatus* and *Ae. aegypti*, respectively. However, during the last few decades, determined efforts led to the development of transposable- element gene vectors as genetic-transformation systems that are applicable to broad groups of insects. Currently, there are four transposable-element gene vectors that have been proven to work in mosquitoes. These are the Minos element from *D. hydei* (Franz and Savakis, 1991) the Hermes element from the house fly *Musca domestica* (Warren et al., 1994) the Mos1 element from *D. mauritiana* (Medhora et al., 1991), and the piggyBac element from the cabbage looper, *Trichoplusia ni* (Fraser et al., 1995).

13.3.2 Transformation Marker System for Transgenic Detection

As a part of transformation operations, marker system is critical to distinguish the rare transformant from the wider non-transformed background. However, the choice of a suitable marker system remains critical for the detection of transformants. The first germline transformation of mosquito utilized the transgene carrying bacterial neomycin phosphotransferase gene, which confers resistance to the neomycin analog G-418 (Miller et al., 1987; Morris et al., 1989). In general, however, selecting for antibiotic resistance in transgenic individuals can be quite difficult owing to the false positives caused most likely by antibiotic breakdown, Furthermore, if the level of transgene expression is insufficient to provide antibiotic resistance to the transgenic individual, true transformants can be lost (Pinkerton et al., 2000).

Later, the identification of genetic transformants of *Ae. aegypti* was facilitated by using a recessive genetic marker. However, this strategy requires a suitable recipient mutant strain and a functional transgene to rescue the phenotype (Osanai-Futahashi et al., 2012). Among mosquito mutants, the *Ae. aegypti* line carrying a recessive mutation in the *kynurenine hydroxylase* gene, which is involved in eye pigment production, has proven to be the most advantageous. This mutation produces a white-eye phenotype (Bhalla, 1968). Therefore, introducing the wild-type allele of the *D. melanogaster* cinnabar (cn^+) gene to complement the white-eye phenotype of the kynurenine hydroxylase-white (kh^w) recipient strain of *Ae. aegypti* results in partial rescue of the wild-type phenotype (Coates et al., 1998). These rescued mosquitoes have red eyes of varied intensities which can be used to distinguish transgenic individuals from others.

When compared to mutant rescue, introducing a new transgene that results in an easily recognized phenotype is preferable since it doesn't require a mutant recipient and can be used with wild-type lines of mosquitoes. Currently, there are several transgenic markers which are widely used to screen transgenic individuals by producing dominant phenotypes. Among them, fluorescent protein genes are the most common biological markers which have been used in a wide variety of studies. These proteins absorb light of a specific wavelength and then emit a photon at a visible wavelength. The first fluorescent marker developed was green fluorescent protein (GFP) gene isolated from the jellyfish, *Aequorea victoria* (Prasher et al., 1992). Even though GFP has been shown to be functional in a number of tissues and species, it has two major drawbacks: the protein is relatively insoluble, and the excitation wavelength may damage live organisms. Therefore GFP has been modified to address these challenges by creating more soluble and redshifted enhanced GFP (EGFP) which can be detected with a longer wavelength of blue light and lower intensity (Higgs and Lewis, 2000; Dilani et al., 2021). Since then GFP has been engineered to produce a number of useful blue, cyan and yellow mutants as well as fluorescent proteins from a range of other species have been identified (Ehrig et al., 1995; Kremers et al., 2011; for example, red fluorescent protein isolated from a sea anemone, *Discosoma striata* (*DsRed*; Fradkov et al., 2000) and novel far-red fluorescent protein (*HcRed*) cloned from the sea anemone *Heteractis crispa* (Cotlet et al., 2006). These marker systems are currently utilized in a variety of mosquito species, including *Ae. aegypti* (Kokoza et al., 2001; Fu et al., 2010), *Ae. fluviatilis* (Rodrigues et al., 2006), *Ae. albopictus* (Labbé et al., 2012), *An. albimanus* (Perera et al., 2002), *An. gambiae* (Ito et al., 2002; Nolan et al., 2002), and *An. stephensi* (Catteruccia et al., 2000; Catteruccia et al., 2003). Among others, the EGFP (Enhanced Green Fluorescent Protein) and DsRed genes appear to be stably integrated into the genome, as they have been passed down through 30 generations with no evidence of losing fluorescent marker or transposon mobilization (Catteruccia et al., 2003) ensuring its potential to serve as universal marker for insect transgensis.

In mosquitoes, several promoters have been employed to express these fluorescent marker genes. This could be tissue-specific or ubiquitous promoters. The artificial 3xP3 promoter, for example, has been created to express in insect eye rhodopsin cells (Berghammer et al., 1999). The tissue-specific nature of this promoter aids in distinguishing marker expression from background fluorescence and minimizing the fitness cost due to the marker gene expression (Dilani et al., 2022). Nevertheless, the promoters which have been designed to express the marker genes in all tissues throughout the development such as Hr5-IE1 (baculovirus IE1 promoter and Hr5 enhancer) are more visible and easier to screen, especially for the insects whose eye pigment masks the expression of 3xP3- driven markers (Rodems and Friesen, 1993; Dilani et al., 2021).

13.3.3 Different Transgenic Strategies for Mosquito Control

With the advancement of microinjection techniques and the availability of a variety of transposable element–based vector systems, promoters, and universal marker genes, now it is easily possible to develop transgenic individuals to control mosquito populations in fight against mosquito-borne diseases (Alphey and Alphey, 2014). The genetic control of mosquito vectors could potentially be achieved in two mechanisms: they either aim to modify the vector to make it refectory to pathogen transmission (population replacement) or reduce the vector population in such ways that the population size will crash below the threshold level necessary to support disease transmission (population reduction/suppression; Gabrieli et al., 2014). Another classification is by the persistence of the modification: Self-sustaining and self-limiting mechanisms. 'Self-sustaining' genetic systems are expected to persist in the wild population indefinitely after the initial release, while 'self-limiting' genetic systems disappear rapidly unless maintained by repeated releases (Alphey and Alphey, 2014).

13.3.3.1 Release of Insects Carrying Dominant Lethal Systems

Recent advances in genetic engineering have pay the way for novel strategies to control natural vector populations. One possibility is to develop transgenic individuals based on traditional SIT by engineering self-limiting genes, which addresses many of the drawbacks of traditional SIT, such as genetic sterilization, genetic sexing, and genetic containment while maintaining the environmentally benign and species-specific. The release of insects carrying dominant lethal (RIDL) systems introduced by Oxitec (Abington, United Kingdom) is such a strategy currently being used to suppress the target population through release of sterile males (Thomas et al., 2000). In this approach, genetic sterility and genetic sexing in mosquitoes are engineered by a repressible dominant lethal gene, where the permissive condition can only be created in the lab and never seen in the wild. In the mass-rearing facility, a dietary additive like tetracycline will suppress the lethal effect, followed by the release of mass-reared mosquitoes into the wild without the need for irradiation (Thomas et al., 2000; Alphey et al., 2006). Following the release, the progeny of the released individuals dies as a consequence of inheriting dominant lethal mutations, and the population tends to decline. Originally, the RIDL was designed to allow the death of all springs, the strategy known as bi-sex RIDL; however, it was later discovered that killing only females is sufficient and more successful than killing both sexes. The latter is known as the female-specific RIDL system (fsRIDL) (Thomas et al., 2000, Fu et al., 2010, Dilani et al., 2021).

13.3.3.1.1 RIDL Mechanism

Genetic sterility in the RIDL system has been engineered in insects mostly by tetracycline-dependent repressible systems (Heinrich and Scott, 2000; Lycett et al., 2004; Phuc et al., 2007; Morrison et al., 2012). These systems can be divided into two main categories: two-component systems (bipartite system) and one-component systems (Figure 13.1). The two-component system contains two sub-components: driver and the effector. The key component of the driver is the tetracycline-repressible transactivator (tTA), which is controlled by a specific promoter. The effector contains the lethal/effector gene placed under the control of a minimal promoter and tetracycline response elements (TREs) upstream of the promoter. The TREs contains several tetracycline operators (tetO) sequences (Gossen and Bujard, 1992; Haghighat-Khah et al., 2019). Tetracycline can bind to the tTA and prevents it from binding to the TREs to repress the expression of lethal gene and allows the insects to develop properly on a tetracycline-supplemented diet. In contrast, tTA binds to the TRE in the absence of tetracycline in the wild, driving the expression of a lethal gene. The variant of this is the one-component positive feedback system containing tTA placed under the control of a minimal promoter and TREs (Gong et al., 2005). In the absence of tetracycline, the basally expressed tTA can bind to TREs, consequently increasing its own expression in a positive feedback manner. The elevated expression of tTA can consequently lead to death, implying that the tTA can serve as both transactivator and the effector (Gong et al., 2005).

13.3.3.1.2 Global Research/Application of RIDL for Mosquito Control

RIDL systems have been tested for several species of mosquitoes, including *Ae. aegypti*, *Ae. Albopictus*, and *An. Stephensi*, and the findings have shown that the late-acting lethality is more effective than conventional SIT at controlling mosquito populations limited by density dependent effects, e.g., limited availability of larval food. The late-acting RIDL *Ae. aegypti* was first

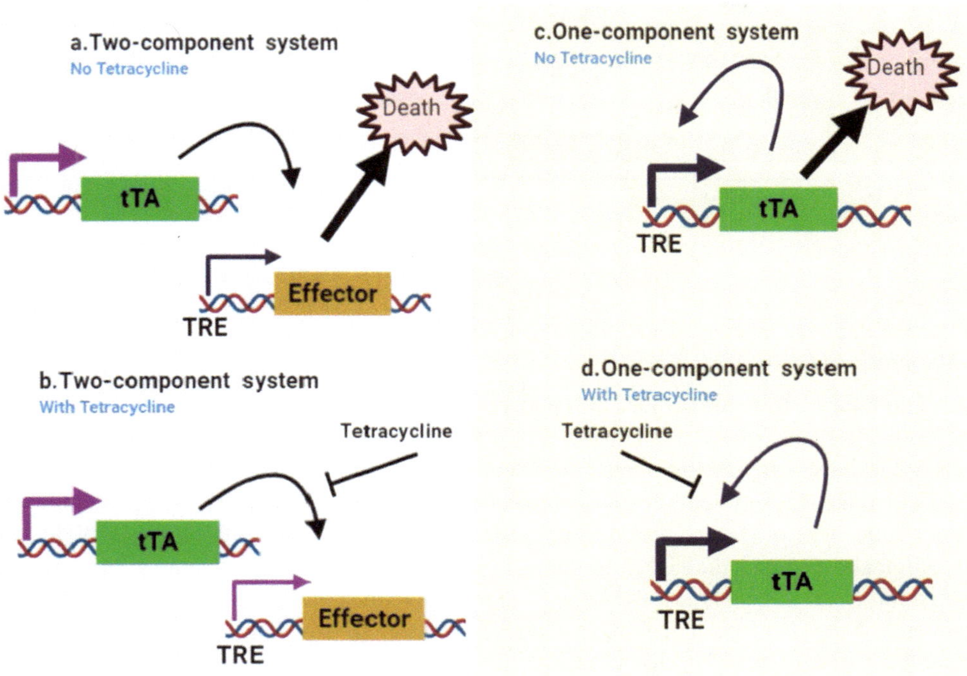

FIGURE 13.1 Schematic diagram of tetracycline-repressible lethal systems. (a, b) Two-component tetracycline-repressible lethal systems; in the absence of tetracycline (Tc), tTA binds TRE and drives the expression of effector gene, leading to death. In the presence of tetracycline, tTA binds tetracycline and represses the expression of the effector and inactivates the lethal system. (c, d) One-component positive feedback system; in the absence of tetracycline, low levels of basally expressed tTA bind TRE to increase the expression of tTA, leading to death. In the presence of tetracycline, tTA preferentially binds to tetracycline to repress the expression of tTA to prevent the lethality (originally drawn for this chapter).

reported in 2007 (Phuc et al., 2007). They employed a lethal positive feedback system to evaluate the tetracycline-repressible lethality and observed late-acting lethality in transgenic line LA513A during the transition from late larvae to early pupae. It was later developed into OX513A RIDL by Oxitec and has been the most successful bi-sex strain by which only males are released to mate with wild females. The progeny of the mating die at the late larvae or pupae stage in the absence of tetracycline. The engineered genetic component of OX513A majorly consist of the variant of the *tTA* gene (*tTAV*) with marker gene, *DsRed2*, and therefore can be readily detected under fluorescent light due to the expression of *DsRed2*. *tTAV* is controlled by its binding site TRE, minimal promoter from *Drosophila* heat-shock protein (Hsp70), and the 30 untranslated region (UTR) sequence of the gene *Dmelfs(1)K10* from *Drosophila* (Figure 13.2; Phuc et al., 2007).

Prior to the release of the OX513A strain into the field, life history characteristics under the controlled laboratory conditions had been compared to its wild counterparts and found lower larval survival and reduced adult longevity (Phuc et al., 2007). Later, the field performance of first-generation RIDL strain of *Ae. aegypti* OX513A males had been assessed in term of mating competitiveness at Grand Cayman, British Overseas Territory in the Caribbean and Brazil where they found that environmental and target-strain differences imposed little impact on the mating success of the OX513A males showing the ability of OX513A to reduce rates of disease transmission through population suppression (Harris et al., 2011; Harris et al., 2012; Carvalho et al., 2015).

Even though OX513A was highly effective in controlling the target population, it is still lacking the efficient genetic sexing mechanism. Therefore, a genetic sexing strain (GSS) of *Ae. aegypti* was

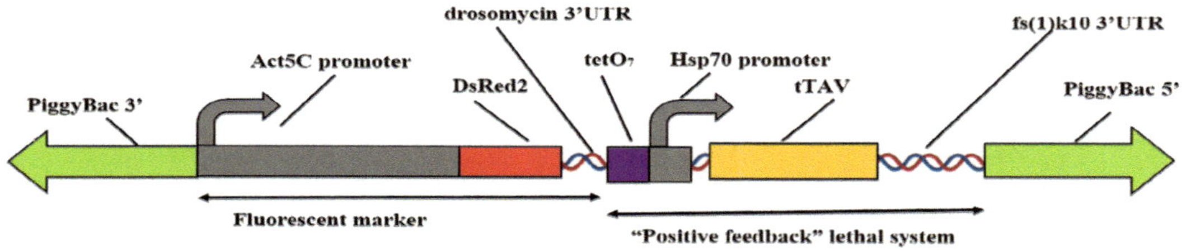

FIGURE 13.2 Schematic representation of transposon LA513. LA513 is an 8.4 kb non-autonomous piggyBac-based transposon, Actin 5-c (***Act5-C***) promoter drives the expression of the marker gene DsRed2. tTAV is controlled by its binding site tetO, the minimal promoter of *Drosophila Hsp70*, and a 30 UTR sequence from *Drosophila fs(1)K10*.

Source: This figure is from the author's own publication, Dilani et al., 2021.

engineered to have a repressible female-specific flightless phenotype by either using single transgene (positive feedback system) or two separate transgenes (two-component system, along with the promoter sequences of *Ae. aegypti* Actin-4 gene (*AeAct-4*), specific to the indirect flight muscle (IFM) of females of which the expression starts at L4 larvae (Fu *et al.*, 2010). For the two-component system, a mosquito line with an *AeAct-4 –tTA* driver component has been crossed with two effector components, Nipp1Dm and micelob x, under the control of the TRE, and it was reported that 69.8%–98.3% of females were flightless. Additionally, some flightless phenotypes have been seen even in the absence of the effector gene, indicating that the RIDL system may be suitable for developing as a positive feedback system. By taking advantage of the *AeAct-4* gene's sex-specific alternative splicing features, the female specificity was further improved to develop fsRIDL strain to have flightless phenotype. However, in the field, inability to fly is likely to be lethal since adult mosquitoes are unable to mate find blood meal, access food resources, and escape predators (Fu et al., 2010). In a trial introducing the fsRIDL strain, OX3604C males into laboratory cages indicated the successful elimination of the targeted population (Wise de Valdez et al., 2011). Later, Oxitec succeeded in improving this to a new second-generation OX5034 self-limiting *Ae. aegypti* fsRIDL system. A recent field trial in Brazil which dealt with releasing of OX5034 fsRIDL mosquitoes has shown high levels of suppression (96%), and the recent progress has finally reached into the open-air release in Florida, United States, for the first time after decades of struggle for regulatory approval and public acceptance (OXITEC, 2019; Caragata et al., 2021; Waltz, 2021).

Later studies reported that the same technology works in related species such as *Ae. albopictus*. The first report of engineering *Ae. albopictus* with Actin-4 (AealbAct-4) and tTA to drive a dominant lethal gene in the IFM resulted in a female-specific flightless phenotype (Labbé et al., 2012). However, it has not been tried out yet for population suppression, and further studies are required to assess their suitability and effectiveness in suppressing wild populations. The technological advances made in relation to *Aedes* mosquitoes were later adapted to the human malaria vector *An. stephensi* to prevent the transmission of malaria parasites. Towards this end, *An. stephensi* was engineered to confer a repressible female-specific phenotype using a similar driver component, *AeAct-4 –tTA*, and effector component (teto-Nipp1Dm). Further experiments by replacing the *AeAct-4 -5'*-end sequence with the orthologous sequence from *An. stephensi* Actin-4 (*AsAct-4*) reported promising results in developing a flightless female-specific phenotype for application in malaria control (Marinotti et al., 2013).

RIDL strains developed so far rely entirely on engineering the cell-autonomous effectors with their activity confined to the tissues where they are expressed typically regulated by specific control elements, such as promoter regionsPhuc et al., 2007; Fu et al., 2010). However, the effectiveness of using regulatory elements with precise spatio-temporal or sex-specific expression patterns depend on the target cells being simultaneously responsive to the effectors. This dependency presents a significant challenge in creating more advanced and versatile RIDL systems. Dissociation of the temporal and spatial expression pattern of the effector offers potential solution to overcome this limitation. In a more recent study, temporal and spatial expression patterns of the effector were dissociated with a novel synthetic effector which was designed to be secreted out of the adult fat body following a blood meal by a female mosquito and to affect the functioning of motor neurons to cause paralysis (Haghighat-Khah et al., 2019). Even though the penetrance of the phenotype was not as strong as anticipated, the outcomes of the study demonstrated that non-cell-autonomous effectors can be used to create more flexible repressible lethal systems.

13.3.3.1.3 RIDL-Aedes aegypti Research in India

Deployment of any new technologies in various scientific fields, including genetically modified organisms, genetically modified crops, pharmaceutical chemicals, etc., requires phased evaluation and is considered and recommended as an appropriate way for risk assessment (UNICEF/UNDP/World Bank/WHO Special Programme for Research and Training in Tropical Diseases, 2015). In this context, the evaluation of genetically engineered mosquitoes involves phased testing, including testing for its potential use and integration in public health tools. The World Health Organization has developed a guidance framework for evaluation well before deployment and implementation of genetically modified (WHO Special Programme for Research and Training in Tropical Diseases 2014). Phase evaluation involves testing of the modern tools and in addition considers addressing the regulatory decisions and ethical, social, and cultural issues at each phase.

OX513A *Ae. aegypti* strain, also referred to as RIDL technology, is a self-limiting technology and is also termed Friendly™ mosquitoes. OX513A consists of a single gene construct insertion consisting of two inherited dominant genes: a self-limiting, repressible dominant lethal gene and a fluorescent (DsRed2) marker gene (Phuc et al., 2007). This repressible, self-limiting gene component confers a lethality in the inherited offspring, in the absence of antidote tetracycline (repressor), causing lethality during the immature stages; hence, offspring fail to develop successfully into adults (Figure 13.3). Monitoring of the OX513A strain offsprings is aided by the presence of fluorescent marker gene encoded for a red fluorescence protein DsRed2, visible under a fluorescence microscope equipped with specific filters at excitation range of 510–550 nm and emission at 600 nm. Thus as sufficient OX513A strain male mosquitoes are released over a sustained period, the local population will crash. As the mosquitoes inherit the gene from OX513A *Ae. aegypti* do not survive in nature due to a lack of tetracycline, and the released OX513A males die a natural death; moreover, the genes from OX513A do not persist in nature, thus making it a self-limiting and non-environmental footprint.

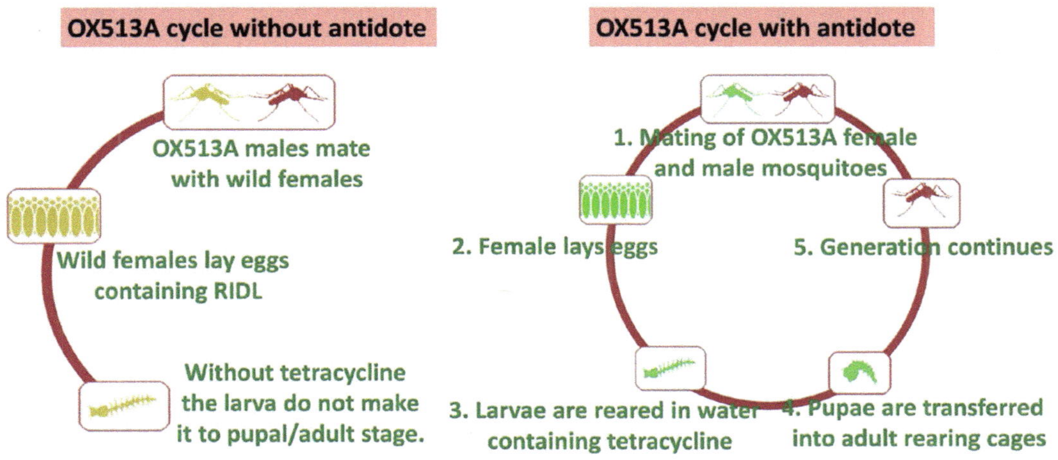

FIGURE 13.3 Effect of offsprings, inheriting the lethal gene from OX513A male adults, reared with and without antidote tetracycline.

Phase wise studies has been conducted by GBIT Limited in India, which involved the laboratory evaluation of the OX513A strain (phase-1 study) under Arthropod Containment Level II facility and physically contained field cage trials (phase 2 study) evaluation of the OX513A strain to drive the technology for the control *Ae. aegypti* mosquito that carry and spread dengue and chikungunya in India.

13.3.3.1.3.1 Phase 1 Laboratory Evaluation of OX513A Strain Laboratory studies were conducted to evaluate OX513A strain for (1) efficacy of lethal gene construct; (2) life table characteristics of the strain, and (3) mating competitiveness.

Evaluation of lethality induced by the gene expression in the immature stages is one of the crucial and important experiments to understand the potential effect of the OX513A strain. The gene expression was studied under homozygous and heterozygous conditions in the offspring for its effectiveness. Progeny of wild-type females mated with OX513A males were used for lethality under heterozygous gene condition. The homozygous and heterozygous progeny were reared in the absence of antidote tetracycline along with control groups of wild type and OX513A strain supplemented with repressor antidote tetracycline. Studies revealed that the homozygous and heterozygous progeny collapsed during the development stages expressing mortality during larval/pupal stages. Progeny under homozygous condition revealed 93.7% mortality during larval stages, and none of the progeny survived into successful adult emergence, while under heterozygous condition mortality was visible maximum during pupal stages leading 83.1% exhibiting total mortality of 98.3% and those adults emerged could not survive long enough to take blood meal for continuation of the life cycle (data unpublished).

Fitness of the strain is an important parameter, which indicates successful effectiveness of the technology under field conditions and its subsequent use in integrated vector control tools (Curtis and Sinkins, 1998; Sinkins and Gould, 2006). Assessment of the biology of the OX513A strain was done in comparison with two wild-type AWD (Aurangabad) and DEL (Delhi) strains) of different regions (Patil et al., 2015). Laboratory studies on the developmental period, adult survivability and population growth parameters (net reproductive rate, mean generation time, intrinsic rate of increase and doubling time) in comparison with wild-type strains have demonstrated OX513A strain fitness to be at par revealing no negative effect on its biology and potential. Among the population growth parameters, successful reproductive rate indicates the fitness of the strain, and studies have shown that net reproductive rate was significantly similar, along with other reproductive parameters to the wild-type strains (Figure 13.4). In addition, survivorship of the male adults is crucial, which are intended for field releases during the implementation programme.

The longevity of male adults was found to be slightly but significantly shorter compared to wild types; however, it has been inferred that decreased longevity could reduce persistence of the transgene and result in selection against released males in an open environment, thereby offering safeguards that may be perceived as a significant biosafety benefit (Figure 13.5). Moreover, males are not expected to persist for longer periods, as open release of male adults is intended for successfully mating with wild-type females.

Mating competitiveness studies were conducted under small adult rearing cages (30cm^3) under laboratory conditions, by releasing equal proportions of male adults of OX513A (42.8%) and wild strains (39.6%) for mating against wild-type female adults. Observations revealed OX513A strain the percent of adult females mated with males of respective strains was significantly similar, in addition females (9.1%) with double mating (females mated with both OX513A and DEL males) was observed. The relative mating index excluding females that were double mated yielded values of 0.52 and 0.48 for OX513A and DEL respectively and were significantly similar (Patil et al. 2015).

13.3.3.1.3.2 Phase 2 Physically Contained Field Cage Trials Phase 2 trials involved field cage studies under physically contained facilities under natural climatic conditions. OX513A strain was evaluated for mating competitiveness with wild-type *Ae. aegypti* strain and to demonstrate suppression of wild-type population with sustained release of male adults of OX513A strain over a period of time. The physically contained facility was set on a large platform with a built-up area of 580 m^2

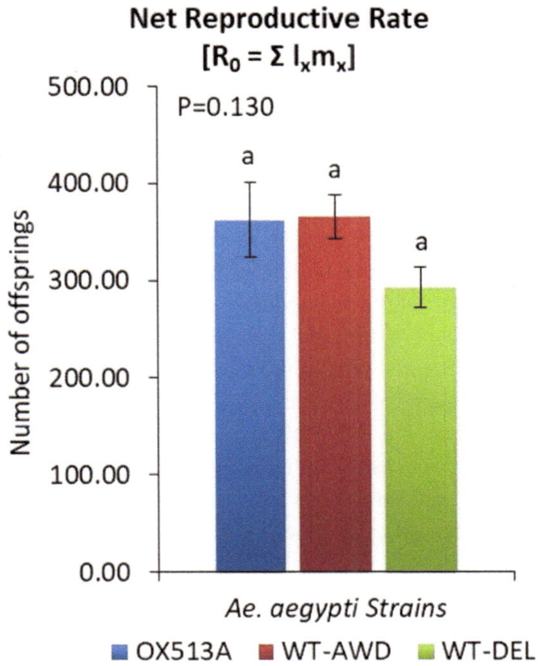

FIGURE 13.4 Reproductive rate of OX513A and wild-type strains of *Ae. aegypti*, where R_0 is the average number of female offspring that a female produces during her lifetime, l_x is the proportion of females surviving to each age (x), and m_x is the average number of female offspring born to the surviving females at each age.

Source: Patil et al. 2015.

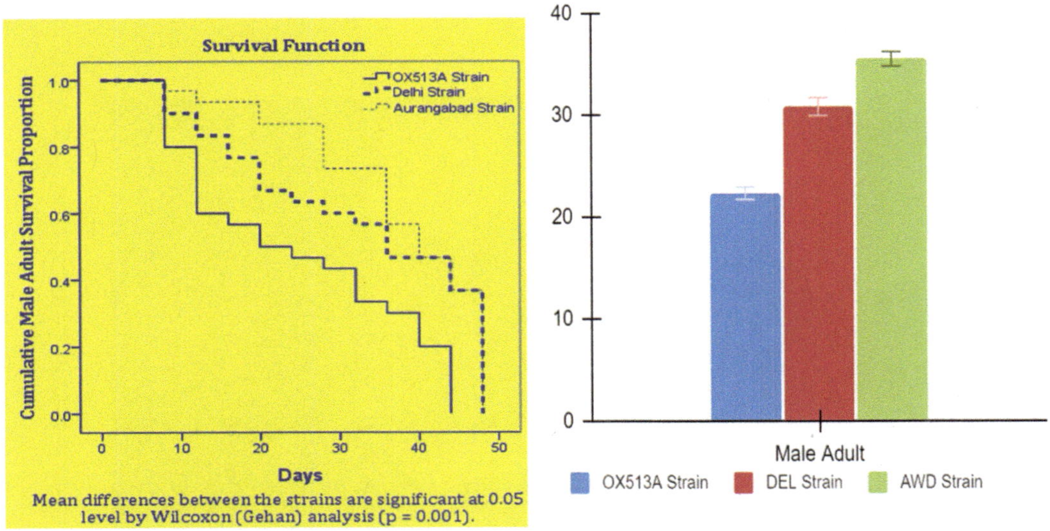

FIGURE 13.5 Male adult survivorship and life span of OX513A and wild-type strains of *Ae. aegypti*.

Source: This data is reproduced from the author's own publication with permission from Patil et al. 2015, © 2014 Gangabishan Bhikulal Investment and Trading Limited. Pest Management Science published by John Wiley & Sons Ltd. on behalf of the Society of Chemical Industry.

(29 m × 20 m) consisting of six paired field cages fastened on a wooden platform elevated 1 m above the ground supported on legs made out of steel (Figures 13.6 and 13.7). Each paired field cage represented one control and one treatment cage for the population suppression experiments (Patil et al., 2022).

13.3.3.1.4 Mating Competitiveness of OX513A Strain Under Physically Contained Field Cage Facility

Mating experiments were conducted to evaluate performance of OX513A strain male adults versus wild-type adults (AWD strain) against wild-type female adults of AWD strain under field cages (Figure 13.8). The release ratio of adults for mating competitiveness was 1:1:1 (OX513A♂: AWD ♂; AWD ♀), with 100 adults of each strain, and experiments were conducted using 3- to 4-day-old adults. Prior to mating experiments, the size of the mosquitoes of respective strains were measured during the pupal stage (measurement of the dorsum of the thorax) and the adult stage (wing measurement was considered to determine the adult size) to determine presence of size difference if any influencing mating competitiveness. Male adults of OX513A and wild-type strains were first introduced into the field cages and allowed to acclimatize for a period of one hour followed by introduction of wild-type female adults. The released adults were allowed mating for a period of 24 hours and were followed with access to blood meal for female adults. Following the blood meal, the female adults were recovered from the respective field cages and allowed oviposition individually under the Arthropod Containment Level 2 Laboratory. The eggs laid from individual female adults of respective cages were hatched and reared until the third instar stage for fluorescence screening to determine the paternity of the offspring, i.e., larvae showing fluorescence of individual female adult indicated mating with Ox513A male, larvae with no fluorescence indicated female mating with wild-type male adult, while the presence of both larvae with fluorescence and no fluorescence indicated double mating of female adults with OX513A and wild-type male adults (Patil et al., 2022).

Mating competitiveness experiments provide crucial insight on the performance under natural conditions in field cages providing data pertinent to the effectiveness of the OX513A male adults for population suppression strategy. Measurements of the pupal size and right wing of adults indicated no significant differences between the two strains, avoiding any suspicion of mating advantages due to differences in the size. Results of the mating experiments revealed wild-type female adults mated

FIGURE 13.6 Physically contained field cage facility for evaluation of transgenic OX513A strain.

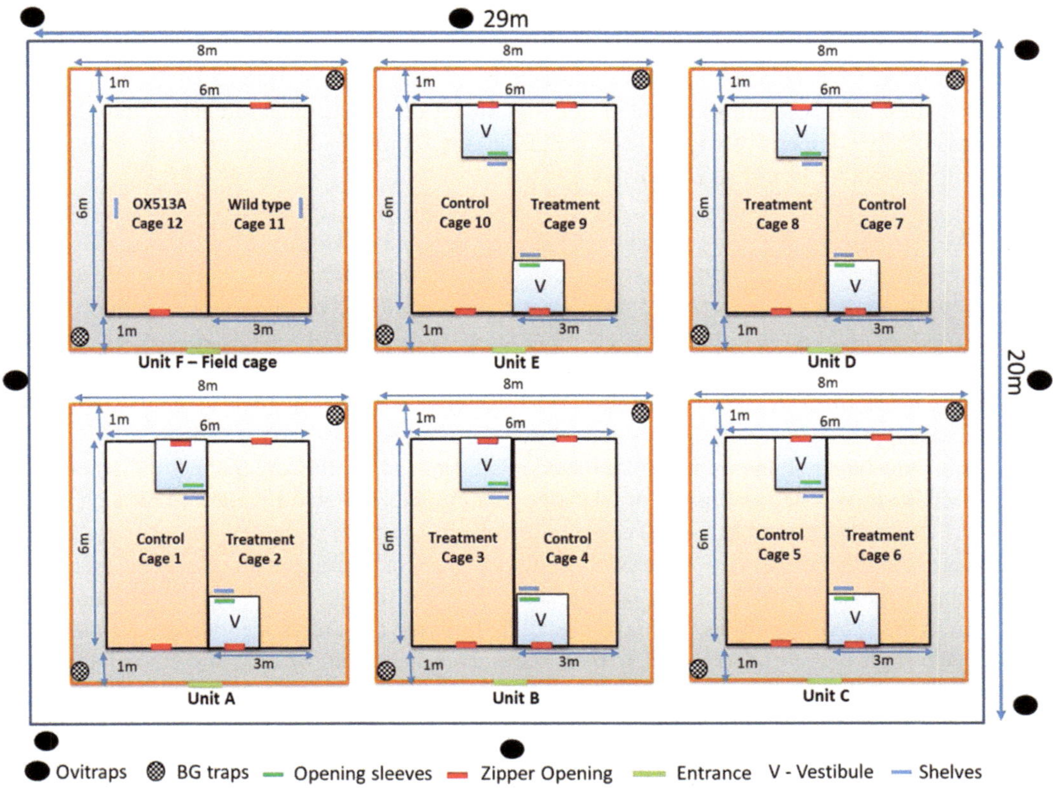

FIGURE 13.7 Layout of field cages under physically contained facility for evaluation of OX513A strain.

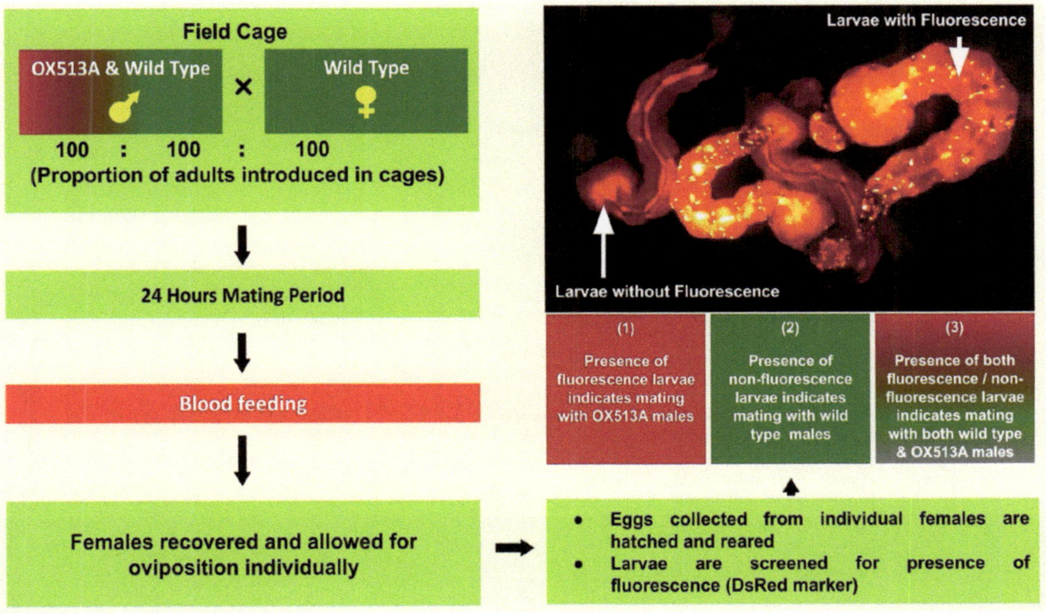

FIGURE 13.8 Illustration of mating competitiveness of OX513A strain under physically contained field cage trials.

with OX513A, and wild-type male adults were 41% and 48% respectively, while 10.9% of the AWD adult females mated with both OX513A and AWD males. The mating proportion including double mating subjected for chi-square analysis against expected mating proportion of 1:1:0 (AWD males: OX513A males: double mating) revealed non-significant deviation, thus indicated that OX513A strain was equally competitive with wild-type male adults. While double mating observed was negligible with respect to the chi-square analysis.

13.3.3.1.4.1 Population Suppression Trials Under Physically Contained Field Cage Facility Population suppression trials were conducted in five paired cages units, with each pair representing a control and a treatment cage assigned randomly. The duration of the experiments were for a period of 34 weeks, consisting of two parts: establishment of a stable wild-type (i.e. AWD strain) population in all the five paired cage units, followed by a population suppression period with sustained releases of OX513A male adults in the designated treatment cage of paired unit cages (Figure 13.9).

A stable wild-type population was established in all the cages, following the method reported by Wise de Valdez et al. The establishment of the population was initiated by the introduction of 400 adults (1:1 male/female ratio) in all 10 cages; adults were provided with 10% sucrose, and female adults were provided access to a blood meal twice a week using membrane feed technique. Oviposition traps were placed for egg laying, and eggs were recovered from the oviposition traps weekly from the respective cages. The larvae hatched from the eggs recovered were reared until the second instar stage and introduced back to their respective cages. Stabilization of the population was determined by weekly egg counts recovered from the respective cages; in addition, the adult population dynamics were determined by adult samples in cages using adult traps (BG Sentinel Traps, Bioagents, Germany) and reintroduced the adult samples in the respective cages post-analysis. The stable population of the wild-type adults in all the cages was achieved over a period of 14 weeks in all the paired cage units. Stable population establishment was determined through analysis of the number of eggs recovered in each cage by one-way ANOVA, which indicated an increase of the mosquito population significantly from week 8 onwards and stabilized by week 14 with consistent

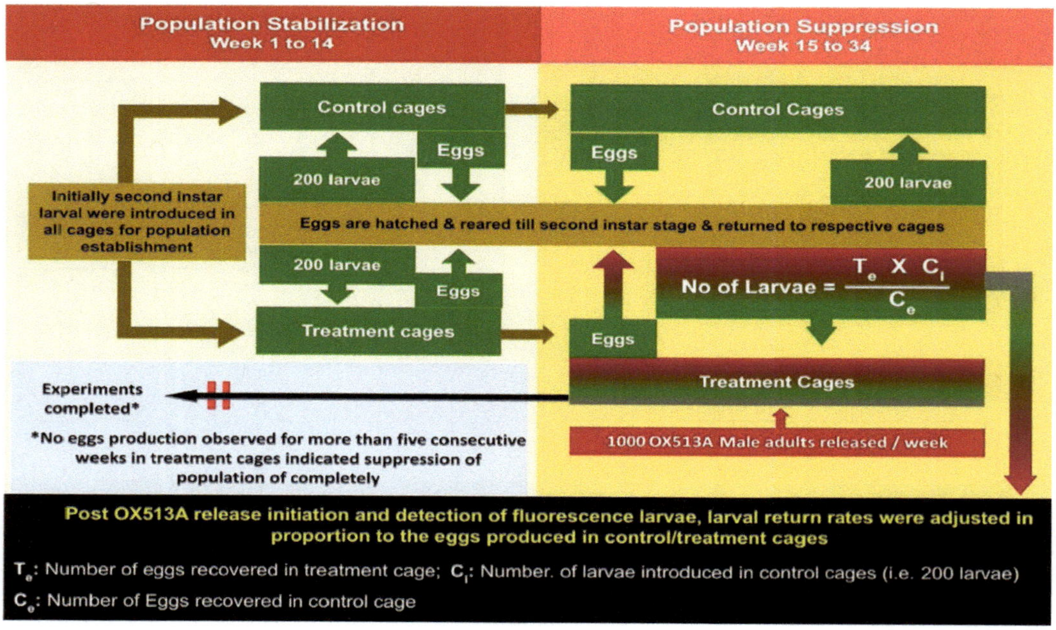

FIGURE 13.9 Illustration of the population suppression trials under a physically contained field cage facility.

egg production. Egg production during the stabilization period has shown no significant difference in the control and treatment cages.

Stabilization of the population was followed by weekly introduction of OX513A male adults in the designated treatment cage of each paired unit cage. During population suppression, a weekly 1,000 OX513A male adults in each treatment cage was introduced, which corresponded to a release of tenfold of wild-type male adult emergence in the control cages, based on the weekly introductory rate of 200 second instar larvae per cage, considering 1:1 male/female emergence ratio. The weekly replenishment rate of 200 second instar larvae from the eggs recovered from respective cages were held constant for control cages throughout the experiential period. For treatment cages, the weekly larval introductory rate was similar for control cages until the first fluorescence larval observation in the respective treatment cages. After initiating the OX513A male adult releases in the treatment cages, the larvae were screened for the presence of DsRed fluorescence. Post-observation of fluorescent larvae, the number of larvae replenished back to the respective treatment cage was adjusted proportionally to reflect the egg productivity and larval introduction rate of the respective paired control cage (Wise de Valdez et al., 2010). For instance, if a control cage produced 1,000 eggs in one week with constant return rate of 200 larvae/week (1/5), and if the treatment cage produced 2,000 eggs in the same week, then a same proportion of the total number of eggs (1/5) were returned as larvae to the cage (i.e., 400 larvae). If a treated cage in the same week produces 500 eggs, then 100 larvae were returned to the cage.

Observations on the mortality due to lethal effect of gene expression during developmental stages in the weekly larvae introduced in treatment cages following OX513A releases were carried out from week 18 onwards. Mortality observed was between 46.8% to 59.4% during the weeks 18 to week 24 and reached 100% by week 29 to 29 in the treatment cages (Figure 13.10). The weekly introduction of OX513A male adults following lethality induced due to gene expression in the

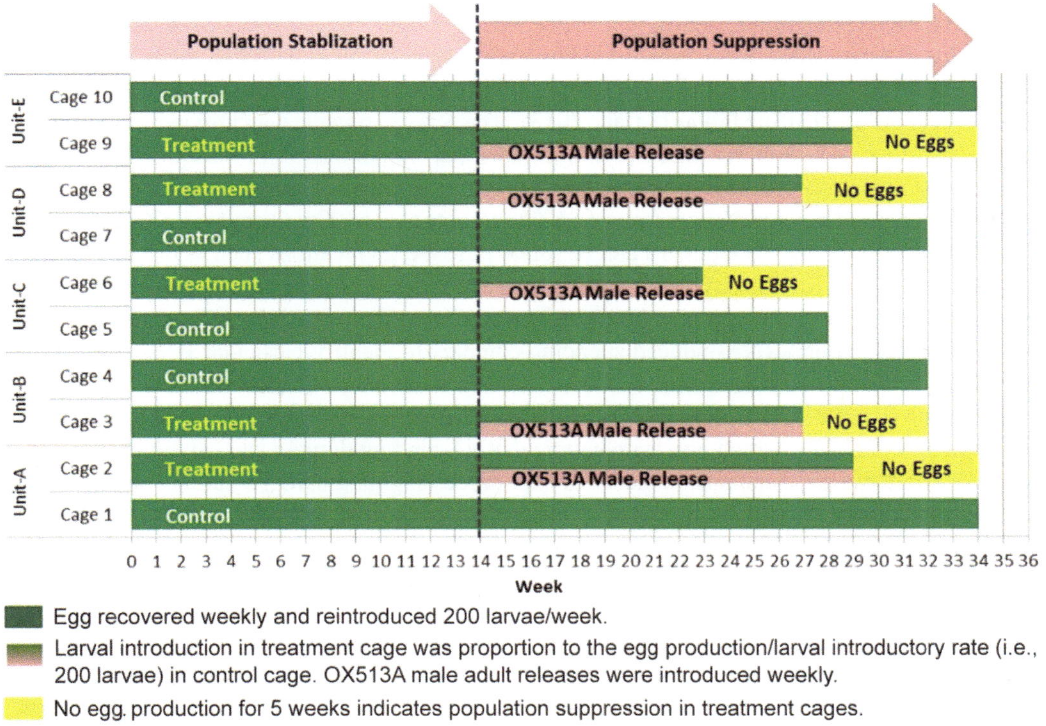

FIGURE 13.10 Illustration of population stabilization and population suppression following sustained releases of OX513A male adults and weekly egg productivity.

weekly introductory larvae led to a decrease in the adult population, leading to a significant decrease in weekly egg production from week 20. Statistical analysis following analysis of covariance (ANCOVA) on the egg productivity in treatment cages revealed significant reduction in egg production in respective treatment cages. The first treatment cage with no egg production was observed during week 24, followed by another two treatment cages from week 28, and remaining two treatment cages from week 30 (Patil et al., 2022). Observations on egg productivity were continued for subsequent weeks; the population was considered eliminated when no egg production was recorded for a period of five consecutive weeks in the treated cages (Figure 13.10).

Such species-specific self-limiting technology could deliver long-term effective protection while minimizing the ecological footprint associated with pest management practices. Thus RIDL technology has no toxin involved, and no adverse effects on human or environmental health, and limited prospect of resistance development – providing true sustainability. This is by far the longest scientific study on GE *Ae. aegypti* in India, whose results are comparable to those of a similar study anywhere in the world.

13.3.3.2 RNA Interference: Insight of Defence

The immune system in cells is capable to eliminate or restrict the growth of pathogenic viruses and the innate immune pathways such as RNAi plays a specific role in RNA viruses-specific antiviral role in insects and mosquitoes (Gammon and Mello, 2015). The term "RNA interference" (RNAi) was introduced by Fire and coworkers after they discovered that introducing dsRNA into nematode (*Caenorhabditis elegans*) resulted in robust and selective inhibition of the endogenous gene identical to the introduced dsRNA (Fire et al., 1998). Lately, a study in the fruit fly *Drosophila melanogaster* revealed anomalies in embryonic patterning that resembled the loss of wingless function following injection of dsRNA into embryos (Kennerdell and Carthew, 1998). Further findings of RNAi pointed out the phenomena known as post-transcriptional gene-silencing/virus-induced silencing described in plant research, which demonstrated the antiviral defence by synthetic dsRNA in plants in 2000 (Baulcombe, 2000).

In insects, RNAi is the most broadly active antiviral mechanism and also plays an important antiviral role in plants (Szittya and Burgyán, 2013), nematodes (Sarkies and Miska, 2013), and mammals (Li et al., 2013; Maillard et al., 2013). The RNAi strategy was utilized as an antiviral defence against mosquito-borne diseases and intracellular immunization of mosquitoes. Past years of transgenic mosquito studies were proven usability and the antiviral resistance of RNAi in flavivirus family viruses of dengue, yellow fever, and Zika (Higgs et al., 1998; Qsim et al., 2017; Buchman et al., 2019). Different RNAi elements can be found in RNAi applications such as dsRNA, miRNA, shRNA, lh-RNA and MicroRNA-based shRNA (miRshRNA) and RNAi mechanism utilizes RNAi-specific elements to generate intermediate product small interference RNA (siRNA), which is subsequently processed into RNA-induced silencing complex to identify particular RNA (Meister and Tuschl, 2004; Carthew and Sontheimer, 2009; Krishnan et al., 2009; Taxman et al., 2010).

Processing of dsRNA or miRshRNA catalyzed by dsRNA-specific RNAase-III-type endonucleases (Dorsha and Dicer), which possess catalytic dsRNA binding domain and catalytic protein subunit RNaseIII domains. Dorsha is not necessary for processing long dsRNA, but it is involved in the long primary transcripts processing which provides precursor-miRNAs (pre-miRNA), and the processed products have 5' phosphates and two nucleotide 3' overhangs. Later, pre-miRNA is exported to the cytoplasm via exportin-5, the nuclear export receptor and in the cytoplasm, the Dicer enzyme further processes pre-miRNA into the miRNA. Cytoplasm reached dsRNA is processed by the Dicer enzyme into small RNA duplexes/ siRNA (~21nt) which is also made up of 5' phosphates and two nucleotide 3' overhangs. The products of miRNA and siRNA duplexes–containing ribonucleoprotein particles (RNPs) are subsequently rearranged into the RNA-induced silencing complex (RISC; Figure 13.11; Meister and Tuschl, 2004).

Argonaute (Ago) RNP proteins act as the heart of RISC also considered key proteins in all gene-silencing pathways (Meister and Tuschl, 2004). Ago proteins are conserved among species, while

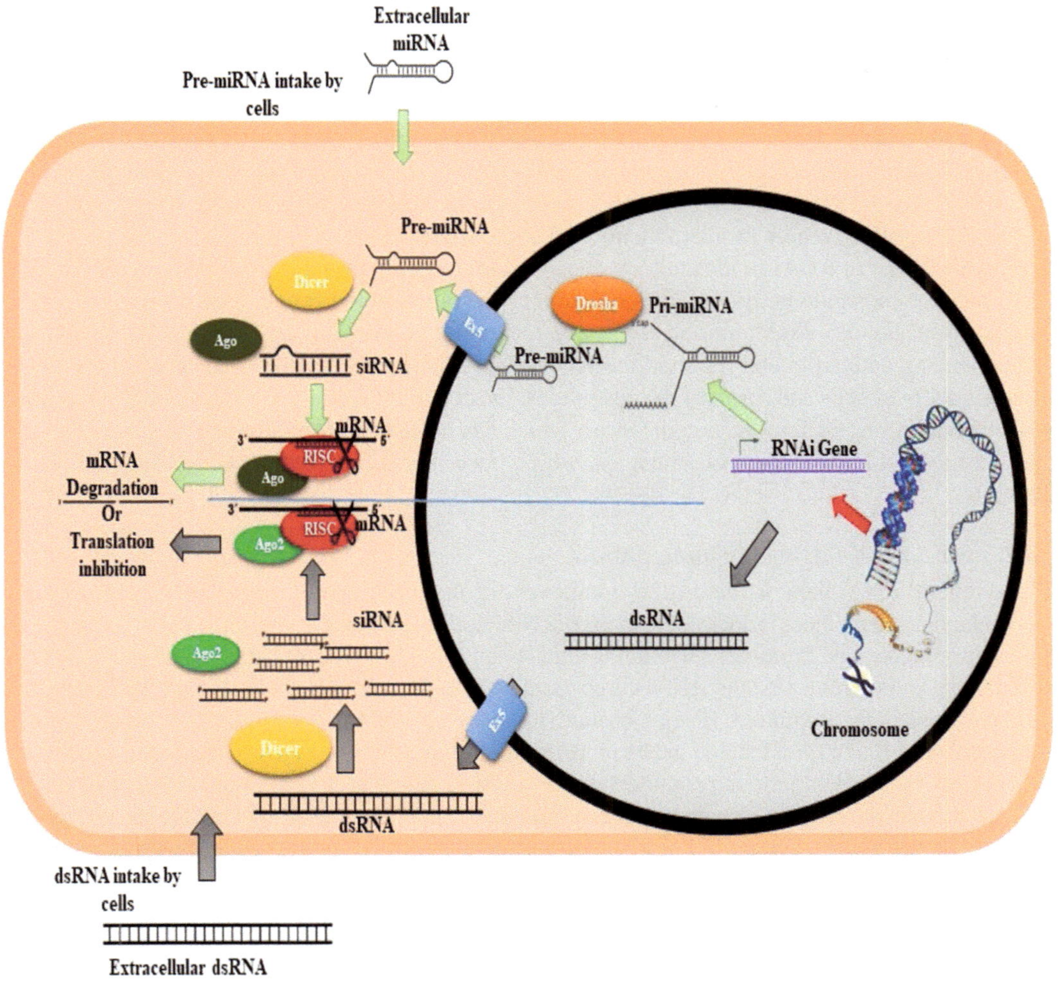

FIGURE 13.11 RNA interference mechanism. RNAi triggering elements of miRNA and dsRNA are processed into siRNA, which subsequently induce RNA binding with the target mRNA to be interfered with, eventually forming the silencing complex (RISC). Ex5 represents exportin 5.

many of them express multiple proteins, *Schizosaccharomyces pombe* expresses one protein, *Drosophila* expresses five proteins, ten proteins in *Arabidopsis* sp., 27 in *Caenorhabditis elegans* and eight proteins in humans. Argo2 protein is involved in the dsRNA-mediated gene-silencing pathway and silencing of Ago2 results in inhibition of RISC complex formation. Decades of research utilized synthetic RNAi elements to suppress specific genes, and the engineering of the RNAi elements were supported by new tools and web applications developed by bioinformaticians and also utilized as a powerful tool to silence endogenous mosquito genes and mosquito-borne pathogen genes in mosquitoes (Arziman et al., 2005; Sánchez-Vargas et al., 2009; Ender and Meister, 2010; Airs and Bartholomay, 2017).

13.3.3.2.1 RNAi-Based Mosquitocides and Virus-Resistant Mosquitoes

RNAi may contribute to the cause by suppressing genes involved in key physiologies that impact mosquito survival, fecundity, behaviour, or vector status. Chemical insecticides can be designed by using targeting the mosquito's essential biological pathways such as mosquito exoskeleton, apoptosis, olfaction receptor proteins, blood-feeding nda digestion, reproduction, embryogenesis, larval and pupal development, detoxification and insecticide Metabolism/Resistance (Airs and Bartholomay, 2017). A previous study based on dsRNA-mediated gene silencing achieved a specific knockdown of chitin synthases A and B (Lopez et al., 2019) and lethality in *Aedes aegypti* larvae. The major drawback of the bioinsecticidal approach is the production cost of RNAi plasmids therefore another study has developed a 3-hydroxykynurenine transaminase RNAi expression plasmid and transformed it into *Chlamydomonas* and *Chlorella* algae and the result reported some level of lethality (Fei et al., 2021) and the transgenic mosquito approach to virus transmission control may solve the cost-effectiveness problem mentioned in the studies.

The discovery of RNAi in mosquitoes was a breakthrough in the fight against flavivirus replication inside the mosquito body, and it's a safe tool to use in arthropod-borne disease vectors. A study conducted on Sendib virus reported resistance to dengue virus serotype 2 by post-transcriptional gene-silencing activation and reported the successful reduction of dengue virus titers in the *Aedes albopictus* C6/36 cells and *Ae. aegypti* mosquitoes. Especially, RNAi will be a vital tool for reducing virus transmissions such as the dengue virus, which has no cure or vaccine and relies solely on traditional mosquito vector control strategies, which have failed in recent decades (Adelman et al., 2001).

13.3.3.2.2 RNAi Gene Designing Basics

The synthetic RNAi gene is the critical component of the RNAi-mediated virus-resistant mosquitoes, which contains host inducible promoter, RNAi coding sequence, and terminator sequence as the basic components. Promoter selection is vital because it allows the RNAi gene to bind host RNA polymerase and to transcribe the RNAi component gene in the mosquito cells. Recent research studies have used both ubiquitous (Poly-Ubiquitin) and tissue-specific promoters (Carboxypeptidase and vitellogenin) to express RNAi genes in mosquitoes and achieve virus resistance and expression of RNAi elements. Ubiquitous promoters express RNAi genes in a variety of tissues, and tissue-specific promoters, such as carboxypeptidase promoter showed expression in mosquito midguts by tissue and time specific expression of RNAi elements, which is expressed when the virus enters the mosquito midguts and recent studies predominately used carboxypeptidase promoters to develop virus-resistant transgenic mosquitoes (Coutinho-Abreu et al., 2010).

The physical and chemical property studies of RNAi elements provide critical information to design RNAi coding sequences and the supportive bioinformatics tools are also available to assist in designing and refining RNAi genes before demonstrating them in mosquito cells *in vivo* (Arziman et al., 2005). In the designing of dsRNA, the sense and antisense sequences of the target are connected with the mosquito-specific intron sequence which is processed into a long double-strand RNA molecule later and then the RNAi machinery processes dsRNA into the RISC complex to fulfil the dengue virus resistance (Sánchez-Vargas et al., 2009). Bioinformatic tools of shRNA can be used to design stem and loop sequences of the shRNA elements and in the miRshRNA designing; the active miRNA sequence of the host-specific pre-miRNA sequence is replaced with the target sequence (~17–27 nt) of the virus genome. Previous transgenic mosquito studies designed miR-shRNA and shRNA genes have shown to be relatively small and efficient compared to dsRNA genes and achieved in the targeting the multiple serotypes and viruses (Yen et al., 2018; Buchman et al., 2019; Ramyasoma et al., 2020).

13.3.3.2.3 Challenges, Advantages and Future of RNAi

The RNAi gene uses the population replacement approach to replace wild mosquitoes with virus-resistant transgenic mosquitoes. Transgenic mosquitoes and wild mosquitoes mate with each other and the balance of the transgenic mosquito and to wild mosquito determines the transmission control ability of the virus (Yen and Failloux, 2020). However, the designed RNAi genes have to face many challenges to achieve the purpose of virus transmission control. One of the challenges in the RNAi genes is the efficiency of its machinery. The efficiency of RNAi machinery varies considerably among insect species, and it is highly efficient in coleopterans (beetles). Dipterans (Mosquitoes),

Lepidopterans, Hymenopterans, and Hemipterans show variable efficiency in RNAi machinery, and the efficiency differences can be observed within the same order due to the availability of the components of the RNAi (Zhu and Palli, 2020). The off-target effect is one of the challenges in the RNAi genes and the following off-targets may be lethal or induce suppression of genes in the mosquito may lead to a significant decrease in the fitness of the mosquito (Murphy et al., 2016). The major challenge in the RNAi gene is the development of an escape mechanism by viruses. These mechanisms counterattack RNAi machinery by mutating the target region, covering the target region with ribonucleoproteins, and encoding viral suppressors and cellular factors. The virus resistance mechanisms can be overcome by introducing multiple targets in a single RNAi gene (Zheng et al., 2005, Leonard et al., 2008). RNAi has a lot of potential as a tool for understanding gene functions and as a therapeutic, and there has been tremendous progress in the understanding of RNAi machinery for more than a decade; the therapeutic implications of the RNAi gene-silencing mechanism are growing every single day.

13.3.3.3 Mosquito Transgenesis Based on CRISPR/Cas

Vector-borne diseases are one of the major problems in the current environmental situations accounting for propagating about 17% of the communicable diseases in the world with a mortality of almost 700,000 every year (WHO, 2017). One of the foremost insect vectors for dengue is the *Aedes aegypti* species that prevails in densely human-populated regions, making it an exclusive vector capable of proliferating extensively (Buchman et al., 2020). This makes it capable of spreading the diseases to large populations, which is a risk-factor that has to be mitigated through effective control strategies.

Strategies to palliate/eradicate this particular vector species have been attempted by several research works via different methods. These include methods like SIT, which works through irradiation for male sterilization, which affects female mosquitoes while breeding (or) a dominant lethal (RIDL) gene, capable of providing reduced fecundity (Alphey et al., 2013). Methods based on the endosymbiotic bacterium 'Wolbachia', which mediates cytoplasmic incompatibility (CI) by affecting male spermatozoon, thus reducing offsprings (Wickramasinghe et al., 2021). Although such methods prevail, these methods are required to have long-term commitments during frequent releases. Hence, there is a requirement of alternate and cost-effective technique to effectively control vector transmission of pathogenic organisms.

Modification/elimination approaches for disease vectors like mosquitoes involve genetically modified mosquitoes, where methods like ZFN, TALENs and CRISPR/Cas systems (Wood et al., 2011; Reyon et al., 2012; Gasiunas et al., 2014) comes to play. These methods provide a double-stranded (or) single-stranded break, which are repaired by either homology-directed repair (HDR) or non-homologous end joining (NHEJ). Due to low specificity of DNA binding proteins like ZFN and TALEN (which are 'context-dependent'; Maeder et al., 2008, Juillerat et al., 2014), a specific method for genome editing was required, which led to the discovery of CRISPR/Cas systems. CRISPR/Cas mechanisms are highly specific genome editing methods which utilize single (or) multiple single-guided RNA sequences, which are capable of identifying specific sequences. With the help of this RNA mechanism and a short PAM sequence, Cas protein utilizes its protein domains like HnH and RuvC for precise double-stranded break, which can be utilized for knock-in/knock out of genes of interest by HDR (Jinek et al., 2012; Figure 13.12).

Although such advanced methods prevail, are the genome variations that we provide are subjected to Mendelian's law of inheritance, where the transgene (gene of interest that is inserted) has reduced possibility of transferring to subsequent generations. This required methods with increased prevalence of the transgene in mosquito populations, which can be efficiently transferred to subsequent generation (Wickramasinghe et al., 2021).

Gene drives (or) selfish genes are DNA sequences that trigger processes like the elimination of sperms without the selfish genes, alteration of DNA replication processes which produce the selfish-DNA systems, etc. These genetic elements are capable of modifying the inheritance pattern by influencing the homologous chromosome and the cell to replicate these DNA elements when a break

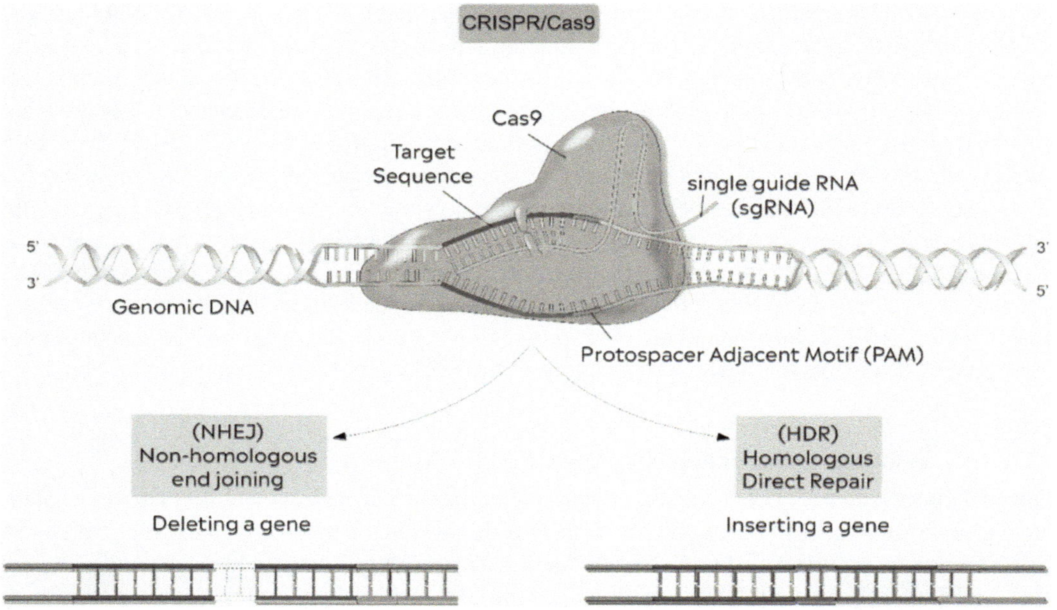

FIGURE 13.12 How CRISPR/Cas9 system work. A DNA double-stranded break (DSB) will be imposed by the Cas protein which can be repaired using either HDR or NHEJ repair mechanism. In the HDR mechanism, the DSB is amended with a template DNA which can be used to introduce a gene, a portion of a gene, or even a point mutation. NHEJ is erroneous and can result in insertions and/or deletions (indels) around the DSB. The latter method could be useful in the formation of a frameshift and an early stop codon, resulting in the deletion of the gene.

is needed. Some of the most notable selfish genetic elements include chromosomal rearrangements (Curtis, 1968) transposons (Skipper et al., 2013) bacterial endosymbiont/parasite Wolbachia (Rasgon, 2007) and CRISPR/Cas9 systems.

Synthetic gene drives are techniques used to alter the genetic makeup of mosquitoes to modify or eliminate vector-based traits. These are mostly applicable to sexually reproducing organisms, where the distribution of genes to offsprings from parents can be increased (above the Mendelian's 50% inheritance) with gene drives (Grunwald et al., 2019).

Gene drives have been successfully implemented in different species including *Drosophila*, and mosquitoes such as *Anopheles* spp., etc (Bassett et al., 2013; Basu et al., 2015; Hammond et al., 2016). Different strategies of genetic control in mosquitoes have also been studied in the recent years of CRIPR/Cas9 development. Some of the ideologies also include prevention of reproduction in females followed by mating with the wild population. To deepen our understanding of mosquito vector biology, Kistler et al. (2015) and others have investigated mutation rates and the inheritance of indel formation in *Ae. aegypti* (*waterwitch*) using the C-NHEJ pathway and single-stranded oligodeoxynucleotide (ssODN) insertion via the HDR pathway. The study reported a 24% and 18.9% of indel inheritance rate in F0 and F1 generations. Gene drive studies using multiple targets were also evaluated with genes involved in DNA repair, RNAi, and sex determination in *Ae. aegypti* (kmo, loqs, r2d2, ku70, lig4, and nix; Basu et al., 2015)

Gene drives are mechanisms that mostly work by cutting specific-site-chromosomes which do not code for the drive sequence, initiating DNA repair, which copies the drive sequence. CRISPR/Cas9 based gene drives can be classified into two types – namely, suppression (of species) and modification (of the target trait). These resemble a 'mutagenic chain reaction (MCR),' which are similar and are new methodologies in the area of selfish-DNA (Gantz and Bier, 2015).

The technique of gene drive is relatively easier in the explanation of Gantz and Bier (2016), who mentioned,

> CRISPR/Cas9-driven autocatalysis that generates insertional mutants can be achieved by a construct consisting of three genetic elements: a central element that encodes Cas9 that can be expressed in both somatic and germline cells; a gRNA sequence encoded by an ubiquitously expressed gene, targeting the desired DNA sequence in the genome; and homology arms designed to flank the genomic sequence immediately adjacent to the DSB which must be placed flanking the Cas9/gRNA cassette.

The cassette insertion is then done by the HDR pathway, which can be utilized for modification/disruption of targeted-trait-gene (or) insertion of transgene into the wild population.

Although genetic control is a good alternative to traditional control techniques, it has numerous drawbacks like natural actions of finding make mosquitoes for mating. Further, it is impossible to track the mutant mosquitoes, owing to the fact that it can travel literally anywhere where our technologies are out of reach. Further, as all genetically modified organisms have bioethical issues and safety checks according to guidelines like WHO, where field trials with modified male mosquitoes are to be performed, to verify the safety and evaluate the effects in releases (Benedict and Robinson, 2003). Since there is no procedure for field deployment and proper roadmap for efficiency analysis, there prevails a challenge for adopting/formulating new methods that proffer safety and efficiency present along with the mutated strains of mosquitoes. This can contribute greatly to the developing guidelines on genetically modified testing. Several challenges also lie ahead, such as genetically modified mosquitoes (GMMs) which are developed for transferring the effect in shorter breeding periods, having their effects filter out after multiple generations (Wickramasinghe et al., 2021). This requires periodic GMM releases, which might not be suitable in several circumstances. Alternate methods for this involve self-sustainment, where dire consequences can be seen when the unintended mishaps become tough to reverse. Since there are endless possibilities of genomic approaches with no proper guidelines, it is of utmost importance to plan countermeasures for containment of accidental spreading of faulty GMM (Akbari et al., 2015).

13.3.4 Perspectives

Mosquito based disease transmission has been a challenging battle to be completely overcome. Due to the insufficiency of resources, most control programmes have become ineffective, and the insects are increasingly developing resistant to pesticides after decades of insecticide application. Diverse tactics have to be employed in controlling the spread of mosquito-borne diseases. Altering the genetics or transgenesis proves to be a promising methodology that would reduce an insect's ability to transmit disease or would induce a population reduction among selected disease-carrying species.

Due to the advancement in the tools of molecular biology, researchers could make specific changes in the mosquito genomes, accurately causing the mosquito populations to be suppressed. The transgenic mosquitoes should also be able to compete with the local/wild strains and to induce the genetic changes for a rapid effect of population suppression. Regardless of which type of transgene is to be inserted into the mosquito genome (anti-pathogen, antiviral, antiprotist gene, a lethal gene), the successfulness of the strategy would be based on the ability to drive that transgene into the wild population. Scientists are already working on with selfish-DNA elements such as transposons, homing endonucleases, and gene drive mechanisms to make the transgenesis more efficient.

Although entomologists consider that the genetic manipulation with super-Mendelian spread is the best and the most efficient means of genetic control, the earliest stable implementation is at least almost a decade away. As the genetically modified mosquito projects move from contained laboratory testing to semi-field testing and small-scale field trials, the requirement to assess them in several stages such as detecting unintended spread of GMM mosquitoes beyond trial sites or presence of any resistance mechanisms are crucial. GMM mosquito monitoring with gene drive mechanisms would be more costly than research, and development. For population suppression systems, detecting rare resistant alleles as well as tracking the population size is mandatory, whereas for population replacement systems, allele frequencies, together with non-functional effector genes, and development of resistance to functional effector genes should also be evaluated. needs for gene drive mosquito projects progressing through the phases of pre-release, release and post-release.

Currently the best methodology would be to use a strategy in utilizing the existing genetic tools and methodologies in momentarily eradicating the mosquito populations.

REFERENCES

Adelman, Z., Blair, C., Carlson, J., Beaty, B. and Olson, K. 2001. Sindbis virus-induced silencing of dengue viruses in mosquitoes. *Insect molecular Biology*, 10, 265–273. 10.1046/j.1365-2583.2001.00267.x

Airs, P.M. and Bartholomay, L.C. 2017. RNA interference for mosquito and mosquito-borne disease control. *Insects*, 8, 4. 10.3390/insects8010004

Akbari, O.S., Bellen, H.J., Bier, E., Bullock, S.L., Burt, A., Church, G.M., Cook, K.R., Duchek, P., Edwards, O.R. and Esvelt, K.M. 2015. Safeguarding gene drive experiments in the laboratory. *Science*, 349, 927–929. 10.1126/science.aac7932

Alphey, L. and Alphey, N. 2014. Five things to know about genetically modified (GM) insects for vector control. *PLoS Pathogens*, 10. 10.1371/journal.ppat.1003909

Alphey, L., Baker, P., Condon, G.C., Condon, K.C., Dafa'alla, T.H., Fu, G., Jin, L., Labbe, G., Morrison, N.M., Nimmo, D.D., O'Connell, S., Phillips, C.E., Plackett, A., Scaife, S., Woods, A., Burton, R.S., Epton, M.J. and Gong, P. 2006. Genetic technologies to enhance the Sterile Insect Technique (SIT). 7th International symposium on fruit flies of economic importance: from basic to applied knowledge, Brazil

Alphey, L., Benedict, M., Bellini, R., Clark, G.G., Dame, D.A., Service, M.W. & Dobson, S.L. 2010. Sterile-insect methods for control of mosquito-borne diseases: an analysis. *Vector-Borne and Zoonotic Diseases*, 10, 295–311. 10.1089/vbz.2009.0014

Alphey, L., Mckemey, A., Nimmo, D., Neira Oviedo, M., Lacroix, R., Matzen, K. & Beech, C., 2013. Genetic control of Aedes mosquitoes. *Pathogens and global health*, 107, 170–179. 10.1179/2047773213Y.0000000095

Ariyaratne, M., Gunasekara, P., Wajirasena, P.H., Rathnayake, D.M., Dilani, D., Chathuranga, T., Gomes, L., Jayatunga, D., Wewita, S. & Meegahage, T., 2022. Comparison of two rapid test kits with real time polymerase chain reaction for early diagnosis of dengue in Sri Lanka. *Journal of Immunoassay and Immunochemistry*, 43, 213–221. 10.1080/15321819.2021.1984252

Aryan, A., Anderson, M.A., Myles, K.M. & Adelman, Z.N., 2013. TALEN-based gene disruption in the dengue vector *Aedes aegypti*. *PLoS One*, 8,. 10.1371/journal.pone.0060082

Arziman, Z., Horn, T. & Boutros, M., 2005. E-RNAi: a web application to design optimized RNAi constructs. *Nucleic acids research*, 33, W582-W588. 10.1093/nar/gki468

Atkinson, P.W., Pinkerton, A.C. & O'Brochta, D.A. 2001. Genetic transformation systems in insects. *Annual review of entomology*, 46, 317. 10.1146/annurev.ento.46.1.317

Bakri, A., Mehta, K., Lance, D.R. 2005. Sterilizing Insects with Ionizing Radiation. In: Dyck, V.A., Hendrichs, J., Robinson, A. (eds) Sterile Insect Technique. Springer. 10.1007/1-4020-4051-2_9

Bassett, A.R., Tibbit, C., Ponting, C.P. & Liu, J.-L., 2013. Highly efficient targeted mutagenesis of *Drosophila* with the CRISPR/Cas9 system. *Cell reports*, 4, 220–228. 10.1016/j.celrep.2013.06.020

Basu, S., Aryan, A., Overcash, J.M., Samuel, G.H., Anderson, M.A., Dahlem, T.J., Myles, K.M. & Adelman, Z.N., 2015. Silencing of end-joining repair for efficient site-specific gene insertion after TALEN/CRISPR mutagenesis in *Aedes aegypti*. *Proceedings of the National Academy of Sciences*, 112, 4038–4043. 10.1073/pnas.150237011

Baulcombe, D.C., 2000. Unwinding RNA silencing. *Science*, 290, 1108–1109. 10.1126/science.290.5494.1108

Becker, N., Langentepe-Kong, S.M., Tokatlian Rodriguez, A., Oo, T.T., Reichle, D., Lühken, R., Schmidt-Chanasit, J., Lüthy, P., Puggioli, A. & Bellini, R., 2022. Integrated control of Aedes albopictus in Southwest Germany supported by the Sterile Insect Technique. *Parasites & vectors*, 15, 1–19. 10.1186/s13071-021-05112-7

Bellini, R., Carrieri, M., Balestrino, F., Puggioli, A., Malfacini, M. & Bouyer, J., 2021. Field competitiveness of *Aedes albopictus* (Diptera: Culicidae) irradiated males in pilot Sterile Insect Technique trials in northern Italy. *Journal of medical entomology*, 58,807–813. 10.1093/jme/tjaa235

Bellini, R., Medici, A., Puggioli, A., Balestrino, F. & Carrieri, M., 2013. Pilot field trials with *Aedes albopictus* irradiated sterile males in Italian urban areas. *Journal of medical entomology*, 50(2), 317–325. 10.1603/ME12048

Benedict, M.Q. & Robinson, A.S. 2003. The first releases of transgenic mosquitoes: an argument for the sterile insect technique. *Trends in parasitology*, 19(8), 349–355. 10.1016/s1471-4922(03)00144-2

Berghammer, A.J., Klingler, M. & Wimmer, A. E. 1999. A universal marker for transgenic insects. *Nature*, 402, 370–371. 10.1038/46463

Bhalla, S.C., 1968. White eye, a new sex-linked mutant of *Aedes aegypti. Mosquito News*, 28, 380–385.

Bhatt, S., Gething, P.W., Brady, O.J., Messina, J.P., Farlow, A.W., Moyes, C.L., Drake, J.M., Brownstein, J.S., Hoen, A.G. & Sankoh, O., 2013. The global distribution and burden of dengue. *Nature*, 496, 504–507. 10.1038/nature12060

Bouyer, J., Culbert, N.J., Dicko, A.H., Pacheco, M.G., Virginio, J., Pedrosa, M.C., Garziera, L., Pinto, A.M., Klaptocz, A. & Germann, J., 2020a. Field performance of sterile male mosquitoes released from an uncrewed aerial vehicle. *Science Robotics*, 5, 10.1126/scirobotics.aba6251

Bouyer, J., Yamada, H., Pereira, R., Bourtzis, K. & Vreysen, M.J., 2020b. Phased conditional approach for mosquito management using sterile insect technique. *Trends in Parasitology*, 36, 325–336. 10.1016/j.pt.2020.01.004

Bouyer, J. 2024. Current status of the sterile insect technique for the suppression of mosquito populations on a global scale. *Infect Dis Poverty*, 13, 68. 10.1186/s40249-024-01242-z

Buchman, A., Gamez, S., Li, M., Antoshechkin, I., Li, H.-H., Wang, H.-W., Chen, C.-H., Klein, M.J., Duchemin, J.-B. & Crowe Jr, J.E., 2020. Broad dengue neutralization in mosquitoes expressing an engineered antibody. *PLoS pathogens*, 16, 10.1371/journal.ppat.1008103

Buchman, A., Gamez, S., Li, M., Antoshechkin, I., Li, H.-H., Wang, H.-W., Chen, C.-H., Klein, M.J., Duchemin, J.-B. & Paradkar, P.N., 2019. Engineered resistance to Zika virus in transgenic Aedes aegypti expressing a polycistronic cluster of synthetic small RNAs. *Proceedings of the National Academy of Sciences*, 116, 3656–3661. 10.1073/pnas.1810771116

Caragata, E.P., Lee, Y. & Buckner, E.A., 2021. Genetically Modified Mosquitoes: ENY2066/IN1326, 8/2021. *EDIS*, 2021. 10.32473/edis-in1326-2021

Carthew, R.W. & Sontheimer, E.J., 2009. Origins and mechanisms of miRNAs and siRNAs. *Cell*, 136, 642–655. 10.1016/j.cell.2009.01.035

Carvalho, D.O., Mckemey, A.R., Garziera, L., Lacroix, R., Donnelly, C.A., Alphey, L., Malavasi, A. & Capurro, M.L., 2015. Suppression of a field population of *Aedes aegypti* in Brazil by sustained release of transgenic male mosquitoes. *PLoS neglected tropical diseases*, 9, 10.1371/journal.pntd.0003864

Catteruccia, F., Godfray, H.C.J. & Crisanti, A., 2003. Impact of genetic manipulation on the fitness of *Anopheles stephensi* mosquitoes. *Science*, 299, 1225–1227. 10.1126/science.1081453

Catteruccia, F., Nolan, T., Loukeris, T.G., Blass, C., Savakis, C., Kafatos, F.C. & Crisanti, A., 2000. Stable germline transformation of the malaria mosquito *Anopheles stephensi. Nature*, 405, 959–962. 10.1038/35016096

Coates, C.J., Jasinskiene, N., Miyashiro, L. & James, A.A., 1998. Mariner transposition and transformation of the yellow fever mosquito, *Aedes aegypti. Proceedings of the National Academy of Sciences*, 95, 3748–3751. 10.1073/pnas.95.7.3748

Cotlet, M., Habuchi, S., Whitier, J.E., Werner, J.H., De Schryver, F.C., Hofkens, J. & Goodwin, P.M., 2006. Single molecule spectroscopic characterization of a far-red fluorescent protein (HcRed) from the Anthozoa coral Heteractis crispaed. *Genetically Engineered Probes for Biomedical Applications*, SPIE, 18–28. 10.1016/j.bpj.2016.05.047

Coutinho-Abreu, I.V., Zhu, K.Y. & Ramalho-Ortigao, M., 2010. Transgenesis and paratransgenesis to control insect-borne diseases: current status and future challenges. *Parasitology International*, 59, 1–8. 10.1016/j.parint.2009.10.002

Curtis, C., 1968. Possible use of translocations to fix desirable genes in insect pest populations. *Nature*, 218, 368–369. 10.1038/218368a0

Dantas, L., Pereira, R., Silva, N., Rodrigues, A. & Costa, R., 2004 The SIT control programme against Medfly on Madeira Islanded. *Proc. 6th International Symposium on Fruit Flies of Economic Importance. Stellenbosch, South Africa. Isteg Scientific Publications, Irene, South Africa*, 127–130.

Degennaro, M., Mcbride, C.S., Seeholzer, L., Nakagawa, T., Dennis, E.J., Goldman, C., Jasinskiene, N., James, A.A. & Vosshall, L.B., 2013. Orco mutant mosquitoes lose strong preference for humans and are not repelled by volatile DEET. *Nature*, 498, 487–491. 10.1038/nature12206

Dilani P.V.D., Gunawardene Y.I.N.S. & Dassanayake, R.S., 2021. Genetic improvements to the sterile insect technique (SIT) for the control of mosquito population. In Tyagi BK (eds) *Genetically Modified and other Innovative Vector Control Technologies*. Springer, 43–65. 10.1007/978-981-16-2964-8_3

Dilani, P.V.D., Dassanayake, R.S., Tyagi, B.K. & Silva Gunawardene, Y.I.N.,2022. The impact of transgenesis on mosquito fitness: A review. *Frontiers in Insect Science*, 38. 10.3389/finsc.2022.957570

Ehrig, T., O'Kane, D.J. & Prendergast, F.G., 1995. Green-fluorescent protein mutants with altered fluorescence excitation spectra. *FEBS Letters*, 367, 163–166. 10.1016/0014-5793(95)00557-p

El Sayed, B.B., Malcolm, C.A., Babiker, A., Malik, E.M., El Tayeb, M.A., Saeed, N.S., Nugud, A.H.D. & Knols, B.G., 2009. Ethical, legal and social aspects of the approach in Sudan. *Malaria Journal*, 8, 1–6. 10.1186/1475-2875-8-S2-S3

Ender, C. & Meister, G., 2010. Argonaute proteins at a glance. *Journal of Cell Science*, 123, 1819–1823. 10.1242/jcs.055210

Fei, X., Zhang, Y., Ding, L., Xiao, S., Xie, X., Li, Y. & Deng, X., 2021. Development of an RNAi-based microalgal larvicide for the control of *Aedes aegypti*. *Parasites & Vectors*, 14, 1–11. 10.1186/s13071-021-04885-1

Fire, A., Xu, S., Montgomery, M.K., Kostas, S.A., Driver, S.E. & Mello, C.C., 1998. Potent and specific genetic interference by double-stranded RNA in *Caenorhabditis elegans*. *nature*, 391, 806–811. 10.1038/35888

Fradkov, A.F., Chen, Y., Ding, L., Barsova, E.V., Matz, M.V. & Lukyanov, S.A., 2000. Novel fluorescent protein from Discosoma coral and its mutants possesses a unique far-red fluorescence. *FEBS letters*, 479, 127–130. 10.1016/s0014-5793(00)01895-0

Franz, G., 2005. Genetic sexing strains in Mediterranean fruit fly, an example for other species amenable to large-scale rearing for the sterile insect technique. *Sterile insect technique*, 427–451. 10.1007/1-4020-4051-2_16

Franz, G. & Savakis, C., 1991. Minos, a new transposable element from Drosophila hydei, is a member of the Tc1-like family of transposons. *Nucleic acids research*, 19, 6646. 10.1093/nar/19.23.6646

Fraser, M.J., Cary, L., Boonvisudhi, K. & Wang, H.-G.H., 1995. Assay for movement of Lepidopteran transposon IFP2 in insect cells using a baculovirus genome as a target DNA. *Virology*, 211, 397–407. 10.1006/viro.1995.1422

Fu, G., Lees, R.S., Nimmo, D., Aw, D., Jin, L., Gray, P., Berendonk, T.U., White-Cooper, H., Scaife, S. & Kim Phuc, H., 2010. Female-specific flightless phenotype for mosquito control. *Proceedings of the National Academy of Sciences*, 107, 4550–4554. 10.1073/pnas.1000251107

Gabrieli, P., Smidler, A. & Catteruccia, F., 2014. Engineering the control of mosquito-borne infectious diseases. *Genome Biology*, 15, 1–9. 10.1186/s13059-014-0535-7

Gammon, D.B. & Mello, C.C., 2015. RNA interference-mediated antiviral defense in insects. *Current Opinion in Insect Science*, 8, 111–120. 10.1016/j.cois.2015.01.006

Gantz, V.M. & Bier, E., 2015. The mutagenic chain reaction: a method for converting heterozygous to homozygous mutations. *Science*, 348, 442–444. 10.1126/science.aaa5945

Gantz, V.M. & Bier, E., 2016. The dawn of active genetics. *Bioessays*, 38, 50–63. 10.1002/bies.201500102

Garg, P., Nagpal, J., Khairnar, P. & Seneviratne, S.L., 2008. Economic burden of dengue infections in India. *Transactions of the Royal Society of Tropical Medicine and Hygiene*, 102, 570–577. 10.1016/j.trstmh.2008.02.015

Gasiunas, G., Sinkunas, T. & Siksnys, V., 2014. Molecular mechanisms of CRISPR-mediated microbial immunity. *Cellular and molecular life sciences*, 71, 449–465. 10.1007/s00018-013-1438-6

Gato, R., Menéndez, Z., Prieto, E., Argilés, R., Rodríguez, M., Baldoquín, W., Hernández, Y., Pérez, D., Anaya, J. & Fuentes, I., 2021. Sterile insect technique: successful suppression of an *Aedes aegypti* field population in Cuba. *Insects*, 12, 469. 10.3390/insects12050469

Gaudinski, M.R., Berkowitz, N.M., Idris, A.H., Coates, E.E., Holman, L.A., Mendoza, F., Gordon, I.J., Plummer, S.H., Trofymenko, O. & Hu, Z., 2021. A monoclonal antibody for malaria prevention. *New England Journal of Medicine*, 385, 803–814. 10.1056/NEJMoa2034031

Gong, P., Epton, M.J., Fu, G., Scaife, S., Hiscox, A., Condon, K.C., Condon, G.C., Morrison, N.I., Kelly, D.W. & Dafa'alla, T., 2005. A dominant lethal genetic system for autocidal control of the Mediterranean fruitfly. *Nature biotechnology*, 23, 453–456. 10.1038/nbt1071

Gossen, M. & Bujard, H., 1992. Tight control of gene expression in mammalian cells by tetracycline-responsive promoters. *Proceedings of the National Academy of Sciences*, 89, 5547–5551. 10.1073/pnas.89.12.5547

Grossman, G., Rafferty, C., Clayton, J., Stevens, T., Mukabayire, O. & Benedict, M., 2001. Germline transformation of the malaria vector, Anopheles gambiae, with the piggyBac transposable element. *Insect molecular biology*, 10, 597–604. 10.1046/j.0962-1075.2001.00299.x

Grunwald, H.A., Gantz, V.M., Poplawski, G., Xu, X.-R.S., Bier, E. & Cooper, K.L., 2019. Super-Mendelian inheritance mediated by CRISPR–Cas9 in the female mouse germline. *Nature*, 566, 105–109. 10.1038/s41586-019-0875-2

Haghighat-Khah, R.E., Harvey-Samuel, T., Basu, S., Stjohn, O., Scaife, S., Verkuijl, S., Lovett, E. & Alphey, L., 2019. Engineered action at a distance: Blood-meal-inducible paralysis in *Aedes aegypti*. *PLoS Neglected Tropical Diseases*, 13, 10.1371/journal.pntd.0007579

Hammond, A., Galizi, R., Kyrou, K., Simoni, A., Siniscalchi, C., Katsanos, D., Gribble, M., Baker, D., Marois, E. & Russell, S., 2016. A CRISPR-Cas9 gene drive system targeting female reproduction in the malaria mosquito vector *Anopheles gambiae*. *Nature biotechnology*, 34, 78–83. 10.1038/nbt.3439

Harris, A.F., Mckemey, A.R., Nimmo, D., Curtis, Z., Black, I., Morgan, S.A., Oviedo, M.N., Lacroix, R., Naish, N. & Morrison, N.I., 2012. Successful suppression of a field mosquito population by sustained release of engineered male mosquitoes. *Nature biotechnology*, 30, 828–830. 10.1038/nbt.2350

Harris, A.F., Nimmo, D., Mckemey, A.R., Kelly, N., Scaife, S., Donnelly, C.A., Beech, C., Petrie, W.D. & Alphey, L., 2011. Field performance of engineered male mosquitoes. *Nature biotechnology*, 29, 1034–1037. 10.1038/nbt.2019

Heinrich, J.C. & Scott, M.J., 2000. A repressible female-specific lethal genetic system for making transgenic insect strains suitable for a sterile-release program. *Proceedings of the National Academy of Sciences*, 97, 8229–8232. 10.1073/pnas.14014269

Helinski, M.E., Parker, A.G. & Knols, B.G., 2006. Radiation-induced sterility for pupal and adult stages of the malaria mosquito *Anopheles arabiensis*. *Malaria Journal*, 5, 1–10. 10.1186/1475-2875-5-41

Hendrichs, J., Vreysen, M., Enkerlin, W. & Cayol, J., 2005. Strategic options in using sterile insects for area-wide integrated pest management. In: Dyck, V.A., Hendrichs, J., Robinson, A. (eds) Sterile Insect Technique. Springer, Dordrecht. 10.1007/1-4020-4051-2_22

Higgs, S. & Lewis, D., 2000. Green fluorescent protein (GFP) as a marker for transgenic insects. *Insect Transgenesis: Methods and Applications*. CRC Press, 93–108.

Higgs, S., Rayner, J.O., Olson, K.E., Davis, B.S., Beaty, B.J. & Blair, C.D., 1998. Engineered resistance in *Aedes aegypti* to a West African and a South American strain of yellow fever virus. *The American journal of tropical medicine and hygiene*, 58, 663–670. 10.4269/ajtmh.1998.58.663

International atomic Energy Agency (IAEA) 2020a. Guidelines for Irradiation of Mosquito Pupae in Sterile Insect Technique Programmes.

International atomic Energy Agency (IAEA) 2020b. Guidelines for mass rearing of *Aedes* mosquitoes.

Ito, J., Ghosh, A., Moreira, L.A., Wimmer, E.A. & Jacobs-Lorena, M., 2002. Transgenic anopheline mosquitoes impaired in transmission of a malaria parasite. *Nature*, 417, 452–455. 10.1038/417452a

Itô, Y., Yamamura, K. 2005. Role of Population and Behavioural Ecology in the Sterile Insect Technique. In: Dyck, V.A., Hendrichs, J., Robinson, A. (eds) *Sterile insect technique*. Springer, Dordrecht. 10.1007/1-4020-4051-2_7

Iyaloo, D.P., Bouyer, J., Facknath, S. & Bheecarry, A., 2020. Pilot Suppression trial of *Aedes albopictus* mosquitoes through an Integrated Vector Management strategy including the sterile insect technique in Mauritius. *bioRxiv*. 10.1101/2020.09.06.284968

Jasinskiene, N., Coates, C.J., Benedict, M.Q., Cornel, A.J., Rafferty, C.S., James, A.A. & Collins, F.H., 1998. Stable transformation of the yellow fever mosquito, Aedes aegypti, with the Hermes element from the housefly. *Proceedings of the National Academy of Sciences*, 95, 3743–3747. 10.1073/pnas.95.7.3743

Jinek, M., Chylinski, K., Fonfara, I., Hauer, M., Doudna, J.A. & Charpentier, E., 2012. A programmable dual-RNA–guided DNA endonuclease in adaptive bacterial immunity. *Science*, 337, 816–821. 10.1126/science.12258

Juillerat, A., Dubois, G., Valton, J., Thomas, S., Stella, S., Maréchal, A., Langevin, S., Benomari, N., Bertonati, C. & Silva, G.H., 2014. Comprehensive analysis of the specificity of transcription activator-like effector nucleases. *Nucleic acids research*, 42, 5390–5402. 10.1093/nar/gku155

Karess, R.E. & Rubin, G.M., 1984. Analysis of P transposable element functions in Drosophila. *Cell*, 38, 135–146. 10.1016/0092-8674(84)90534-8

Kennerdell, J.R. & Carthew, R.W., 1998. Use of dsRNA-mediated genetic interference to demonstrate that frizzled and frizzled 2 act in the wingless pathway. *Cell*, 95, 1017–1026. 10.1016/S0092-8674(00)81725-0

Kistler, K.E., Vosshall, L.B. & Matthews, B.J., 2015. Genome engineering with CRISPR-Cas9 in the mosquito *Aedes aegypti*. *Cell reports*, 11, 51–60. 10.1016/j.celrep.2015.03.009

Kittayapong, P., Kaeothaisong, N.-O., Ninphanomchai, S. & Limohpasmanee, W., 2018. Combined sterile insect technique and incompatible insect technique: sex separation and quality of sterile *Aedes aegypti* male mosquitoes released in a pilot population suppression trial in Thailand. *Parasites & Vectors*, 11, 73–83. 10.1186/s13071-018-3214-9

Klassen, W., 2009. *Introduction: development of the sterile insect technique for African malaria vectors.* Springer, 1–4. 10.1186/1475-2875-8-S2-I1

Klassen, W., Curtis, C.F. 2005. History of the Sterile Insect Technique. In: Dyck, V.A., Hendrichs, J., Robinson, A. (eds) *Sterile Insect Technique*. Springer. 10.1007/1-4020-4051-2_1

Knipling, E., 1985 Sterile insect technique as a screwworm control measure: the concept and its developmented. *Symposium on eradication of the screwworm from the United States and Mexico. Misc. Pub. Entomol. Soc. Am* (pp. 4–7).

Knipling, E.F., 1979. *The basic principles of insect population suppression and management.* US Department of Agriculture.

Kokoza, V., Ahmed, A., Wimmer, E. & Raikhel, A., 2001. Efficient transformation of the yellow fever mosquito Aedes aegypti using the piggyBac transposable element vector pBac [3xP3-EGFP afm]. *Insect biochemistry and molecular biology*, 31, 1137–1143. 10.1016/s0965-1748(01)00120-5

Koyama, J., Kakinohana, H. & Miyatake, T., 2004. Eradication of the melon fly, Bactrocera cucurbitae, in Japan: importance of behavior, ecology, genetics, and evolution. *Annual Reviews in Entomology*, 49, 331–349. 10.1146/annurev.ento.49.061802.123224

Kremers, G.-J., Gilbert, S.G., Cranfill, P.J., Davidson, M.W. & Piston, D.W., 2011. Fluorescent proteins at a glance. *Journal of Cell Science*, 124, 157–160. 10.1242/jcs.072744

Krishnan, P., Babu, P.G., Saravanan, S., Rajendran, K. & Chaudhari, A., 2009. DNA constructs expressing long-hairpin RNA (lhRNA) protect Penaeus monodon against white spot syndrome virus. *Vaccine*, 27, 3849–3855. 10.1016/j.vaccine.2009.04.011

Labbé, G.M., Scaife, S., Morgan, S.A., Curtis, Z.H. & Alphey, L., 2012. Female-specific flightless (fsRIDL) phenotype for control of *Aedes albopictus. PLoS neglected tropical diseases*, 6, e1724. 10.1371/journal.pntd.0001724

Lee, H., Halverson, S. & Ezinwa, N., 2018. Mosquito-borne diseases. *Primary Care: Clinics in Office Practice*, 45, 393–407.

Lees, R.S., Carvalho, D. & Bouyer, J., 2021. Potential impact of integrating the sterile insect technique into the fight against disease-transmitting mosquitoes. In: Dyck, V.A., Hendrichs, J., Robinson, A. (eds) *Sterile Insect Technique*. Springer.

Leonard, J.N., Shah, P.S., Burnett, J.C. & Schaffer, D.V., 2008. HIV evades RNA interference directed at TAR by an indirect compensatory mechanism. *Cell Host & Microbe*, 4, 484–494. 10.1016/j.chom.2008.09.008

Li, Y., Lu, J., Han, Y., Fan, X. & Ding, S.-W., 2013. RNA interference functions as an antiviral immunity mechanism in mammals. *Science*, 342, 231–234. 10.1126/science.1241911

Lopez, S.B.G., Guimarães-Ribeiro, V., Rodriguez, J.V.G., Dorand, F.A., Salles, T.S., Sá-Guimarães, T.E., Alvarenga, E.S., Melo, A.C.A., Almeida, R.V. & Moreira, M.F., 2019. RNAi-based bioinsecticide for Aedes mosquito control. *Scientific Reports*, 9, 1–13. 10.1038/s41598-019-39666-5

Lycett, G.J., Kafatos, F.C. & Loukeris, T.G., 2004. Conditional expression in the malaria mosquito *Anopheles stephensi* with Tet-On and Tet-Off systems. *Genetics*, 167, 1781–1790. 10.1534/genetics.104.028175

Maeder, M.L., Thibodeau-Beganny, S., Osiak, A., Wright, D.A., Anthony, R.M., Eichtinger, M., Jiang, T., Foley, J.E., Winfrey, R.J. & Townsend, J.A., 2008. Rapid "open-source" engineering of customized zinc-finger nucleases for highly efficient gene modification. *Molecular cell*, 31, 294–301. 10.1016/j.molcel.2008.06.016

Maillard, P., Ciaudo, C., Marchais, A., Li, Y., Jay, F., Ding, S. & Voinnet, O., 2013. Antiviral RNA interference in mammalian cells. *Science*, 342, 235–238. 10.1126/science.1241930

Mains, J.W., Brelsfoard, C.L., Rose, R.I. & Dobson, S.L., 2016. Female adult *Aedes albopictus* suppression by Wolbachia-infected male mosquitoes. *Scientific reports*, 6, 1–7. 10.1038/srep33846

Marinotti, O., Jasinskiene, N., Fazekas, A., Scaife, S., Fu, G., Mattingly, S.T., Chow, K., Brown, D.M., Alphey, L. & James, A.A., 2013. Development of a population suppression strain of the human malaria vector mosquito, *Anopheles stephensi. Malaria journal*, 12, 1–10. 10.1186/1475-2875-12-142

Mcdonald, J.F., 1993. Evolution and consequences of transposable elements. *Current opinion in genetics & development*, 3, 855–864. 10.1016/0959-437X(93)90005-A

Mcgrane, V., Carlson, J.O., Miller, B.R. & Beaty, B.J., 1988. Microinjection of DNA into *Aedes triseriatus* ova and detection of integration. *The American journal of tropical medicine and hygiene*, 39, 502–510. 10.4269/ajtmh.1988.39.502

Medhora, M., Maruyama, K. & Hartl, D., 1991. Molecular and functional analysis of the mariner mutator element Mos1 in Drosophila. *Genetics*, 128, 311–318. 10.1093/genetics/128.2.311

Meister, G. & Tuschl, T., 2004. Mechanisms of gene silencing by double-stranded RNA. *Nature*, 431, 343–349. 10.1038/nature02873

Miller, L.H., Sakai, R.K., Romans, P., Gwadz, R.W., Kantoff, P. & Coon, H.G., 1987. Stable integration and expression of a bacterial gene in the mosquito Anopheles gambiae. *Science*, 237, 779–781. 10.1126/science.3039658

Morris, A.C., Eggleston, P. & Crampton, J.M., 1989. Genetic transformation of the mosquito *Aedes aegypti* by micro-injection of DNA. *Medical and veterinary entomology*, 3, 1–7. 10.1111/j.1365-2915.1989.tb00467.x

Morrison, N.I., Simmons, G.S., Fu, G., O'Connell, S., Walker, A.S., Dafa'alla, T., Walters, M., Claus, J., Tang, G. & Jin, L. 2012. Engineered repressible lethality for controlling the pink bollworm, a lepidopteran pest of cotton. *PLoS One*, 7, 10.1371/journal.pone.0050922

Murphy, K.A., Tabuloc, C.A., Cervantes, K.R. & Chiu, J.C., 2016. Ingestion of genetically modified yeast symbiont reduces fitness of an insect pest via RNA interference. *Scientific reports*, 6, 1–13. 10.1038/srep22587

Nazni, W.A., Teoh, G.-N., Shaikh Norman Hakimi, S.I., Muhammad Arif, M.A., Tanusshni, M., Nuradila, M.A., Nurfarahin Hanini, A., Shazia, I.A., Tan, A.-M. & Rabizah, H. 2021. *Aedes* Control Using Sterile Insect Technique (SIT) in Malaysia. In: Tyagi, B.K. (eds) Genetically Modified and other Innovative Vector Control Technologies. Springer. 1007/978-981-16-2964-8_8

Nolan, T., Bower, T.M., Brown, A.E., Crisanti, A. & Catteruccia, F., 2002. piggyBac-mediated germline transformation of the malaria mosquito Anopheles stephensi using the red fluorescent protein dsRED as a selectable marker. *Journal of Biological Chemistry*, 277, 8759–8762. 10.1074/jbc.C100766200

O'Connor, L., Plichart, C., Sang, A.C., Brelsfoard, C.L., Bossin, H.C. & Dobson, S.L., 2012. Open release of male mosquitoes infected with a Wolbachia biopesticide: field performance and infection containment. *PLoS neglected tropical diseases*, 6, 10.1371/journal.pntd.0001797

Osanai-Futahashi, M., Ohde, T., Hirata, J., Uchino, K., Futahashi, R., Tamura, T., Niimi, T. & Sezutsu, H., 2012. A visible dominant marker for insect transgenesis. *Nature Communications*, 3, 1–9. 10.1038/ncomms2312

Oxitec 2019. https://www.oxitec.com/en/news/oxitec-successfully-completes-first-field-deployment-of-2nd-generation-friendly-aedes-aegypti-technology (Accessed March 22).

Patterson, R., Weidhaas, D., Ford, H. & Lofgren, C., 1970. Suppression and elimination of an island population of Culex pipiens quinquefasciatus with sterile males. *Science*, 168, 1368–1370. 10.1126/science.168.3937.1368

Perera, O., Harrell Ii, R. & Handler, A., 2002. Germ-line transformation of the South American malaria vector, Anopheles albimanus, with a piggyBac/EGFP transposon vector is routine and highly efficient. *Insect molecular biology*, 11, 291–297. 10.1046/j.1365-2583.2002.00336.x

Phuc, H.K., Andreasen, M.H., Burton, R.S., Vass, C., Epton, M.J., Pape, G., Fu, G., Condon, K.C., Scaife, S. & Donnelly, C.A., 2007. Late-acting dominant lethal genetic systems and mosquito control. *BMC biology*, 5, 1–11. 10.1186/1741-7007-5-11

Pinkerton, A., Michel, K., O'Brochta, D. & Atkinson, P., 2000. Green fluorescent protein as a genetic marker in transgenic Aedes aegypti. *Insect Molecular Biology*, 9, 1–10. 10.1046/j.1365-2583.2000.00133.x

Prasher, D.C., Eckenrode, V.K., Ward, W.W., Prendergast, F.G. & Cormier, M.J., 1992. Primary structure of the Aequorea victoria green-fluorescent protein. *Gene*, 111, 229–233. 10.1016/0378-1119(92)90691-h

Qsim, M., Ashfaq, U.A., Yousaf, M.Z., Masoud, M.S., Rasul, I., Noor, N. & Hussain, A., 2017. Genetically modified Aedes aegypti to control dengue: a review. *Critical Reviews™ in Eukaryotic Gene Expression*, 27. 10.1615/CritRevEukaryotGeneExpr.2017019937

Ramyasoma, H.P.B.K.D., Dassanayake, R.S., Hapugoda, M., Capurro, M.L. & Silva Gunawardene, Y.I.N., 2020. Multiple dengue virus serotypes resistant transgenic *Aedes aegypti* fitness evaluated under laboratory conditions. *RNA biology*, 17, 918–929. 10.1080/15476286.2020.1735210

Ranjha, R. & Sharma, A., 2021. Forest malaria: the prevailing obstacle for malaria control and elimination in India. *BMJ global health*, 6, 10.1136/bmjgh-2021-005391

Rasgon, J., 2007. Population replacement strategies for controlling vector populations and the use of Wolbachia pipientis for genetic drive. *Journal of Visualized Experiments*. 10.3791/225

Reyon, D., Tsai, S.Q., Khayter, C., Foden, J.A., Sander, J.D. & Joung, J.K., 2012. FLASH assembly of TALENs for high-throughput genome editing. *Nature biotechnology*, 30, 460–465.

Rivera, J.P., Hernández, E., Toledo, J., Salvador, M. & Gómez, Y., 2012. Optimización del proceso de cría de Anastrepha ludens Loew (Diptera: Tephritidae) utilizando una dieta larvaria a base de almidón pregelatinizado. *Acta zoológica mexicana*, 28, 102–117.

Robinson, A., Vreysen, M., Hendrichs, J. & Feldmann, U., 2009. Enabling technologies to improve area-wide integrated pest management programmes for the control of screwworms. *Medical and veterinary entomology*, 23, 1–7. 10.1111/j.1365-2915.2008.00769.x

Rodems, S.M. & Friesen, P.D., 1993. The hr5 transcriptional enhancer stimulates early expression from the Autographa californica nuclear polyhedrosis virus genome but is not required for virus replication. *Journal of Virology*, 67, 5776–5785. 10.1128/jvi.67.10.5776-5785.1993

Rodrigues, F.G., Oliveira, S.B., Rocha, B.C. & Moreira, L.A., 2006. Germline transformation of Aedes fluviatilis (Diptera: Culicidae) with the piggyBac transposable element. *Memórias do Instituto Oswaldo Cruz*, 101, 755–757. 10.1590/s0074-02762006000700008

Rubin, G.M. & Spradling, A.C., 1982. Genetic transformation of Drosophila with transposable element vectors. *Science*, 218, 348–353. 10.1126/science.6289436

Sampath, K.S. & Puttaraju, H., 2012. Improvised microinjection technique for mosquito vectors. *The Indian journal of medical research*, 136, 971.

Sánchez-Vargas, I., Scott, J.C., Poole-Smith, B.K., Franz, A.W., Barbosa-Solomieu, V., Wilusz, J., Olson, K.E. & Blair, C.D., 2009. Dengue virus type 2 infections of *Aedes aegypti* are modulated by the mosquito's RNA interference pathway. *PLoS pathogens*, 5 10.1371/journal.ppat.1000299

Sarkies, P. & Miska, E.A., 2013. RNAi pathways in the recognition of foreign RNA: antiviral responses and host–parasite interactions in nematodes. *Biochemical Society Transactions*, 41, 876–880. 10.1042/BST20130021

Shirk, P.D. & Bossin, H., 1991. Development of piggyBac transposon-derived gene vectors and their utilization for transgenic insects.

Siriyasatien, P. & Thavara, U., 2006. Genetically modified mosquitoes: A new strategy to control mosquito borne diseases. *The Thai Journal of Veterinary Medicine*, 36, 9–19.

Skipper, K.A., Andersen, P.R., Sharma, N. & Mikkelsen, J.G., 2013. DNA transposon-based gene vehicles-scenes from an evolutionary drive. *Journal of biomedical science*, 20, 1–23. 10.1186/1423-0127-20-92

Smidler, A.L., Terenzi, O., Soichot, J., Levashina, E.A. & Marois, E., 2013. Targeted mutagenesis in the malaria mosquito using TALE nucleases. *PloS one*, 8. 10.1371/journal.pone.0074511

Szittya, G. & Burgyán, J., 2013. RNA interference-mediated intrinsic antiviral immunity in plants. *Intrinsic Immunity*, 153–181. 10.1007/978-3-642-37765-5_6

Taxman, D.J., Moore, C.B., Guthrie, E.H. & Huang, M.T.-H., 2010. Short hairpin RNA (shRNA): design, delivery, and assessment of gene knockdown. *RNA Therapeutics*, 139–156. 10.1007/978-1-60761-657-3_10

Thomas, D.D., Donnelly, C.A., Wood, R.J. & Alphey, L.S., 2000. Insect population control using a dominant, repressible, lethal genetic system. *Science*, 287, 2474–2476. 10.1126/science.287.5462.2474

Tur, C., Almenar, D., Benlloch-Navarro, S., Argilés-Herrero, R., Zacarés, M., Dalmau, V. & Pla, I., 2021. Sterile Insect Technique in an Integrated Vector Management Program against Tiger Mosquito Aedes albopictus in the Valencia Region (Spain): Operating Procedures and Quality Control Parameters. *Insects*, 12, 272. 10.3390/insects12030272

Vreysen, M.J., Abd-Alla, A.M., Bourtzis, K., Bouyer, J., Caceres, C., De Beer, C., Oliveira Carvalho, D., Maiga, H., Mamai, W. & Nikolouli, K., 2021. The Insect Pest Control Laboratory of the Joint FAO/IAEA Programme: Ten Years (2010–2020) of Research and Development, Achievements and Challenges in Support of the Sterile Insect Technique. *Insects*, 12, 346. 10.3390/insects12040346

Vreysen, M.J., Klassen, W. & Carpenter, J.E., 2016. Overview of technological advances toward greater efficiency and efficacy in sterile insect-inherited sterility programs against moth pests. *Florida Entomologist*, 99, 1–12. 10.1653/024.099.sp102

Vreysen, M.J., Saleh, K.M., Ali, M.Y., Abdulla, A.M., Zhu, Z.-R., Juma, K.G., Dyck, V.A., Msangi, A.R., Mkonyi, P.A. & Feldmann, H.U., 2000. Glossina austeni (Diptera: Glossinidae) eradicated on the island of Unguja, Zanzibar, using the sterile insect technique. *Journal of economic entomology*, 93, 123–135. 10.1603/0022-0493-93.1.123

Waltz, E., 2021. First genetically modified mosquitoes released in the United States. *Nature*, 593, 175–176. 10.1038/d41586-021-01186-6

Wang, G.-H., Gamez, S., Raban, R.R., Marshall, J.M., Alphey, L., Li, M., Rasgon, J.L. & Akbari, O.S., 2021. Combating mosquito-borne diseases using genetic control technologies. *Nature communications*, 12, 1–12. 10.1038/s41467-021-24654-z

Warren, W.D., Atkinson, P.W. & O'Brochta, D.A., 1994. The Hermes transposable element from the house fly, Musca domestica, is a short inverted repeat-type element of the hobo, Ac, and Tam3 (hAT) element family. *Genetics Research*, 64, 87–97. 10.1017/s0016672300032699

WHO 2017. Vector-borne diseases. https://www.who.int/news-room/factsheets/detail/vector-borne-diseases (Accessed 15 October 2020)

WHO 2020. Guidance framework for testing the sterile insect technique as a vector control tool against aedes-borne diseases.

WHO 2021. *World malaria report 2021*. World Health Organization.

Wickramasinghe, P., Silva, G., Silva Gunawardene, Y. & Dassanayake, R., 2021. Advances in Aedes Mosquito Vector Control Strategies Using CRISPR/Cas9. *Genetically Modified and other Innovative Vector Control Technologies*. Springer, 67–87. 10.1007/978-981-16-2964-8_4

Wijewickrama, A., Fernando, S., Jayerathne, G.S.B., Perera, P.A., Abeynaike, S.A., Gomes, L., Jeewandara, C., Tippalagama, R., Dilani, D. & Rajendran, A., 2018. Emergence of a dengue virus serotype 2 causing the largest ever dengue epidemic in Sri Lanka. *BioRxiv*, 329318. 10.1101/329318

Wilke, A.B.B. & Marrelli, M.T., 2012. Genetic control of mosquitoes: population suppression strategies. *Revista do Instituto de Medicina Tropical de Sao Paulo*, 54, 287–292. 10.1590/s0036-46652012000500009

Wise De Valdez, M.R., Nimmo, D., Betz, J., Gong, H.-F., James, A.A., Alphey, L. & Black Iv, W.C., 2011. Genetic elimination of dengue vector mosquitoes. *Proceedings of the National Academy of Sciences*, 108, 4772–4775. 10.1073/pnas.1019295108

Wood, A.J., Lo, T.-W., Zeitler, B., Pickle, C.S., Ralston, E.J., Lee, A.H., Amora, R., Miller, J.C., Leung, E. & Meng, X., 2011. Targeted genome editing across species using ZFNs and TALENs. *Science*, 333, 307–307. 10.1126/science.1207773

Yen, P.-S. & Failloux, A.-B., 2020. A review: Wolbachia-based population replacement for mosquito control shares common points with genetically modified control approaches. *Pathogens*, 9, 404. 10.3390/pathogens9050404

Yen, P.-S., James, A., Li, J.-C., Chen, C.-H. & Failloux, A.-B., 2018. Synthetic miRNAs induce dual arboviral-resistance phenotypes in the vector mosquito *Aedes aegypti*. *Communications Biology*, 1, 1–9. 10.1038/s42003-017-0011-5

Zheng, X., Zhang, D., Li, Y., Yang, C., Wu, Y., Liang, X., Liang, Y., Pan, X., Hu, L. & Sun, Q., 2019. Incompatible and sterile insect techniques combined eliminate mosquitoes. *Nature*, 572, 56–61. 10.1038/s41586-019-1407-9

Zheng, Z.M., Tang, S. & Tao, M., 2005. Development of resistance to RNAi in mammalian cells. *Annals of the New York Academy of Sciences*, 1058, 105–118. 10.1196/annals.1359.019

Zhu, K.Y. & Palli, S.R. 2020. Mechanisms, applications, and challenges of insect RNA interference. *Annual Review of Entomology*, 65. 10.1146/annurev-ento-011019-025224

Section IV

Brief Review of Mosquitoes in Other Countries of the Indian Subcontinent

14 Mosquitoes of Bangladesh

Mohammad Nazrul Islam
Hamdard University Bangladesh, Gazaria, Bangladesh

14.1 INTRODUCTION

One of the fastest developing countries, Bangladesh (24°N to 27°N latitude and 88°E to 92°E longitude) is surrounded by the Indian states of West Bengal, Meghalaya, Assam, Mizoram, Tripura, and Burma on the east and the Bay of Bengal on the south. The Tropic of Cancer divides the country into two halves: (1) the northern and north-western part where the climate is hot and humid (average rainfall 2,000–3,000 mm) and (2) the south-eastern and south-western zone where the climate is humid and tempered, with an average rainfall of 1,500 to 1,800 mm (Banglapedia 2021). Bangladesh (area: 148,460 sq. km) is the eighth-most populous country in the world (>165 million; Bangladesh CIA, 2021). It has three distinct seasons: (i) pre-monsoon hot season (summer) from March to May, (ii) rainy monsoon season from June to October, and (iii) cool and dry winter season from November to February. The environmental conditions and other factors are conducive for a preponderance and diversity of mosquitoes in Bangladesh.

14.2 CATALOGUE OF MOSQUITOES OF BANGLADESH

There are 123 species of mosquitoes in Bangladesh (Covell, 1927, 1931; Christophers, 1933; Barraud, 1934; Puri, 1936, 1948; Ahmed, 1987; Irish et al., 2016 Table 14.1).

TABLE 14.1
An Updated Catalogue of Mosquitoes of Bangladesh

FAMILY CULICIDAE

I. Subfamily Anophelinae

GENUS ANOPHELES (MEIGEN, 1818)

Subgenus Anopheles (Meigen, 1818)

1. *An. (An.) aitkenii* (James, 1903)
2. *An. (An.) baileyi* (Edwards, 1929)
3. *An. (An.) barbirostris* (van der Wulp, 1884) (s.l.)
4. *An. (An.) barbumbrosus* (Strickland and Chowdhury, 1927)
5. *An. (An.) bengalensis* (Puri, 1930)
6. *An. (An.) gigas* (Giles, 1901)
7. *An. (An.) nigerrimus* (Giles, 1900)
8. *An. (An.) peditaeniatus* (Leicester, 1908)
9. *An. (An.) umbrosus* (Theobald, 1903)

Subgenus Cellia (Theobald, 1902)

10. *An. (Cel.) aconitus* (Dönitz, 1902)
11. *An. (Cel.) annularis* (van der Wulp, 1884) (s.l.)
12. *An. (Cel.) baimaii* (Sallum and Peyton, 2005)
13. *An. (Cel.) culicifacies* (Giles, 1901) (s.l.)
14. *An. (Cel.) fluviatilis* (James, 1902) (s.l.)

(*Continued*)

DOI: 10.1201/9781003327028-18

TABLE 14.1
(Continued)

15. *An. (Cel.) jamesii* (Theobald, 1901)
16. *An. (Cel.) jeyporiensis* (James, 1902)
17. *An. (Cel.) karwari* (James, 1903)
18. *An. (Cel.) kochi* (Dönitz, 1901)
19. *An. (Cel.) maculatus* (Theobald, 1901)
20. *An. (Cel.) majidi* (Young and Majid, 1928)
21. *An. (Cel.) minimus* (Theobald, 1901) (s.l.)
22. *An. (Cel.) nivipes* (Theobald, 1903)
23. *An. (Cel.) pallidus* (Theobald, 1901)
24. *An. (Cel.) philippinensis* (Ludlow, 1902)
25 *An. (Cel.) pseudojamesi* (Strickland and Chowdhury, 1927)
26. *An. (Cel.) pseudowillmori* (Theobald, 1910) [Al-Amin, in preparation]
27. *An. (Cel.) splendidus* (Koidzumi, 1920)
28. *An. (Cel.) stephensi* (Liston, 1901)
29. *An. (Cel.) subpictus* (Grassi, 1899) (s.l.)
30. *An. (Cel.) sundaicus* (Rodenwaldt, 1925) (s.l.)
31. *An. (Cel.) tessellatus* (Theobald, 1901)
32. *An. (Cel.) theobaldi* (Giles, 1901)
33. *An. (Cel.) turkhudi* (Liston, 1901)
34. *An.* (*Cel.*) *vagus* (Dönitz, 1902)
35. *An.* (*Cel.*) *varuna* (Iyengar, 1924)
36. *An.* (*Cel.*) *willmori* (James, 1903)

GENUS AEDEOMYIA (THEOBALD, 1901)

Subgenus Aedeomyia (Theobald, 1901)

37. *Ad.* (*Ady.*) *catastica* (Knab, 1909)

GENUS AEDES (MEIGEN, 1818)

Subgenus Aedimorphus (Theobald, 1903)

38. *Ae.* (*Adm.*) *caecus* (Theobald, 1901)
39. *Ae.* (*Adm.*) *pallidostriatus* (Theobald, 1907)
40. *Ae.* (*Adm.*) *punctifemoris* (Ludlow, 1921)
41. *Ae.* (*Adm.*) *vexans* (Meigen, 1830)

Subgenus *Christophersiomyia* (Barraud, 1923)

42. *Ae.* (*Chr.*) *thomsoni* (Theobald, 1905)

Subgenus *Collessius* Reinert (Harbach and Kitching, 2006)

43. *Ae.* (*Col.*) *pseudotaeniatus* (Giles, 1901)

Subgenus *Dendroskusea* (Edwards, 1929)

44. *Ae.* (*Dsk.*) *reginae* (Edwards, 1922)

Subgenus *Downsiomyia* (Vargas, 1950)

45. *Ae.* (*Dow.*) *albolateralis* (Theobald, 1908)
46. *Ae.* (*Dow.*) *niveus* (Ludlow, 1903)

Subgenus *Edwardsaedes* (Belkin, 1962)

47. *Ae.* (*Edw.*) *imprimens* (Walker, 1861)

Subgenus *Fredwardsius* (Reinert, 2000)

48. *Ae.* (*Fre.*) *vittatus* (Bigot, 1861)

Subgenus *Hulecoeteomyia* (Theobald, 1904)

49. *Ae.* (*Hul.*) *chrysolineatus* (Theobald, 1907)

(Continued)

TABLE 14.1
(Continued)

Subgenus *Kenknightia* (Reinert, 1990)
50. *Ae.* (*Ken.*) *dissimilis* (Leicester, 1908)
Subgenus *Mucidus* (Theobald, 1901)
51. *Ae.* (*Muc.*) *scatophagoides* (Theobald, 1901)
Subgenus *Neomelaniconion* (Newstead, 1907)
52. *Ae.* (*Neo.*) *lineatopennis* (Ludlow, 1905)
Subgenus *Petermattinglyius* (Reinert, Harbach, and Kitching, 2009)
53. *Ae.* (*Pet.*) *iyengari* (Edwards, 1923)
Subgenus *Phagomyia* (Theobald, 1905)
54. *Ae.* (*Phg.*) *assamensis* (Theobald, 1908)
55. *Ae.* (*Phg.*) *khazani* (Edwards, 1922)
56. *Ae.* (*Phg.*) *lophoventralis* (Theobald, 1910)
Subgenus *Stegomyia* (Theobald, 1901)
57. *Ae.* (*Stg.*) *aegypti* (Linnaeus, 1762)
58. *Ae.* (*Stg.*) *albopictus* (Skuse, 1895)
59. *Ae.* (*Stg.*) *annandalei* (Theobald, 1910)
60. *Ae.* (*Stg.*) *desmotes* (Giles, 1904)
61. *Ae.* (*Stg.*) *gardnerii imitator* (Leicester, 1908)
62. *Ae.* (*Stg.*) *w-albus* (Theobald, 1905)
GENUS *ARMIGERES* (THEOBALD, 1901)
Subgenus *Armigeres* (Theobald, 1901)
63. *Ar.* (*Arm.*) *kesseli* (Ramalingam, 1987)
64. *Ar.* (*Arm.*) *kuchingensis* (Edwards, 1915)
65. *Ar.* (*Arm.*) *malayi* (Theobald, 1901) [Al-Amin, in preparation]
66. *Ar.* (*Arm.*) *subalbatus* (Coquillett, 1898)
67. *Ar.* (*Arm.*) *theobaldi* (Barraud, 1934)
Subgenus *Leicesteria* (Theobald, 1904)
68. *Ar.* (*Lei.*) *annulitarsis* (Leicester, 1908)
69. *Ar.* (*Lei.*) *dentatus* (Barraud, 1927)
70. *Ar.* (*Lei.*) *digitatus* (Edwards, 1914)
71. *Ar.* (*Lei.*) *flavus* (Leicester, 1908)
72 *Ar.* (*Lei.*) *inchoatus* (Barraud, 1927)
73. *Ar.* (*Lei.*) *magnus* (Theobald, 1908)
74. *Ar.* (*Lei.*) *omissus* (Edwards, 1914)
GENUS *COQUILLETTIDIA* (DYAR, 1905)
Subgenus *Coquillettidia* (Dyar, 1905)
75. *Cq.* (*Coq.*) *crassipes* (van der Wulp, 1881)
76. *Cq.* (*Coq.*) *ochracea* (Theobald, 1903)
GENUS *CULEX* (LINNAEUS, 1758)
Subgenus *Culex* (Linnaeus, 1758)
77. *Cx.* (*Cux.*) *annulus* (Theobald, 1901)
78. *Cx.* (*Cux.*) *fuscocephala* (Theobald, 1907)
79. *Cx.* (*Cux.*) *gelidus* (Theobald, 1901)
80. *Cx.* (*Cux.*) *hutchinsoni* (Barraud, 1924)
81. *Cx.* (*Cux.*) *mimulus* (Edwards, 1915)

(*Continued*)

TABLE 14.1
(Continued)

82. *Cx.* (*Cux.*) *pseudovishnui* (Colless, 1957)
83. *Cx.* (*Cux.*) *quinquefasciatus* (Say, 1823)
84. *Cx.* (*Cux.*) *sitiens* (Wiedemann, 1828)
85. *Cx.* (*Cux.*) *theileri* (Theobald, 1903)
86. *Cx.* (*Cux.*) *tritaeniorhynchus* (Giles, 1901)
87. *Cx.* (*Cux.*) *vagans* (Wiedemann, 1828)
88. *Cx.* (*Cux.*) *vishnui* (Theobald, 1901)
89. *Cx.* (*Cux.*) *whitei* (Barraud, 1923)
90. *Cx.* (*Cux.*) *whitmorei* (Giles, 1904)

Subgenus *Culiciomyia* (Theobald, 1907)

91. *Cx.* (*Cui.*) *fragilis* (Ludlow, 1903)
92. *Cx.* (*Cui.*) *nigropunctatus* (Edwards, 1926)
93. *Cx.* (*Cui.*) *pallidothorax* (Theobald, 1905)
94. *Cx.* (*Cui.*) *pullus* (Theobald, 1905)

Subgenus *Eumelanomyia* (Theobald, 1909)

95. *Cx.* (*Eum.*) *brevipalpis* (Giles, 1902)
96. *Cx.* (*Eum.*) *malayi* (Leicester, 1908)

Subgenus *Lophoceraomyia* (Theobald, 1905)

97. *Cx.* (*Lop.*) *minutissimus* (Theobald, 1907)

Subgenus *Oculeomyia* (Theobald, 1907)

98. *Cx.* (*Ocu.*) *bitaeniorhynchus* (Giles, 1901)
99. *Cx.* (*Ocu.*) *epidesmus* (Theobald, 1910)
100. *Cx.* (*Ocu.*) *infula* (Theobald, 1901)
101. *Cx.* (*Ocu.*) *sinensis* (Theobald, 1903)

GENUS *FICALBIA* (THEOBALD, 1903)

102. *Fi. minima* (Theobald, 1901)

GENUS *HEIZMANNIA* (LUDLOW, 1905)

Subgenus *Heizmannia* (Ludlow, 1905)

103. *Hz.* (*Hez.*) *covelli* (Barraud, 1929)

GENUS *LUTZIA* (THEOBALD, 1903)

Subgenus *Metalutzia* (Tanaka, 2003)

104. *Lt.* (*Mlt.*) *fuscana* (Wiedemann, 1820)
105. *Lt.* (*Mlt.*) *halifaxii* (Theobald, 1903)

GENUS *MALAYA* (LEICESTER, 1908)

106. *Ml. genurostris* (Leicester, 1908)
107. *Ml. jacobsoni* (Edwards, 1930)

GENUS *MANSONIA* (BLANCHARD, 1901)

Subgenus *Mansonioides* (Theobald, 1907)

108. *Ma.* (*Mnd.*) *annulata* (Leicester, 1908)
109. *Ma.* (*Mnd.*) *annulifera* (Theobald, 1901)
110. *Ma.* (*Mnd.*) *dives* (Schiner, 1868)
111. *Ma.* (*Mnd.*) *indiana* (Edwards, 1930)
112. *Ma.* (*Mnd.*) *uniformis* (Theobald, 1901)

GENUS *MIMOMYIA* (THEOBALD, 1903)

Subgenus *Etorleptiomyia* (Theobald, 1904)

113. *Mi.* (*Eto.*) *luzonensis* (Ludlow, 1905) [Al-Amin, in\ preparation]

(*Continued*)

TABLE 14.1
(Continued)

Subgenus *Mimomyia* (Theobald, 1903)
114. *Mi.* (*Mim.*) *chamberlaini* (Ludlow, 1904)
115. *Mi.* (*Mim.*) *hybrida* (Leicester, 1908)
GENUS *ORTHOPODOMYIA* (THEOBALD, 1904)
116. *Or. anopheloides* (Giles, 1903)
GENUS *TOXORHYNCHITES* (THEOBALD, 1901)
Subgenus *Toxorhynchites* (Theobald, 1901)
117. *Tx.* (*Tox.*) *bengalensis* (Rosenberg and Evenhuis, 1985)
118. *Tx.* (*Tox.*) *splendens* (Wiedemann, 1819)
GENUS *TRIPTEROIDES* (GILES, 1904)
Subgenus *Rachionotomyia* (Theobald, 1905)
119. *Tp.* (*Rah.*) *aranoides* (Theobald, 1901)
Genus *Uranotaenia* Lynch (Arribálzaga, 1891)
Subgenus *Pseudoficalbia* (Theobald, 1912)
120. *Ur.* (*Pfc.*) *novobscura* (Barraud, 1934)
Subgenus *Uranotaenia* Lynch (Arribálzaga, 1891)
121. *Ur.* (*Ura.*) *campestris* (Leicester, 1908)
122. *Ur.* (*Ura.*) *rampae* (Peyton and Klein, 1970)
GENUS *VERRALLINA* (THEOBALD, 1903)
Subgenus *Neomacleaya* (Theobald, 1907)
123. *Ve.* (*Nma.*) *andamanensis* (Edwards, 1922)

Source: Irish et al., 2016.

14.3 ANNOTATED BIO-ECOLOGY OF IMPORTANT DISEASE VECTORS IN BANGLADESH

14.3.1 Genus Anopheles

A total of 35 Anopheles species have been recorded in Bangladesh (Jannat and Ahmed, 2006; Bashar et al., 2012; Elias et al., 1982; Alam et al., 2010), of which seven species were found to transmit malaria in the country. Out of these seven vectors, four species, *Anopheles dirus, An. philippinensis, An. Sundaicus*, and *An. Minimus*, are considered important malaria vectors. The other three species, *An. aconitus, An. Annularis*, and *An. vagus* have also been infrequently incriminated recently in places where malaria outbreaks took place. *Anopheles jeyporiensis* is the most abundant species (32.9%), followed by *An. kochi* (12.27%), *An. philippinensis* (11.45%), *An. karwari* (11.4%), and *An. vagus* (9.06%; Alam et al. 2010). Only four species (*An. philippinensis, An. minimus, An. baimai*, and *An. annularis*) have been collected using human bait, alluding to their anthropophilic nature (Bashar et al. 2012; Bashar and Tuno, 2014).

i. ***Anopheles jeyporiensis*:** The species is abundantly found in Bandarban. It has a unique covering of short, oblong white scales on the scutum, extending all the way back to the scutellum. It is the only unplaced member of the Funestus Group. The species is highly variable, especially in morphological characters of the wing; palps and four chromosomal forms (A, B, C, D) are identified (Mohammad Shafiul Alam et al., 2012).

ii. ***Anopheles dirus* (*An. baimaii*):** This species is transmitting malaria in hilly forested and foot-hill areas in the eastern part of Bangladesh. This wild species is largely an outdoor resting mosquito and bites mainly in the middle of the night. It is difficult to find adults in human dwellings during the daytime. Members of this species seem to rest a while before feeding and leave soon after feeding in outdoor shelters (Baimai et al., 1988).
iii. ***Anopheles minimus*:** This species is usually found in forested hills and foot-hill areas and considered as the primary malaria vector. It starts feeding just after sunset but attacks more after midnight. It is generally considered a domestic mosquito. Recently, there has been signs of reappearance of this species in low density, probably due to the long absence of any insecticidal spray (Banglapedia, 2021).
iv. ***Anopheles philippinensis*:** This species is more abundant in the flood-plain deltaic region of Bangladesh. Its favourable breeding places are marshy land, ground pools, ditches, tanks, ponds, etc. This species is now mainly seen resting in cattle sheds or on vegetation around outdoor shelters, rarely found indoors (Banglapedia 2021).
v. ***Anopheles sundaicus*:** This species is transmitting malaria in coastal areas of Bangladesh. It virtually disappeared after DDT spraying. Presently, it is found in low numbers in coastal areas, including Cox's Bazar. It breeds in brackish water (Banglapedia, 2021).

Other important species include *Anopheles annularis, A. vagus*, and *A. aconitus*, which are found in some particular flood-plain areas of Bangladesh. These species are generally considered cattle feeders and occasionally bite man. *Anopheles annularis* is usually found in large numbers and represents as a malaria vector in some localities. *Anopheles aconitus* is not common but observed as a secondary vector of malaria in some particular areas of Bangladesh. *Anopheles vagus*, a fairly common species of mosquito in all parts of Bangladesh, has recently been incriminated as a malaria vector. These species are playing a vital role in malarial transmission due to changes in the ecological situation. It is still endemic in the northern and eastern parts of bordering India and Myanmar. Outbreaks of malaria are also being reported from those areas, as well as from some pockets of flood-plain deltoid areas of Bangladesh (Alam et al., 2010). Species such as *An. annularis, An. baimaii, An. barbirostris, An. jeyporiensis, An. karwari, An. minimus, An. philippinensis, An. umbrosus, An. Wilmori*, and *An.vagus* are the dominant species present almost throughout the year with highest peaks in March and smallest peaks in September, but *An. baimaii* and *An. willmori* were found during monsoon season (July–September) only. Rainfall and relative humidity significantly influenced *An. baimaii, An. willmori, An. Vagus*, and *An. subpictus* density in the Kumari area of Bangladesh.

14.3.2 Genus Aedes

i. ***Aedes aegypti*:** It is the most important vector of dengue and *chikungunya fever* in Bangladesh, although *Ae. albopictus* also plays a major role. It is noteworthy that the mosquito spreads yellow fever virus (YFV) in Africa and Latin America, although a few imported YFV cases were recently detected in China (Zettel and Kaufman, 2010; Wasserman et al., 2016). The mosquito (4–7 mm long and dark coloured) can be recognized by black-and-white markings on its legs and a marking in the form of a *lyre* on the upper surface of its *thorax*. *Aedes* mosquitoes are visually distinctive because they have noticeable black-and-white markings on their bodies and legs. Unlike most other mosquitoes, they are active and bite only *during the daytime*. The peak biting periods are early in the morning and in the evening before dusk. *Aedes aegypti* can be mixed up with *Aedes albopictus* without a magnifying glass; the latter has a white stripe on the top of the mid thorax. Male mosquitoes feed on nectar, and female mosquitoes feed on blood for maturing their eggs. The female mosquito is attracted to chemical compounds emitted by mammals, including *ammonia* (Geier et al., 1999), *carbon dioxide* (Ghaninia et al., 2019), *lactic acid*, and *octenol* (Bohbot et al., 2013). Research has shown that the specific chemical structure of octenol attracts the mosquito to its host and found that the mosquito has a preference for "right-handed" (*dextrorotatory*) octenol molecules (Dennis, 2010). The preference for biting humans is dependent on expression of the *odorant receptor* AaegOr4 (McBride et al., 2014). The white eggs are laid separately into water and not together, unlike most other mosquitoes, and soon turn black. The larvae feed bacteria and grow over a period of weeks until they reach the pupa stage. The lifespan of an adult Ae. *aegypti* is two to four weeks depending on conditions, but the eggs can be viable for over a year in a dry state, which allows the mosquito to re-emerge after a cold winter or dry spell (Mortimer, 2010). They inflict an innocuous bite, usually on the back of the neck or the ankles

ii. ***Aedes albopictus*:** It is also known as the tiger mosquito or forest mosquito, which is native to the tropical and subtropical areas of South-East Asia. The presence of *Aedes* mosquitoes is significantly ($p < 0.05$) higher in low socio-economic zones of Dhaka (Paul et al., 2018). It has been observed that temperate populations of *Aedes albopictus* are more affected in biology. That may be due to seasonal temperature, geographical area, and photoperiodicity. They usually lay their eggs during late summer or early autumn when daylight hours are reducing (Carrieri et al., 2003). *Aedes albopictus* has been shown to be able to survive a cold spell of −10°C, whereas eggs of tropical *Ae. albopictus* could only survive −2°C. The drought-resistant eggs are laid above the water line. Larval/pupal development takes three to eight weeks and is continuous throughout the year. Adult females can survive over three weeks (Calzolari et al., 2012). *Aedes albopictus* is an opportunistic feeder (Richards et al., 2006; Tsetsarkin et al., 2007). Blood hosts include humans, domestic and wild animals, reptiles, birds, and amphibians. It has been observed that host availability and abundance has a direct impact on the feeding activities of *Ae. albopictus*. Adult women, most of whom spend the majority of their day at or very near the home, experienced the highest attack rates in outbreak (Khatun et al., 2015)

Aedes albopictus has the ability to breed in natural and artificial habitats, some of which include tyres, barrels, rainwater gulley catch basins, and drinking troughs. Natural habitats consist of phytotelms (water bodies held by terrestrial plants, e.g., tree holes) and rock pools (Delatte et al., 2008). They are not known to breed in brackish or salt water (Diallo, Laganier, and Nangouma, 2010). *Aedes albopictus* is said to be superior in competing for food resources with *Ae. triseriatus* and *Ae. japonicus* (Leisnham and Juliano, 2012). *Aedes albopictus* is currently considered a serious biting nuisance for humans (Vazeille et al., 2008); it is significantly reducing the quality of life in infested areas. Adult females bite aggressively, usually during the day and preferably outdoors. However, there are reports that *Ae. albopictus* is becoming partially endophilic and is found to be biting indoors (Drago, 2003). Another study reported that *Ae. albopictus* females develop indoors in containers such as flower vases, empty paint cans, and sinks. The mosquitoes' survival time was long enough to complete a gonotrophic cycle and to allow the development of transmissible arboviruses within the vector (Qualls et al., 2013). A summer temperature of 25°C–30°C is required for optimum development (Straetemans, 2008). However, there are reports of populations establishing in areas with lower mean temperatures (5°C–28.5°C) and lower rainfall (290 mm annually) than previously suspected. There is generally little adult activity below 9°C, but adults do seek warmer microclimates indoors (Roiz et al., 2010). Flight range (and hence dispersal on the wing) is limited to ~200 m (Turell et al., 2005). Other species of *Aedes* found in Bangladesh are *Ae. caecus, Ae. pallidostriatus, Ae. punctifemoris, Ae. vexans, Ae. thomsoni, Ae. pseudotaeniatus, Ae. reginae, Ae. albolateralis, Ae. niveus, Ae. imprimens Ae. vittatus, Ae. chrysolineatus, Ae. scatophagoides, Ae. lineatopennis, Ae. iyengari, Ae. assamensis, Ae. khazani, Ae. lophoventralis, Ae. annandalei, Ae. desmotes, Ae. gardnerii* (botanical garden in Dhaka), and *Ae. albus*.

14.4 GENUS CULEX

14.4.1 Culex Mosquitoes

The mosquito is commonly known as the southern house mosquito. It is a medium-sized mosquito found in tropical and subtropical regions. It is a vector of *Wuchereria bancrofti*, filariasis, avian malaria, and arboviruses, including St Louis encephalitis virus, Western equine encephalitis virus, Zika virus, and West Nile virus (Hill and Connelly, 2009). It is taxonomically a member of the *Culex pipiens* species complex (Harbach, 2012). It has 18,883 protein-coding genes.

An adult of *Cx. quinquefasciatus* is brown in colour. The body is about 3.96 to 4.25 mm long. While the main body is brown, the proboscis, thorax, wings, and tarsi are darker than the rest of the body. The head is light brown, with the lightest portion in the centre. The antennae and the proboscis are about the same length, but in some cases, the antennae are slightly shorter than the proboscis. The flagellum has 13 segments that may have few or no scales. The scales of the thorax are narrow and curved. The abdomen has pale, narrow, rounded bands on the basal side of each tergite. Males can be differentiated from females in having large palps and feathery antennae. The larva has a short and stout head. The mouth brushes have long yellow filaments used for filtering organic materials. The abdomen consists of eight segments, the siphon, and the saddle. Each segment has a unique setae pattern. The siphon is on the dorsal side of the abdomen and is four times longer than its breadth. The siphon has multiple setae tufts. The saddle is barrel-shaped and located on the ventral side of the abdomen, with four long anal papillae protruding from the posterior end (Hill and Connelly, 2009).

Mature *Cx. quinquefasciatus* females fly at night to nutrient-rich standing water to lay eggs. They breed profusely in dirty water collections, including stagnant drains, cesspools, septic tanks with leaks, burrow pits, and almost all organic polluted water collections. A single female can lay up to five rafts of eggs in a lifetime, with each raft containing thousands of eggs. The exact number varies depending on climatic conditions (Hill and Connelly, 2009). The larvae feed on organic material in the water and require between five and eight days to complete their development at 30°C (86°F). The larvae pass through four larval instars, and towards the end of the fourth instar, they stop eating and undergo moulting to give rise to pupae. After 36 hours at 27°C (81°F), adults emerge. The exact timing of development can vary depending on temperature. In optimum temperature and humidity, the lifecycle will be completed in seven days, passing through the egg, larval, pupal, and adult stages. Both male and female adults take sugar meals from plants. After mating, the female seeks a blood meal from a mammal or bird, as ingested blood is necessary for egg development. *Culex quinquefasciatus* shows a preference for the blood of birds but will also commonly bite humans. The southern house mosquito is a principal vector of numerous pathogens, transmitting the phleboviruses such as Rift Valley fever virus, St Louis encephalitis virus, filarial worms, and avian malarial parasites (Guagliard and Levine, 2021). It transmits zoonotic diseases that affect humans and wild and domestic animals and is also a vector of the Zika virus (Garvey, 2016). It causes infection through biting during the blood meal. It acts as an intermediate host for the helminth parasite by harbouring the larval stages (Albuquerque et al., 1999). It is the definitive host for the malarial parasite, as it harbours the sexual cycle (Farajollahi et al., 2011). Other species of Culex mosquitoes found in Bangladesh are *Cx. fuscocephala, Cx. tritaeniorhynchus, Cx. gelidus, Cx. whitmorei, Cx. vishnui, Cx. whitei, Culex annulus*.

14.5 CONCLUSION

Mosquitoes in Bangladesh are diverse and many (123 species). They transmit different life-threating diseases like dengue, malaria, chikungunya, filariasis, yellow fever, etc. (Shirin et al., 2019). Bangladesh is a fast-developing country where unplanned urbanization, improper drainage, poor budget, and lack of knowledge on mosquito play vital roles in deleting either vector mosquitoes *perse*, or the various mosquito-borne diseases.

REFERENCES

Ahmed, T.U. (1987). Checklist of the mosquitoes of Bangladesh. *Mosquito Systematics*, 19, 187–200.

Alam, M.S., Chakma, S., Khan, W.A., Glass, G.E., Mohon, A.N., Elahi, R., Norris, L.C., Podder, M.P., Ahmed, S., Haque, R., Sack, D.A., Sullivan, D.J. Jr., Norris, D.E. (2012). Diversity of anopheline species and their *Plasmodium* infection status in rural Bandarban, Bangladesh. *Parasites and Vectors*, 5, 150. https://doi.org/10.1186/1756-3305-5-150

Alam, M.S., Khan, M.G., Chaudhury, N., Deloer, S., Nazib, F., Bangali, A.M. and Haque, R. (2010). Prevalence of anopheline species and their Plasmodium infection status in epidemic-prone border areas of Bangladesh. *Malaria Journal*, 9, 15.

Albuquerque, C.M.R., Cavalcanti, V.M.S., Melo, M.A.V., Verçosa, P., Regis, L.N. and Hurd, H. (1999). Bloodmeal microfilariae density and the uptake and establishment of Wuchereria bancrofti infections in *Culex quinquefasciatus* and *Aedes aegypti*. *Memórias do Instituto Oswaldo Cruz*, 94(5), 591–596.

Baimai, V., Thu. M.M., Paing, M. and Maheswary, N.P. (1988). Distribution and chromosomal polymorphism of the malaria vector *Anopheles dirus* species D. *Southeast Asian Journal of Tropical Medicine and Public Health*, 19, 661–665.

Banglapedia (2021): https://en.banglapedia.org/index.php/Malaria. Bangladesh. The World Factbook. CIA, 2021

Barraud, P.J. (1934). *The fauna of British India, including Ceylon and Burma. Diptera. vol. 5, family Culicidae. Tribes Megarhinini and Culicini*. Taylor and Francis.

Bashar, K., Tuno, N., Ahmed, T.U. and Howlader, A.J. (2012). Blood-feeding patterns of Anopheles mosquitoes in a malaria-endemic area of Bangladesh. *Parasites Vectors*, 5, 39. https://doi.org/10.1186/1756-3305-5-39

Bashar, K. and Tuno, N. (2014). Seasonal abundance of *Anopheles* mosquitoes and their association with meteorological factors and malaria incidence in Bangladesh. *Parasites and Vectors*, 7, 442.

Bohbot, J.D., Durand, N.F., Vinyard, B.T. and Dickens, J.C. (2013). Functional development of the octenol response in *aedes aegypti*. *Frontiers in Physiology*, 4, 39. https://doi.org/10.3389/fphys.2013.00039

Calzolari, M., Gaibani, P., Bellini, R., Defilippo, F., Pierro, A., Albieri, A., Maioli, G., Luppi, A., Rossini, G., Balzani, A., Tamba, M., Galletti, G., Gelati, A., Carrieri, M., Poglayen, G., Cavrini, F., Natalini, S., Dottori, M., Sambri, V., Angelini, P. and Bonilauri, P. (2012). Mosquito, bird and human surveillance of West Nile and Usutu viruses in Emilia-Romagna Region (Italy) in 2010. *PLoS One*, 7(5), e38058. https://doi.org/10.1371/journal.pone.0038058

Carrieri, M., Bacchi, M., Bellini, R. and Maini, S. (2003). On the competition occurring between *Aedes albopictus* and *Culex pipiens* (Diptera: Culicidae) in Italy. *Environmental Entomology*, 32(6), 1313–1321.

Christophers, S.R. (1933). *The fauna of British India, including Ceylon and Burma. Diptera Vol. 4, Family Culicidae. Tribe Anophelini*. Taylor and Francis.

Covell, G. (1927). The distribution of anopheline mosquitoes of India and Ceylon. *Indian Medical Research Memoirs* 5, 1–85.

Covell G. (1931).The distribution of anopheline mosquitoes in India and Ceylon. Additional records, 1926–1930. *Records of Malaria Survey of India* 2, 225–268.

Delatte, H., Dehecq, J.S., Thiria, J., Domerg, C., Paupy, C., Fontenille, D. (2008). Geographic distribution and developmental sites of Aedes albopictus (Diptera: Culicidae) during a Chikungunya epidemic event. *Vector Borne Zoonotic Diseases*, 8(1), 25–34. https://doi.org/10.1089/vbz.2007.0649

Dennis, O'Brien (2010). *ARS study provides a better understanding of how mosquitoes find a host*. U.S. Department of Agriculture.

Diallo, M., Laganier R, and Nangouma A. (2010). First record of *Ae. albopictus* (Skuse 1894), in Central African Republic. *Tropical Medicine and International Health*, 15(10), 1185–1189.

Drago, A. (2003). Education, information and public awareness in vector control. *14th European Conference of the Society of Vector Ecology*. Bellinzona, Switzerland.

Elias, M., Dewan, Z.A.R. and Ahmed, R. (1982). Vectors of malaria in Bangladesh. *Journal of Preventive and Social Medicine*, 1, 20–28.

Farajollahi, A., Fonseca, D.M., Kramer, L.D. and Marm Kilpatrick, A. (2011). Bird biting- mosquitoes and human disease: A review of the role of *Culex pipiens* complex mosquitoes in epidemiology. *Infection, Genetics and Evolution*, 11(7), 1577–1585.

Garvey, K.K. (2016). *Are Culex mosquitoes potential vectors of the zika virus?* Bug squad, UC-ANR.

Geier, M., Bosch, O.J., and Boeckh, J. (1999). Ammonia as an attractive component of host odour for the yellow fever mosquito, *Aedes aegypti*. *Chemical Senses*, 24(6), 647–653.

Ghaninia, M., Majeed, S., Dekker, T., Hill, S.R. and Ignell, R. (2019). Hold your breath - Differential behavioral and sensory acuity of mosquitoes to acetone and carbon dioxide. PLoS *One*, 14(12), e0226815. https://doi.org/10.1371/journal.pone.0226815

Guagliard, S.A.J. and Levine, R.S. (2021). Etymologia: *Culex quinquefasciatus. Emerging Infectious Diseases*, 27(8), 2041.

Harbach, R.E. (2012). *Culex pipiens*: species versus species complex – taxonomic history and perspective. *Journal of the American Mosquito Control Association*, 28(4), 10–23.

Hill, S. and Connelly, R. (2009). *Features Creatures: Southern house mosquito*. University of Florida. https://en.wikipedia.org/wiki/Aedes_aegypti

Irish, S.R., Al-Amin, H.M., Alam, M.S. and Harbach, R.E. (2016). A review of the mosquito species (Diptera: Culicidae) of Bangladesh. *Parasites and Vectors*, 9, 559. https://doi.org/10.1186/s13071-016-1848-z

Jannat, K.N.E., and Ahmed, T.U. (2006). The mosquito fauna, including new records of St. Martin's Island, Bangladesh. *Bangladesh Journal of Entomology*, 16(1), 67–81.

Khatun, S., Chakraborty, A., Mahmudur, R., Banu, N.N., Rahman, M.M., Hasan S.M.M., Luby, S.P. and Gurley Emily S. (2015). An Outbreak of Chikungunya in Rural Bangladesh, 2011. *PLoS Neglected Tropical Diseases*, 9(7), e0003907. https://doi.org/10.1371/journal.pntd.0003907

Leisnham, P.T. and Juliano, S.A. (2012). Impacts of climate, land use, and biological invasion on the ecology of immature *Aedes* mosquitoes: implications for La Crosse emergence. *EcoHealth*, 9(2), 217–228.

McBride, C.S., Baier, F., Omondi, A.B., Spitzer, S.A., Lutomiah, J., Sang, R., Ignell, R., Vosshall, L.B. (2014). Evolution of mosquito preference for humans linked to an odorant receptor. *Nature*, 515(7526), 222–227. https://doi.org/10.1038/nature13964

Paul, K.K., Dhar-Chowdhury, P., Haque, C.E., and Al-Amin, H.M. (2018). Risk factors for the presence of dengue vector mosquitoes, and determinants of their prevalence and larval site selection in Dhaka, Bangladesh. *PLoS ONE* 13(6), e0199457. https://doi.org/10.1371/journal.pone.0199457

Shirin, T., Muraduzzaman, A.K.M., Alam, A.N., Sultana, S., Siddiqua, M., Khan, M.H., Akram, A., Sharif, A.R., Hossain, S., Flora, M.S. (2019). Largest dengue outbreak of the decade with high fatality may be due to reemergence of DEN-3 serotype in Dhaka, Bangladesh, necessitating immediate public health attention. *New Microbes New Infect.* 16;29:100511. https://doi.org/10.1016/j.nmni.2019.01.007

Puri, I.M. 1936. The distribution of anopheline mosquitoes in India: additional records, 1931–1935. *Records of Malaria Survey of India*, 6, 177–211.

Puri, I.M. (1948). The distribution of anopheline mosquitoes in India, Pakistan, Ceylon and Burma. Part V: additional records 1936–1947. *Indian Journal of Malariology*, 2, 67–107.

Qualls, W.A., Xue, R.D., Beier, J.C. and Muller, G.C. (2013). Survivorship of adult *Aedes albopictus* (Diptera: Culicidae) feeding on indoor ornamental plants with no inflorescence. *Parasitological Research*, 12(6), 2313–2318.

Roiz, D., Rosa, R., Arnoldi, D. and Rizzoli, A. (2010). Effects of temperature and rainfall on the activity and dynamics of host-seeking *Aedes albopictus* females in northern Italy. *Vector Borne and Zoonotic Diseases*, 10(8), 811–816.

Richards, S.L., Ponnusamy, L., Unnasch, T.R., Hassan, H.K. and Apperson, C.S. (2006). Host-feeding patterns of Aedes albopictus (Diptera: Culicidae) in relation to availability of human and domestic animals in suburban landscapes of central North Carolina. *Journal Medical Entomology*, 43(3), 543–551. https://doi.org/10.1603/0022-2585(2006)43[543:hpoaad]2.0.co;2

Sean Wasserman, Paul Anantharajah Tambyah, Poh Lian Lim,(2016). Yellow fever cases in Asia: primed for an epidemic. *International Journal of Infectious Diseases*, 48, 98–103. https://doi.org/10.1016/j.ijid.2016.04.025

Straetemans, M. (2008). Vector-related risk mapping of the introduction and establishment of *Aedes albopictus* in Europe. *European Surveillance*, 13(7), 8040.

Tsetsarkin, K.A., Vanlandingham, D.L., McGee, C.E. and Higgs, S. (2007). A single mutation in Chikungunya virus affects vector specificity and epidemic potential. *PLoS Pathogens*, 3(12), e201. https://doi.org/10.1371/journal.ppat.0030201.3 (12)

Turell, M.J., Dohm, D.J., Sardelis, M.R., Oguinn, M.L., Andreadis, T.G. and Blow, J.A. (2005). An update on the potential of north American mosquitoes (Diptera: Culicidae) to transmit West Nile virus. *Journal of Medical Entomology*, 42(1), 57–62.

Vazeille, M., Jeannin, C., Martin, E., Schaffner, F. and Failloux, A.B. (2008). Chikungunya: a risk for Mediterranean countries? *Acta Tropica*, 105(2), 200–202.

Zettel, C., and Kaufman, P. (2010). *Yellow fever mosquito, Aedes aegypti.* University of Florida, Institute of Food and Agricultural Sciences.

15 Mosquitoes of Bhutan

B.K. Tyagi
Chandigarh University, Mohali, India

15.1 INTRODUCTION

Mosquitoes and some mosquito-borne infections abound in Bhutan – a relatively smaller Himalayan Kingdom of Bhutan (latitudes 26° and 29 °N, and longitudes 88° and 93 °E; population 800,000). It is a landlocked country, situated between northern India and southern Tibet in the lap of the Eastern Himalayan Mountains (United Nations, Department of Economic and Social Affairs, Population Division, 2017). The country is organized into 20 districts (Figure 15.1), with Thimphu being its capital. Physiographically, the country has a mélange of steep and high mountains, rising from 100 m in the southern foothills to more than 7,000 m in the north, intersected by a network of torrentially flowing rivers merging into deep valleys before draining into the Indian plains in the south. There are plains areas along the Indian border, though limited in. Bhutan's climate alters with elevation, from subtropical in the south to temperate in the highlands, and with polar-like climate in the north with snow throughout the year. The country is very rich in natural forests (60% of the land) with a great diversity of plants and animals, with mosquitoes galore.

15.2 MOSQUITO RESEARCH IN BHUTAN

Bhutan, despite its tiny appearance, boasts a sizeable quantum of species belonging to both Anophelinae and Culicinae. So far, 132 species of mosquitoes have been enlisted in Bhutan, including sibling species such as, for example, that of *Anopheles (Anopheles) lindesayi Giles* (Namgay et al., 2020). As a result, vector-borne diseases such as malaria, dengue, chikungunya, and Japanese encephalitis have been occurring and/or recurring for decades. https://thebhutanese.bt/the-threat-of-vector-borne-diseases-in-bhutan/). Malaria is particularly rampant in Bhutan, being endemic in seven districts bordering India, viz., Samtse, Chhukha, Dagana, Sarpang, Zhemgang, Pemagatshel, and Samdrup Jongkhar (Tobgay et al., 2011). A total of 34 Anopheles species were identified. It should be noted that their distributions are based on a limited number of collection sites.

Recently, several new mosquito species have been discovered and described in Bhutan, such as, for example, *Anopheles bhutanensis, An. himalayensis, An. thimphuensis, Culex bhutanensis, Aedes (Hul.) bhutanensis*, and *Uranotaenia bhutanensis*. Obviously, some of these seem to be endemic to the country (Somboon et al., 2022).

15.3 MAJOR DISEASE VECTOR AND PEST MOSQUITOES IN BHUTAN

15.3.1 Malaria

Plasmodium falciparum and *P. vivax* are the main malaria parasite species in Bhutan, and the main vector species are *Anopheles pseudowillmori* and *An. culicifacies* (Tobgay et al., 2011, Wangdi et al., 2016a, b). Bhutan has made great strides in almost eliminating malaria by 99% between 2000 and 2020. Local malaria transmission is successfully contained in most of the districts, imported malaria in particular is posing a major threat to its elimination efforts along the southern border (Kinley et al., 2023; Penjor et al., 2023).

DOI: 10.1201/9781003327028-19

FIGURE 15.1 Bhutan; location and geography.

Source: Free Wikipedia.

15.3.2 Dengue

In the recent past, dengue has been reported in some districts where *Aedes aegypti* exists (Tsheten et al., 2021), and the dengue outbreak was particularly explosive in Bhutan when the disease spread from six subdistricts (with a history of sporadic cases of dengue outbreaks) to a total of 44 subdistricts. So far, DENV1-3 has been reported in Bhutan, and only DENV is evasive.

15.3.3 Japanese Encephalitis

Japanese encephalitis, transmitted by Culex vishnui subgroup mosquitoes, is sporadically occurring in Bhutan, as evidenced by a nationwide surveillance which reported only 20 cases of JE between 2011 and 2018 (Wangchuk et al., 2020).

REFERENCES

Kinley, P., Zangpo, U., Tshering, D., Ley, B., Price, R.N. and Wangdi, K. (2023). Imported malaria and its implication to achievement of malaria-free Bhutan. *Journal of Travel Medicine*, 30(3). https://doi.org/10.1093/jtm/taad044

Namgay, R., Pemo, D., Wangdi, T., Thanari, P., Harbach, R.E. and Somboon, P. (2020). Molecular and morphological evidence for sibling species within *Anopheles (Anopheles) lindesayi* Giles (Diptera: Culicidae) in Bhutan. *Acta Tropica*, 207, 0001–706X. https://doi.org/10.1016/j.actatropica.2020.105455

Penjor, K., Zangpo, U., Tshering D., Ley, B., Price, R. N., and Wangdi, K. (2023). Imported malaria and its implication to achievement of malaria-free Bhutan. *Journal of Travel Medicine*, 30(3), taad044. https://doi.org/10.1093/jtm/taad044

Somboon, P., Namgay, R. and Harbach, R.E. (2022). *Uranotaenia (Pseudoficalbia) bhutanensis* (Diptera: Culicidae), A New Species From Bhutan. *Journal of Medical Entomology*, 59(4), 1303–1307. https://doi.org/10.1093/jme/tjac061

Tobgay, T., Torres, C.E. and Na-Bangchang K. (2011). Malaria prevention and control in Bhutan: successes and challenges. *Acta Tropica*, 117(3), 225–228. https://doi.org/10.1016/j.actatropica.2010.11.008

Tsheten, T., Mclure, A., Clements, A.C.A., Gray, D.J., Wangdi, T., Wangchuk, S. and Wangdi, K. (2021). Epidemiological Analysis of the 2019 Dengue Epidemic in Bhutan. *International Journal of Environmental Research & Public Health*, 18(1), 354. https://doi.org/10.3390/ijerph18010354

Wangchuk, S., Tamang, T.D., Darnal, J.B., Pelden, S., Lhazeen, K., Mynak, M.L., Letson, G.W., Khare, S., Leader, B.T., Marfin, A.A. and Hills, S.L. (2020). Japanese Encephalitis Virus as Cause of Acute Encephalitis, Bhutan. *Emerging Infectious Diseases*, 26(9), 2239–2242. https://doi.org/10.3201/eid2609.200620

Wangdi, K., Banwell, C., Gatton, M. L., Kelly, G. C., Namgay, R., Clements, A. C. A. (2016a). Malaria burden and costs of intensified control in Bhutan, 2006–14: an observational study and situation analysis. *The Lancet Global Health*, 4(5), E336–E343. https://doi.org/10.1016/S2214-109X(16)00083-8

Wangdi, K., Banwell, C., Gatton, M.L., Kelly, G.C., Namgay, R. and Clements, A.C. (2016b). Malaria burden and costs of intensified control in Bhutan, 2006-14: an observational study and situation analysis. *Lancet Global Health*, 4(5), e336–43. https://doi.org/10.1016/S2214-109X(16)00083-8

16 Mosquitoes of the Maldives

Akib Jahir
Entomologist, Soneva Fushi, Kunfunadhoo Island, Baa Atoll, Republic of Maldives

Bart G. J. Knols
Medical Entomologist, K&S Consulting, Dodewaard, The Netherlands

16.1 INTRODUCTION

The Republic of Maldives is an island nation located southwest of India (Figure 16.1). While it is the smallest country in Asia by land size (296 km^2), it is uniquely characterized by the vast 14,300 km^2 of sparkling blue Indian Ocean which surrounds it. Known globally for gorgeous sandy white beaches, azure waters, and colourful coral reefs, the Maldives comprises over a thousand coral islands and sandbanks within 20 atolls. Of these islands, only about two hundred are inhabited with local populations, private resorts, industry, or agriculture. As of 2018, the population was reported to be 500,000 residents with high concentration in urban areas; nearly a fifth of the population reside in the island cities of Malé, Addu City, and Fuvahmulah.

In past the Maldives had been experiencing disease burden due to malaria and human lymphatic filariasis, both of which have been declared eliminated from the country. However, the rising climate temperatures of both land and sea pose a serious threat to sustain these successes, on one hand, and continue effective vector control, on the other. The country's highest natural point is only eight feet above sea level, making it highly susceptible to the effects of climate change.

Spread across the equator, the tropical temperatures of the Maldives range from 24°C to 33°C, and the country receives an average annual rainfall of 2,540 millimetres in the northern atolls and 3,810 millimetres in the southern atolls. The two key seasons of the Maldives, the dry season and the rainy season, are shaped by the winter north-east monsoon and the summer south-west monsoon respectively. Maldives' unique climate and geography, that of tropical isolated islands separated by ocean, have led to unique mosquito populations, mosquito behavioural patterns, and spread of mosquito-borne diseases. Although both malaria and lymphatic filariasis were eliminated from the country's population, residents of the Maldives remain at risk for dengue fever, chikungunya, and Zika virus.

The Republic of Maldives is a small island nation with limited mosquito research history and documentation. Additionally, due to the coral islands being separated by ocean, the mosquito species have adapted and evolved into their isolated island's specific conditions. Thus, mosquito species populations present in the Maldives are likely to change and diverge over time.

The World Health Organization (WHO) identified the yellow fever mosquito *Aedes aegypti* and the Asian tiger mosquito *Ae. albopictus* as native to the Maldives in its 2018 report on dengue fever. These mosquitoes are known to transmit a suite of arboviral diseases, such as dengue fever, chikungunya, and Zika virus. Other mosquito genera, such as *Culex* and *Mansonia*, are present in the Maldives, but information about their spatio-temporal distribution is limited.

DOI: 10.1201/9781003327028-20

FIGURE 16.1 The Republic of Maldives in the Indian Ocean.

16.2 CHECKLIST OF MOSQUITOES OF MALDIVES

Mosquitoes in the Maldives pose a significant public health risk because to the tropical atmosphere that fosters their proliferation. The Maldives, characterised by numerous islands, coastal regions, and wetlands, offers ample breeding habitats for mosquitoes. According to Iyengar and Menon (1955), 15 different species of mosquitoes can be found in the Maldives, as follows:

(1) SUBFAMILY ANOPHELINAE
1. *Anopheles (Myzomyia) subpictus* (Grassi)
2. *Anopheles (Myzomyia) tessellatus* (Theo.)

(2) SUBFAMILY CULICINAE
3. *Harpagomyia genurostris* (Leic.)
4. *Armigeres (Armigeres) obturbans* (Wlk.)
5. *Aedes (Stegomyia) aegypti* (L.)
6. *Aedes (Stegomyia) albopictus* (Skuse)
7. *Aedes (Stegomyia) sp. in scutellaris* (Wlk.) group
8. *Culex (Lophoceraomyia) minutissimus* (Theo.)
9. *Culex (Lophoceraomyia) infantulus* (Edw.)
10. *Culex (Mochthogenes) malayi* (Leic.)
11. *Culex (Culiciomyia) spathifurca* (Edw.)
12. *Culex (Culex) tritaeniorhynchus* (Giles)
13. *Culex (Culex) vishnui* (Theo.)
14. *Culex (Culex) sitiens* (Wied.)
15. *Culex (Culex) pipiens fatigans* (Wied.) = *Culex quinquefasciatus* (Say)

16.3 *AEDES ALBOPICTUS* AND *AEDES AEGYPTI* – IMPORTANT VECTORS AND PEST SPECIES IN MALDIVES

Amongst the variety of mosquito species found in the Maldives at present, *Aedes aegypti* and *Ae. albopictus* are the two most important species, as both can carry and transmit a plethora of vector-borne diseases. Additionally, both are linked specifically in settings that have a high concentration of people and human activities. Both *Ae. aegypti* and *Ae. albopictus* are container breeders and frequently engage in resource competition, indicating that they have comparable biological and behavioural traits. The primary distinction between *Ae. aegypti* and *Ae. albopictus* mosquitoes is that while the former species has two longitudinal white stripes and a white lyre-shaped marking laterally on the thorax, the latter has white scales forming a median longitudinal line on the head extending on to the dorsum of the thorax (Näslund et al., 2021). Both species are mostly active in the two hours after sunrise and before dusk and are aggressive daytime biters. *Aedes aegypti* will linger in proximity to humans, and its bite often goes unnoticed. Both *Ae. aegypti* and *Ae. albopictus* can naturally vector arboviruses like dengue, chikungunya and Zika viruses in several nations in the Indian Ocean, including the Maldives, besides Central Africa, Americas and Europe (Rezza, 2014).

Very few entomological studies have been conducted in the Maldives to determine the locations where *Aedes* mosquitoes reproduce. However, a group of WHO representatives did a survey as recently as 2018 in seven different islands in the Maldives to observe how various water-holding containers, such as those used for domestic purposes and abandoned containers, aid in the reproduction of *Ae. aegypti* in the Maldives. Earlier records of dengue cases led to the selection of the seven islands for the entomological survey. They looked for *Aedes* mosquito breeding in homes, mosques, hospitals, schools, rehabilitation facilities, police stations, parks, building sites, disused boats laying on beaches, etc. While *Ae. albopictus* was discovered breeding in natural habitats, including tree holes (Phytotelmata), solid trash, abandoned boats, coconut shells, and tarpaulins in open places, *Ae. aegypti* was discovered breeding in household containers, such as storage tanks, buckets, and other plastic containers.

During the recent larval breeding ground survey of *Aedes albopictus* and *Ae. aegypti* by WHO in 2018, around 30 different types of breeding places were reported that demonstrated tremendous lack in efficient and proper mosquito breeding surveillance (WHO Dengue Bulletin, 2018). In the past several years, dengue cases have increased in the Maldives, and dengue outbreaks have become more common. Effective community interventions are required for dengue control. *Aedes* mosquito breeding site recognition, methods for reducing larval sources and treatment, and inspiring people to take charge of mosquito management in and around their homes are some critical elements of community engagement. Both the sociocultural significance of the varied container habitats and the larval ecology of these mosquitoes should be considered when developing integrated community-based vector control programmes.

16.4 MOSQUITO-BORNE DISEASES FOUND IN THE MALDIVES

More than one million people die from vector-borne illnesses each year, which affect more than one billion people and are mostly transmitted by mosquitoes (WHO, 2014). Given that both *Aedes* species are effective carriers of human arboviral infections like Zika (Thangamani et al., 2016), dengue (Jansen and Beebe, 2010), chikungunya (Burt et al., 2012), and yellow fever (Aitken et al., 1979), they have drawn a lot of interest from all around the world. A growing worldwide health issue is being raised by the aforementioned arboviral infections because of their quick geographic expansion and significant disease burden. Due to the worldwide dispersion of their vectors and an increase in commerce and travel, especially during the past 30 years, the dissemination and public health impact of these arboviruses have grown substantially (Girard et al., 2020). Notably, WHO has recently certified the Maldives for successfully eliminating lymphatic filariasis as a public health problem and malaria as an endemic disease (Samir et al., 2021; WHO, 2018).

Despite being a world class tourist destination, the Republic of Maldives is currently afflicted with three notorious and neglected tropical diseases: Zika, dengue fever, and chikungunya fever.

16.4.1 Dengue Fever

The first dengue epidemic in the Maldives occurred in 1979, and since then there have been sporadic outbreaks in 1988, 1998, 1999, 2006, and 2011. The Maldives became a nation with an endemic dengue epidemic in 2004. In the Maldives, a total of 2,768 dengue cases and 10 fatalities were reported in 2006, with a 52% rise in the number of infected persons. Maldives recorded 920 dengue cases in 2010, including two fatalities. The year 2011 had the biggest dengue outbreak on record, with 2,909 cases and 12 fatalities (WHO, 2018). In 2015, there were eight fatalities, while in 2016, there were four. In 2017, a newborn with congenital dengue died. The first four months of 2019 saw a 200% increase in dengue cases with 1,912 cases as compared to the same period the previous year where only 539 cases were reported.

According to the Health Protection Agency (HPA), the Maldives witnessed more than 1,600 cases of dengue fever in 2022. Of these, 939 cases were from the atolls and 695 cases reported were isolated to the Greater Malé Region; HPA attributed five deaths to dengue fever in the Maldives for that year. These cases from the Greater Malé Region were disseminated as follows: 362 cases from Malé City, 294 cases from the neighbouring island of Hulhumale, and 39 cases from Villimale. In terms of atolls, the central and southern atolls recorded the greatest number of instances. Raa Atoll and Laamu Atoll had the highest number of cases reported, 218 and 99 cases, respectively. Additionally, more than 50 dengue cases were recorded by the atolls of Kaafu, Alifu Dhaalu, Meemu, Dhaalu, Gaafu Alifu, and Gaafu Dhaalu.

16.4.2 Chikungunya Fever

Chikungunya has spread globally since 2006, with epidemics happening in the Indian Ocean, Oceania, Southern Europe, and, starting in 2013, the Americas (Dudouet et al., 2020). Chikungunya was first noted in the Maldives in 2006. In Malé and adjacent Maldives Islands, an outbreak marked by fever, arthralgia, and myalgia started in December of that year. Clinical signs and the ongoing chikungunya pandemic in the Indian Ocean region led to the chikungunya virus being considered as the cause of the sickness. Blood samples from 67 individuals were obtained as part of an inquiry by the Department of Public Health; 21% tested positive for chikungunya by IgM and 96% by Polymerase chain reaction (PCR). By using an enzyme immunosorbent assay, 6% of people tested positive for dengue, and 4% of people exhibited dual infection symptoms.

Before this 2006 outbreak, chikungunya fever had never been identified in the Maldives. By the conclusion of the pandemic in April 2007, there had been a total of 11,879 suspected and confirmed cases, impacting 61% of inhabited islands, with attack rates on specific islands ranging from 0% to 72% (Yoosuf et al., 2009). Up until 2019, there was little evidence of chikungunya virus (CHIKV) circulation in the Maldives. Over the years 2009 to 2018, only five imported cases of chikungunya were reported to the EuroTravNet/GeoSentinel networks. Local health officials in the Maldives documented a total of 1,736 confirmed cases in 2019, of which 277 cases were noted only in the month of December. During that time, visitors to the Maldives would have been more susceptible to contracting CHIKV. This greatest known CHIKV epidemic in the Maldives is believed to have started in November or December of 2018, and by the beginning of 2019, all atolls save for a few islands had been infected.

All age groups, including infants, are included in the affected population. During this epidemic in the Maldives, a sizeable uncounted number of refugees from the Indian subcontinent were impacted. The Maldives' 2019 CHIKV outbreak reached epidemic proportions and spread to every atoll in the nation. The responsible isolate was discovered to be indigenous and to have a unique lineage that is exclusive to the Maldives. Analysis of a sample of patients from this epidemic showed that encephalitis only manifested in children, and that adults who were infected were more likely to experience arthralgia and arthritis than infected children (Imad et al., 2021).

16.4.3 Zika Fever

The Zika virus (ZIKV) has been responsible for multiple epidemics in Pacific Ocean islands since 2007: in 2007 on Yap Island (Federated States of Micronesia; Duffy et al., 2009) and from 2013 to 2014 in French Polynesia (Musso et al., 2014; Campos et al., 2015). In the South American countries of Brazil and Columbia, outbreaks have been documented for the first time since 2015.

The National IHR Focal Point of the Maldives informed WHO of a ZIKV infection case in a patient returning from Finland on 7 January 2016, with the commencement of symptoms in June 2015. The patient was a 37-year-old Finnish man who spent several months in the Maldives before travelling back to Finland on 16 June 2015. The patient first experienced symptoms on June 18 (moderate fever, rash on the face and trunk, eye discomfort, and arthralgia), but his symptoms soon disappeared. The patient contacted a workplace medical centre, and serum samples were taken on June 24, originally due to dengue suspicion. These tested negative for dengue virus (DENV) NS1 antigen but positive for IgG and IgM dengue antibodies. On June 25, urine samples were collected and sent for RT-PCR flavivirus RNA detection. The identified ZIKV was identical to Asian lineage strains found during earlier Zika outbreaks (Easter Island in 2014, French Polynesia and Thailand in 2013), according to sequencing of the RNAs recovered from the serum and urine samples. This was the first ever reported ZIKV case in the Maldives (Korhonen, 2016).

In 2016 and 2017, Maldives conducted ongoing Zika testing and monitoring. A total of 651 instances were examined during this time, and 3 of those cases tested positive for ZIKV. There have been no confirmed cases in the Maldives since January 2018.

16.5 FUTURE OF MOSQUITO CONTROL IN THE MALDIVES

A third of the world's population is at risk from mosquito-borne arboviruses, which are growing because of human interference with natural ecosystems, globalization of trade and travel, and climate change (Kraemer et al., 2019). Chemical control of mosquitoes continues to be the most common technique to stop transmission because there are currently no viable vaccines or medications specifically designed to treat many of these illnesses, such as DENV, CHIKV, or ZIKV. However, environmental effects and growing pesticide resistance undercut their effectiveness and necessitate the use of new, more eco-friendly techniques (Achee et al., 2019).

In the Maldives, conventional mosquito control technologies typically include the following measures: larviciding, source reduction, insecticide spraying, and population awareness and training. Currently, this insecticide-based method continues to be used widely across the Maldives on both tourist islands and neighbouring local islands. The long-term usage of harsh, chemical insecticides not only causes damage to the non-target terrestrial species of beneficial insects but also causes detrimental effects to coral reefs and other marine life. Since the inception of tourism in the Maldives, it is a trend for each resort to hire a locally based pest control company. These companies enact harsh chemical insecticides on the natural ecosystem without understanding the detrimental effect the chemicals might cause to the terrestrial and aquatic environments. On average, there were 66% fewer species of insects recorded in the Maldives on islands with resorts than in the rest of the archipelago; this demonstrates the irresponsible usage of pesticides and the detrimental effects they have had against vital, beneficial insect species from the resort island ecosystems.

Since 2011, more than 20 different kinds of chemicals that have been registered for use in the Maldives. Over the years, these pesticides have been imported more frequently into the nation. In 2013 alone, 10.3 tonnes and 18.3 m^3 of pesticides were imported into the nation. Some of these chemicals are part of the vector management initiatives of the local islands and resorts.

Jahir et al. (2022) developed a new way to significantly reduce mosquito populations on small, tropical islands in the Maldives. The pioneering strategy employed was based on extensive larval source management (LSM) and mass trapping where, for the first time, mosquito populations on two small islands in the Maldives were nearly eliminated, and *Aedes* mosquitoes were eliminated from a third island for more than one year now. This technique utilizes human odours as bait in mosquito traps and extensive elimination of mosquito larvae in breeding sites.

Such suppression levels protect tourism, a vital source of revenue for the small island nation and its government, by eliminating the possibility of arboviral disease spreading to nearby communities. Additionally, the restoration of insect biodiversity, coral reefs, and marine life in these small, vulnerable island ecosystems would improve both human and environmental health when intensive pesticide usage (chemical fogging of insecticides) is stopped. The combination of mass trapping and intensive LSM offers an effective substitute for chemical control and achieves impact levels equivalent to modern genetic control techniques (which are vastly more expensive). However, given the successes in other parts of the world with the use of *Wolbachia*, the sterile insect technique or a combination of these tools is likely to eliminate mosquito populations from small islands in a short period of time, thereby permanently solving the risk of disease transmission, as well as nuisance biting. We strongly encourage the Maldives government to consider researching such options, as these may contribute to the fulfilment of its Strategic Action Plan (SAP) to significantly reduce the use of pesticides in the country.

The Maldives' unique geography, that of isolated tropical islands surrounded by the Indian Ocean, offers a unique opportunity for research, development, and implementation of these pioneering practices. Each community has the opportunity to employ sustainable and impactful techniques for mosquito suppression that support the biodiversity native to the island's ecosystem and eliminate the threat of mosquito-borne diseases from their island homes.

ACKNOWLEDGEMENTS

We would like to thank Julia May Neeson for her support in editing this chapter, and Mr. Sonu Shivdasani for providing us with the opportunity to conduct mosquito research on Soneva properties in the Maldives.

REFERENCES

Achee, N.L., Grieco, J.P., Vatandoost, H., Seixas, G., Pinto, J., Lee Ching, N.G., Ademir, J. M., Juntarajumnong, W., Corbel, V., Gouagna, C., Jean-Philippe, D., Logan, J.G., Orsborne, J., Marois, E., Devine, G.J. and Vontas J. (2019). Alternative strategies for mosquito-borne arbovirus control. *PLOS Neglected Tropical Diseases*, 13(3), e0007275. https://doi.org/10.1371/journal.pntd.0007275

Aitken, T.H.G., Tesh, R.B. and Beaty, B.J. (1979). Rosen Transovarial transmission of yellow fever virus by mosquitoes (*Aedes aegypti*). *American Journal of Tropical Medicine & Hygiene*, 28, 119–121.

Burt, F.J., Rolph, M.S., Rulli, N.E., Mahalingam, S. and Heise, M.T. (2012). Chikungunya: A re-emerging virus. *The Lancet*, 379, 662–671.

Campos, G.S., Bandeira, A.C. and Sardi, S.I. (2015). Zika virus outbreak, Bahia, Brazil. *Emerging Infectious Diseases*, 21(10), 1885–1886. https://doi.org/10.3201/eid2110.150847

Dudouet, P., Gautret, P., Larsen, C.S., Díaz-Menéndez, M., Trigo, E., von Sonnenburg, F., Gobbi, F., Grobusch, M.P., Malvy, D., Field, V., Asgeirsson, H., Souto, I.O., Hamer, D.H., Parola, P. and Javelle, E. (2020). Chikungunya resurgence in the Maldives and risk for importation via tourists to Europe in 2019–2020. A GeoSentinel case series. *Travel Medicine & Infectious Diseases*, 36, 101814. https://doi.org/10.1016/j.tmaid.2020.101814

Duffy, M.R., Chen, Tai-Ho., Thane Hancock, W., Powers, A.M., Kool, J.L., Lanciotti, R.S., Pretrick, M., Marfel, M., Holzbauer, S., Dubray, C., Guillaumot, L., Griggs, A. et al. (2009). Zika Virus Outbreak on Yap Island, Federated States of Micronesia. *The New England Journal of Medicine*, 360(24), 2536–2543. https://doi.org/10.1056/NEJMoa0805715

Girard, M., Nelson, C.B., Picot, V. and Gubler, D.J. (2020). Arboviruses: A global public health threat. *Vaccine*, 38(24), 3989–3994. https://doi.org/10.1016/j.vaccine.2020.04.011

Imad, H.A., Phadungsombat, J., Nakayama, E.E., Suzuki, K., Ibrahim, A.M., Afaa, A., Azeema, A., Nazfa, A., Yazfa, A., Ahmed, A., Saeed, A., Waheed, A., Shareef, F., Islam, M.M., Anees, S.M., Saleem, S., Aroosha, A., Afzal, I., Leaungwutiwong, P., Piyaphanee, W., Phumratanaprapin, W. and Shioda, T. (2021). Clinical Features of Acute Chikungunya virus infection in children and adults during an outbreak in the Maldives. *American Journal of Tropical Medicine & Hygiene*, 105(4), 946–954. https://doi.org/10.4269/ajtmh.21-0189

Iyengar, M.O.T. and Menon, M.A.U. (1955). Mosquitos of the Maldives Islands. *Bulletin of entomological Research*, 46(1), 1–10. https://doi.org/10.1017/s0007485300030716

Jahir, A., Kahamba, N.F., Knols, T.O., Jackson, G., Patty, N.F.A., Shivdasani, S., Okumu, F.O., Knols, B.G.J. (2022). Mass trapping and larval source management for mosquito elimination on Small Maldivian Islands. *Insects*, 13, 805. https://doi.org/10.3390/insects13090805

Jansen, C.C. and Beebe, N.W. (2010). The dengue vector Aedes aegypti: what comes next. *Microbes and Infection*, 12(4), pp. 272–279. doi:https://doi.org/10.1016/j.micinf.2009.12.011

Korhonen, E. M. (2016). Zika virus infection in a traveler returning from the Maldives, June 2015. Euro surveillance. *European Communicable Disease Bulletin*, 21(2), 5 pp. https://doi.org/10.2807/1560-7917.ES.2016.21.2.30107

Kraemer, M.U.G., Robert, C.R. Jr., Oliver, J.B., Jane, P.M., Gilbert, M., Pigott, D.M., Yi, D., Johnson, K., Earl, L., Marczak, L.B., Shirude, S., Weaver, N.D., Bisanzio, D., Perkins, T.A., Lai, S., Lu, X., Jones, P., Coelho, G.E., Carvalho, R.G., Wim Van Bortel, Marsboom, C., Hendrickx, G., Schaffner, F., Moore, C.G., Nax, H.H., Bengtsson, L., Wetter, E., Tatem, A.J., Brownstein, J.S., Smith, D.L., Lambrechts, L., Cauchemez, S., Linard, C., Faria, N.R., Pybus, O.G., Scott, T.W., Liu, Q., Yu, H., William, W.G.R., Hay Simon, I. and Golding, N. (2019). Past and future spread of the arbovirus vectors *Aedes aegypti* and *Aedes albopictus*. *Nature Microbiology*, 4(5), 854–863. https://doi.org/10.1038/s41564-019-0376-y

Musso, D., Nilles, E. J. and Cao-Lormeau, V.-M. (2014). Rapid spread of emerging Zika virus in the Pacific area. *Clinical Microbiology and Infection*, 20(10), O595–6. https://doi.org/10.1111/1469-0691.12707

Näslund, J. et al. (2021). Emerging Mosquito-Borne Viruses Linked to Aedes aegypti and Aedes albopictus: Global Status and Preventive Strategies. *Vector Borne and Zoonotic Diseases*, 21(10), 731–746. https://doi.org/10.1089/vbz.2020.2762

Rezza, G. (2014) Dengue and chikungunya: long-distance spread and outbreaks in naïve areas. *Pathogens and Global Health*. [Online] 108(8), 349–355. [online]. http://doi.org/10.1179/2047773214y.0000000163

Samir, H., Mohamed, N., Saleem, S., and Tjandra, A. (2021). Unburdening the Poor: Elimination of Lymphatic Filariasis in the Maldives. https://doi.org/10.1007/978-981-16-5566-1_5

Thangamani, S. et al. (2016). Tesh Vertical transmission of Zika virus in *Aedes aegypti* mosquitoes. *American Journal of Tropical Medicine & Hygiene*, 95, 1169–1173.

WHO (2014). Zika virus outbreaks in the Americas. *Weekly Epidemiological Records*, 90(45), 609–610.

WHO SEARO (2018) Elimination of Lymphatic Filariasis - Maldives. Youtube. Available at: https://www.youtube.com/watch?v=UypQKyRTJ58 (Accessed: May 23, 2023).

Yoosuf, A.A., Ibrahim Shiham, Ahmed Jamsheed Mohamed, Geela Ali, J.M. Luna, R. Pandav, G.N. Gongal, Ananda Nisaluk, Richard G. Jarman, Robert V. Gibbons (2009). First report of chikungunya from the Maldives. *Transactions of the Royal Society of Tropical Medicine and Hygiene*, 103(2), 192–196. https://doi.org/10.1016/j.trstmh.2008.09.006

17 Mosquitoes of Pakistan

B.K. Tyagi
Department of Biosciences, University Institute of Biotechnology, Chandigarh University, Mohali, India

17.1 INTRODUCTION

Pakistan, situated between latitudes 24°–37° north and between the longitudes 62°–75° east, occupies a rather newly crafted land position in the north-west of India. Much before Pakistan was still to be born in 1947, as a nation in two parts out of the Indian mainland, Christophers (1933) and Barraud (1934) had already catalogued and taxonomically described comprehensively mosquitoes from British India, including Ceylon (now Sri Lanka) and Burma (now Myanmar), in the world famous series *The Fauna of British India*, which, despite being incomplete and out-of-date as of now, and besides containing many unclear interpretations, still remains the standard reference work of mosquito identification for a serious researcher (Tribe Anophelini included 43 species with 9 varieties, while Tribes Megarhinini and Culicini possessed 252 species with 17 varieties of the mosquitoes). After Bangladesh emerged as a new country out of Pakistan, leaving behind only western Pakistan as its only land, in 1971, the number of species described in *The Fauna of British India* volumes actually shrank to merely 34 species and 1 variety of anophelines and 66 species, 3 sub-species, and 1 variety of culicine mosquitoes distributed in localities now constituting Pakistan. Save for the anopheline mosquitoes, the other mosquito fauna of Pakistan is poorly known (Amerasinghe et al., 2002; Manzoor et al., 2020). At present, there are 134 species of mosquitoes found in Pakistan belonging to Anophelinae and Culicinae subfamilies (Aslamkhan, 1971,1972).

17.2 MOSQUITO CHECKLIST OF PAKISTAN

Aslamkhan (1971, 1972) inventorized *de novo* the extant taxa of mosquitoes in Pakistan and listed 134 species of which 24 nominal taxa were described from the present-day Pakistan. Of this number, 14 were at that time considered valid species and sub-species and 10 were understood to be synonyms. This however lacked modern taxonomic interpretations and identification keys to species of the major genera, except for regional keys to female Anopheles that include 20 species and keys and descriptions to adult and larvae of subgenus Culex that include 10 species from Pakistan (Glick 1992; Harbach and Knight, 1980). In case of anophelines, too, despite of a few local reports, there are no recently published keys to all the known anopheline mosquitoes of that country, albeit malaria being an important public health problem and the fact that an unpublished pictorial key to female anophelines of Pakistan developed by Bashir Ahmad (a senior technician at the malaria entomology unit in Lahore) are still available. Whatsoever, keys to the immature larval stages were never prepared for the Pakistani anophelines before and Amerasinghe et al. (2002) arduously filled that gap.

17.3 MOSQUITO-BORNE DISEASES IN PAKISTAN

The main mosquito-borne diseases in Pakistan are malaria, dengue, chikungunya, and West Nile fever.

DOI: 10.1201/9781003327028-21

17.3.1 Malaria

Malaria in Pakistan has been a major public health issue because every tear thousands of people fall sick and hundreds of them die due to the infection, particularly the resistant *P. falciparum* malaria. Major mosquito species involved in transmission of disease include *An. culicifacies* s.l. Giles in rural Pakistan, and *An. stephensi* Liston in mostly urban though rural areas are also affected by this species (Amerasinghe et al., 2002). In the absence of long-term studies and evidence-based documentation, some other mosquitoes such as *An. fluviatilis* s.l. James, *An. annularis* s.l. van der Wulp, *An. subpictus* s.l., *An. superpictus*, and *An. pulcherrimus* Theobald are still suspected to be vectors in different parts of the country. Therefore, the need for up-to-date taxonomic keys hardly needs to be re-emphasized in Pakistan. It is noteworthy that many of the species in Pakistan have species complexes, hence the need for a well-documented key become all the more significant.

17.3.2 Dengue

Dengue is common in Pakistan, with seasonal outbreaks taking a huge toll, such as during the year 2019 when more than 47 people died and more than 30,000 were infected, or in 2021, when, according to the World Health Organization (WHO), 48,906 cases, including 183 deaths, were reported in Pakistan (Khan and Azeem, 2022). The dengue situation is highly alarming in Pakistan due to the rising trend of cases (Arshad et al., 2022). Transmitted by *Aedes aegypti* and *Ae. albopictus*, all the four serotypes, DENV1-4, exist in different areas of the country (Ali et al., 2017).

17.3.3 Chikungunya

The chikungunya virus (CHIKV) was first reported in Pakistan in 1983 in four rodents and one human being (Darwish et al., 1983). However, the first human case of chikungunya in Pakistan was reported in 2011 in Lahore (Afzal et al., 2015). Frequent outbreaks have been reported in Karachi, Pakistan, between 2016 and 2017 (Mallhi et al., 2017).

17.3.4 Japanese Encephalitis

Japanese encephalitis virus (JEV) transmission in Pakistan was documented in 1980s onwards, although it is reported to be occurring in fewer than 1% of JEV infections, often with catastrophic sequelae including neuropsychiatric disability and eventually death (Imran et al., 2022). *Culex tritaeniorhynchus* and related species are reported to be vectors for the JEV (Fatima et al., 2020).

17.3.5 West Nile

West Nile virus (WNV), a serious arboviral infection, is getting increasingly serious in Pakistan where seropositivity for the virus has been detected as high as 65% of the samples collected (Zohaib et al., 2014). In recent times, Pakistan has experienced many outbreaks of WNV (Waseem et al., 2022; Khan et al., 2018). This is one emerging mosquito-borne infection in Pakistan which warrants urgent investigation, with particular reference mosquito vector species.

17.4 FINAL REMARKS

Pakistan shares the highest burden of major vector-borne diseases. such as malaria, dengue, chikungunya, Japanese encephalitis, and West Nile. It is noteworthy that malaria, which is at the verge of elimination in many countries including India, is still continuing to be a leading cause of morbidity and mortality in Pakistan under the impact of changing climate (Jabeen et al. 2022). The poor health facilities and health-care delivery system in Pakistan are a major reason for the failure to eliminate various vector-borne diseases, which need special attention with focused research.

REFERENCES

Afzal, M.F., Naqvi, S.Q., Sultan, M.A. and Hanif, A. (2015). Chikungunya fever among children presenting with nonspecific febrile illness during an epidemic of dengue fever in Lahore, Pakistan. *Merit Research Journal of Medical Science*,3(3),69–73.

Ali, H., Alvi, A., Fatima, S., Zafar, F., Naveed, S., Khan, K., Ali, U., Tariq, A., Naqvi, G.R. and Mallick, N. (2017). Dengue fever in Pakistan, episodes of epidemic to endemic: treatment challenges, prevention and current facts. *Journal of Bioequivalent Availablity*, 9, 473–476.

Amerasinghe, F.P., Mukhtar, M. and Herrel, N. (2002). Keys to the Anopheline Mosquitoes (Diptera: Culicidae) of Pakistan, *Journal of Medical Entomology* 39(1): 28–35. https://doi.org/10.1603/0022-2585-39.1.28

Arshad, T., Wajahat, A., Jabeen, A., Ali, S.H. (2022). Malaria and dengue outbreaks during a national disaster in Pakistan: A rising concern for public health. *Journal of Global Health*, 3,12:03076. https://doi.org/10.7189/jogh.12.03076

Aslamkhan, M., 1971. The Mosquitoes of Pakistan. I. A Checklist. *Mosquito Systematics Newsletter* 3(4): 147–159.

Aslamkhan, M. (1972). The Mosquitoes of Pakistan. II. Mosquitoes originally described from Pakistan. *Mosquito Systematics* 4(4): 98–102.

Barraud, P.J. (1934). *The fauna of British India, including Ceylon and Burma. Diptera: Family Culicidae; Tribes Megarhinini and Culicini. vol. 5.* Taylor & Francis, London, 463 pp.

Christophers, S.R. (1933). *The Fauna of British India, including Ceylon and Burma. Diptera: Family Culicidae; Tribe Anophelini. vol. 4.* Taylor & Francis, London, 271 pp.

Darwish, M.A., Hoogstraal, H., Roberts, T.J., Ahmed, I.P., Omar, F. (1983). A sero-epidemiological survey for certain arboviruses (Togaviridae) in Pakistan, *Transactions of The Royal Society of Tropical Medicine and Hygiene*, 77(4), Pages 442–445. https://doi.org/10.1016/0035-9203(83)90106-2

Fatima, T., Rais, A., Khan, E., Hills. S.L., Chambers. T.V., Hotwani, A., Qureshi, S., Shafquat, S., Malik, S., Qamar, F., Mir, F., Marfin, A.A., Zaidi, A., Khowaja, A.R. and Shakoor, S. 2020. Investigation of Japanese encephalitis virus as a cause of acute encephalitis in southern Pakistan, April 2015-January 2018. *PLoS One*, 15(6), e0234584. https://doi.org/10.1371/journal.pone.0234584

Glick, J. I. 1992. Illustrated key to the female Anopheles of southwestern Asia and Egypt (Diptera: Culicidae). *Mosq. Syst.* 24: 125Ð153

Harbach, R.E. and Knight, K.L. (1980) *Taxonomists' Glossary of Mosquito Anatomy*. Plexus Publishing, Marlton, New Jersey.

Imran, M., Ye, J., Saleemi, M.K., Iqra Shaheen, Ali Zohaib, Zheng Chen, Shengbo Cao. 2022. Epidemiological trends of mosquito-borne viral diseases in Pakistan. *Animal Diseases*, 2, 5. https://doi.org/10.1186/s44149-021-00034-4

Jabeen, A., Ansari, J.A., Ikram, A., Khan, M.A. and Safdar, M., 2022. Impact of climate change on the epidemiology of vector-borne diseases in Pakistan. *Global Biosecurity*, 4(1). https://doi.org/10.31646/gbio.163

Khan, E., Barr, K.L., Farooqi, J.Q., Prakoso, D., Abbas, A., Khan, Z.Y., Ashi, S., Imtiaz, K., Aziz, Z., Malik, F., Lednicky, J.A. and Long, M.T. (2018). Human West Nile Virus Disease Outbreak in Pakistan, 2015–2016. Front. *Public Health (Sec. Infectious Diseases): Epidemiology and Prevention*, 6(20), 1–11. https://doi.org/10.3389/fpubh.2018.00020

Khan, U. and Azeem, S. (2022). The rising toll of dengue cases in Pakistan every year: An incipient crisis. *Annals of Medicine & Surgery (London)*, 76, 103549. http://doi.org/10.1016/j.amsu.2022.103549

Mallhi, T.H., Khan, Y.H., Khan, A.H., Tanveer, N. and Qadir, M.I. (2017). First chikungunya outbreak in Pakistan: a trail of viral attacks. *New Microbes New Infections*, 19:13–14. http://doi.org/10.1016/j.nmni.05.008

Manzoor, F., Shabbir, R., Sana, M., Nazir, S. and Khan, M.A. (2020). Determination of Species Composition of Mosquitoes in Lahore, Pakistan. *Journal of Arthropod Borne Diseases*, 14(1), 106–115. http://doi.org/10.18502/jad.v14i1.2717

Waseem, S., Shaikh, T.G., Ahmed, S.H., Usama, R.M., Asghar, M.S. and Tahir, M.J. (2022). West Nile virus outbreak in Pakistan: Another potential public health threat in the pipeline? *Annals of Medicine & Surgery (London)*, 77, 103663. http://doi.org/10.1016/j.amsu.2022.103663

Zohaib, A., Saqib, M., Beck, C., Hussain, M.H., Lowenski, S., Lecollinet, S., Sial, A., Asi, M.N., Mansoor M.K., Saqalein, M., Sajid, M.S., Ashfaq, K., Muhammad, G., Cao, S. (2014). High prevalence of West Nile virus in equines from the two provinces of Pakistan. *Epidemiology & Infection*, 143(9), 1931–5. http://doi.org/10.1017/S0950268814002878

18 Mosquitoes of Nepal

Meghnath Dhimal
Nepal Health Research Council, Kathmandu, Nepal

Ishan Gautam
Tribhuvan University, Swayambhu, Kathmandu, Nepal

Buddhi Ram Oli
Institute of Science and Technology, Tribhuvan University, Kathmandu, Nepal

Pramod Shrestha
Nepal Health Research Council, Kathmandu, Nepal

Reshma Tuladhar
Institute of Science and Technology, Tribhuvan University, Kathmandu, Nepal

Ruth Mueller
Institute of Tropical Medicine, Entomology Unit, Antwerp, Belgium

18.1 INTRODUCTION

Nepal, a mountainous country with altitudinal variations ranging from the lowlands 60 m above mean sea level (asl) to the peak of Mount Everest (8,848 m asl), is now endemic to at least six major vector-borne diseases (VBDs). These include malaria, lymphatic filariasis, Japanese encephalitis, visceral leishmaniasis, chikungunya, and dengue fever, all caused by pathogens transmitted by insect vectors. The pathogens of those VBDs, except visceral leishmaniasis, are transmitted to humans by infected mosquitoes. In contrast, visceral leishmaniasis is transmitted by infected sandflies. The first mosquito survey in Nepal was initiated in 1952 to identify the vectors of malaria. Recent entomological studies in Nepal showed the presence of disease-transmitting vectors at an elevation of at least 2,100 m asl. In this chapter, we describe the mosquito fauna in Nepal with reference to their (i) inventory, (ii) the reported evidence of vector species for medically relevant pathogens, (iii) VBDs, and (iv) the impact of climate change.

18.2 A BRIEF INVENTORY OF MOSQUITOES IN NEPAL

A systematic study of mosquitoes in Nepal began in the 1950s with the first record of some anopheline mosquitoes by Puri (1955) and Peters et al. (1955) from the Rapti valley area of Nepal. Thereafter, Peters and Dewar (1956) recorded some culicine species for the first time from Nepal,

DOI: 10.1201/9781003327028-22

including *Aedes albopictus*. The number of anopheline species increased to 31 species by Brydon et al. (1961). Joshi et al. (1964) contributed to the mosquito inventory of a new anopheline species, i.e., *Anopheles kochi*, and in the following year, Joshi et al. (1965) reported 59 culicine species with 28 new country records including *Aedes albopictus, Culex quinquefasciatus, Cx. fuscocephala, Cx. tritaeniorhynchus, Cx. vishnui, Cx. fuscocephala, Cx. gelidus, Cx. bitaeniorhynchus*, and *Cx. whitmorei*. Additional reports of *Aedes albopictus* from Nepal were added by Shrestha (1966). Shrestha (1966) reported a total of 97 mosquito species from Nepal, including 36 anophelines and 61 culicines. With 13 new country records of mosquito species belonging to the genera *Aedes*, *Culex*, *Mimoyia* and *Uranotaenia*, 70 culicine mosquito species were recorded in Nepal by Pradhan and Darsie (1989). Additional mosquito fauna was not recorded until 1989, when Pradhan and Darsie (1989) discovered 13 new country records. Also, *Aedes pulchriventer, Ae. subalbopictus, Anopheles fragilis, An. dravidicus, Armigeres durhami, Culex infula*, and *Cx. pseudovishnui* were partially the result of new collections reported by Shreedhar Prasad Pradhan in 1988. An extensive review of the mosquito fauna of Nepal was published by Darsie and Pradhan (1990), with 132 mosquito species belonging to 15 genera and 26 subgenera registered, along with their identification key with biology and bionomics. Darsie et al. (1991) recorded four new mosquito species for Nepal, and they revised the key of Darsie and Pradhan (1990) and deleted two species, along with some taxonomic modifications of both. The work was followed by Darsie et al. (1992), and they declared 18 species as new country records. Then, the number of mosquito species in Nepal reached 150 belonging to 15 genera and 27 subgenera.

Similarly, Darsie et al. (1993) added a genus *Aedeomyia* and two subgenera, *Aedeomyia* and *Tripteroides*, and 14 new mosquito species, with a total record of 164 mosquito species encompassing 16 genera and 29 subgenera. After a year, Darsie et al. (1994) made a substantial contribution to the mosquito inventory. They added four *Aedes* species, one *Anopheles* species, and five *Culex* species from the higher altitude (3,738 m asl) for the first time in Nepal. The species were *Aedes dissimilis, Ae. pseudotaeniatus, Ae. pulchriventer, Ae. shortti, Anopheles baileyi, Culex jacksoni, Cx. theileri, Cx. vagans, Cx. pallidothorax*, and *Cx. viridiventer* collected from the Mustang district. After a couple of years, six new mosquito species were added to the mosquito inventory of Nepal by Darsie et al. (1996); among them, five species belonging to the genus *Aedes* and one to *Heizmannia*, respectively. In previous works, the genera *Aedeomyia* and *Tripteroides* remained unidentified at the species level, and *Aedes aegypti* did not occur in Nepal then (Scholte et al., 2012; Walther et al., 2017). Therefore, based on previously published records, 16 genera, 29 subgenera, and more than 181 species are known in Nepal to date.

This is the complete information based on Darsie and Pradhan (1990), Darsie et al. (1991, 1992, 1993, 1996), and Darsie et al. (1994) regarding mosquito species previously reported in Nepal. Given the number of genera, subgenera, and species, the existing knowledge on this field is, however, still limited. Further work is needed to update this information on mosquito biodiversity, which is of particular importance in light of the climate crisis that provokes alterations in the distribution patterns of mosquito fauna and the establishment of new vector mosquito species in Nepal. Morphologically based taxonomy coupled with molecular methods for species identification will explicitly depict the updated picture in the years to come.

18.3 MOSQUITO VECTORS OF MEDICALLY RELEVANT PATHOGENS

Historically, several mosquito species are known as vectors for arboviruses and parasites of human and veterinary medical relevance. A large number of mosquito species have been listed in Nepal as vectors/potential vectors of human and animal pathogens by Darsie and Pradhan (1990). The distribution of vector mosquito species, which are regularly observed in Nepal, is summarized in Table 18.1.

The vectors of chikungunya and dengue viruses are *Ae. aegypti* and *Ae. albopictus*, which have been reported from at least 2,100 m above mean sea level (asl) in Nepal (Dhimal et al., 2015). *Aedes aegypti* and *Ae. albopictus* are widely recognized to be responsible for the spread of the dengue virus.

TABLE 18.1
Mosquito Species in Nepal with Reported Evidence to Be a Vector of Medically Relevant Pathogens

Sl. No.	Genera	Subgenera	Species	Reported Evidence to Be a Vector of Medically Relevant Pathogens	Geographic Distribution in Nepal	First Record in Nepal	Sources
1.	*Aedes* Meigen	*Aedimorphus* Theobald	*Aedes pallidostiatus* (Theobald)	Females have been found naturally infected with larvae of *Brugia malayi* (Reinert, 1973).	Janakpur: Mahotttari, Pipara; Mahakali: Kanchanpur, Haldu Khal; Kosi: Morang, Lohana	Pradhan and Darsie (1989)	Darsie and Pradhan (1990); Reinert (1973)
2.			*Aedes pipersalatus* (Giles)	Larvae of *Burgia malayi* have been recovered from females (Reinert, 1973).	Mahakali: Kanchanpur, Badaipur, Musepani	Pradhan and Darsie (1989)	Darsie and Pradhan (1990) Reinert (1973)
3.		*Stegomyia* Theobald	*Aedes aegypti* (Linnaeus)	Primary vector of dengue viruses	Terai regions of Nepal, Kathmandu valley	Gautam et al. (2009)	WHO (2006) Malla et al. (2008) Gautam et al. (2009)
4.			*Aedes albopictus* (Skuse)	Important vectors of dengue viruses	Kosi: Sunsari, Dharan; Morang, Biratnagar; Kathmandu, Halchowk; Swayambhunath; Mahakali: Kanchanpur, Musepani, Khutepani; Janakpur: Sindhuli, Ranibas; Narayani: Makawanpur, Hetaura	Peters and Dewar (1956)	Peters and Dewar (1956) Darsie and Pradhan (1990)
5.	*Anopheles*	*Anopheles*	*Anopheles barbirostris* Van der Wulp	Recognized as vector of human filariasis in Indonesia and has been found naturally infected with JE virus in West Bengal (Rao, 1981).	Kosi: Sunsari, Tarahara	Brydon et al. (1961)	Darsie and Pradhan (1990) Brydon et al. (1961) Rao (1981)
6.		*Cellia* Theobald	*Anopheles aconitus* Doenitz	This species is an important vector of malaria-causing parasites in Indonesia (Rao, 1981).	Kosi: Sunsari, Tarahara	Peters et al. (1955)	Darsie and Pradhan (1990)
7.			*Anopheles annularis* van der Wulp	This species has been found naturally infected with malaria parasites in the Terai of Nepal (Shrestha and Parajuli, 1980).	Abundant in the cultivated areas of the Terai and inner Terai	Puri (1955)	Darsie and Pradhan (1990)
8.			*Anopheles culicifacies* Giles	It is a major vector of malaria in the lowlands of India and a suspected vector in the cultivated Terai of Nepal (Shrestha et al., 1988).	Common in Terai and inner Terai areas of the country	Puri, 1955	Darsie and Pradhan (1990)

(*Continued*)

TABLE 18.1
(Continued)

Sl. No.	Genera	Subgenera	Species	Reported Evidence to Be a Vector of Medically Relevant Pathogens	Geographic Distribution in Nepal	First Record in Nepal	Sources
9.			*Anopheles dirus* Peyton and Harrison	It is an important vector of malaria parasites (Darsie and Pradhan, 1990).	Mechi: Jhapa, Suryanagar	Shrestha (1966)	Darsie and Pradhan (1990)
10.			*Anopheles fluviatilis* James	This species is one of the most efficient vectors of Plasmodium species in the Indian subcontinent. It is one of the principal malaria vectors in the foothills, lower mountains, and high mountain valleys up to 1,344 m asl in Nepal (Darsie and Pradhan, 1990).	Kosi: Tarahara; Narayani: Makawanpur, Hetaura; All Terai and inner Terai districts and up the mountain valleys to at least 1,344 m	Puri (1955)	Darsie and Pradhan (1990)
11.			*Anopheles jeyporiensis* James	The species is an important vector of malaria in Indochina. It has also been found infected with *Wuchereria bancrofti* (Harrison, 1980).	Throughout the Terai, inner Terai and lower hills	Brydon et al. (1961)	Darsie and Pradhan (1990)
12.			*Anopheles maculatus* Theobald	It is an important vector of malaria in Malaysia (Rao, 1981).	Occurs more commonly in the mountains all over the country and much less abundant in the lowland areas	Puri (1955)	Darsie and Pradhan (1990)
13.			*Anopheles minimus* Theobald	*Anopheles minimus* has been a primary vector of malaria in Nepal (Darsie and Pradhan, 1990).	This species is confined to the forest and forest fringe in the Terai and inner Terai belts across the whole country	Puri, 1955	Darsie and Pradhan (1990)
14.			*Anopheles pallidus* Theobald	It was found infected with malaria parasites and is considered a secondary malaria vector in Bihar State, India (Rao, 1981).	Widely distributed in the lowland but never abundant	Brydon et al. (1961)	Darsie and Pradhan (1990)
15.			*Anopheles philippinensis* Ludlow	There is convincing evidence that this species serves as a vector of malaria parasites but not in all parts of its range (Rao, 1981).	Kosi: Morang, Khokse, Narayani: Bara, Nizghard; Janakpur: Sindhuli, Khutepani. Terai region – Rupandehi east to Jhapa	Brydon et al. (1961)	Pradhan, and Brydon, (1960) Darsie and Pradhan (1990)

16.			*Anopheles stephensi* Liston	This species is a major vector of malaria in India, Pakistan, Iran and Iraq (Darsie and Pradhan, 1990).	Tanahu, Makawanpur, Mahottari, and Morang	Brydon et al. (1961)	Darsie and Pradhan (1990)
17.			*Anopheles varuna* Iyengar	It is considered a secondary vector of malaria parasites in parts of India (Rao, 1981).	Widely distributed in the Terai and inner Terai (Shrestha, 1966)		
18.			*Anopheles willmori* (James)	The altitude in this valley was 3,170 m, and this is the highest known record of malaria transmission. Pradhan et al. (1970).	Janakpur, Sindhuli, Ranibas, Kanchanpur, Musepani (Darsie and Pradhan, 1990)		
19.	*Culex* Linnaeus	*Culex* Linnaeus	*Culex fuscocephala* Theobald	Japanese encephalitis has been isolated from this mosquito in Thailand (Gould et al., 1962) and is a suspected vector in Nepal (Darsie and Pradhan, 1990).	Bagmati: Laliptur, Pharping, Taudaha, Kathmandu, Swyamphunath, Kalimati; Kosi: Sunsari, Tarahara; Narayani: Makawanpur, Hetaura	Peters and Dewar (1956)	Darsie and Pradhan (1990)
20.			*Culex gelidus* Theobald	JE virus has been isolated from *Cx. gelidus* in Thailand (Gould et al., 1962).	Bagmati: Kathmandu, Guheswari; Kosi: Morang, Murgatola, Baukajhora, Tarahara; Narayani: Bara, Nizgarh; Makawanpur, Hetaura	Peters and Dewar (1956)	Darsie and Pradhan (1990)
21.			*Culex quinquefasciatus* Say	The species is the principal vector of *Wuchereria bancrofti* in Nepal, which is within the endemic zone of filariasis (Jung, 1973).	Bagmati: Lalitpur, Taudaha, Kathmandu, Dilli Bazar, Kalimati; Kosi: Sunsari, Dharan, Tarahara; Narayani: Makawanpur, Chisapani, Hetaura; Janakpur: Sindhuli, Khutepani	Peters and Dewar (1956)	Darsie and Pradhan (1990)
22.			*Culex theileri* Theobald	This species has been found naturally infected with Sindbis and West Nile viruses (McIntosh et al., 1967).	Bagmati: Lalitpur, Taudaha, Swayambhunath	Joshi et al. (1965)	Darsie and Pradhan (1990)

(*Continued*)

TABLE 18.1
(Continued)

Sl. No.	Genera	Subgenera	Species	Reported Evidence to Be a Vector of Medically Relevant Pathogens	Geographic Distribution in Nepal	First Record in Nepal	Sources
23.			*Culex tritaeniorhynchus* Giles	This species is a major vector of Japanese encephalitis virus in many parts of the Oriental region, including Nepal (Hammon et al., 1949; Hale et al., 1957; Buescher et al., 1959; Reuben, 1971a, 1971b; Leake et al., 1986).	Narayani: Bara, Nizgarh, Chitwan, Tarpti Valley; Bagmati: Kathmandu, Kalimati, Kathmandu; Kosi: Sunsari, Tarahara	Joshi et al. (1965)	Darsie and Pradhan (1990)
24.			*Culex vishnui* Theobald	Japanese encephalitis virus has been isolated from this species (Reuben, 1971a, 1971b; Sirivanakarn, 1976).	Bagmati: Kathmandu, Balaju, Kalimati; Narayani: Chitwan, Rapti valley, Makawanpur, Hetaura; Kosi: Sunsari, Tarahara	Peters and Dewar (1956)	Darsie and Pradhan (1990)
25.	*Mansonia* Blanchard	*Mansonioides* Theobald	*Mansoniaannulifera* (Theobald)	This species is an important vector of *Brugiamalayi* in India (Burton, 1960).	Lumbini: Kapilvastu, Taulihawa; Janakpur: Mahottari, Pipara	Joshi et al. (1965)	Darsie and Pradhan (1990)
26.			*Mansoniauniformis* (Theobald)	Major vector of filariasis in many parts of its range, i.e., Japan, Ryukyu Islands, Afrotropical, Oriental, and Australasian regions (Delfinado, 1966).	Kosi: Morang, Baukajhora; Lumbini: Kapilvastu, Taulihawa; Janakpur: Mohattari, Pipara; Narayani: Makawanpur, Nayagaon	Peters and Dewar (1956)	Darsie and Pradhan (1990)

The primary vector of the dengue virus, *Ae. aegypti*, which has its origin in tropical Africa, was identified in five major urban areas of the Terai region bordering India, i.e., Biratnagar (Morang), Birganj (Parsa), Bharatpur (Chitwan), Tulsipur (Dang), and Nepalganj (Banke), during the entomological surveillance in Japanese encephalitis endemic districts after the first dengue outbreak in 2006 in Nepal (WHO, 2006; Malla et al., 2008). From the Kathmandu valley, *Ae. aegypti* was recorded for the first time in 2009 (Gautam et al., 2009). The secondary vector of the dengue virus, the Asian tiger mosquito *Ae. albopictus*, was described in Nepal initially from Hetauda (Makawanpur) (Peters and Dewar, 1956) and in the following years from Dharan, Biratnagar, Janakpur, Sindhuli, Ranibas, Hetaura, Halchowk, Swayambhunath, Kathmandu, Kanchanpur, Musepani, and Khutepani (Joshi et al., 1965; Shrestha, 1966; Huang, 1979). *Aedes albopictus* was collected by Shreedhar Prasad Pradhan in 1961, 1963, 1987, 1988, and 1989 and by Richard F. Darsie in 1987. Nowadays, the chikungunya and dengue disease and its vectors *Ae. albopictus and Ae. aegypti*, are found in 77 districts of Nepal, including the mountain districts. In general, both *Aedes* species are commonly observed up to an altitude of 1,350 m asl and are sparsely distributed from 1,700 m asl to 2,100 m asl in Nepal (Dhimal et al., 2015), depicting the ongoing expansion of their species distribution ranges towards the highlands of Nepal.

Certain species of *Anopheles* and *Culex* have been documented as vectors for diseases such as malaria, the Japanese encephalitis, and lymphatic filariasis (Table 18.1). *Anopheles minimus*, *An. fluviatilis*, *An. maculatus*, *An. Willmori* [as *maculatus* var. *willmori*], and *An. annularis* have been incriminated in malaria transmission. In Nepal, four members of the *An. maculatus* complex (*maculatus*, *willmori*, *pseudowillmori*, and *dravidicus*) occur (Rattanarithikul and Green, 1987).

The mosquito species *Culex tritaeniorhynchus* has been reported to be the principal vector of the Japanese encephalitis virus in many parts of Asia, including Nepal, and it was first recorded in Nepal in 1965 with a distribution ranging from the lowland to hill regions, including the Kathmandu valley. The mosquito species *Culex quinquefasciatus*, the principal vector of *Wuchereria bancrofti* microfilaria in South Asia, was first recorded in Nepal in 1956 and found to occur within the lymphatic filariasis endemic zones of this country (almost throughout the country). The vectors of malaria, lymphatic filariasis, Japanese encephalitis, dengue, and chikungunya are reported from at least 2,100 m asl in Nepal (Dhimal et al., 2014).

Table 18.1 shows mosquito species in Nepal with reported evidence to be vectors of medically relevant pathogens. Next to basic taxonomic information on a given mosquito species, field reports on the infection with or transmission of arboviruses and parasites of medical relevance, the first detection of a given vector species, and its geographic distribution in Nepal are additionally given.

Source: This table was originally formed by the authors. *Note*: The source of information is adopted from Darsie and Pradhan (1990).

Reported evidence to be a vector of medically relevant pathogens are additionally given in brackets. References for first records for Nepal are provided in the separate column.

18.4 MOSQUITO-BORNE DISEASES IN NEPAL

The spatio-temporal distribution of mosquito-borne diseases in Nepal, as well as the Hindu-Kush Himalayan (HKH) region and their characteristics, are described elsewhere (Dhimal et al., 2015; Phuyal et al., 2020).

A heterogeneous spatial distribution and declining trend of malaria incidence is observed in Nepal. Though a higher incidence was observed in southern districts bordering India in the past, in the later years, sporadic outbreaks of malaria were reported from the hilly districts of Nepal (Dhimal et al., 2014). The distribution of the disease, which was previously confined to the forest and forest fringe regions of the Terai lowlands and the so-called Inner Terai valleys and hills (<1,200 m asl), is now observed to have extended to the hills and mountains (>2,000 m asl), and frequent outbreaks of malaria are reported from districts of hills and mountains. Forty-four species of mosquitoes have been recorded in Nepal, and out of these, seven have been incriminated in malaria transmission

(Dhimal et al., 2014). During the last decade, the incidence of confirmed malaria cases has declined significantly in Nepal following the introduction of the free distribution of long-lasting insecticide-impregnated bed nets (LLINs), insecticide residual spray, and artemisinin combination therapy (ACT) for the treatment of *Plasmodium falciparum*- associated malaria.

Malaria risk stratification was conducted in 2021 to design a programme intervention tailored to the evolving epidemiology of malaria in the country. The findings revealed that malaria is concentrated within some wards of rural municipalities, while other wards remain relatively free of malaria. This stratification shows that the 22 wards at high risk were scattered across 9 districts, and 69 wards at moderate risk were across 12 districts. Malaria transmission is concentrated in the provinces Sudurpaschim and Karnali and accounts for approximately 81% of the high-risk burden and around 94% of the moderate risk burden. Malaria transmission has reached a low level of endemicity in most of the Terai regions (plain lands), but malaria infection is increasingly being detected even in upper hilly river valleys, which were earlier classified as "No Malaria" areas in Nepal. Nepal is also part of the global E-2025 countries with the aim of attaining "Malaria Elimination in Nepal by 2025."

The first case of dengue was reported in Nepal in 2004, and the first local transmission of dengue virus (DENV) in Nepal was confirmed at the beginning of an outbreak in 2006 (August–November) in lowland urban areas of 11 districts (EDCD, 2019). A rapid expansion of DENV being witnessed in Nepal with reports of more than 15,000 cases in 2019 and more than 55,000 cases throughout the country in 2022. In 2022, approximately 55,000 cases and 88 deaths from dengue were reported. The majority of cases of dengue in the 2022 outbreak were reported from Bagmati Province, followed by Lumbini Province. The majority of cases of dengue in the 2024 outbreak were also reported from Bagmati Province. Studies conducted in Nepal show the circulation of all four serotypes of DENV in Nepal (Malla et al., 2008).

The first case of chikungunya was reported in Nepal in 2013 (Pun et al., 2014), and subsequent studies confirmed the circulation of the chikungunya virus in Nepal (Pandey et al., 2015, 2017). However, there has not been systematic diagnosis and regular reporting of chikungunya cases in Nepal.

Japanese encephalitis virus was first recorded in Nepal in 1965. Although most reported Japanese encephalitis cases in Nepal were initially confined to 24 districts in the lowland Terai, Japanese encephalitis transmission has now been established in hill and mountain districts of Nepal, including the Kathmandu valley, which were previously considered non-endemic to this disease. Moreover, there are reports of spatial clusters of Japanese encephalitis incidences, with a shift from the Terai lowlands to hill and mountain regions after 2005, and the risk of Japanese encephalitis was also associated with paddy field configuration at the landscape level. Nepal introduced the Japanese encephalitis vaccine to accelerate Japanese encephalitis control, and the number of Japanese encephalitis cases is less than 30 per year.

Nepal is among the 73 countries listed by the WHO as endemic to lymphatic filariasis, and this disease has been detected in different topographical areas, ranging from an altitude of 80 m asl (the plain terrain) to 2,100 m asl (high hill areas; Dhimal et al., 2015). Comparatively, more lymphatic filariasis cases are reported from the Terai than from the hills, but the valleys and river basin areas of hilly districts have a high disease burden as well. Nepal is among the countries that has started mass drug administration in all endemic districts and is on track to achieve elimination status.

18.5 IMPACT OF GLOBAL WARMING ON THE EXPANSION OF MOSQUITO-BORNE DISEASES AND THEIR VECTORS IN NEPAL

The geographic range and life cycle of mosquitoes are sensitive to temperature changes and thus, ultimately, to climate change. The consequences of climate change and other non-climatic factors, such as rapid urbanization, poor waste disposal systems, expansion of roads and increasing vehicular movement, and travel of people, are the contributing factors for expanding the distribution of vectors and disease transmission towards previously considered non-endemic hill and mountain regions. The expansion is projected to shift to higher elevations. A systematic review shows that both the observed and projected climate are favourable for the transmission of mosquito-borne pathogens in the mountain regions of Nepal, which were previously considered non-endemic for these diseases (Dhimal et al., 2015). Short-term data reveal a clear association between climatic factors and VBDs; however, more robust studies are necessary to explore the long-term effects of climate change in the face of rapid environmental and socio-economic changes. Notably, shifts in the spatio-temporal distribution of VBDs have been observed in the HKH region as well (Phuyal et al., 2020), with climate change linked to the expansion of VBDs and their vectors in the area (Dhimal et al., 2021).

18.6 CONCLUDING REMARKS AND RECOMMENDATIONS FOR FURTHER RESEARCH

Studies on the mosquito inventory in Nepal provide valuable insights into the composition of mosquito fauna, with a primary focus on their geographical distribution as well as aspects of their morphology, physiology, ecology, and bionomics. However, the majority of studies carried out so far are based on the identification of mosquito species via distinct morphological traits, which requires extensive entomological training and expertise. To support the reliable identification of vector species by health professionals in the biodiverse country Nepal, a molecular barcoding pipeline for mosquito surveillance applicable in the field has been recently developed by Hartke et al. (2022). Combining morphological approaches with molecular barcoding and phenotyping will certainly improve our understanding species diversity, functional diversity, and phylogenetic diversity which will be new dimensions to mosquito biodiversity in Nepal. Further investigation on correlations between the genetic divergence and morphological characteristics is needed to uncover overlooked species, bridging the gap between what is previously known and what can be expected to be found. Furthermore, molecular approaches help us to better understand the adaptive potential of mosquitoes to climate warming (Kramer et al., 2023) and help to unravel the insecticide resistance and the vectoral capacity of mosquitoes that are yet to be explored in the vector species of Nepal.

Despite previous solid studies, the recent countrywide dengue outbreaks in Nepal highlight the need to fully register the countrywide mosquito and arboviral biodiversity, explore the vector competence to transmit pathogens of medically and veterinary relevance in multiple mosquito species, and identify the underlying ecological, environmental, and climatic factors of the vectorial capacity of mosquitoes. There is still scarce information available on the introduction and establishment of vector mosquitoes, which has increased the threat of emerging and re-emerging mosquito-borne diseases. *Aedes aegypti* and *Ae. albopictus* especially have increased their occurrence and geographical range and spread into new areas, and climate adaptation is ongoing (Kramer et al., 2021, 2023). In order to improve the assessment of mosquito-borne disease transmission dynamics in Nepal, the trends of geographical expansion (specifically into remote mountainous regions) in the context of environmental and climate change definitely merit further research, allowing a large step forward in resolving and unveiling the shifting of behavioural characteristics of vector species (e.g., host preference, diurnal activity pattern) using valuable, integrative approaches. Therefore, we highly recommend interdisciplinary research to develop new approaches for vector control to ultimately be able to eliminate mosquito-borne diseases from Nepal and elsewhere.

ACKNOWLEDGEMENTS

The authors wish to thank especially Keshav Luitel, Sunita Baral and all staff members of the Nepal Health Research Council, Kathmandu, Nepal, for their cooperation.

FUNDING: The authors were financially supported by the Directorate-General for Development Cooperation and Humanitarian Aid (DGD), Belgium, within the FA5 country programme Nepal, project,

Evidence-Based Vector-Borne Disease Management and Elimination of Visceral Leishmaniasis in Nepal. This research was further part of the project CLIMB: *The impact of rapid CLIMate change on the Biodiversity-health interface*, funded by the SOFI programme (Structural Research Funding) of the Institute of Tropical Medicine Antwerp supported by the Flemish Government, Science and Innovation, and the DGD FA5 Synergy Fund.

REFERENCES

Brydon, H.W., Joshi, G. and Pradhan, S. (1961). Anopheles distribution and its relation to the malaria eradication program in Central Nepal. *Mosquito News*, 21:102–105.

Buescher, E.L., Scherer, W.F., Rosenberg, M.Z., Gresser, I., Hardy, J.L. and Bullock, H.R. (1959). Ecological studies of Japanese encephalitis virus in Japan II. Mosquito infections. *American Journal of Tropical Medicine and Hygiene*, 8:651–677.

Burton, G.J. (1960). *Studies on the bionomics of mosquito vectors which transmit filariasis in India*. U.S. Tech. Coop. Mission to India, New Delhi, 145 pp.

Darsie, Jr. R. F. and Pradhan, S.P. (1990). The mosquitoes of Nepal: Their identification, distribution and biology. *Mosquito Systematics*, 22(2):69–130.

Darsie, R. F. Jr., Pradhan, S.P. and Vaidya, R.G. (1991). Notes on the Mosquitoes of Nepal I. New Country Records and Revised *Aedes* Keys (Diptera, Culicidae). *Mosquito Systematics*, 23(1):39–49.

Darsie Jr., R. F. Pradhan, S.P. and Vaidya, R. G. (1992). Notes on the Mosquitoes of Nepal: II. New Species Records From 1991 Collections. *Mosquito Systematics*, 24(1):23–28.

Darsie Jr, R. F., Courtney, G.W. and Pradhan, S. P. (1993). Notes on the Mosquitoes of Nepal: III. Additional New recodes in 1992 (Diptera: Culicidae). *Mosquito Systematics*, 25(3):186–191.

Darsie Jr, R.F., Courtney, G.W. and Pradhan, S.P. (1994). The mosquitoes of Mustang (Diptera, Culicidae). *Proceedings of the Entomological Society of Washington*, 96(2):230–235.

Darsie, Jr. R.F., Courtney, G.W. and Pradhan, S.P. (1996). Notes on the mosquitoes of Nepal. IV. Results of the 1994 collecting in the Midwestern Region, including new country records and voucher confirmation (Diptera, Culicidae). *Journal of the American Mosquito Control Association*, 12(1):130–134.

Delfinado, M.D. (1966). The culicine mosquitoes of the Philippines, tribe Culicini (Diptera, Culicidae). *Memoirs of the American Entomological Institute*, 7:1–252.

Dhimal, M., Ahrens, B., Kuch, U. (2014). Species composition, seasonal occurrence, habitat preference and altitudinal distribution of malaria and other disease vectors in eastern Nepal. *Parasites & Vectors*,7(1):1–1.

Dhimal, M., Gautam, I. Joshi, H.D., O'Hara, R.B., Ahrens, B. and Kuch, U. (2015). Risk factors for the presence of chikungunya and dengue vectors (*Aedes aegypti* and *Aedes albopictus*), their altitudinal distribution and climatic determinants of their abundance in central Nepal. *PLoS Neglected Tropical Disease*, 9(3):e0003545. https://doi.org/10.1371/journal.pntd.0003545

Dhimal, M., Kramer, I. M., Phuyal, P., Budhathoki, S. S., Hartke, J., Ahrens, B., ... Müller, R. (2021). Climate change and its association with the expansion of vectors and vector-borne diseases in the Hindu Kush Himalayan region: a systematic synthesis of the literature. *Advances in Climate Change Research*, 12(3):421–429.

EDCD (2019). Dengue Updates [Internet]. Available at: http://edcd.gov.np/news/download/dengue-updates1

Gautam, I., Dhimal, M., Shrestha, S.R. and Tamrakar, A.S. (2009). Firs record of *Aedesaegypti* (L.) vector of dengue virus from Kathmandu, Nepal. *Journal of Natural History Museum*, 24:156–164.

Gould, D. J., Barnett, H.C. and Suyemoto, W. (1962). Transmission of Japanese encephalitis virus by *Culex gelidus* Theobald. *Transactions of Royal Society of Tropical Medicine and Hygiene*, 56:429–435.

Hale, J.H., Colless, D.H. and Kim, K.A. (1957). Investigation of the Malaysian form of *Culex tritaeniorhynchus* as a potential vector of Japanese B encephalitis virus on Singapore Island, *Annals of Tropical Medicine and Parasitology*, 51:17–25.

Hammon, W.M., Tigertt, W.D., Salter, G. and Schenker, H. (1949). Isolations of Japanese B encephalitis virus from naturally infected *Culex tritaeniorhynchus* collected in Japan. *American Journal of Hygiene*, 50(1), 50–56.

Harrison, B.A. (1980). Medical entomology studies - XIII. The Myzomyia series of *Anopheles* (*Celliia*) in Thailand, with emphasis on intra-interspecific variations (Diptera: Culicidae). *Contributions of the American Entomological Institute (Ann Arbor)*, 17(4):1–195.

Hartke, J., Reuss, F., Kramer, I.M., Magdeburg, A., Deblauwe, I., Tuladhar, R., Gautam, I., Dhimal, M., Müller, R. (2022). A barcoding pipeline for mosquito surveillance in Nepal, a biodiverse dengue-endemic country, *Parasites & Vectors*, 15:145.

Joshi, G., Shrestha, S.L. and Darsie Jr., R.F. (1964). First record of *Anopheles kochi* Doenitz, 1901 in Nepal (Diptera: Culicidae). *Bulletin of the Indian Society for Malaria and other Communicable Diseases*,1:135–139.

Joshi, G., Pradhan, S. and Darsie, Jr. F. R. (1965) Culicine sabethine and toxorhynchitine mosquitoes of Nepal including new country records, *Proceedings of the Entomological Society of Washington*, 67:137–146.

Jung, R.K. (1973) A brief study on the epidemiology offilariasis in Nepal, *Journal of Nepal Medical Association*, 11:155–166.

Kramer, I.M., Pfeiffer, M., Steffen, O., Schneider, F., Greger, V., Phuyal, P., Braun, M., Magdeburg, A., Ahrens, B., Groneberg, D.A., Kuch, U., Dhimal, M. and Müller, R. (2021) The ecophysiological plasticity of *Aedes aegypti* and *Aedes albopictus* concerning overwintering in cooler ecoregions is driven by local climate and acclimation capacity. *Science of the Total Environment*, 778:146128.

Kramer, I. M., Pfenninger, M., Feldmeyer, B., Dhimal, M., Gautam, I., Shreshta, P., Baral, S., Phuyal, P., Hartke, J., Magdeburg, A., Groneberg, D. A., Ahrens, B., Müller, R., and Waldvogel, A.M. (2023). Genomic profiling of climate adaptation in Aedes aegypti along an altitudinal gradient in Nepal indicates nongradual expansion of the disease vector, *Molecular Ecology*, 32:350–368.

Leake, C. J., Ussery, M.A., Nisalak, A., Hoke, C.H., Andre, R.G. and Burke, D.S. (1986). Virus isolations from mosquitoes collected during the 1982 Japanese encephalitis epidemic in northern Thailand. *Transactions of the Royal Society of Tropical Medicine and Hygiene*, 80:831–837.

Malla, S., Thakur, G.D., Shrestha, S.K., Banjeree, M.K., Thapa, L.B., Gongal, G., Ghimire, P., Upadhyay, B.P., Gautam, P., Khanal, S., Nisaluk, A., Jarman, R.G., Gibbons, R.V. (2008). Identification of all dengue serotypes in Nepal. *Emerging Infectious Diseases*, 14(10):1669–1670.

Mclntosh, B.M., Jupp, P.G., Dickinson, D.B., McGillivray, G.M. and Sweetman, J. (1967) Ecological studies on Sindbis and West Nile viruses in South Africa. I. Viralactivity as revealed by infection of mosquitoes and sentinal fowls. *South African Journal of Medical Science*, 32:1–14.

Pandey, B.D., Neupane, B., Pandey, K., Tun, M.M. and Morita, K. (2015). Detection of chikungunya virus in Nepal. *The American Journal of Tropical Medicine and Hygiene*, 93(4):697.

Pandey, K., Pandey, B.D., Chaurasiya, R.R., Thakur, M., Neupane, B., Shah, Y., Ngwe Tun, M.M. and Morita, K. (2017). Evidence of Chikungunya virus circulation in the Terai region of Nepal in 2014 and 2015. *Transactions of The Royal Society of Tropical Medicine and Hygiene*, 111(7):294–299.

Phuyal, P., Kramer, I.M., Klingelhöfer, D., Kuch, U., Madeburg, A., Groneberg, D.A., Wouters, E., Dhimal, M. and Müller, R. (2020). Spatiotemporal Distribution of Dengue and Chikungunya in the Hindu Kush Himalayan Region: A Systematic Review. *International Journal of Environment Research and Public Health*, 17(18):6656.

Pradhan, J.M., Shrestha, S.L. and Vaidya, R.G. (1970).Malaria transmission in high mountain valleys of west Nepal including first record of *Anopheles maculatus willmori* (James) as a third vector of malaria. *Journal of Nepalese Medical Association* 8:89–97.

Pradhan, S.P. and Brydon, H.W. (1960). A record of *Anopheles filipinae* in Nepal. *Mosquito News* 20:319–320.

Pradhan, S.P. and Darsie Jr., R.F. (1989). New mosquito records for Nepal. *Journal of the American Mosquito Control Association*, 5:21–24.

Peters, W., Dewar, S.C., Das Balla, B. and Manadhar, T.L. (1955). A preliminary note on the Anophelini of the Rapti valley area of the Nepal, Terai. *Indian Journal of Malariology*, 9(3):207–212.

Peters, W. and Dewar, S. (1956). A preliminary record of the megarhine and culicine mosquitoes of Nepal with notes on their taxonomy (Diptera: Culicidae). *Indian Journal of Malariology*, 10:37–51.

Pun, S.B., Bastola, A. and Shah, R. (2014). First report of Chikungunya virus infection in Nepal. *The Journal of Infection in Developing Countries*, 8(6):790–792.

Puri, I.M. (1955). The Distribution of Anopheline Mosquitoes in India. Health Bulletin No. 17; Malaria Bureau No. 7, Government of India. 104 pp.

Rao, R.T. (1981). *The anophelines of India*. Indian Council of Medical Research, New Delhi, 594 pp.

Rattanarithikul, R. and Green, C.A. (1987). Formal recognition of the species of the *Anopheles maculatus* group (Diptera: Culicidae) occurring on Thailand, including the descriptions of two new species and a preliminary key to females. *Mosquito Systematics*, 18(1986):246–278.

Reinert, J.F. (1973) Contributionsto the mosquitofaunaofSoutheast Asia XVI. Genus *Aedes* Meigen. Subgenus *Aedimolphus* Theobald in Southeast Asia. *Contributions of the American Entomological Institute (Ann Arbor)*, 9(5):1–218.

Reuben, R. (1971a). Studies on the mosquitoes of NorthArcot District, Madras State, India. Part 5. Breeding Places of the *Culex vishnui* group of species. *Journal of Medical Entomology*, 8:363–366.

Reuben, R. (1971b). Studies on the mosquitoes of NorthArcot District, Madras State, India. Part 6. Seasonal prevalence of the *Culex vishnui* group of species. *Journal of Medical Entomology*, 8(4):367–371.

Scholte, E.J., Dik, M., Ibañez-Justicia, A., Hartog, W.D., Schoelitsz, B., Brooks, M., Braks, M. and Steeghs, M.(2012) Findings and control of two invasive exotic mosquito species, *Aedes albopictus* and *Ae. atropalpus* (Diptera: Culicidae) in the Netherlands, 2011. *European Mosquito Bulletin*, 30:1–14.

Shrestha, S.L. (1966) Mosquitoes. In: *Wild is Beautiful*. pp. 51–66.

Shrestha, S.L. and Parajuli, M.B. (1980). Reappearance of malaria in terai areas of Nepal and incrimination of *An. annularis* Van der Wulp. *Journal of Nepal Medical Association*, 18:11–18.

Shrestha, S.L., Pradhan, J.P.B., Shrestha, J.D., Shrestha, Y., Rajbhandari, G.L., Shrestha, T., Swar, B., Nushii, M.K. and Reisen, W.K. (1988) Observations on anopheline and malaria ecology in the far western region of Nepal, 1986. *Bulletin of the Society for Vector Ecology*, 13:332–342.

Sirivanakarn, S. (1976) Medical entomology studies. III. A revision of the subgenus *Culex* in the Oriental Region (Diptera: Culicidae),*Contributions of the American Entomological Institute* (Ann Arbor), 12(2):1–272.

Walther, D., Scheuch, D.E. and Kampen, H. (2017), The invasive Asian tiger mosquito *Aedes albopictus* (Diptera: Culicidae) in Germany: local reproduction and overwintering. *Acta Tropica*, 166:186–192.

WHO (2006). *Outbreak investigation of dengue Fever in Nepal*. World Health Organization South East Asia Region Office. Available at: http://www.searo.who.int/LinkFiles/Dengue_dengue_Nepal.pdf

19 Mosquitoes of Sri Lanka

Tharaka Ranathunge
Department of Zoology, Eastern University of Sri Lanka, Chenkaladi, Sri Lanka

19.1 INTRODUCTION

Vector-borne diseases have been one of the major health concerns in Sri Lanka for decades, with particular reference to mosquito-borne infections, such as, for example, malaria, chikungunya, Japanese B-encephalitis, yellow fever, filariasis, and dengue. Alarmingly, outbreaks caused by some of these mosquito-borne diseases are increasingly becoming rampant across the country. That is due to natural environmental changes as well as human-induced changes in ecological landscapes in the country (Chandra et al., 2008).

The main purpose of the chapter is to offer an authentic checklist of all the extant taxa of mosquitoes in Sri Lanka as a first step towards conserving knowledge on the mosquito taxonomy/ An annotated catalogue will facilitate comprehension about the intricate organization of various species, with an extensive bibliography.

19.2 ECO-PHYSIOGRAPHY OF SRI LANKA

Sri Lanka is located in the Indian Ocean between latitudes 5°55' and 9°50' north and 79°42' and 81°52' east. The island is situated inside the monsoon belt, where the major wind directions from May to September are from the south-west and from December to February are from the north-east. The island also experiences various weather conditions during the inter-monsoon seasons. The island is divided between a "wet zone" in the southwest, an "intermediate zone," and a "dry zone" in the remainder of the island by climate and topography. This explains the distribution of numerous flora types, including forests in the central mountain range, moist tropical evergreen forests in the southwest, tropical dry savannah forests in the north, and several types of grasslands (Lambrecht, 1974).

19.3 MOSQUITO RESEARCH IN SRI LANKA

The first list of mosquitoes of Sri Lanka was published by Green (1901), including 20 species of which 18 are valid taxa to date. Theobald (1901) listed 54 mosquito species he had described between 1905 and 1910. The research done by Christophers (1933) and Barraud (1934) under the anopheline and culicine mosquito genera, led to expand the list to 92 mosquito species. Using previous records, as well as the investigations made in various parts of the island, Carter (1950) indicated ten anopheline mosquito species, and created a thorough database of mosquito species that included 125 species under 14 genera. Chow et al. (1954) and Jayasekera and Chelliah (1982) later modified this list. In a later update, Amerasinghe (1990, 1991, 1992) added 140 species under 16 genera, updating the inventory of Sri Lankan mosquito species and offered a guide to identification of the Sri Lankan species of mosquitoes. Finally, Gunathilaka (2018) inventorized a total of 159 mosquito species, including sibling species complexes, belonging to 19 genera in Sri Lanka. This annotated checklist provides concise information on taxonomic information of the mosquito fauna in Sri Lanka.

DOI: 10.1201/9781003327028-23

19.4 CONTROL OF MOSQUITO VECTORS AND MOSQUITO-BORNE DISEASES IN SRI LANKA

In the past, from time to time, Sri Lanka has been threatened with mosquito-borne disease outbreaks and the island nation of Sri Lanka prepared systematically against such onslaughts of the diseases. Therefore, Sri Lanka has a strong entomological surveillance system, dating back to the 1930s, which has evolved based on the changing epidemiology of vector-borne diseases in the country. The state institutions such as the Anti-Malaria Campaign (AMC), National Dengue Control Unit (NDCU), Medical Research Institute (MRI), and the Anti-Filariasis Campaign (AFC) are engaged in entomological surveillances to controlling mosquito-borne diseases in the country through their well-established public health networks (Gunathilaka, 2018). In addition, various research groups in national universities and research institutions in Sri Lanka are also conducting entomological surveillance to fill the knowledge gaps on vector ecology and distribution patterns of medically important mosquitoes in the country. Therefore, the number of species has increased considerably compared to previous checklists, particularly the last published by Gunathilaka (2018). Later, Rangana et al. (2022) and Wickramasinghe et al. (2022) published two research articles reporting two mosquito species – namely, *Aedes (Hulecoeteomyia) japonicus* and *Culex (Lophoceraomyia) cinctellus* – which were not reported from the country earlier, hence records for the nation. Interestingly, *Anopheles stephensi* was discovered as an invasive species in Sri Lanka in 2017 (Gayan et al., 2017).

19.5 AN UPDATED CATALOGUE OF MOSQUITO SPECIES IN SRI LANKA

Subfamily: Anophelinae
GENUS: ANOPHELES MEIGEN
Subgenus: *Anopheles* Meigen

1. *Anopheles aitkenii* James, 1903
2. *Anopheles barbirostris* Van der Wulp, 1884
3. *Anopheles barbumbrosus* Strickland and Choudhury, 1927
4. *Anopheles gigas refutans* Alcock, 1913
5. *Anopheles interruptus* Puri, 1929
6. *Anopheles nigerrimus* Giles, 1900
7. *Anopheles peditaeniatus* Leicester, 1908
8. *Anopheles peytoni* Kulasekera, Harrison, and Amerasinghe, 1988
9. *Anopheles reidi* Harrison, 1973

Subgenus: *Cellia* Theobald

10. *Anopheles aconitus* Donitz, 1902
11. *Anopheles annularis* Van der Wulp, 1884
12. *Anopheles culicifacies* Giles, 1901
13. *Anopheles elegans* James, 1903
14. *Anopheles jamesii* Theobald, 1903
15. *Anopheles jeyporiensis* James, 1902
16. *Anopheles karwari* James, 1902
17. *Anopheles maculatus* Theobald, 1901
18. *Anopheles mirans* Sallum and Peyton, 2005
19. *Anopheles pallidus* Theobald, 1901
20. *Anopheles pseudojamesi* Strickland and Chowdhury, 1927
21. *Anopheles subpictus* Grassi, 1899
22. *Anopheles stephensi* Liston, 1901
23. *Anopheles tessellatus* Theobald, 1901
24. *Anopheles vagus* Donitz, 1902
25. *Anopheles varuna* Iyengar, 192

Subfamily: Culicinae
GENUS: *AEDES* MEIGEN
Subgenus: *Aedimorphus* Theobald
26. *Aedes alboscutellatus* Theobald, 1905
27. *Aedes argenteoscutellatus* Carter and Wijesundera, 1948
28. *Aedes jamesi* Edwards, 1914 Barraud
29. *Aedes pallidostriatus* Theobald, 1907
30. *Aedes pipersalatus* Giles, 1902
31. *Aedes stenoetrus* Theobald, 1907
32. *Aedes taeniorhynchoides* Christophers, 1911
33. *Aedes vexans* Meigen, 1830

Subgenus: *Bruceharrisonius* Theobald
34. *Aedes greenii* Theobald, 1903

Subgenus: *Cancraedes* Edwards
35. *Aedes simplex* Theobald, 1903

Subgenus: *Christophersiomyia* Barraud
36. *Aedes annulirostris* Theobald, 1905
37. *Aedes thomsoni* Theobald, 1905

Subgenus: *Collessius* Reinert, Harbach, and Kitching
38. *Aedes macdougalli* Edwards, 1922
39. *Aedes Collessius pseudotaeniatus* Giles, 1901

Subgenus: *Danielsia* Theobald
40. *Aedes albotaeniatus* Leicester, 1904

Subgenus: *Dendroskusea* Edwards
41. *Aedes ramachandrai* Reuben, 1967
42. *Aedes reginae* Edwards, 1922

Subgenus: *Diceromyia* Theobald
43. *Aedes periskelatus* Giles, 1902

Subgenus: *Downsiomyia* Vargas
44. *Aedes albolateralis* Theobald, 1908
45. *Aedes mohani* Knight, 1969

Subgenus: *Finlaya* Theobald
46. *Aedes aureostriatus* Doleschall, 1857
47. *Aedes niveus* Ludlow, 1903

Subgenus: *Fredwardsius* Theobald
48. *Aedes vittatus* Bigot, 1861

Subgenus: *Hulecoeteomyia* Theobald
49. *Aedes chrysolineatus* Theobald, 1907
50. *Aedes harveyi* Barraud, 1923
51. *Aedes japonicus* Theobald, 1901

Subgenus: *Mucidus* Theobald
52. *Aedes quasiferinus* Mattingly, 1961
53. *Aedes scatophagoides* Theobald, 1901

Subgenus: *Neomelaniconion* Newstead
54. *Aedes lineatopennis* Ludlow, 1905

Subgenus: *Paraedes* Edwards
55. *Aedes ostentatio* Leicester, 1908
56. *Aedes chrysoscuta* Theobald, 1910

Subgenus: *Phagomyia* Theobald
57. *Aedes gubernatoris* Giles, 1901

Subgenus: *Rhinoskusera* Edwards

58. *Aedes wardi* Reinert, 1976 Reinert 1976

Subgenus: *Stegomyia* Theobald

59. *Aedes aegypti* Linnaeus, 1762
60. *Aedes albopictus* Skuse, 1894
61. *Aedes krombeini* Huang, 1975
62. *Aedes mediopunctatus* Theobald, 1905
63. *Aedes novalbopictus* Barraud, 1931
64. *Aedes scutellaris* Walker, 1858
65. *Aedes w-albus* Theobald, 1905

GENUS: *AEDEOMYIA* THEOBALD

Subgenus: *Aedeomyia* Theobald

66. *Aedeomyia catasticta* Knab, 1909

GENUS: *ARMIGERES* THEOBALD

Subgenus: *Armigeres* Theobald

67. *Armigeres aureolineatus* Leicester, 1908
68. *Armigeres durhami* Edwards, 1917
69. *Armigeres obturbans* Walker, 1859
70. *Armigeres subalbatus* Coquillett, 1898

Subgenus: *Leicesteria* Theobald

71. *Armigeres mangus* Theobald, 1908
72. *Armigeres omissus* Edwards, 1914

GENUS: *CULEX* LINNAEUS

Subgenus: *Culex* Linnaeus

73. *Culex fuscocephala* Theobald, 1907
74. *Culex gelidus* Theobald, 1901
75. *Culex hatchinsoni* Barraud, 1924
76. *Culex jacksoni* Edwards, 1934
77. *Culex mimulus* Edwards, 1934
78. *Culex perplexus* Leicester, 1908
79. *Culex pseudovishnui* Colless, 1957
80. *Culex quinquefasciatus* Say, 1823
81. *Culex sitiens* Wiedemann, 1828
82. *Culex tritaeniorhynchus* Giles, 1901
83. *Culex vishnui* Theobald, 1901
84. *Culex whitmorei* Giles, 1904

Subgenus: *Culiciomyia* Theobald

85. *Culex bahri* Edwards, 1914
86. *Culex bailyi* Barraud, 1934
87. *Culex fragilis* Ludlow, 1903
88. *Culex nigropunctatus* Edwards, 1926
89. *Culex pallidothorax* Theobald, 1905
90. *Culex scanloni* Bram, 1967
91. *Culex spathifurca* Edwards, 1915

Subgenus: *Eumelanomyia* Theobald

92. *Culex brevipalpis* Giles, 1902
93. *Culex campilunati* Carter and Wijesundera, 1948
94. *Culex castrensis* Edwards, 1922
95. *Culex foliates* Brug, 1932
96. *Culex malayi* Leicester, 1908
97. *Culex pluvialis* Barraud, 1924 Sirivanakarn 1977b

Subgenus: *Lophoceraomyia* Theobald

98. *Culex bicornutus* Theobald, 1910
99. *Culex cinctellus* Edwards, 1922
100. *Culex infantulus* Edwards, 1922
101. *Culex lasiopalpis* Sirivanakarn, 1977
102. *Culex mammilifer* Leicester, 1908
103. *Culex minutissimus* Theobald, 1907
104. *Culex quadripalpis* Edwards, 1914
105. *Culex rubithoracis* Leicester, 1908
106. *Culex uniformis* Theobald, 1905
107. *Culex wardi* Sirivanakarn, 1977

Subgenus: *Oculeomyia* Theobald

108. *Culex bitaeniorhynchus* Giles, 1901
109. *Culex infula* Theobald, 1901
110. *Culex sinensis* Theobald, 1903

GENUS: *COQUILLETTIDIA* DYAR

Subgenus: *Coquillettidia* Dyar

111. *Coquillettidia crassipes* Van der Wulp, 1881

GENUS: *FICALBIA* THEOBALD

112. *Ficalbia minima* Theobald, 1901

GENUS: *HEIZMANNIA* LUDLOW

Subgenus: *Heizmannia* Ludlow

113. *Heizmannia greenii* Theobald, 1905
114. *Heizmannia carteri* Amerasinghe, 1993

GENUS: *HODGESIA* THEOBALD

115. *Hodgesia bailyi* Barraud, 1929
116. *Hodgesia malayi* Leicester, 1908

GENUS: *LUTZIA* THEOBALD

Subgenus *Metalutzia* Tanaka

117. *Lutzia fuscanus* Wiedemann, 1820
118. *Lutzia halifaxi*i Theobald, 1903
119. *Lutzia vorax* Edwards, 1921 Bram 1967

GENUS: *MALAYA* LEICESTER

120. *Malaya genurostris* Leicester, 1908

GENUS: *MANSONIA* BLANCHARD

Subgenus: *Mansonioides* Theobald

121. *Mansonia annulifera* Theobald, 1901
122. *Mansonia indiana* Edwards, 1930
123. *Mansonia uniformis* Theobald, 1901

GENUS: *MIMOMYIA* THEOBALD

Subgenus: *Etorleptiomyia* Theobald

124. *Mimomyia nluzonensis* Ludlow, 1905

Subgenus: *Mimomyia* Theobald

125. *Mimomyia chamberlaini* Ludlow, 1904
126. *Mimomyia hybrid* Leicester, 1908
127. *Mimomyia intermedia* Barraud, 1929

GENUS: *ORTHOPODOMYIA* THEOBALD

128. *Orthopodomyia anopheloides* Giles, 1903
129. *Orthopodomyia flavithorax* Barraud, 1927

GENUS: *TRIPTEROIDES* GILES

Subgenus: *Rachionotomyia* Theobald

130. *Tripteroides affinis* Edwards, 1913
131. *Tripteroides ceylonensis* Theobald, 1905
132. *Tripteroides dofleini* Gunther, 1913

Genus: Topomyia

** Species name is not yet confirmed.

GENUS: *TOXORHYNCHITES* THEOBALD

Subgenus: *Toxorhynchites* Theobald

133. *Toxorhynchites amboinensis* Doleschall, 1857
134. *Toxorhynchites minimus* Theobald, 1905
135. *Toxorhynchites splendens* Wiedmann, 1819

GENUS: *URANOTAENIA*

Subgenus: *Pseudoficalbia* Theobald

136. *Uranotaenia bicolor* Leicester, 1908
137. *Uranotaenia gouldi* Peyton and Klein, 1970
138. *Uranotaenia nivipleura* Leicestor, 1908
139. *Uranotaenia obscura* Edwards, 1915
140. *Uranotaenia novobscura* Barraud, 1934
141. *Uranotaenia srilankensis* Payton, 1974

Subgenus: *Uranotaenia* Lynch Arribalzaga

142. *Uranotaenia campestris* Leicester, 1908
143. *Uranotaenia lateralis* Ludlow, 1905
144. *Uranotaenia rutherfordi* Edwards, 1922

GENUS: *VERRALLINA* THEOBALD

145. *Verrallina butleri* Theobald, 1901
146. *Verrallina indicus* Theobald, 1907
147. *Verrallina lankaensis* Stone and Knight, 1958
148. *Verrallina lugubris* Barraud, 1928
149. *Verrallina pteroelephantus* Wijesundara, 1951
150. *Verrallina pseudomediofasciatus* Theobald, 1910
151. *Verrallina secullatus* Menon, 1950
152. *Verrallina spermathecus* Wijesundara, 1951
153. *Verrallina srilankensis* Reinert, 1977
154. *Verrallina yerburyi* Edwads, 1917
155. *Verrallina yasafi* Barraud, 1931

19.6 FINAL REMARKS

Mosquito diversity in Sri Lanka is well known (Christophers, 1933; Barraud, 1934 Wilkerson et al. 2015). Mosquito-borne diseases are both fatal and debilitating (Ong et al., 2022). In addition to diseases that affect humans, mosquitoes can also spread a number of illnesses and parasites that dogs and horses are particularly vulnerable to. These include West Nile virus, dog heart worms, and eastern equine encephalitis (Haynes et al., 2022). Due to the movement of people, animals, and things on a worldwide scale, as well as climate changes, mosquitoes and/or mosquito-borne diseases are being reported increasingly and frequently globally, and Sri Lanka is no exception (Weaver and Reisen, 2010). These circumstances also give invasive and frequently alien vector species a chance to spread throughout the world (Becker et al., 2012; Gayan et al., 2017). Their potential to spread locally amid human populations and their wind-borne, long-distance dispersal over suitable habitats highlight the need to fill in the knowledge gaps about the distribution of vector species (Tyagi, 2023). Monitoring the species composition in the country is one of the most crucial measures in preventing the spread of diseases caused by mosquitoes, and Sri Lanka with a mosquito wealth of 155 extant taxa has to remain alert both in time and space.

REFERENCES

Amerasinghe, F. P. (1990). A guide to the identification of the anopheline mosquitoes (Diptera: Culicidae) of Sri Lanka. I. Adult females. *Ceylon Journal of Science (Biological Science)*, 21, 1–16.

Amerasinghe, F. P. (1991). *A catalog of the mosquitoes in Sri Lanka*. Natural Resources, Energy and Science Authority of Sri Lanka, 1–23 pp.

Amerasinghe, F. P. (1992). A guide to the identification of the anopheline mosquitoes (Diptera: Culicidae) of Sri Lanka. *Ceylon Journal of Science (Biological Science)*, 22(1), 1–13.

Barraud, P. J. (1934). *The Fauna of British India including Ceylon and Burma. Diptera. vol. 5. Family Culicidae. Tribes Megarhinini and Culicini*. Taylor and Francis, London, 463 pp.

Becker, N., Pluskota, B., Kaiser, A. and Schaffner, F. (2012). Exotic mosquitoes conquer the world. In: Mehlhorn H., editor. *Parasitology research monographs 3*. Berlin: Springer, pp. 31–60.

Carter H. F. (1950). Ceylon mosquitoes: Lists of species and names of mosquitoes recorded from Ceylon. *Ceylon Journal of Science*, 24(2), 85–115.

Chandra, G., Bhattacharjee, I., Chatterjee, S.N. and Ghosh A. (2008). Mosquito control by larvivorous fish. *Indian Journal of Medical Research*, 127, 13–27.

Chow, C.Y., Thevasagayam E.S. and Tharumarajah, K. (1954). Insects of public health importance in Ceylon. *Revista Ecuatoriana de Entomologia y Parasitologia*, 2(1–2), 115–119.

Christophers S.R. (1933). *The fauna of British India, including Ceylon and Burma. Diptera. Family Culicidae, Tribe Anophelini, Vol. 3*. Taylor and Francis, London.

Gayan, D.A.G., Perera A.Y., Harishchandra J. et al. (2017). First record of *Anopheles stephensi* in Sri Lanka: A potential challenge for prevention of malaria reintroduction. *Malaria Journal*, 16, 326. https://doi.org/10.1186/s12936-017-1977-7

Green, E. E. (1901). Mosquitoes and malaria. *Circular of Royal Botanical Gardens, Ceylon*, 1, 345–368.

Gunathilaka, H.N. (2018). Annotated checklist and review of the mosquito species (Diptera: Culicidae) in Sri Lanka. *Journal of Insect Biodiversity*, 007, 038–050. https://doi.org/10.12976/jib/2018.07.3.1

Haynes, E., Cleveland, C.A., Garrett, K.B., Grunert, R.K., Bryan, J.A., Sidouin, M., Oaukou, P.T., Ngandolo, B.N.R. and Yabsley, M.J. (2022). Characterization of the genetics and epidemiology of *Brugia* sp. in domestic dogs in Chad, Africa. *Veterinary Parasitology*, 35, 100784. https://doi.org/10.1016/j.vprsr.2022.100784

Jayasekera, N. and Chelliah, R.V. (1982). An annotated checklist of mosquitoes of Sri Lanka. Issue 8 of Publication (UNESCO, Man and the Biosphere National Committee for Sri Lanka). National Science Council of Sri Lanka. 16 pp.

Lambrecht, F.L. (1974). Entomological aspects of filariasis control in Sri Lanka. *Bulletin of the World Health Organization*, 51(2), 133.

Ong, S.Q., Pauzi, M.B.M. and Gan, K.H. (2022). Text mining in mosquito-borne disease: a systematic review. *Acta Tropica*, 231, 106447. https://doi.org/10.1016/j.actatropica.2022.106447

Rangana, U.L.P., Kumarasinghe, K.G.D.S.G.S., Jayasinghe, H.P.N.T., Refai, F.W., Udayanga, N.W.B.A.L. and Ranathunge, R.M.T.B. 2022. First report of *Culex (Lophoceraomyia) cinctellus* in Sri Lanka. International Conference on Applied and Pure Sciences, 2022, Faculty of Science, University of Kelaniya, Sri Lanka, 4 pp.

Theobald, F. V. 1901. *A Monograph of the Culicidae or Mosquitoes. Vol 1*, British Museum (Natural History), London, 424 pp.

Tyagi, B.K. 2023. *Desert Malaria: An emerging malaria paradigm and its global impact on disease elimination*. Springer-Nature, Singapore, 416 pp. https://doi.org/10.1007/978-981-19-7693-3

Weaver, S.C. and Reisen, W.K. (2010). Present and future arboviral threats. *Antiviral Research*, 85(2), 328–345.

Wickramasinghe, T.K.C., Udayanga, N.W.B.A.L., Manamperi, A.A.P.S., Hapugoda, M.D., Karunarathna, K.H.T. and Ranathunge, R.M.T.B. 2022. Use of morphological and molecular methods in identification of Asian bush mosquito, *Aedes japonicus* (Theobald, 1901), in Sri Lanka. *International Conference on Applied and Pure Sciences, Kelaniya, Sri Lanka*, p 2.

Wilkerson, R. C., Linton, Y. M., Fonseca, D. M., Schultz, T.R., Price, D.C. and Strickman D. A. (2015). Making mosquito taxonomy useful: A stable classification of tribe aedini that balances utility with current knowledge of evolutionary relationships. *Plos One*, 10(7), e0133602. https://doi.org/10.1371/journal.pone.0133602

Section V

Mosquito-Borne Infections and Biosafety, Regulatory, and Ethical Considerations

20 Biosafety Approaches for Deployment of Genetically Modified *Aedes* Mosquitoes

S. K. Dasgupta
Ex-Tata Institute of Genetics and Society, New inStem Building, Bengaluru, India

Prabhakargouda B. Patil
Maharashtra Hybrid Seeds Company Pvt. Ltd., Dawalwadi, Badnapur, Jalna, India

Jhansi Charles
Rajaji Medical College & Hospital, Madurai, India

Vibha Ahuja
Biotech Consortium India Limited, Anuvrat Bhawan, Deen Dayal Upadhyaya Marg, New Delhi, India

B.K. Tyagi
Chandigarh University, Mohali, India

20.1 INTRODUCTION

Aedes aegypti L, also commonly known as the yellow fever mosquito, is a primary vector for transmission of the dengue virus as well as several other incapacitating arboviral diseases such as chikungunya, yellow fever and the newly emerging Zika virus worldwide (Brady et al., 2012; Bhatt et al., 2013; Bhardwaj et al., 2017). The incidence of dengue has grown exponentially around the world recently, with cases reported to WHO increased from 505,430 cases in 2000 to 5.2 million in 2019. The numbers available could be much higher since many cases are underreported, as they could be asymptomatic or misdiagnosed as febrile illnesses. One modelling estimate indicates 390 million dengue virus infections per year, of which 96 million manifests clinically. Another study on the prevalence of dengue estimates that 3.9 billion people are at risk of infection with dengue viruses. Dengue is now endemic in more than 100 countries in the WHO Regions of Africa, the Americas, the Eastern Mediterranean, South-East Asia, and the Western Pacific. The Americas, South-East Asia, and Western Pacific regions are the most seriously affected, with Asia representing around 70% of the global disease burden (WHO, 2023).

Aedes aegypti, the primary vector of dengue, is mainly an urban vector, but recent studies indicate the geographical spread of the vector from urban to rural areas has further raised the risk of epidemic

DOI: 10.1201/9781003327028-25

outbreaks and increased its economic burden on the health-care system. This alarming increase in the situation is due to global warming, population growth and migration, and urbanization, as well as including lacunae in mosquito control strategies and tools (Kay and Vu, 2005).

20.2 DENGUE IN INDIA

Disease due to dengue virus (DENV) has been considered of prime importance among the mosquito-borne diseases not only due to the severity of the disease and fatal nature but also the lack of vaccine and specific drugs. Dengue infection is caused by any of the four antigenically similar viruses (DENV, 1–4 serotypes) and is the most important arthropod-borne viral infection globally in tropical and subtropical regions (Mutheneni et al., 2017). Infection with any of the strain is thought to confer immunity towards other variants of respective DENV serotype; however, immunity against other serotypes is partial and for a transient period of two to three months. It is also reported that the possibility of a severe form of dengue infection increases during subsequent re-infection with other DENV serotypes (Halstead, 2007). The first epidemiology of dengue in India was reported in Chennai (formerly Madras) in 1780, and the first outbreak was in Kolkata (formerly Calcutta) in 1963 (Ramakrishnan et al., 1964; Chaturvedi and Nagar, 2008). The total number of dengue cases has significantly increased over the past ten years. The number of reported cases during the year 2021 remained close to 2 lakhs, with a note that underreporting exists which when considered can be much higher (Shepard et al., 2014).

In one of the estimates, the global economic burden due to dengue illness was US$8.9 billion per year. Forty percent of this estimate was due to what are known as productivity costs productivity loss from both paid and unpaid work, treatment, or premature death (Hung et al., 2020; Koopmanschap and van Ineveld, 1992).

20.3 CHIKUNGUNYA IN INDIA

Chikungunya virus (CHIKV) belongs to the family Togaviridae, genus Alphavirus; has a single-stranded RNA genome, a 60–70 nm diameter capsid, phospholipid envelope; and was first isolated in 1953 in Tanzania (Robinson, 1955; Pialoux et al., 2007). The word "chikungunya," the term for both the virus and the disease, means "to walk bent over" in east African languages, is due to the characteristic symptom of joint pain, in addition to fever, rash, and incapacitating arthralgia (Enserink, 2006). In India, the first major epidemic was reported in Kolkata in 1963 followed by those in Puducherry, Tamil Nadu, Andhra Pradesh, Madhya Pradesh, and Maharashtra (Sudeep and Parashar, 2008). The CHIKV infection outbreak re-emerged in 2006 after a gap of 32 years from the year 1973 and was reported in 13 states infecting all age groups. In 2006, more than 1.4 million suspected cases for CHIKV infection were reported in India, and subsequently, frequent infections were reported every year in all states (Ravi, 2006; Cecilia, 2014; Figures 20.1 and 20.2).

20.4 *AEDES* MOSQUITO CONTROL

Aedes aegypti and *Ae. albopictus* are the most important mosquito vectors of dengue fever viruses (Knudsen, 1995; Mousson et al., 2005). *Aedes aegypti* is the principal vector of dengue fever and dengue haemorrhagic fever in almost all countries (Guillena et al., 2010; WHO 2019; Were, 2012). *Aedes aegypti* and *Ae. albopictus* prefer laying their eggs in containers like broken crockery, discarded vehicle tires, buckets, flowerpots, etc. (Gubler, 1998; Knudsen, 1995; Vezzani, 2007). *Aedes aegypti* is the most efficient vector for arboviruses because it is a highly anthropophilic, frequent biter and thrives in close proximity to humans (WHO, 2019). The adult *Ae. aegypti* prefers to rest indoors and feed on humans during daylight hours. Its peak biting periods are early in the morning and before dark in the evening (Gubler, 1998). Once it has contracted the virus, the mosquito remains infected during its entire life and may transmit the virus during blood meals (de Figueiredo, 2012).

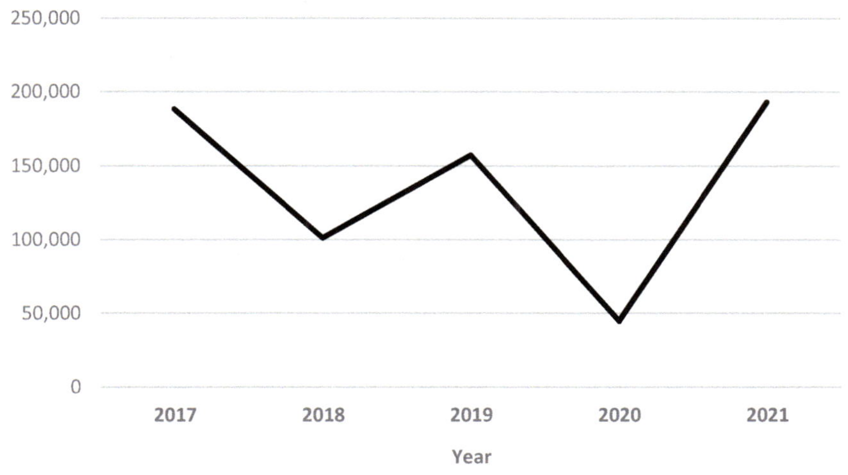

FIGURE 20.1 Number of dengue cases in India.

Source: Adapted from NVBDCP, https://ncvbdc.mohfw.gov.in/index4.php?lang=1&level=0&linkid=431&lid=3715.

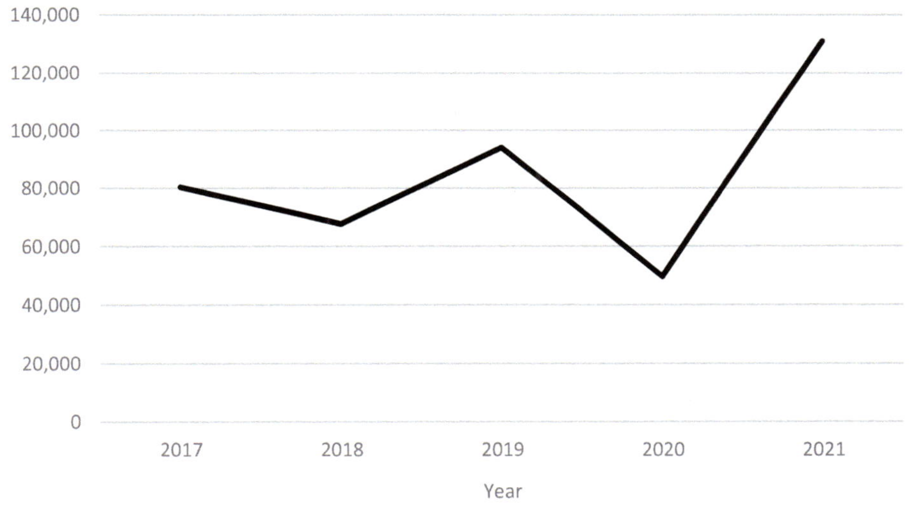

FIGURE 20.2 Number of suspected chikungunya cases in India.

Source: NVBDCP, 2022.

The potential for disease transmission by *Ae. aegypti* is a major factor for disease spread. In one of the studies, at a threshold level as low as two, adult mosquitoes emerging per day in a locality of 100 people has the potential to transmit disease (Lee et al., 2008). Awareness among the communities and proper waste management to eliminate potential breeding sites particularly during changes in temperature and humidity affecting population density and distribution would complement the implementation of integrated vector management to curb the transmission of diseases spread by *Aedes* mosquito.

Various types of vector-control strategies are used to curb the breeding of *Aedes* spp. The principal intervention categories are as follows (Weeratunga et al., 2017).

- Chemical interventions: insecticides, chemical larvicides
- Habitat management
- Non-chemical larvicides, e.g., larvivorous fish, oil coating, and mass trapping of larvae
- Population replacement methods like *Wolbachia*-infected mosquito
- Genetic techniques like the sterile insect technique

Detailed discussion of each of them is beyond the scope of this review.

20.5 GENETICALLY MODIFIED MOSQUITOES

> If major advances are to be made in coping with most of the major arthropod pest problems, then the tactics and strategies for managing such insects, ticks and mites must change. They must change from the current, limited scale, reactive, broad-spectrum measures to preventive measures that are target-pest specific and rigidly applied on an area-wide basis.
>
> (Knipling, 1992)

Mosquitoes will continue to challenge mankind, forcing a review of existing management tools and strategies, and stimulating the development and need for new innovative control strategies. Investing in research for new innovative technologies, including the use of genetically modified mosquitoes (GMMs), could be an integral part of strategies to combat the frequent epidemic disease outbreaks. GMMs are one of the emerging technologies and have several advantages; however, their use and implementation in integrated management for vector control yet need public and scientific discourse. GMM strains are primarily being developed for their use in area-wide control programmes and with other vector-control tools to control and reduce the rate of disease transmission. GMMs have also been reported to be compatible with other disease control methods, such as the use of drugs and or vaccines (Alphey et al., 2013; Bhatt et al., 2013). For instance, the inclusion of GMMs in an integrated approach to control disease transmission could improve the protective potential of new vaccines by reducing the number of infectious mosquitoes. During the last decade, one such transgene-based sterile insect technique being tested for its effectiveness in many countries is the OX513A strain, a genetically engineered (GE) *Ae. aegypti* mosquito strain developed by Oxitec Ltd, Abingdon, United Kingdom. Open field trials conducted in Brazil have shown effective suppression of the natural population of *Ae. aegypti* by sustained releases of OX513A male adults over a period, and the strain has been deregulated for its use in management and controlling the *Ae. aegypti* population (Harris et al., 2012, 2011; Carvalho et al., 2015).

The phased testing pathway is considered an appropriate approach to risk assessment for the implementation of any such new technologies in various scientific fields, including GMOs (genetically modified organisms), GM crops, pharmaceuticals chemicals, etc. WHO has proposed a guidance framework (WHO, 2021) for testing and implementation of GMMs, as illustrated in Figure 20.3. The phased testing also includes consideration of ethical, social, and cultural issues in each phase.

20.5.1 Genetic Technologies for Mosquito Control

Four types of mosquito technologies (some of which are not necessarily GM technologies) are currently available or have been designed.

20.5.1.1 Radiation-Based Sterile Insect Technology with the Release of Sterile Males

This involves release of radiation or chemically induced sterile males in wild for suppressing mosquito population with repeated release. Sterile insect technology (SIT) has been used in *Aedes* and experimental field releases have been done in some instances (Dame et al., 2009; Gato et al., 2021).

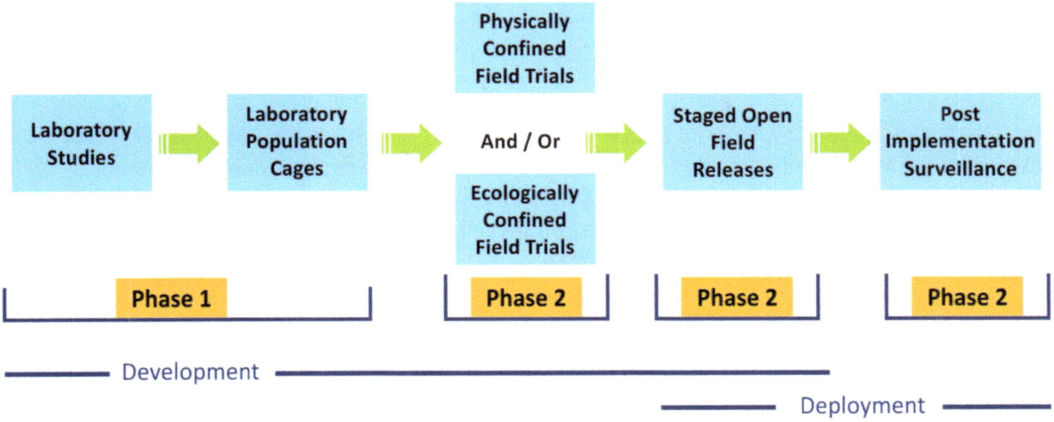

FIGURE 20.3 Schematic diagram illustrating the unidirectional pathway activities pertaining to development of basic data and deployment of a GMO in the field.

Source: WHO 2021.

20.5.1.2 Self-Limiting Insects Carrying Dominant Lethal Gene (RIDL)

The technology involves release of laboratory reared GM male mosquitoes carrying a dominant lethal gene. These mosquitoes mate with wild females, and the eggs do not hatch into adults. Thus, a localized population gets suppressed over time with repeated release of the GM male mosquitoes. RIDL technology has been approved for commercial release in Brazil and is being tested in open environments at several locations globally (Carvalho et al., 2015).

20.5.1.3 Endosymbiont *Wolbachia* Mediated Incompatible Males

Incompatible insect technique (IIT) involves the release of male mosquitoes infected with *Wolbachia*, a maternally inherited endosymbiotic bacterium. Due to cytoplasmic incompatibility, mating between *Wolbachia*-infected males and wild-type females not infected with the same *Wolbachia* strain yields non-viable eggs. Repeated releases of *Wolbachia*-infected males are thus expected to suppress wild-type mosquito populations and reduce disease transmission. This is being tested with open release in some countries in Latin America, Asia, and Oceania (Zheng et al., 2019; Crawford et al., 2020).

20.5.1.4 CRISPR-Based, Precision-Guided Sterile Insect Technology (pgSIT)

A simple CRISPR-based approach is used to generate sterile males to disrupt genes for male fertility and female viability (i.e., flight, which is essential for mating). Separate lines expressing Cas9 and guide RNAs are crossed to generate sterile males and short-lived, flightless females. The release of sterile males results in population suppression. Presently, the technology is being tested on the laboratory level (Li et al., 2021).

Several other techniques like those involving 'gene drives' (Gantz et al., 2015; Terradas et al., 2023) are in the process of development and are mostly in their laboratory stages, providing an attractive option to control either by suppression or replacement of mosquito population.

The aforementioned strategies under development and or developed strategies have their own pros and cons. All these strategies have to pass stringent assessment for field implementation, which involves biosafety studies as and when developed by the respective countries where they are to be implemented. The following discussion revolves around biosafety study requirements for deployment of such technology, with specific regulations applied in India.

20.6 GENERAL CONSIDERATIONS AND REGULATORY APPROVAL FOR THE NEXT STAGE OF RESEARCH

The phased biosafety studies pertaining to GMMs as proposed by WHO are widely accepted by many countries (Figure 20.3). With minor modifications as may be needed on a case-by-case basis based on the development of technology. Each of these phases are evaluated and monitored by regulatory authorities as per the country's regulatory framework to move to the next step. Note that with the development of gene drive–modified mosquitoes (GDMMs), the testing pathway may have to be modified to suit the requirements (WHO, 2021).

Activities involving GMOs, also referred to as living modified organisms (LMOs) are also governed by international treaties such as the Convention on Biological Diversity (CBD), Cartagena Protocol on Biosafety, Nagoya Kuala Lumpur Supplementary Protocol on Liability and Redress and Nagoya Protocol on Access and Benefit Sharing etc. Hence, it is important to understand applicable national regulations and international obligations while undertaking research and development of GMMs.

20.7 GLOBAL REGULATORY FRAMEWORK

Different regulatory frameworks exist in various countries with their defined scope of jurisdiction. As GM crops have taken the centre stage of GMOs, the regulatory frameworks globally have been developed with basic considerations of GM crops and mostly based on such hierarchy. However, not all countries where research or release is being undertaken follow a similar route of approval. For example, in the United States experimental field releases of GMMs was given by the Environmental Protection Agency (EPA, 2022) in 2022, while in Brazil, the National Biosafety Technical Committee (CTNBio) grants permission for field trials and commercialization (Andrade et al., 2018).

India has a robust, well-defined regulatory structure for evaluating and monitoring the biosafety aspects of research with GE organisms, that currently holds good for GMMs also. All activities are regulated as per the "Rules for Manufacture, Use/Import/Export and Storage of Hazardous Microorganisms/Genetically Engineered Organisms or Cells, 1989" (Rules, 1989) as required by the Ministry of Environment, Forest, and Climate Change (MoEF&CC), Government of India under the Environment (Protection) Act (EPA, 1986). Under these rules the mandate and functions of the competent authorities are given next.

20.8 COMPETENT AUTHORITIES IN INDIA: MANDATE AND FUNCTIONS

20.8.1 Recombinant DNA Advisory Committee

The Recombinant DNA Advisory Committee (RDAC) functions in the Department of Biotechnology (DBT), Ministry of Science and Technology, and is responsible for the review of developments in biotechnology at the national and international levels, and recommends suitable and appropriate safety regulations for India in recombinant research, use, and applications from time to time.

20.8.2 Institutional Biosafety Committee

An Institutional Biosafety Committee (IBSC) is constituted by an occupier or any person including research institutions, handling hazardous microorganisms/genetic engineered organisms at the research and development level. The occupier or any person, including research institutions, shall prepare, with the assistance of the IBSC, an up-to-date onsite emergency plan according to the manuals/guidelines of the Review Committee on Genetic Manipulation (RCGM) and make available copies for the District-Level Committees (DLC)/State Biotechnology Coordination Committees (SBCC) and RCGM/Genetic Engineering Appraisal Committee (GEAC).

20.8.3 Review Committee on Genetic Manipulation

RCGM functions in the DBT to monitor the safety related aspects in respect of ongoing research projects and activities involving GE organisms/hazardous microorganisms. It shall bring out manuals of guidelines specifying procedures for regulatory processes with respect to activities involving GE organisms in research, use, and applications, with a view towards ensuring environmental safety. RCGM shall lay down procedures restricting or prohibiting production, sale, importation and use of such GE organisms of cells.

20.8.4 Genetic Engineering Appraisal Committee

The GEAC functions in the MOEF&CC and is responsible for the appraisal of activities involving large-scale use of hazardous microorganisms and recombinants in research and industrial production from the environmental angle. The committee is also responsible for the appraisal of proposals relating to the release of GE organisms and products into the environment, including experimental field trials.

In addition, there are provisions for SBCCs and DLCs for the monitoring of activities.

20.9 BIOSAFETY AND RISK ASSESSMENT STUDIES RELATED TO GENETICALLY ENGINEERED MOSQUITOES

20.9.1 Risk Assessment

Risk can be defined as the likelihood that harm will occur from a particular action, and the level of risk is estimated to be the probability at which this harmful event will occur, along with consequences or the impact of the event. Risk assessment (RA) is a methodological approach for defining the level of risk. Risk management (RM) is the process to develop strategies to avoid and reduce risk to acceptable levels. Risk analysis encompasses RA and RM, as well as risk communication. With proper risk analysis, the focus areas and risk acceptability can be communicated to the developers, decision-makers, and public.

The principal outcome of risk analysis and assessment after identifying the defined hazards are the potential routes of exposure; their measurement; their effect on target organism, non-target organism, and human health; the level of concern about the negative impacts; and the planned action to mitigate the risks. The planned action identified for both standard and responsive measures should be acceptable to regulators and possibly to the affected community through risk communication needed to determine that RM is also acceptable to a wider community.

The evaluation of risk should be set against the benefits of GMMs for improving human health on a case-by-case basis. Cost-benefit or cost-effectiveness analyses can provide the framework under which the appropriate (economic, health, social) returns of a GMMs-release programme may be quantified and provide a context for decision-making about the level of acceptable risk. RA of novel technologies should be set against the risk of relevant alternatives, such as the risk of no action or the risk of conventional control methods. For example, "causes more harm" than current practice is a reasonable comparator for RA of GMMs-based vector-control systems.

On evaluation, risk in some cases may be judged as negligible when the probability of a harmful event will occur is determined to be very low or the consequences of an event occurring would be minimal. Moreover, in many cases, despite potentially harmful events being identified, the practical level of risk to which the public is exposed can be reduced to acceptable levels by effective management. The identification of potential hazards does not in itself indicate an unacceptable risk (WHO, 2021).

20.9.2 Biosafety Evaluation

It is accomplished in different phases for which the following standard operating procedures are to be followed, and trained personnel are required in both the laboratory and field in nature.

20.9.2.1 Development of Standard Operating Procedures and Training

Detailed standard operating procedures (SOPs) must be developed for rearing, release, monitoring, and efficacy studies, determining the expression level in GMMs. The following are suggested for such studies:

i. Baseline studies involving wild species and GE mosquitoes
ii. The number of mosquitoes to be released should be given in mosquitoes per unit area
iii. Source of food for insects to be provided
iv. Mating competitiveness against mosquito strains having a wild genetic constitution: The detailed description of actual number and ratios of GE male/female and wild-type male/female to wild-type male/female mosquitoes should be provided
v. Frequency of GMMs that express the desired characteristic and the level of expression in strains containing the wild genetic population
vi. Appropriate insect traps should be used for collecting the insect sample. The GMM should be identified for the desired gene by PCR (polymerase chain reaction). These data should be correlated to determine the frequency of GMMs that express the desired characteristic
vii. For population suppression and replacement strategies, the rate of suppression and replacement, respectively, against wild types in cage trials should be provided
viii. Mass rearing of both wild-type and GE insect should be carried out at similar large field containment. Strict quality control mechanisms should be in place to release only quality GMM. The weekly data of the wild-type insect population should be monitored and included in the report
ix. Population study to be done includes egg hatching rates, survival age, mating patterns, behaviour, male biology, population size and structure, mechanisms of population regulation, fitness and phenotypic effects of colonization and mass production, and other parameters as required to establish the efficacy and biosafety. This will develop appropriate contained semi-field systems to improve understanding of the biology of (transgenic) mosquitoes
x. The frequent monitoring of inside and outside the cage for (1) GMMs, (2) mosquito predators and vertebrate insectivores, (3) blood hosts for mosquitoes, and (4) other non-target vectors should be carried out and reported
xi. Appropriate statistical analysis should be done before interpretation of results

20.9.2.2 Phase 1 Testing

Phase 1 testing is essentially contained within the laboratory of Arthropod Containment Level 2 (ACL 2) nature (ACL, 2003). In India laboratory work with GMMs is required to be carried out in Insect Biosafety Level 2 (IBSL 2) laboratories. The parameters governing such laboratories are in minor modification from ACL 2 type. In Phase 1, testing choice of appropriate comparators and the data are to be generated to determine the end points that should be established. The efficacy and stability of the transgene at the population level with the evaluation of partial transgene function needs to be determined. In the case where multiple transgenes are incorporated, the effect of them on the phenotype needs to be evaluated rather than studying the genes in isolation. Wherever only one type of sex-separated GMMs are deployed, the methodology of sex separation and possible impact in the case of any mixture needs to be assessed. Also important is the methodology for the detection of transgenes in the wild population, monitoring of gene integrity maintenance, and trial determination of trial end points. With the release of GMMs, the effect on non-target organisms (NTOs) and ecological processes needs to be analysed.

20.9.2.3 Phase 2 Testing

Phase 2 testing could be either in physically contained cages or in ecologically confined areas, such as islands separated adequately from the mainland or locations which are geographically and ecologically isolated so that it is not possible for the GMMs to mix with the neighbouring population. The choice depends upon the type of GMMs to be deployed. For example, testing of mosquitoes containing gene drives in confined cages would not provide any meaningful end points which can only be achieved in a dynamic population.

Contained Phase 2 trials with GMMs needs to be conducted in a facility such as large cages in open environmental conditions. Under contained conditions, there is a physical barrier or barriers that contain organisms being tested so there is virtually no direct contact of viable GMOs with the environment. Activities carried out within such contained facilities are performed with specific biosafety guidelines.

General information that may be required to be furnished before undertaking the trials are as follows:

1. Name of the developer with contact details of the proposed activities
2. Objectives of proposed activities
3. Name of the GMMs
4. Intended use of the proposed product
5. Summary of the product characteristics and process of development
6. Details of prior safety studies and protocols for the proposed product approved if conducted anywhere outside the proposed country
7. Containment measures put in place for the proposed activities
8. Decontamination and disposal mechanisms at the research locations
9. RM (emergency plans)
10. Technical and scientific details, including the following:
    i. The scientific and common name(s) for the parent species
    ii. Biology document
    iii. Details of genetic modification including all steps of the genetic modification with complete nucleic acid sequence and origin of vectors, regulatory elements, markers and genes, and descriptions of all methods used
    iv. Information on the scheme for developing the final modified organism
    v. Details of new trait(s) in the resulting GMM should include the final modified sequences that are a part of the genome of the modified organism, the traits encoded or altered by the modified sequences and phenotypic characteristics that are different from the parental species. Also, the rate and level of expression on mating with the non-GM counterpart needs to be quantified
    vi. Mode of action of the expressed traits, including potential for allergenicity, toxicity, or pathogenicity
    vii. Important information is provided on the copy number (complete and partial copies), including any effects (functional, insertional, or positional; intentional or unintentional) the modifications may have on the final product; only intended sequences that are integrated into the genome of the final GM product are present, and unintended sequences (such as vector backbone, site-specific recombination recognition sequences, or selectable markers) are absent
    viii. Additional information may also be required to be provided on the specific breeding requirements for maintaining subsequent generation, stability of GM products in subsequent generations, or detection methods of the GM organism with limit of detection

20.9.2.3.1 Contained Trials in Cages

Several studies have been done with contained trials in cages (Facchinelli et al., 2011; Ferguson et al., 2008; Patil et al., 2022). The design and methodology may be adopted for an effective evaluation. Essentially, cage or caged refers to conditions where organisms, GE or otherwise, are physically segregated from the field. Fully caged conditions closely simulate natural habitats which can serve as an intermediate research stage between controlled laboratory conditions and field trials involving release to the environment. Some of the essential features that the cages must have to ensure physical isolation and prevention of the transgene leaking into the environment are as follows:

i. A semi-field system may consist of large caging units sufficient to provide free movement of the species under test. The experimental cage units used in successful evaluation measured 6 × 6 × 2 m (L × W × H).
ii. The number of units depends on the experimental requirement; however, the caging units should be divided into half by mesh walls where one half is used for the control experiments.
iii. One caging unit is to be used as a field laboratory for larval rearing and adult mosquito handling.
iv. Each caging unit needs to be further covered with another cage leaving a buffer zone of 1 metre on sides and top to provide additional containment. This space is to be used for monitoring personnel to walk between layers to check the cage integrity and set traps to monitor for escaped mosquitoes. BG (BioGents) traps can be placed in this buffer zone to capture any escapees. Zippered access to inner experimental cages is to be provided. Each pair of the experimental cages may be provided with openings to access inside of cage for example two vestibules that allow access to cages through sleeves, opening to shelves inside the cage and an opening to the cage floor. The platform of the cages are to be covered with white plastic sheets.
v. Cages need to be made with white tricot mesh (120–200 mesh/square inch) reinforced with white fabric and are located on a platform elevated 1 m above the ground, with ~3–5 m between the top of each cage and the roof covering the platform. The mesh selected should be resistant to ripping, rusting, tearing, and stretching. A solid plastic roof with sunscreen around the edges should cover all cages to provide shade and protection from direct sunlight.
vi. The whole structure should be isolated from the ground by ant traps (i.e., metal buckets filled with water and soap) under the legs of the platform, which is intended to minimize ground access to the cage by arthropods and small vertebrates.
vii. A plastic sheet laid on the ground below the platform prevents grass and other plants from growing under the cage and eventually touching the platform, which could be a means for arthropods and small animals to climb onto the platform and eventually into the cage.
viii. Items like a potted plant, clay bricks and clay pots with soil can be placed inside the cage to provide refuge for mosquitoes.
ix. SOPs should contain a plan for replacement of mesh before it becomes worn and in such a way that containment is not compromised.
x. Measures should be taken to protect the cages from the intrusion of unwarranted entry of human or animals.
xi. The large field cage experiment protocol should also include procedures for detection of the potential escape of mosquitoes to the open environment. ovitraps at different locations and mosquito traps should be distributed on the ground around the cage platform.

20.9.2.3.2 Ecologically Confined Trials

i. In case of ecologically confined trials, appropriate site selection criteria should be made (Lanzaro et al., 2011), bearing in mind the spatial location, timing, and duration of the trials.

ii. Spatial extent of the trial, including potential risks in areas outside the designated trial site(s).
iii. Development of detailed SOPs to ensure that rearing, release, and monitoring are carried out consistently with the relevant assumptions made in RA, with clear lines of responsibility and reporting, and RM strategies for field trials.
iv. Potential for unanticipated effects on disease burden.
v. Non-target species assessments, if appropriate, for confined field trials.

20.9.2.4 Phase 3 Testing

Following the successful completion of the Phase 2 field trials, the next phase of testing (Phase 3) is staged open field releases. The following information may be required from the applicant seeking approval to conduct the trials.

i. Collection of surveillance data: Surveillance data includes incidence and intensity or prevalence of the target insect at a specific point of time or season at a given site or location (Patil et al., 2020). Surveillance data would benefit from the availability of baselines before release (such as the level and seasonal pattern of disease burden, the past abundance of the vector population, effects of conventional vector-control methods). Assessment of population size, age structure, and/or sex ratio should also be considered. The applicant may have to produce supportive epidemiological vector-borne diseases data from the release sites and/or surrounding areas before open field release
ii. Epidemiological data on vector-borne diseases in the proposed experimental sites/villages
iii. Information on vectorial capacity of the wild-type and GMMs
iv. Annual meteorological data of the proposed site
v. Data on probable direct and indirect feeding of dead tissues by non-target animals available in and around the trial site
vi. Insecticidal spray details
vii. Data on release parameters, i.e., release density, frequency, release time, weather condition, height of release
viii. A detailed map of the trial site covering more than the available flight range of the mosquito species and location of ovitraps and sentinel traps to be used during the trial
ix. Characterization of local target mosquito ecology as required to set appropriate trial end points, including impact on human health and the wider environment
x. Methods for evaluating GMM's success through population-level assessments
xi. Appropriate RM plans for any potential resistance to the genetic modification, designating the lines of responsibility for managing this risk
xii. Proportionate assessment and management of non-target and off-target effects and the likely risk of transgenic gene flow
xiii. Proportionate assessment and management of risks associated with the mass production of mosquitoes

20.9.2.5 Phase 4 Activities

When a decision is made to deploy GMMs broadly as a public health tool, there may be a need for post-implementation quality control and surveillance to monitor for effectiveness and development of specific risks identified by post-release assessment. Biosafety considerations in Phase 4 should include the following:

i. Methods available for ongoing monitoring of the epidemiological impact of GMMs on human health
ii. Methods available for ongoing monitoring of safety for the environment and human health (in a manner analogous to pharmacovigilance, the monitoring applied to medicines after introduction to market)

iii. Available mitigation methods in the case that a negative effect is observed
iv. Risk implications and management of the movement of GMMs across borders

20.10 ETHICAL CONSIDERATIONS AND COMMUNITY ENGAGEMENT

As GMM technology is not limited to laboratory alone and involves communities where the tests are to be undertaken, several considerations, such as nature and scope of the obligations and kinds of protection, need to be provided. It should be kept in mind that ethical considerations and engagement have a much broader outreach than what it is thought of from the ones that has been mandated by policy- or lawmakers. Debates and discussion from transparent communication to the stakeholder community on the goals, the methods to be applied and the potential risks and benefits arising out of this technology should be continued and their voices heard by scientists and decision-makers. Informed consent will be required whenever human or household level identifiable data or clinical samples are collected. Intensive communication is also required for any community which may not be directly associated with the research but nonetheless affected by such research, and an agreement should be reached before progressing with the research. Equally important is to consider and respond to the views of people who are not associated with the research but have interest in the conduct or outcome of the research. GMM projects should incorporate a communications/public engagement strategy that not only includes information about the goal and methods but also provides opportunities for follow-on discussion.

Research must be led by in-country scientists in a co-development approach along with the public. This would in turn promote the community to interact with the scientists and promote further progress of the trials. Also, engagement must be continuous, beginning from the stage where research moves out of the laboratory. The basic aim of such engagement is not only to provide information on technology, risk and benefits but also not to provide any misleading assurances that may lead to the dropping of the current practices to protect themselves.

Community engagement and undertaking activities through consensus will increase as the research progresses to Phases 3 and 4. Implementation of GMMs as a public health initiative will happen after considering legal, regulatory, and political decisions. However, the need for public engagement activities is likely to continue.

A point to note is that it may not be sufficient for the scientific team to undertake communication exercises at different levels. Specialized skills of communication experts and social scientists may be required. It may be highlighted at the end that independent, ongoing safety reviews during testing are recommended, covering relevant aspects of environmental monitoring and human health. This may be accomplished through existing institutional- or national-level biosafety committees or through the establishment of new review bodies focused on GMM activities. The strengthening of biosafety oversight capabilities within disease endemic countries should be encouraged. National biosafety laws and regulations developed primarily to regulate GM crops may need to be reinterpreted for GMMs or additional guidance provided.

REFERENCES

Alphey, L., Andrew, McKemey, Nimmo, D., Oviedo, M.N., Lacroix, R., Matzen, K. and Beech, C. (2013) Genetic control of Aedes mosquitoes. *Pathogens and Global Health*, 107(4): 170–179. 10.1179/2047773213Y.0000000095

Andrade, P.P., Ferreira M.A. da Silva, Muniz, M.S. and Lira-Neto, A. de Casto (2018). GM insect pests under the Brazilian regulatory framework: development and perspectives. *BMC Proceedings* 12(Suppl. 8): 16. 10.1186/s12919-018-0107-z

Arthropod Containment Levels (ACL) (2003). *Vector Borne and Zoonotic Diseases*, 3, 75–90.

Bhardwaj, S., Gokhale, M.D.and Mourya, D.T. (2017). Zika virus: Current concerns in India. *Indian Journal of Medical Research*, 146(5), 572–575. 10.4103/ijmr.ijmr_1160_17

Bhatt, S., Gething, P.W., Brady, O.J., Messina, J.P., Farlow, A.W., Moyes, C.L., Drake, J.M., Brownstein, J.S., Hoen, A.G., Sankoh, O., Myers, M.F., George, D.B., Jaenisch, T., Wint, G.R., Simmons, C.P., Scott, T.W., Farrar, J.J. and Hay, S.I. (2013) The global distribution and burden of dengue. *Nature*, 496, 504–507. 10.1038/nature12060

Brady, O.J., Gething, P.W., Bhatt, S., Messina, J.P., Brownstein, J.S., Hoen, A.G., Moyes, C.L., Farlow, A.W., Scott, T.W. and Hay, S.I. (2012). Refining the global spatial limits of dengue virus transmission by evidence-based consensus. *PLOS Neglected Tropical Diseases*, 6(8): e1760. 10.1371/journal.pntd.0001760

Carvalho, D.O., Mckemey, A.R., Garziera, L., et al (2015) Suppression of a field population of *Aedes aegypti* in Brazil by sustained release of transgenic male mosquitoes. *PLoS Neglected Tropical Diseases* 9(7): e0003864. 10.1371/journal.pntd.0003864

Cecilia, D. (2014) Current status of dengue and chikungunya in India. WHO South East Asia. *Journal of Public Health* 3(1): 22–26. 10.4103/2224-3151.206879

Chaturvedi, U.C. and Nagar, R. (2008) Dengue and dengue haemorrhagic fever: Indian perspective. *Journal of Biosciences*, 33, 429–441. 10.1007/s12038-008-0062-3

Crawford, J.E., Clarke, D.W., Criswell, V. et al. (2020). Efficient production of male Wolbachia-infected *Aedes aegypti* mosquitoes enables large-scale suppression of wild populations. *Nature Biotechnol* 38, 482–492. https://doi.org/10.1038/s41587-020-0471-x

Dame, D. A., Curtis, C. F., Benedict, M. Q., Robinson, A. S. and Knols, B. G. J. (2009) Historical applications of induced sterilisation in field populations of mosquitoes. *Malaria Journal*, 8, S2.

De Figueiredo, R.M.P. (2012) Molecular Characterization of Dengue Virus Circulating in Manaus, the Capital City of the State of Amazonas, Brazil. In: *Current Topics in Tropical Medicine*. Rodriguez-Morales, Ed., chapter 6, pp. 81–90. InTech. 10.5772/26085

Enserink, M. (2006). Infectious diseases. Massive outbreak draws fresh attention to little-known virus. *Science* 311: 1085. 10.1126/science.311.5764.1085a

EPA (1986). Environmental Protection Act, No. 29. Government of India Notification No. G.S.R. 1198(E) published in the Gazette of India No. 525 dated 12-11-86.

EPA (2022). Environmental Protection Agency experimental use permit amended for 93167-eup-2 to allow releases of OX5034 *Aedes aegypti* in Florida and California. https://www.regulations.gov/document/EPA-HQ-OPP-2019-0274-0470

Facchinelli L, Valerio L, Bond JG, Wise de Valdez MR, Harrington LC, Ramsey JM, Casas-Martinez M, Scott TW. (2011). Development of a semi-field system for contained field trials with *Aedes aegypti* in southern Mexico. *American Journal of Tropical Medicine and Hygiene*, 85(2), 248–56. 10.4269/ajtmh.2011.10-0426

Ferguson, H.M., Ng'habi, K.R., Walder, T., Kadungula, D., Moore, S.J., Lyimo, I., Russell, T.L., Urassa, H., Mshinda, H., Killeen, G.F. and Knols, B.G.J. (2008) Establishment of a large semi-field system for experimental study of African malaria vector ecology and control in Tanzania. *Malaria Journal*, 7(158): 1–15.

Gato, R., Menéndez, Z., Prieto, E., Argilés, R., Rodríguez, M., Baldoquín, W., Hernández, Y., Pérez, D, Anaya, J., Fuentes, I., Lorenzo, C., González, K., Campo, Y. and Bouyer, J. (2021) Sterile Insect Technique: Successful Suppression of an *Aedes aegypti* Field Population in Cuba. *Insects*, 12(5), 469. 10.3390/insects12050469

Gantz, V.M., Jasinskiene, N., Tatarenkova, O., Fazekas, A., Macias, V.M., Bier, E. and James, A.A. (2015). Highly efficient Cas9-mediated gene drive for population modification of the malaria vector mosquito *Anopheles stephensi*. *Proceedings of the National Academy of Sciences* 112, E6736–E6743.

Gubler, D. J. (1998), Dengue and dengue haemorrhagic fever. *Clinical Microbiology Reviews*, 11(3), 480–496.

Guillena, J. B., Opena, E. L. and Baguio, M. L. (2010). Prevalence of dengue fever (DF) and dengue hemorrhagic fever (DHF): A description and forecasting. *Proceedings of the 11th National Convention on Statistics* (NCS 11), p. 16.

Halstead, S.B. (2007) Dengue. *The Lancet*, 370, 1644–1652. 10.1016/S0140-6736(07)61687-0

Harris, A.F., McKemey, A.R., Nimmo, D., Curtis, Z., Black, I., Morgan, S.A., Oviedo M.N., Lacroix, R., Naish, N., Morrison, N.I., Collado, A., Stevenson, J., Scaife, S., Dafa'alla, T., Fu, G., Phillips, C., Miles, A., Raduan, N., Kelly, N., Beech, C., Donnelly, C.A., Petrie, W.D. and Alphey, L. (2012). Successful suppression of a field mosquito population by sustained release of engineered male mosquitoes. *National Biotechnology* 30(9): 828–830. 10.1038/nbt.2350

Harris, A.F., Nimmo, D., McKemey, A.R., Kelly, N., Scaife, S., Donnelly, C.A., Beech, C., Petrie, W.D., and Alphey, L. (2011) Field performance of engineered male mosquitoes. *National Biotechnology* 29(11): 1034–1037. 10.1038/nbt.2019

Hung, T.M., Shepard, D.S., Bettis, A.A., Nguyen, H.A., McBride, A., Clapham, H.E. and Turner, H.C. (2020) Productivity costs from a dengue episode in Asia: a systematic literature review. *BMC Infectious Diseases*, 20(1):393. 10.1186/s12879-020-05109-0

Kay, B. and Vu, S.N. (2005) New strategy against *Aedes aegypti* in Vietnam. *The Lancet* 365(9459): 613–617. 10.1016/S0140-6736(05)17913-6

Knipling, E.F. (1992). Brief talk on the occasion of receiving the World Food Prize on 12th October 1992 at Des Moines, Iowa, USA (*Unpublished document*).

Koopmanschap, M.A., and van Ineveld, B.M. (1992). Towards a new approach for estimating indirect costs of disease. *Social Science & Medicine*, 34(9),1005–1010.

Knudsen, A.B., (1995). Global distribution and continuing spread of *Aedes albopictus. Parassitologia*, 37(2–3), 91–97.

Lanzaro, G.C., Campos, M., Crepeau, M., Cornel, A., Estrada, A., Gripkey, H., Haddad, Z., Kormos, A. and Palomares S. (2011) Selection of sites for field trials of genetically engineered mosquitoes with gene drive. *Evolutionary*, *14*(9), 2147–2161. 10.1111/eva.13283

Lee, H., Vasan, S., Nazni, W. and Shanaz M. (2008). *Scientific report on the innovative application of Aedes aegypti RIDL-sterile insect technique to combat dengue and chikungunya in Malaysia*. Institute of Medical Research, Kuala Lumpur, Malaysia.

Li, M., Yang, T., Bui, M., Gamez, S., Wise, T., Kandul, N.P., Liu, J., Alcantara, L., Lee, H., Edula, J.R., Raban, R., Zhan, Y., Wang, Y., DeBeaubien, N., Chen, J., Sánchez C.H.M., Bennett, J.B., Antoshechkin, I., Montell, C., Marshall, J. M. and Akbari, O.S. (2021) Suppressing mosquito populations with precision guided sterile males. *Nature Communications* 12, 5374. 10.1038/s41467-021-25421-w

Mousson, L., Dauga, C., Garrigues, T., Schaffner, F., Vazeille, M. and Failloux, A.B. (2005). Phylogeography of *Aedes (Stegomyia) aegypti* (L.) and *Aedes (Stegomyia) albopictus* (Skuse) (Diptera: Culicidae) based on mitochondrial DNA variations. *Genetical Research*, 86(1), 1–11.

Mutheneni, S.R., Morse, A.P., Caminade, C. and Upadhyayula S.M. (2017) Dengue burden in India: recent trends and importance of climatic parameters. *Emerging Microbes & Infections*, 6(8), e70. 10.1038/emi.2017.57

Pialoux, G., Gaüzère, B.A., Jauréguiberry, S. and Strobel, M. (2007) Chikungunya, an epidemic arbovirosis. *Lancet Infectious Diseases* 7(5): 319–327. 10.1016/s1473-3099(07)70107-x

Patil, P.B., Dasgupta, S.K., Seshu Reddy, K.V., Char, B., Zehr, U.B. and Barwale, S.R. (2020). Surveillance of *Aedes* mosquito species in villages of Jalna district, Maharashtra. *Journal of Entomology and Zoology Studies*, 8(2), 669–677.

Patil, P.B., Dasgupta, S.K., Gorman, K., Pickl-Herk, A., Puinean, M., McKemey, A., Char, B., Zehr, U.B., Barwale, S.R. (2022). Elimination of a closed population of the yellow fever mosquito, *Aedes aegypti*, through releases of self-limiting male mosquitoes. *PLOS Neglected Tropical Diseases* 16(5): e0010315. 10.1371/journal.pntd.0010315

Ramakrishnan, S.P., Geljand, H.M., Bose, P.N., Mukherjee, R.N. et al, (1964). The epidemic of acute haemorrhagic fever, Calcutta, 1963; epidemiological inquiry. *Indian Journal of Medical Research 52*: 633–650.

Ravi, V. (2006). Re-emergence of chikungunya virus in India. *Indian Journal of Medical Microbiology* 24: 83–84. 10.4103/0255-0857.25175

Robinson, M.C. (1955). An epidemic of virus disease in Southern Province, Tanganyika Territory, in 1952-53. I. Clinical features. *Transactions of Royal Society of Tropical Medicine & Hygiene*, 49(1), 28–32. doi: 10.1016/0035-9203(55)90080-8

Shepard, D.S., Halasa, Y.A., Tyagi, B.K., Adhish, S.V., Nandan, D., Karthiga, K.S., Chellaswamy, V., Gaba, M. and Arora N.K., (2014). The Inclen Study Group. Economic and disease burden of dengue illness in India. *American Journal of Tropical Medicine and Hygiene*, 91(6), 1235–1242. 10.4269/ajtmh.14-0002

Sudeep, A.B. and Parashar, D. (2008). Chikungunya: an overview. *Journal of Biosciences* 33:443–449. 10.1007/s12038-008-0063-2

Terradas, G., Bennett, J.B., Li, Z., Marshall, J.M. and Bier E. (2023) Genetic conversion of a split-drive into a full-drive element. *Nature Communications* 14, 191. 10.1038/s41467-022-35044-4

Vezzani, D. (2007) Review: artificial container-breeding mosquitoes and cemeteries: a perfect match. *Tropical Medicine and International Health*, 12 (2), 299–313.

Were, F. (2012). The dengue situation in Africa. *Paediatrics and International Child Health*, 32, Suppl. (s1), 18–21. 10.1179/2046904712Z.00000000048

WHO (2019), *Dengue Guidelines for Diagnosis, Treatment, Prevention and Control*. World Health Organization, Geneva, Switzerland.

WHO (2023). *Factsheet on dengue and severe dengue*. World Health Organization, Geneva, Switzerland.

Weeratunga, P., Rodrigo, C., Fernando, S.D. and Rajapakse S. (2017) Control methods for *Aedes albopictus* and *Aedes aegypti*. *Cochrane Database of Systematic Reviews*, 2017(8), CD012759. 10.1002/14651858

WHO (2021) *Guidance framework for testing genetically modified mosquitoes*, 2nd edition. Geneva: World Health Organization.

Zheng, X., Zhang, D., Li, Y. et al. (2019) Incompatible and sterile insect techniques combined eliminate mosquitoes. *Nature* 572, 56–61. 10.1038/s41586-019-1407-9

21 Regulatory Policy, Guidelines and Ethical Aspects in Genetically Modified Mosquito Research and Their Application in Disease Control Programmes, with Special Reference to India

S. K. Dasgupta
Ex-Tata Institute of Genetics and Society, New inStem Building, Bengaluru, India

Prabhakargouda B. Patil
Maharashtra Hybrid Seeds Company Pvt. Ltd., Jalna, India

Vibha Ahuja
Biotech Consortium India Limited, Anuvrat Bhawan, Deen Dayal Upadhyaya Marg, New Delhi, India

B.K. Tyagi
Chandigarh University, Mohali, India

21.1 INTRODUCTION

Testing and further deployment of any type of genetically modified (GM) mosquitoes is controlled through national laws and regulations in a country. Most countries have biosafety regulatory frameworks; however, the guidelines are mostly tailored to GM crops. With the application of genetic engineering technology to other types of organisms, new guidelines are being adopted, or existing laws and guidelines are being tweaked and revised to be suitable for GM mosquitoes. Apart from laws and guidelines of a particular country, several international agreements like Cartagena

DOI: 10.1201/9781003327028-26

Protocol for Biosafety (CPB) an agreement under the Convention on Biological Diversity (CBD), to which most countries in the world are parties, may also be relevant to the testing and use of GM mosquitoes.

Environmental releases of GM mosquitoes in various countries in the world (in some cases at a very large scale) have been undertaken, and they function mostly with a target of population suppression by inundating the population with GM mosquitoes at regular intervals. The released GM mosquitoes do not persist in nature (Kofler et al., 2018). However, the regulation of such necessitates the development of a clear guideline before conducting research. To date, research on and the environmental release of GM mosquitoes has been restricted to a few countries only, and some clear guidelines have been developed while few others are still in the process of development. Most of country specific policies have been developed based on policies and guidelines framed by multilateral international organizations.

21.2 INTERNATIONAL ORGANIZATIONS

Highlighted next are some of the policies and guidelines of global organizations that are responsible for key policy development in the governance of research and testing of GM mosquitoes.

21.2.1 The Convention on Biological Diversity

The United Nation CBD in its policy framework addresses the use of genetically modified organisms (GMOs). The CBD entered into force on 29 December 1993 with three main objectives (https://www.cbd.int/doc/legal/cbd-en.pdf):

i. The conservation of biological diversity
ii. The sustainable use of the components of biological diversity
iii. The fair and equitable sharing of the benefits arising out of the utilization of genetic resources

A total of 196 countries are parties to this convention spread across the globe. The convention also includes an obligation to regulate, manage or control GMOs or living modified organisms (LMOs) that may have an adverse effect on biodiversity. It also includes a legal basis for a protocol setting out the safe transfer, handling, and use of any LMO resulting from biotechnology where required. The Conference of the Parties to the CBD adopted a supplementary agreement on 29 January 2000 known as the Cartagena Protocol on Biosafety (https://bch.cbd.int/protocol/background/). The Cartagena Protocol establishes procedures for the safe transfer, handling, and use of LMOs and provides a risk assessment framework for LMOs. One hundred seventy-three countries are parties to this protocol. A supplementary protocol to the Cartagena Protocol the Nagoya – the Kuala Lumpur Supplementary Protocol, entered into force in March 2018 (https://bch.cbd.int/protocol/supplementary/about/) – was developed on liability and redress to specify measures to be taken in the event of damage to biodiversity resulting from LMOs and includes provisions in relation to civil liability.

21.2.2 World Health Organization

The World Health Organization (WHO) founded in 1948 is the United Nations agency that connects nations, partners, and people to promote health, keep the world safe, and serve the vulnerable. As mosquito-borne diseases are one of the greatest global health problems, WHO has a large programme of work around this disease. The World Malaria Programme through its world malaria report (WHO, 2022) provides statistics on malaria cases. WHO assesses new tools for malaria control, such as insecticides and vaccines, using an evidence-based framework. Tools for vector control pass through three WHO processes:

i. The Malaria Policy Advisory Group (MPAG) gives policy recommendations for all tools.
ii. The Vector Control Advisory Group (VCAG) assesses public health value for novel vector control tools based on epidemiological evidence.
iii. The prequalification team for vector control (PQTVC) assesses the safety, quality, and efficacy of new tools. Gene drive mosquitoes would most likely be assessed using the same processes as other vector control tools.

As early as 1991, WHO initiated discussions on genetic modification of insects that are vectors of human diseases with experts and initiated training courses. WHO has publicly acknowledged the value of GM mosquitoes as innovative vector control tools through its position statement on the evaluation and use of GM mosquitoes for the control of vector-borne diseases (WHO, 2020a).

WHO has issued guidance and recommendations in connection with testing and allied activities of GM mosquitoes.

i. The guidance framework for testing GM mosquitoes (WHO, 2021) aims to foster quality and consistency among processes for testing and regulating new genetic technologies by proposing standards of efficacy and safety testing comparable to those used for trials of other new public health tools.
ii. Recommendations for efficacy-testing of traps for control of Aedes mosquito vectors (WHO, 2018) supports product developers, programmers, and testing institutions to generate robust entomological evidence of the efficacy of vector traps for control and, for a first-in-class vector trap, evidence of the public health impact from reducing arboviral diseases.
iii. Guidance and recommendations on designing vector control efficacy trials (WHO, 2017a) for the phase III vector control field trial design supports designing and conducting phase III epidemiological field trials of new vector control interventions. A framework of steps is set out, as well as concerns to be considered when designing and conducting an epidemiological trial to assess the public health value of a new vector control intervention; these steps include defining the research question, choosing the study's design, randomizing the interventions, and calculating the sample size.
iv. Data requirements and methods to support the evaluation of new vector control products (WHO, 2017b): WHO reviewed data requirements associated with the evaluation of new vector control interventions. The aim is to ensure that these can be deployed as soon as possible while also ensuring that policy recommendations to guide deployment remain evidence based.
v. Considerations of ethical issues associated with vector-borne diseases is important and ethical issues associated with vector-borne diseases highlighting environmental, and social determinants of health, the ethics of vector control (including new technologies), relevant aspects of ethics in surveillance and research, and the ethics of mass public health interventions should be mapped (WHO, 2017c).

21.2.3 Commonwealth Scientific and Industrial Research Organisation

The Commonwealth Scientific and Industrial Research Organisation (CSIRO) is an Australian Government agency responsible for scientific research. Their Data 61 Ecological and Environmental Risk Assessment (DEERA) team are leading the implementation of probabilistic risk assessments for GM mosquitoes and for gene drive mammals.

CSIRO published a report (Hayes et al. 2015) on risk assessment for controlling mosquito vectors with engineered nucleases by controlled field release for sterile male construct and assessed the human health and environmental risks associated with a proposed field release of transgenic dominant sterile male *Anopheles gambiae* mosquitoes from insectaries in Western Africa.

21.2.4 Office of the Gene Technology Regulator (Australia)

Office of the Gene Technology Regulator under the Australian Government, Department of Health, issued guidelines in 2019 for regulatory requirements for contained research with GMOs containing engineered gene drives. These guidelines provide guidance for Institutional Biosafety Committees and researchers on notification, licensing and physical containment requirements. The guideline state that if gene technology is used to introduce or create a gene drive in an organism, the resulting organism will be a GMO and subject to regulation under the Gene Technology Act 2000.

21.2.5 European Food Safety Agency

The European Food Safety Agency (EFSA) is the agency of the European Union (EU) set up in 2002 that serves a source of scientific advice to risk managers. They are a key authority in providing the scientific basis for laws and regulation to protect European consumers from food related risks.

EFSA published a guidance document on the environmental risk assessment of GM animals in 2012 (EFSA, 2012) which includes recommendations for the environmental risk assessment (ERA) of animals, including insects. EFSA recommends assessing potential effects of GM animals on animal and human health and the environment and the rationales for data requirements for a comprehensive ERA. The document underlines that ERA should be carried out on a case-by-case basis, following a step-by-step assessment approach and highlights the six sequential steps for the assessment (1) problem formulation including hazard and exposure identification; (2) hazard characterization; (3) exposure characterization; (4) risk characterization; (5) risk management strategies, and (6) an overall risk evaluation.

In 2020 (EFSA 2020), EFSA published a report, at the request of the EU Commission, assessing whether their current guidelines for the risk assessment of GM animals (EFSA 2012, 2013), including insects. They concluded that the current framework was largely sufficient but that additional guidelines could be needed in some areas, such as persistence and invasiveness, use of modelling, and molecular tools for detection.

21.2.6 The International Union for Conservation of Nature

The International Union for Conservation of Nature (IUCN) is a membership Union uniquely composed of both government and civil society organizations. By harnessing the experience, resources, and reach of its more than 1,400 member organizations and the input of some 15,000 experts, IUCN is the global authority on the status of the natural world and the measures needed to safeguard it. IUCN has observer and consultative status at the United Nations and plays a role in the implementation of several international conventions on nature conservation and biodiversity, including the CBD. In 2016, the IUCN resolution "Development of IUCN Policy on Biodiversity Conservation and Synthetic Biology" was adopted to examine the impacts of the production and use of the products resulting from synthetic biology on the conservation and sustainable use of biodiversity to recommend how IUCN could engage in ongoing discussions and deliberations with the synthetic biology community and to develop guidance on the topic. This led to the formation of a task force and technical subgroup, and the publication of a technical assessment in 2019, "Genetic Frontiers for Conservation" (Redford et al. 2019), but an official IUCN policy was not formally adopted. Since then, a further resolution "Towards the Development of an IUCN Policy on Synthetic Biology in Relation to Nature Conservation" was made in 2019; the policy is still in development.

21.2.7 Organisation of Economic Development and Co-operation

The Organization for Economic Development and Co-operation (OECD) is an international economic non-governmental organization with a membership of 38 countries spanning from North and South America to Europe and Asia-Pacific. The organization works to build better policies for better lives with a goal is to shape policies that foster prosperity, equality, opportunity and well-being for all. They have produced 68 science-based consensus documents widely used by regulators. The science-based consensus documents collate information for use during the regulatory risk assessment of biotechnology products, i.e., transgenic organisms (plants, animals, microorganisms) when intended for release in the environment. Volume 8 of the series "Safety Assessment of Transgenic Organisms in the Environment" includes documents on the biology of *Aedes aegypti* mosquitoes (OECD, 2018) issued by the OECD Working Group on the Harmonisation of Regulatory Oversight in Biotechnology. This is the first 'consensus document' to deal with the biology of an insect, the mosquito *Aedes aegypti*, for supporting biosafety assessment. As *Ae. Aegypti* mosquito is vectoring yellow fever, dengue, Zika and chikungunya diseases in tropical and subtropical regions worldwide, biotechnological applications are being developed to control the mosquito population and reduce virus transmission. The consensus biology document provides information on Ae. aegypti taxonomy, morphology, life cycle, reproductive biology, genetics, ecology, interactions with other species and the environment. The mosquito effects on human and animal health, and the control strategies/specific programmes to limit its development are also summarized.

21.3 SELECTIVE NATIONAL LEGISLATION AND REGULATION PERTAINING TO GM MOSQUITOES

21.3.1 India

In India, activities involving GMOs are regulated as per the "Rules for Manufacture, Use/Import/Export and Storage of Hazardous Microorganisms/Genetically Engineered Organisms, or Cells, 1989" (GOI, 1989) notified under the Environment (Protection) Act, 1986. The "Rules, 1989" essentially cover the entire spectrum of activities involving GMOs and products derived from them, including their sale, storage, exportation, importation, production, manufacturing, packaging, etc. These rules cover areas of research, as well as large-scale applications of GMOs and their products. These rules are implemented by the Ministry of Environment, Forest and Climate Change (MoEFCC), Department of Biotechnology (DBT) under the Ministry of Science and Technology (DBT, 2017, 2020, 2022). There are half a dozen competent authorities: Recombinant DNA Advisory Committee (RDAC), Institutional Biosafety Committee (IBSC), Review Committee on Genetic Manipulation (RCGM), Genetic Engineering Appraisal Committee (GEAC), State Biotechnology Coordination Committee (SBCC), and District Level Committee (DLC). Of these IBSC, RCGM, and GEAC are responsible for regulatory approvals.

A series of guidelines to be followed at various stages of GMO development have been adopted under the "Rules, 1989" from time to time. The new guidelines "Regulations and Guidelines on Biosafety of Recombinant, DNA Research and Biocontainment, 2017" and "Guidelines for Establishment of Containment Facilities: Biosafety Level 2 (BSL2) & 3 (BSL 3) and Certification of BSL 3 Facility 2020" provide containment levels for microorganisms, animals, plants, insects and aquatic organisms. Insect biosafety levels (IBSLs) have been prescribed with details of facilities in order to prevent escape and establishment of the experimental arthropods in the natural environment and ensure the safety of laboratory personnel in the facility. Arthropods to be considered include but are not limited to insects (*Lepidoptera, Coleoptera, Diptera* [e.g., mosquitoes, fruit flies], *Hemiptera*), *Blattodea*, and *Arachnida* (ticks and mites). GM arthropods, or non-GM arthropods that are challenged/infected with GM organisms, are covered in the guidelines. RCGM has approved contained trials of GM mosquitoes in large cages (Patil et al., 2022).

In 2023, DBT issued "Guidelines and Standard Operating Procedures for Research on Genetically Engineered Insects" (DBT, 2023), which includes mosquitoes. The guidelines and standard operating procedures (SOPs) provide a regulatory roadmap and describe SOPs and data requirement for conducting research with GE insects under contained facilities. The guidelines are also applicable to GM mosquitoes.

21.3.2 Brazil

Brazil is the only country to have approved commercial release of GM mosquitoes (CTNBio, 2014a). In Brazil, the principal legal framework for biotechnology is governed by Federal Law No. 11105/2005 providing safety regulation and inspection tools for activities concerning GMOs and their by-products. The National Biosafety Council (CNBS) created under this law provided a new format for the National Biosafety Technical Commission (CTNBio; http://ctnbio.mctic.gov.br/inicio) and established a framework through the National Biosafety Policy (PNB).

CTNBio belongs to the Ministry of Science and Technology of the Federal Government of Brazil, whose purpose is to provide technical advisory support and advice to the Federal Government in formulating, updating and implementing the National Biosafety Policy related to GMOs, as well as the establishment of technical safety standards and technical opinions regarding the protection of human health, living organisms and the environment, for activities involving construction, experimentation, cultivation, handling, transport, marketing, consumption, storage, release and disposal of GMOs and derivatives.

All organizations (university, research institution, and industry) must have an internal biosafety commission (CIBio) that is responsible for ensuring biosafety within their areas. Any activity in containment using risk class 1 GMOs is assessed and eventually authorized by the CIBio. For all other risk classes or for activities in confinement (field releases), the CIBio assesses the proposal, but CTNBio gives the final approval. Research labs and other facilities dealing with GMOs must get a Certificate in Biosafety (CQB) given by CTNBio.

The requirements for approval of commercial products acceptance may take years. However, with the advent of new rules, the waiting time for commercialization could be expedited. After approval, the executing organization is required to monitor for adverse effects if the pre-approval risk assessment identifies a risk.

The Brazilian regulatory framework has been applied to GM mosquitoes (de Andrade et al., 2016) and Brazil has approved the use of GM mosquitoes. These mosquitoes do not contain a gene drive construct; they express a conditional lethal gene and a fluorescent marker. An example risk assessment is available through the BCH (CTNBio 2014b).

21.3.3 Malaysia

To regulate the release, import, export, and contained use of LMOs and the release of their products, with the objective of protecting human, plant, and animal health, environment and biological diversity under the Biosafety Act 2007 (Act 678; Biosafety Act, 2007) of the National Biosafety Board was established. The board decides on all applications; monitors activities relating to LMOs and the products of such organisms; promotes research, development, education, and training activities related to biosafety; and establishes mechanisms to facilitate the collection, storage and dissemination of data related to LMOs, the products of such organisms and biosafety. The Genetic Modification Advisory Committee has been established to provide scientific, technical and other relevant advice to the director general of the board.

An application is submitted to the director general, in a prescribed format, for the approval of any release activity and/or any importation of LMOs, along with a risk assessment, risk management report, and emergency response plan.

GM mosquitoes were released for testing in Malaysia under regulatory approval in 2010 (Subramaniam et al., 2012). The publication describes in detail the regulatory process applied for this release.

21.3.4 United States of America

The USA uses its existing national legislation and agencies to regulate GMOs under the Coordinated Framework for Regulation of Biotechnology developed in 1986 (OSTP, 1986). The Coordinated Framework for Regulation of Biotechnology is guidance and not law in the USA. An update to the Coordinated Framework (OSTP, 2017) was published to clarify the current roles and responsibilities of the primary agencies, i.e., the Food and Drug Administration (FDA), Environmental Protection Agency (EPA), and United States Department of Agriculture (USDA).

The roles of the EPA and FDA with respect to mosquito-related products were further clarified by guidance issued in 2017 (FDA, 2017). According to this guidance, FDA regulates mosquito-related products intended to reduce the virus/pathogen load within a mosquito and/or reduction in virus/pathogen transmissibility from mosquitoes to humans or prevent mosquito-borne disease in humans or animals. The EPA regulates mosquito-related products intended to reduce the population of mosquitoes either by killing them or by interfering with their reproduction or development.

For any action that may significantly affect the environment on release of a GMO an environmental assessment (EA) or a broader environmental impact statement (EIS) will have to be prepared. The National Environmental Policy Act (NEPA) imposes procedural requirements, including an open public comment phase announced in the USA Federal Register for all federal agencies to prepare such analysis prior to making a decision before release (FDA, 2016). The EPA issued an Experimental Use Permit in 2020 to allow the release of GM mosquitoes at US sites (https://www.regulations.gov/document/EPA-HQ-OPP-2019-0274-0470).

21.3.5 Sri Lanka

In Sri Lanka, GMOs/LMOs are regulated as per the National Biosafety Framework of Sri Lanka (NBF), 2005. In 2021, the Ministry of Environment published the "Guidelines for Testing of GM Mosquitoes" (MOE, 2021), which have been adapted from the guidance framework for testing of GM mosquitoes by WHO, and a phased testing of GM mosquitoes has been proposed.

21.3.6 Other Countries

***38.3.6.1 Mexico*:** The regulatory process for the conduct of physically confined outdoor testing of GM mosquitoes has been documented (Ramsey et al., 2014)

***38.3.6.2 Burkina Faso*:** The use of GM mosquitoes falls under environmental law and implementing regulations and requires environmental and social impact assessment. Active GM mosquitoes research is currently ongoing in Burkina Faso. Burkina Faso authorized the contained use of GM mosquitoes in 2016 (ABNE, 2017) and subsequently the limited field release of GM mosquitoes in 2019 (Target Malaria, 2019)

21.3.6.3 Panama and Cayman Islands *(British overseas territory)* have also issued permits to undertake field trials with GM mosquitoes (Gorman et al., 2016) but no specific regulatory structure and guidelines have been developed here.

21.4 ETHICAL CONSIDERATIONS

With the ongoing research and plans for releasing GM mosquitoes for vector-borne disease control programmes, there is a need for thorough consideration of ethical issues from the very beginning, starting from the conception of an idea. There should be clear communication with all stakeholders, including the communities and public in general, about its purpose, goal, effect on the environment if any, evaluation and management of risk and benefits. Communication should be transparent and made at all stages of development. Respect for communities should be the main ethical goal in GM mosquito research. Community engagement will clarify any concern for affected communities and fulfilling ethical responsibilities to them and as appropriate; individual- or household-level informed consent must be taken. GM mosquito research should also recognize ethical responsibilities that extend beyond standard compliance criteria. Engagement with a broader set of stakeholders and publics is important for realizing research goals, especially in the development of new technologies. Properly articulated engagement will reduce the chance of misunderstanding of the science needed to meet the goals. Planned repetitive community and stakeholder engagement should be undertaken at a very early stage of research and sustained throughout the process.

As the technology involving GM mosquitoes are novel (WHO, 2017a-c), and there remains a sense of uncertainty on the projected risk and benefits, communities may fear and distrust its introduction, which can garner a lot of social concern (Juma, 2016). Researchers should bear in mind that complying with the regulatory requirements as per any guidelines is not adequate for addressing ethical responsibilities. Laws and procedures do exist for any research related to human subjects mandating an ethical clearance; however, not all laws or guidelines can satisfy all ethical issues and a researcher's responsibility.

21.5 ETHICS ISSUES

Addressing the ethical questions that will arise is an obligation for scientist doing research with GM mosquitoes as we move forward with each phase of the testing pathway. Of these, primary to all issues is having a clear idea about the purpose and value of the research from the proof-of-concept stage. Communication of purpose and value of the research with relevant stakeholders helps in a proper experimental design and build trust at from the very onset of development.

Scientists doing research in areas of public health and disease control have a strong social motivation. However, the motivation should aim at broader health agendas like disease elimination (WHO, 2015) and achieving Sustainable Development Goals (SDGs) of elimination of poverty (SDG1), promotion of health and well-being (SDG3), access to education (SDG4), gender equality (SDG5), and reduced inequities (SDG10), particularly with technologies that aim to control vector-borne diseases.

One of the priority ethical areas that needs to be addressed is how genetic engineering research is going to affect the environment. Anthropocentrism and ecocentrism are two ways of understanding an extension of ethics to nature. In an anthropocentric ethic, nature deserves moral consideration because how nature is treated affects humans. In an ecocentric ethic, nature deserves moral consideration because nature has intrinsic value (Kortenkamp and Moore, 2001). Both GM mosquito strategies (population suppression and population replacement) test the relationship between humans and the environment. Population suppression theoretically permits the elimination of vector species, an outcome many find objectionable owing to the belief that all species have intrinsic value (Bouyer et al., 2019). Population replacement may permanently alter the genome of the vector species, inducing the kind of evolutionary change some find unnatural (Pugh, 2016). Keeping all this in mind in the context of GM mosquito research, two important aspects need to be considered. Firstly, the interventions should not be judged on the basis of natural or non-natural origin (WHO, 2020b) but on the origin of these interventions' potential of harm or benefits. Secondly, communities affected by such interventions should have the maximum priority in making the decision (Kofler et al., 2018; Resnik, 2019).

Understanding potential risks (WHO, 2020a, b) that would affect individuals and communities should be considered, as this is a population level intervention. Potential benefits should outweigh potential risks before proceeding with research on GM mosquitoes. Phased development of such interventions helps in understanding the potential of risk and designing means to mitigate them.

Progressive studies give opportunity for incremental exposure of the risk and only after fulfilling the safety, efficacy and acceptability criteria the next phase is initiated. Risks should be evaluated on a case-by-case basis, considering health, environmental, and socio-economic risks.

In an ethical assessment, there is the need to carefully consider the benefits of GM mosquito release. How a research activity is intended to benefit parties exposed to the risks needs to be evaluated (Emanuel et al., 2004). Assessment of health, environmental, and socio-economic benefits should be undertaken and considered alongside the risks.

Decision-makers should avoid decisions based on assumptions of the safety of any new technology, which is different from a familiar intervention on the basis of its known risks (12). Any improper decisions affect the lives of people and community at large (Krimsky and Golding, 1992). Data sharing and modelling help in making such decisions (James et al., 2018).

For trials involving open release of GM mosquitoes, community engagement plays an integral part in addressing concerns for the community where they are planned for release. Though consent for such release may not be applicable, informed consent would be required when interaction with the community gives rise to individual or household identifiable data (Kolopack and Lavery, 2017).

Informed consent leads to another area of ethics wherein authorization of the community as a whole for carrying forward the research would be required. Community engagement plays a vital role in gaining community authorization. Community authorization can be thought of as representation of the informed consent protecting the interests of those who will be affected by the research. The goal of communication should achieve assurance from the community to move forward with the research. Community authorization can be an outcome of group decisions through common consensus, voting, or through leaders and decision-makers in the host communities.

Transparency about the research as appropriate is important for the stakeholders to take a proper decision and for gaining public confidence. Not being transparent about the data creates an impression of mistrust with a feeling that the data possibly does not match what the scientist is claiming or there is a lacuna in the data which the scientist is hiding. However, it is important there is a need to control or restrict disclosure of data that could result in harm to the research team or host communities. Privacy protections should also be observed for identifiable persons and for groups where appropriate.

REFERENCES

ABNE, (2017). *African biosafety network of expertise. Burkina Faso country profile.* New Partnership for African Development. http://nepad-abne.net/contry_report/burkina-faso/. Accessed on 4 January 2020.

Biosafety Act (2007). s. https://www.aimst.edu.my/ibc/pdf/Guidelines/A.%20Biosafety%20Act%202007%20(Act%20678).pdf

Bouyer, J., Carter, N.H., Batavia, C. and Nelson, M.P. (2019). The ethics of eliminating harmful species: The case of the tsetse fly. *BioScience*, 69,125–35. doi:10.1093/biosci/biy155

CTNBio (2014a). Technical opinion no. 3964/2014. Brasília: National Technical Biosafety Commission. http://bch.cbd.int/database/attachment/?id=14514

CTNBio (2014b). *Risk assessment of Aedes aegypti strain OX513A*. National Technical Biosafety Commission. https://bch.cbd.int/database/record.shtml?documentid=105831

DBT (2017). *Regulation and guidelines for recombinant DNA research and biocontainment.* Department of Biotechnology.

DBT (2020). *Guidelines for Establishment of containment facilities: Biosafety Level 2 (BSL2) & 3 (BSL 3) and certification of BSL 3 facility.* Department of Biotechnology.

DBT (2022). *Guidelines for safety assessment for genome edited plants.* Department of Biotechnology.

DBT (2023). *Guidelines and standard operating procedures for research on genetically engineered insects.* Department of Biotechnology.

de Andrade, P., Aragão, F. J., Colli, W., Dellagostin, O. A., Finardi-Filho, F., Hirata, M. H., Lira-Neto, A. C., Almeida de Melo, M., Nepomuceno, A. L., Gorgônio da Nóbrega, F., Delfino de Sousa, G., Valicente, F. H., & Zanettini, M. H. (2016). Use of transgenic *Aedes aegypti* in Brazil: Risk perception and assessment. *Bulletin of the World Health Organization, 94*(10), 766–71. https://doi.org/10.2471/BLT.16.173377

EFSA (2012) Scientific Opinion on the Guidance on the risk assessment of food and feed from genetically modified animals and animal health and welfare aspects. *EFSA Journal*, 10(1), 2501. doi:10.2903/j.efsa.2012.2501

EFSA (2013) European Union report on pesticide residues in food. *EFSA Journal*, 13(3):4038, doi:10.2903/j.efsa.2015.4038

EFSA, (2020). GMO Panel (EFSA Panel on Genetically Modified Organisms), Scientific Opinion on the adequacy and sufficiency evaluation of existing EFSA guidelines for the molecular characterisation, environmental risk assessment and post-market environmental monitoring of genetically modified insects containing engineered gene drives. *EFSA Journal*, 18(11), 6297, 90. doi:10.2903/j.efsa.2020.6297

Emanuel, E.J., Wendler, D., Killen, J. and Grady, C. (2004). What makes clinical research in developing countries ethical? The benchmarks of ethical research. *Journal of Infectious Diseases*, 189, 930–7. doi:10.1086/381709

FDA (2016). Environmental assessment for investigational use of *Aedes aegypti* OX513A. Silver Spring: United States Food and Drug Administration. https://www.fda.gov/media/99722/download

FDA (2017). CVM GFI #236 Clarification of FDA and EPA jurisdiction over mosquito-related products: Guidance for industry. Silver Spring: US Department of Health and Human Services Food and Drug Administration. https://www.fda.gov/regulatory-information/search-fda-guidance-documents/cvm-gfi-236-clarification-fda-and-epa-jurisdiction-over-mosquito-related-products

GOI (1989). *Government of India Use/import/export and storage of hazardous microorganisms/genetically engineered organisms or cells (Rules, 1989).* Government of India, Ministry of Environment and Forests).

Gorman, K., Young, J., Pineda, L., Márquez, R., Sosa, N., Bernal, D., Torres, R., Soto, Y., Lacroix, R., Naish, N., Kaiser, P., Tepedino, K., Philips, G., Kosmann, C. and Cáceres, L. (2016). Short-term suppression of *Aedes aegypti* using genetic control does not facilitate *Aedes albopictus*. *Pest Management Science*, 72: 618–28. doi:10.1002/ps.4151

Hayes, K., Barry, S., Beebe, N., Dambacher, J., Ford, J., Hosack, G., Peel, D., Ferson, S., Goncalves da Silva, A. and Thresher, R. (2015). *Risk assessment for controlling mosquito vectors with engineered nucleases, Part I: Sterile male construct final report.* CSIRO. https://doi.org/10.25919/5c017aa7231b2

James, S.L., Collins, F.H., Welkhoff, P.A., Emerson, C., Godfray, H.C.J., Gottlieb, M., Greenwood, B., Lindsay, S.W., Mbogo, C.M., Okumu, F.O., Quemada, H., Savadogo, M., Singh, J.A., Tountas, K.H. and Touré, Y.T. (2018). Pathway to deployment of gene drive mosquitoes as a potential biocontrol tool for elimination of malaria in sub-Saharan Africa: Recommendations of a scientific working group. *American Journal of Tropical Medicine and Hygiene*, 98(Suppl 6), 1–49. doi:10.4269/ajtmh.18-0083

Juma, C. (2016). *Innovation and its enemies: Why people resist new technologies.* Oxford University Press.

Kofler, N., Collins, J., Kuzma, J., Marris, E., Esvelt, K., Nelson, M.P., Newhouse, A., Rothschild, L.J., Vigliotti, V.S., Semenov, M., Jacobsen, R., Dahlman, J.E., Prince, S., Caccone, A., Brown, T. and Schmitz, O.J. (2018). Editing nature: Local roots of global governance. *Science*, 362:527–9. doi:10.1126/science.aat4612

Kolopack, P.A. and Lavery, J.V. (2017). Informed consent in field trials of gene-drive mosquitoes. *Gates Open Research*, 1, 14. doi:10.12688/gatesopenres.12771.1

Kortenkamp, K.V. and Moore, C.F. (2001). Ecocentrism and anthropocentrism: Moral reasoning about ecological commons dilemmas. *Journal of Environmental Psychology*, 21(3), 261–72. doi:10.1006/jevp.2001.0205

Krimsky, S. and Golding, D. editors. (1992). *Social theories of risk.* Praeger Publishers.

MOE (2021). *Guidelines for testing of ethics of eliminating harmful species: The case.* Ministry of Environment.

OECD (2018). Safety assessment of transgenic organisms in the environment, Volume 8: OECD consensus document of the biology of mosquito Aedes aegypti, Harmonisation of Regulatory Oversight in Biotechnology, OECD Publishing. doi:10.1787/9789264302235-en

OSTP (1986). *Coordinated framework for regulation of biotechnology.* Executive Office of the President, Office of Science and Technology Policy. https://www.aphis.usda.gov/brs/fedregister/coordinated_framework.pdf

OSTP (2017). *Modernizing the regulatory system for biotechnology products: Final version of the 2017 update to the Coordinated Framework for the regulation of biotechnology.* Executive Office of the President, Office of Science and Technology Policy. https://www.epa.gov/sites/production/files/2017-01/documents/2017_coordinated_framework_update.pdf

Patil, P.B., Dasgupta, S.K., Gorman, K., Pickl-Herk, A., Puinean, M., McKemey, A., Char, B., Zehr, U.B., & Barwale, S.R. (2022). Elimination of a closed population of the yellow fever mosquito, *Aedes aegypti*, through releases of self-limiting male mosquitoes. *PLoS Neglected Tropical Diseases*, 16(5), e0010315. https://doi.org/10.1371/journal.pntd.0010315

Pugh, J. 2016. Driven to extinction? The ethics of eradicating mosquitoes with gene-drive technologies. *Journal of Medical Ethics*, 42, 578–81. doi:10.1136/medethics-2016-103462

Ramsey, J.M., Bond, J.G., Macotela, M.E., Facchinelli, L., Valerio, L., Brown, D.M., Scott, T.W. and James, A.A. (2014). A regulatory structure for working with genetically-modified mosquitoes: Lessons from Mexico. *PLoS Neglected Tropical Diseases*, e2623. doi:10.1371/journal.pntd.0002623

Redford, K.H., Brooks, T.M., Macfarlane, N.B.W. and Adams, J.S. (eds.) 2019. *Genetic frontiers for conservation: An assessment of synthetic biology and biodiversity conservation. Technical assessment.* IUCN. xiv + 166pp.

Resnik, D.B. (2019). Two unresolved issues in community engagement for field trials of genetically modified mosquitoes. *Pathogens & Global Health*, 113:238–45. doi:10.1080/20477724.2019.1670490

Subramaniam, T.S., Lee, H.L., Ahmad, N.W. and Murad, S. (2012). Genetically modified mosquito: The Malaysian public engagement experience. *Biotechnology Journal*, 7:1323–7. doi:10.1002/biot.201200282

Target Malaria (2019). Target Malaria proceeded with a small-scale release of genetically modified sterile male mosquitoes in Bana, a village in Burkina Faso. https://targetmalaria.org/target-malaria-proceeded-with-a-small-scale-release-of-genetically-modified-sterile-male-mosquitoes-in-bana-a-village-in-burkina-faso/

WHO (2015). *Global technical strategy for malaria 2016–2030*. World Health Organization. https://apps.who.int/iris/bitstream/handle/10665/176712/9789241564991_eng.pdf

WHO (2017a). *How to design vector control efficacy trials, guidance on phase III vector control field trial design*. World Health Organization. https://apps.who.int/iris/handle/10665/259688

WHO (2017b). *Data requirements and methods to support the evaluation of new vector control products*. World Health Organization. https://apps.who.int/iris/handle/10665/259536

WHO (2017c). *Ethical issues associated with vector-borne diseases*. Report of a WHO scoping meeting, Geneva, 23–24 February 2017. World Health Organization. https://www.who.int/publications/i/item/WHO-HTM-NTD-VEM-2017.07

WHO (2018). *Efficacy-testing of traps for control of Aedes spp. Mosquito vectors (WHO/CDS/NTD/VEM/2018.6)*. World Health Organization. https://www.who.int/publications/i/item/WHO-CDS-NTD-VEM-2018.06

WHO (2020a). *Position statement on genetically modified mosquitoes*. World Health Organization. https://www.who.int/news/item/14-10-2020-who-takes-a-position-on-genetically-modified-mosquitoes#:~:text=Position%20statement,-In%20recent%20years&text=According%20to%20the%20WHO%20statement,diseases%20carried%20by%20Aedes%20mosquitoes

WHO (2020b) *Ethics and vector-borne diseases: WHO guidance*. World Health Organization. https://apps.who.int/iris/bitstream/handle/10665/336075/9789240012738-eng.pdf

WHO (2021). *Guidance framework for testing genetically modified mosquitoes*, 2nd edition. World Health Organization. https://www.who.int/publications/i/item/9789

WHO (2022). *World malaria report 2022*. World Health Organization. https://www.who.int/teams/global-malaria-programme/reports/world-malaria-report-2022

Section VI

Mosquitoes and Society

22 Mosquitoes in Human Psyche

B.K. Tyagi
University Institute of Biotechnology, Chandigarh University, Mohali, India

Sajal Bhattacharya
Asutosh College, Kolkata, India

B. Reddya Naik
Osmania University, Hyderabad, India

22.1 INTRODUCTION

Mosquitoes are a kind of true fly. Generally, in principle, they are vegetarian (plant sap- and nectar-feeding) but females of most species are non-vegetarian (blood-feeding). While male mosquitoes are, on one hand, entirely innocuous and are rather beneficial as either biocontrol agents or pollinators, the females are obnoxious and, more importantly, vectors to several deadly and/or debilitating diseases causing pathogens (Gubler, 2002; Hayes, 2009; Jones et al., 2017; CDC, 2024; Tyagi, 2023). Therefore, there are numerous reasons to generate scientific interest in mosquitoes; the mosquito has been branded man's deadliest foe on Earth; the mosquito is rated the deadliest predator on our planet; the mosquito is blamed as the economy's enemy number one (Bynum and Overy, 1998; Spielman and D'Antonio, 2001; Tyagi, 2004; Winegard, 2020). Through their bites, however, mosquitoes levy unbearable uneasiness, ill temperament, and loss of sleep, thereby affecting human psychology in attending to important jobs at hand. A mosquito bite *per se* may not bring instant havoc, but it is capable of causing disasters if it carries disease pathogens. Now, since there is both fear and awareness about them, it is only reasonable to put the whole subject in its right perspective to help comprehend mosquito-related fears (mosquito phobia), especially in context with the Indian people's psyche. Mosquitoes were referred to in the Indian scriptures of old times, the Vedas, from 1500 to 1600 BC, and, in fact, they were classified on the basis of size, colour, and behaviour, and even related to intermittent fevers (Rao, 1984). The fact that the mosquito has been known by as many as 14 different names is a grand testimony to its being known to folklore in India in a much-diversified manner.

22.2 MOSQUITOES IN THE INDIAN SCRIPTURES OF YORE

Mosquitoes have been known to the inhabitants of India since ancient times. Even though the medical practitioners of yore did not pinpoint a particular mosquito for being responsible for transmitting disease pathogens; nevertheless, there existed several important scriptures implicating mosquitoes for both the nuisance and unspecific human maladies and scourges, particularly in context with malarial fever, through their painful bites. Some of the earliest scholarly works, such as the Atharva Veda, around 1500 BC, described and classified malarial fever. There is even evidence of the usage of mosquito nets during, and possibly much before, the thirteenth century in some parts of India. Mosquitoes have been known by different names in various Indian states, often depicting their relationship with some disease or particular social situation (Rao, 1984). A few of these vernacular

DOI: 10.1201/9781003327028-28

TABLE 22.1
Names of Mosquitoes in Different Indian Languages

Sl. No.	Languages	Vernacular Names
1	Sanskrit	*Masha, Mashakh*
2	Assamese	*Masha, Mashak*
3	Bengali	*Masha, Mashak*
4	Gujarati	*Dās*
5	Hindi	*Machchar*
6	Kannad	*Solle*
7	Kashmiri	*Moh*
8	Malayalam	*Kothuku*
9	Marathi	*Dās*
10	Oriya	*Mashā, Mashak*
11	Punjabi	*Machchad*
12	Tamil	*Koshu*
13	Telugu	*Dommalu*
14	Urdu	*Machchad, Koohu*

names are presented in Table 22.1. *Susruta Samhita*, a Sanskrit work of ancient India, which identified five types of mosquitoes (Mashakah), has referred to mosquitoes as follows:

> Their bite is as painful as that of serpents, and causes diseases resulting from the three humours joined together; the bite as if burnt with caustic or fire, is red, yellow, white or pink colour accompanied by fever, pain in limbs, hair standing on end, pain, vomiting, diarrhoea, thirst, heat, giddiness, yawning, shivering, hiccup, burning sensation, intense cold, vesicles or pustules increasing, swellings, knots under the skin, circles etc.

Mosquitoes referred to in this scripture of yore were 'Samudrah' (marine or coastal mosquito), 'Parimandalah' (very small mosquito), 'Hastimashakah' (elephant mosquito), 'Krishnah' (Black or dark mosquito), and 'Parwatiyah' (hill mosquito). Their bites were identified to have caused "swelling as well as itching" on the part bitten.

The Atharvaveda also suggests certain mosquito control measures, such as sunlight, yagya (sacrificial fire), vegetarian diet, and self-control, as well as the use of herbal plants, such as *kustha, asvagandha, guggulu, ajasrngi*, and *pramandani* (Kaur and Singh, 2017).

22.3 THE 'MOSQUITO FACTOR' IN INDIVIDUALS, NATIONAL HEALTH, AND THE ECONOMY

Mosquitoes can be an annoying and serious problem in human's lives. In fact, the animal responsible for the most human deaths worldwide is the mosquito (Tyagi and Tilak, 2022). They interfere with work and spoil hours of leisure time. Their attacks on farm animals can cause loss of weight and decreased milk production. Some mosquitoes are capable of transmitting diseases, such as malaria, yellow fever, dengue, and filariasis, as well as various types of encephalitides, e.g., St Louis encephalitis (SLE), Western equine encephalitis (WEE), Eastern equine encephalitis (EEE), La Crosse encephalitis (LAC), Japanese encephalitis (JE), West Nile virus, to humans and/or animals. Malaria, the severest of all, is also humanity's greatest bane triggered by the mosquito. Throughout the world, more than 40% of the population is exposed to varying levels of risk of malaria in 91 countries.

Of a total world population of approximately 6 billion, 1.7 billion (35%) live in areas where endemic malaria has been considerably reduced or even eradicated but where transmission has begun again, and the situation is now unstable or deteriorating. In some areas, severe malaria problems developed following major ecological or social changes, particularly the agricultural or the exploitation of forests or socio-political interests. Each year, 300–500 million cases of malaria, 90% of them in Africa, and between 1.5 and 2.7 million deaths, are reported worldwide. Severe malaria morbidity and mortality are linked to areas endemic for *Plasmodium falciparum*, the predominant species in Africa. According to a report, malaria kills a child somewhere in the world every 30 seconds (Alnwick, 2000), whereas it has been estimated that 800,000 children die each year of malaria in Africa alone. Mortality is lower but nevertheless important in the Americas and South-East Asia, especially where *P. falciparum* drug resistance is widespread. The South-East Asia region embodying India is of particular interest to us, as approximately 3.54 million cases occur in the region every year.

Once seen as little more than a small sacrifice for the lush vistas of tropical climes, mosquitoes, and the diseases they transmit, have put international agencies and the medical community up in arms. Mosquito-borne diseases exert unfathomable economic losses to humans. These losses are not merely monetary but relate to health as well. Diseases like malaria, filariasis, dengue, and JE incur annual losses worth several trillions of US dollars all over the world, if only expressed in terms of disability adjusted life years (DALYs; Table 22.2). In sub-Saharan Africa, a major malaria-infested region, the malaria-related annual economic burden in 1987 was to the tune of US$0.8 billion (0.6% of gross domestic product (GDP)), which was expected to further hike to US$1.7 billion (1.0% of GDP) by 1995.

Economic losses tend to rise with the national income and improved socio-economic conditions. Also, there are a number of direct and indirect factors which are considered when calculating the economics of malaria. Often, the related documentation is faulty and grossly incomplete. In spite of this, even if only the recorded facts are taken into consideration, the economic loss due to malaria in India came to somewhere between Rs. 15,777.85 and 29,934.85 million at the 1991 level. This estimate obviously did not consider major areas like enormous losses in agriculture, industry, mining, railways, and other important productive sectors in the country that are susceptible to malaria-related perils, in addition to indiscernible yet apparent losses to the physical and intellectual growth of the population living in endemic areas. Keeping in mind some factors like estimates of morbidity and

TABLE 22.2
Major Mosquito-Borne Diseases and DALYs

Vector Species	Disease	Dalys	Distribution
Anopheles spp. (*An. gambiae*)	**Malaria**	40,213,000[a] 56,201,200[b]	Tropics, sub-Saharan Africa
Culex spp. (*Cx, quinquefasciatus*)	**Filariasis** Lymphatic filariasis	5,549,000[a] 1,189,000	Tropical urban areas
Culex spp.	**Encephalitides** Japanese encephalitis	426,000[a]	South and South-East Asia, linked to irrigated rice/pigs
Aedes spp. (*Ae. aegypti*)	**Arboviral diseases** Dengue Yellow Fever	433,000[a] 2,946,900[b] 374,000[a]	Global, urban tropics Africa, Central and South Americas
Aedes spp. (*Ae. albopictus*)	**Arboviral diseases** Chikungunya	25, 588[c]	Asia, Pacific, Europe

Notes: Those with
[a] are at 2,000 level and those with
[b] are by GBD (Global Burden of Diseases) (2017); figures for chikungunya
[c] are for India by Krishnamoorthy et al. (2009)

mortality due to malaria expressed as DALYs, an overall loss is postulated to be falling in the range of Rs. 16,192–Rs. 20,192 million, provided that the prevalence figure of only two million malaria cases is considered (which, according to the WHOSEARO, is probably at least six to seven times lower than the actual incidences in the country!). The aforementioned is a clear indication of the enormity of the seriousness of malaria disease, which needs to be dealt with in India. It needs be understood emphatically that the management of mosquito-borne diseases in India takes away more than half of the financial layout earmarked for the entire health sector!

22.4 SHIFT IN HUMAN PSYCHE ABOUT MOSQUITO

Mosquitoes and malarial disease were both prolifically rampant and devastating during the nineteenth century; the mystery surrounding the Anopheles mosquito's role in transmitting malaria, however, would still take several years to be solved when Ronald Ross entered the scene! Accounts of Ross's encounters with these blood-sucking creatures during his painfully cumbersome and never-ceasing extensive service-bound peregrinations find repeated references in his memoirs (Ross, 1923). He understood the significance of mosquito nets since those who defied them fell sick almost instantaneously. The British government in India manoeuvred sporadic, infrequent, and half-hearted mosquito control operations in selected areas on a small scale, but these were largely evanescent without any appreciable output. Indian folklore during the British regime was always suffering with mosquito menace. Mostly locally available means, such as the neem leaves (*Azadirachta indica*), lemongrass oil (*Cymbopogon citratus*), and smouldering dung cakes, were available for prevention of mosquito bites. Mosquitoes had become an inseparable part of human life, particularly in rural areas, and the use of locally available means to repel them was ritually practiced. Now we know that mosquitoes are the most dangerous animals in the world, killing an estimated two to three million people every year by way of various deadly infections. Even though the mosquito-borne infections had in the remotely distant past a history of worldwide distribution, most of them are presently restricted in their epidemic proportions to tropical countries, including India where over 300 species and subspecies of mosquitoes exist.

When Ross struck the 'mosquito malaria theory' in 1897, mosquitoes instantly became the target of control with more enthusiasm and zest. In the years that followed, beginning from the end of the nineteenth century to nearly throughout the twentieth century, some very impressive attempts were made to abate mosquito vectors in different parts of the country under the redoubtable stewardship of such great malariologists as Christophers, Barraud, Covell, Viswanathan, and Ramachandra Rao, to name a few. Still, by and large, Indian peoples could not really isolate themselves from their foremost foe, the mosquito, despite countrywide national malaria and filariasis eradication schemes running for the past nearly five decades. In the process, the mosquito has become a subject of day-to-day tête-à-tête, so much so that when the sun would lower, the mosquito was the first thing to be remembered, to be cautioned about!

Over the last century, since the epoch-making discovery by Ross (1897), malariologists have made several important advances in research about the mosquito, with which our familiarity has improved a great deal. We must, however, be absolutely cautious about making claims since even after a century of intensive investigations, many a mystery remains, and an enigma continues to haunt the genius of the human mind. Why does a mosquito look for human blood at all when other more 'bare-skinned' alternatives persist? Is it only maturing of eggs in females for which the mosquito risks to engorge on human blood? How does a mosquito scan and assort blood vessels deep under the skin to draw blood from them? What kinds of cues does a mosquito use to identify one host from another before landing for feeding, particularly in pitch-black nights? What type of sucking force does a mosquito develop to pull blood through a feeding tube with a diameter that is so infinitesimal that the resulting friction should make it impossible? How does a salivary pump work through even finer salivary tubules? Why do certain types of mosquitoes support the development of one kind of pathogen and not the other? Is the mosquito is aware of the impending parasitism or pathogenicity to her body following engorgement on human or other animal blood? These and many other baffling and arcane inquiries make a mosquito the most sought-after organism in the animal kingdom.

In its enormous variety and adaptive ingenuity, the ubiquitous mosquito offers us a microscopic view of nature's logic. For those mosquitoes which share humans' living environment and subsist on their blood, their dependency appears to be their strength. Like weeds, vector mosquitoes, too, tend to thrive best where their environment is disturbed, which had so often been done to bring a large number of otherwise innocuous species in contact with humans and to feed on their blood eventually. Having thus been driven away from their original sylvan homes, mosquitoes have been forced to earn their livelihood against an enormously dangerous environment. Yet, these seemingly simple and feeble-bodied organisms bestowed with gigantic powers continue to thrive. Everything about the mosquito's body, life, and behaviour is a perfect example of great ingeniousness, as it commences all of its jobs from feeding to egg laying with exquisite precision and economy, on one hand, and keeps meeting high challenges, such as humans' continuous use of insecticides as a weapon against it, deprivation from its original habitats, and, on top of all that, a cadence of inhospitable climate, on the other. It is indeed surprising to see the speed with which the mosquito, which was subjected to all kinds of adversaries by being pushed into the alien human environment, has adapted diversely and successfully to live with man.

22.4.1 Malaria, Ross, and Christophers' Indian '*Anopheles stephensi*'

Even destiny seems to have been planned in India's favour by pushing the greatest of twentieth-century research activities onto its local mosquito, *Anopheles stephensi* (Ross, 1897; Christophers, 1933; Tyagi, 2023). Ross's discovery that the *Anopheles* mosquito transmitted malaria was carried out in India as if to re-emphasize a grand tradition of knowledge seeking in this great country. Mysteries surrounding malaria, prolifically rampant and devastating in the warmer countries, were being solved in many countries during the nineteenth century, both in the tropic and temperate regions, much before the entry of Ronald Ross in the arena. Many European nations, especially Britain, France, Spain, and Holland, had under their rule several countries in the fragmented Africa, Americas, and Asia, but the mosquito-borne diseases, such as malaria and filariasis, made them dread making their effective governance and living in those regions. Dr Patrick Manson had already determined the role of mosquitoes in transferring filarial worms from human to human in China in the early 1880s; a team of great scientists like Prof Grassi, Dr Golgi, and Dr Bignami were already working on malaria (and the possible role of mosquitoes) for several years in Italy around the same period – the seed for which was sown about 100 years ago by Giovanni Lancisi, and even the most plausible interpretation on the possible mosquito association with malaria dissemination was already sounded by Dr Daniel Beauperthuy in Venezuela and Dr Josiah Nott in Alabama more than two decades earlier; still the 'mosquito malaria theory' was destined to be written in India by Ronald Ross, who, despite his total ignorance about mosquito systematics and biology, and fragmentary surface knowledge of the malaria parasite (which also gained the courtesy of Dr Patrick Manson in England), proved through dint of his perseverance and application the real cradle of great malaria invention in this country. The Indian scientific community took upon themselves this moment of great discovery as a matter of pride and dignity, as well as a grand research legacy. There was no surprise, therefore, when the Indian scientists, under the banner of the Indian Society of Parasitology, logically organized an international symposium, with malaria as the central theme, in 1997 at Hyderabad to commemorate a hundred years of Ross's discovery of the malaria mosquito in 1897. Not only this, but very recently a malaria seminar was organized by the Godrej Sara Lee Pvt. Ltd., a great commercial name in manufacturing mosquito-repelling products, on 20 August 2003 in Mumbai (old name Bombay) to coincide with 'Mosquito Day,' coined by the great malariologist Dr Ross. Ever since Ross's famous 1897 discovery, India has been at the forefront of malaria-related research, ably contributed to internationally famous malariologists or culicidologists like Drs Christophers, Barraud, Sinton, Covell, McDonald, Viswanathan, Puri, Ramachandra Rao, Rajagopalan (V.P.), and Sharma, to name a few. Thus, the mosquito continues to be incredibly closer to the Indian human psyche by virtue of its place both in its people's lives and the research programmes aimed at the country. Interestingly, in Indian life, one is aware of quite a few sayings referable to the mosquito for its bearing on humans: for example, when a human is extremely emaciated, he or she is referred to a resembling a 'mosquito'; when somebody is annoyingly exacerbated with a hateful opponent, the former reminds the latter of being crushed like a 'mosquito'; when a person is uncontrollably, but secretively, talking close to another person's ear, the former is urged not to whine like a 'mosquito'; when an already big human population continues to grow unchecked, they are often said to be increasing like flies and 'mosquitoes.'

22.4.2 Malaria-*Numero Uno* Amongst All Mosquito-Borne Diseases

Malaria offers perhaps the best example for determining the economic burden levied by any arthropod-borne disease in a country since (i) among all the vector-borne diseases, malaria continues to be one of the biggest contributors to disease burdens in terms of death and suffering, and from a public health point of view, it is undoubtedly the most devastating parasitic disease in the tropics, and (ii) it is definitely the most extensively studied subject. By undermining the health and capacity to work of hundreds of millions of people, it is closely linked to poverty and contributes significantly to stunting social and economic development. Nearly four decades ago, there was hope that malaria could be eradicated in a fairly short time, mainly through the intensive use of insecticides. However, malaria still remains a serious threat in endemic countries. Those at greater risk of dying from the disease are children under age 5 in malaria-endemic areas; pregnant women; people moving from non-malarious to malarious zones for reasons of work, migration, refuge, war, or tourism; and travellers who visit endemic countries and return home with the disease. In many parts of the country, malaria is becoming even a greater problem than it has been before. Epidemics are recurring in areas where transmission has been interrupted, and they are generally associated with deteriorating social and economic conditions.

22.4.3 Economics of Malaria in India

In India, before a malaria control programme was launched in 1953, nearly 100 million people suffered from the disease annually, and close to 0.1 million people died every year (Sharma and Mehrotra, 1986; Tyagi, 1994). The latter figure clearly appears to be a gross underestimate since during frequent epidemics, so common in the country in those days, the overall number would have been much higher (Sinton, 1935), as in 1931, when the number of deaths indirectly or directly due to malaria exceeded two million. Besides, malaria affected the national as well as individual development as economic losses of the order of Rs. 2936 million per year were forced on India.

Soon after the countrywide National Malaria Eradication Programme (NMEP) was launched in 1958 (in fact, the formal approval for the initiative was accorded by the Government of India as early as late 1957), malaria cases reduced to less than 0.1 million in 1965–66 without a single death, and it appeared that malaria would be wiped out in India before very long! It is noteworthy that the decline in cases was in *P. vivax* and not in *P. falciparum*. Within the next ten years, malaria epidemics were widespread, and the disease, recording the highest ever number of malaria cases (6.47 million) by 1976, invaded even those areas which were earlier not known for malaria transmission or were once freed from its scourges. Subsequently, during the 1980s and early 1990s, following a series of critical reviews of the malaria eradication programme, malaria incidences were brought down to less than two million but soon increased to nearly three million in the mid-1990s, as reported by the NMEP. Ironically, incidences of malaria and deaths reported by NMEP were based on parasitologically proven cases, thus excluding a large number of asymptomatic carriers and/or those who approached other medicinal systems for cure for the disease or even resorted to self-medication.

As a consequence, both morbidity and mortality were apparently underestimated. While the NMEP reported 2.5 to 3 million malaria cases and about 1,000 malarial deaths annually, the WHO South-East Asia Regional Office put the estimates at 15 million malaria cases (nearly 6 times) and 19,500 deaths due to malaria (Sharma, 1999). Therefore, it appears safer to assume that the malaria incidences reported offered at most the disease's trend and not its true number of incidences, the knowledge of which, for logical reasons, is inevitably indispensable for effective implementation of malaria control in the country.

22.4.4 Control of Malaria in India

Malaria control in India is presently hampered for two reasons: (i) drug resistance development in malarial parasites and (ii) insecticide resistance development in malaria vectors. *Plasmodium vivax* and *P. falciparum* are dominant species in India, with the former responsible for 60%–65% cases and the latter for 35%–40% cases (Sharma, 1999). Among the six major vector species prevalent in India, *An. culicifacies*, the main transmitter of rural and peri-urban malaria in the peninsular India, is alone responsible for 65% of malaria and 55% *P. falciparum* annually. This mosquito is resistant to DDT and HCH (hexachlorocyclohexane) all over the country, and to malathion in Maharashtra, Gujarat, and certain other states. *Anopheles stephensi* transmits nearly 12% of malaria cases in the country, mostly in the urban and industrial areas. This species, too, has developed multiple resistances against various insecticides, many of which are used as larvicides. Yet another vector species, *An. fluviatilis*, which, like the two aforementioned species, is common in Rajasthan, transmits 15% of the total malaria cases and 30% *P. falciparum* in the country.

The strategy of malaria control in India has been to cause interruption in disease transmission by spraying residual insecticides in the rural areas (351.8 million population) and by source reduction and larviciding in urban areas (62.1 million population). Because most of the primary malaria vectors in India are endophilic and/or endophagic in behaviour, indoor residual spraying of insecticides in the rural areas and anti-larval operations in the urban set-ups have been the major means to control malaria in the country. Led by DDT's widespread usage in controlling vector population, now largely jeopardized owing to the development of resistance in vector mosquitoes, other insecticides like HCH (later banned in 1997), malathion, and certain synthetic pyrethroids such as cyfluthrin, deltamethrin, and lambdacyhalothrin have also been selectively employed in the control of malaria, although spray targets were rarely achieved due to logistic and operational constraints.

Major anti-malaria activities launched in India since its independence can be summed up briefly as follows:

i. *National Malaria Control Programme (NMCP)*: It was launched in 1953 with comprehensive indoor residual spraying with DDT. By 1957, 165.7 million people were effectively protected.
ii. *NMEP*: The success achieved under the NMCP was so spectacular and encouraging that the Government of India launched NMEP in 1958, with the objective of eradicating the disease from the country. This brought down malaria cases to 0.1 million, and no deaths due to malaria were reported in the country in 1964. The euphoria of victory, however, proved to be ephemeral as malaria incidences began to escalate in the late '60s owing to multiple factors, including financial, logistic, administrative, and technical constraints, on one hand, and a feeling of complacency due to a near victory over the disease in the country on the other. Consequently, malaria resurged with a vengeance and reached its peak in 1976, when about 6.47 million cases of malaria were recorded in the country.

The NMEP originally began as a completely centrally sponsored health programme and, until 1979, the entire expenditure on malaria control throughout the country was borne by the centre. Thereafter, the programme was implemented on a 50:50 cost-sharing basis with the states, with the exception of the seven north-eastern states, which, from December 1994 onwards, were brought under 100% central assistance as special cases. All other states in the country, while meting out the entire operational costs for malaria control as their part of the cost-sharing, continued to receive 50% of the costs from the centre in the form of required equipment and material. The Indian government spends nearly Rs. 350 crores on malaria control annually. Additionally, the World Bank also recently provided a sum of Rs. 1075 crores for five years for the Enhanced Malaria Control Programme.

iii. *Modified Plan Operation (MPO)*: The successes and failures under the NMEP were reviewed in depth, and a MPO was introduced in the year 1977, with three main objectives: (i) prevention of deaths due to malaria, (ii) reduction of morbidity due to malaria, and (iii) maintenance of Industrial and Green Revolutions due to freedom from malaria, as well as retention of achievement gained so far. Under the MPO, a new strategy was adopted from the eradication of malaria to the containment of the disease in the country, i.e., prevention of death and reduction in morbidity due to malaria. As a strategy under this plan, areas with two annual parasite incidences (API) or more were earmarked for regular rounds of spray. Surveillance activities in the whole country were decentralized to Primary Health Centres (PHC), along with laboratory services. People's participation was solicited by involving volunteers in the distribution of antimalarials through Drug Distribution Centres (DDC) and Fever Treatment Depots (FTD). Malaria control in urban areas (<40,000 population) is carried out under the Urban Malaria Scheme (UMS) of the NMEP, launched in 1971/72. This led to a significant reduction in malaria incidences in the country to a level of around two million cases by 1987 for one decade. Since 1994, however, focal outbreaks of malaria have been continuously reported from different parts of country, particularly the north-eastern states and the Thar Desert region in Rajasthan in north-west India. This unwarranted situation resulted in a resurgence of malaria, generally pronounced by *P. falciparum* dominance, as well as deaths due to malaria in the country.

iv. *Plasmodium falciparum* Containment Programme (PfCP): The *P. falciparum* problem attained new dimension when a focus of chloroquine-resistant strain was discovered in Assam in 1973. Subsequently, new foci of chloroquine resistance were traced in different north-eastern states: Orissa, Madhya Pradesh, Maharashtra, Rajasthan, etc. This situation led to the creation of PfCP in 1977 to tackle the rising threat of *P. falciparum.*

v. *Malaria Action Programme Based on Expert Committee Report – 1995*: An expert committee that took cognizance of the availability of appropriate technologies for control of malaria in different epidemiological situations or paradigms in the country attributed the various setbacks in the malaria eradication programme, particularly the recurrence of periodic epidemics and high mortality due to various different imminent factors like administrative indifference, organizational weakness, low prioritization of malaria under health services and apathy of middle-level and peripheral workers in the states. Accordingly, the Expert Committee recommended intensification of malaria control activities throughout the country with a focus on high-risk areas through an integrated malaria control strategy, with following mentioned components: (a) early case detection and prompt treatment; (b) selective vector control; (c) promotion of personal protection methods; (d) early detection and containment of epidemics; (e) information, education, and communication towards personal prevention and community participation; and (f) institutional and management capacity building, trained manpower development, and efficient management information system.

vi. *Enhanced Malaria Control Project (EMCP)*: It was launched in September 1997 with the help of World Bank and covered, in addition to 19 towns in 10 states, 62.2 million high-risk tribal populations living in 100 districts and 1,045 PHCs in the 7 peninsular states, viz., Andhra Pradesh, Bihar, Gujarat, Madhya Pradesh, Maharastra, Orissa, and Rajasthan. These PHCs, with a tribal population of more than 25%, contributed 65.5% of the malaria cases and 78.60% of the *P. falciparum* cases in 1997. The benefits of the programme are conceived to reach other malaria-endemic areas as well since the strengthening of the components of the Information, Education, and Communication (IEC) Trained Manpower Development and Management Information System (MIS) covers the whole country.

vii. *Roll Back Malaria (RBM)*: In January 1998, the World Health Organization proposed the Roll Back Malaria initiative, taking into consideration large-scale morbidity and mortality due to malaria in developing countries, and observing the changing epidemiological scenario and the experience of the past control efforts. The basic concepts of the RBM initiative are not different from the existing control strategy in India, but it lays more emphasis on social mobilization, intersectoral collaboration, effective partnerships, strengthening of primary health care, investment in the development of effective methods of control, and shaping existing tools by operational research, and it is expected that this will result in sustained reduction in disease burden by the year 2010.

viii. *National Anti-Malaria Programme (NAMP)*: The eradication concept was changed to Anti-Malaria Programme and, thus, the programme was once again renamed National Anti-Malaria Programme.

ix. *National Vector Borne Disease Control Programme (NVBDCP)*: The vertical anti-malaria control programme transformed into a horizontal one with the establishment of a wide spectrum National Vector Borne Disease Control Programme in 2003 for synchronizing efforts to control/eliminate six vector-borne diseases. It was erected as an umbrella programme for prevention and control of malaria and other vector-borne diseases, viz., lymphatic filariasis, Kala-azar, JE, chikungunya, and dengue with special focus on the vulnerable groups of the society – namely, children, women, scheduled castes (SC), and scheduled tribes (ST).

x. *The Global Fund to Fight AIDS, Tuberculosis, and Malaria (GFATM)*: It is an international financing and partnership organization that aims to "attract, leverage and invest additional resources to end the epidemics of HIV/AIDS, tuberculosis and malaria" to support attainment of the Sustainable Development Goals established by the United Nations. Founded in 2005, it aims to extend support and additional inputs in 106 districts of 10 states.

xi. *World Bank Assisted Project on Malaria Control and Kala-Azar Elimination (2008).*

xii. *National Framework for Malaria Elimination (2016–30) and Integrated Vector Management (IVM) 2016* were launched to provide a roadmap for eliminating malaria from India (Nema et al., 2020).

xiii. *Operational Manual for Malaria Elimination*: It was launched in 2016.

xiv. *National Strategic Plan (2017–22) for Malaria Elimination by 2030*: It was launched in 2017.

xv. *Mosquito and Other Vector Control Response (MVCR)*: It was launched in 2020.

xvi. *National Centre for Vector Borne Diseases Control (NCVBDC)*: The long-standing Directorate of NVBDCP was rechristened the National Centre for Vector Borne Diseases Control in 2021.

22.4.5 Early Malaria Epidemics

In the nineteenth century, tropical diseases, particularly vector-borne infections, among which malaria ranked first, were a major concern for the colonial British government as its empire reached around the globe. In several African and Asian countries, including India, colonial officers and their families were terrified by a host of illnesses, which claimed lives, slowed the march of conquest, and exacted an enormous financial price (Spielman and D'Antonio, 2001). Unfortunately, no systematic countrywide studies were performed prior to 1961, and details are available through scattered surveys in different parts of India. Records show just how devastating malaria was in the early days of the British occupation of India, particularly the five-river-bathed Punjab, which had remained impacted by malaria throughout. It is not difficult to comprehend how much the State of Punjab, one of the greenest and most watered in the country, suffered in maintaining health, especially of children and women, the two most vulnerable segments of the population, as well as economic and social activities. Mortality rates due to malaria were not enumerated, albeit being stupendous, until the initiation of independent India's specific office of NMCP in 1953 and later NMEP in 1958.

Malaria epidemics occurred throughout India with varied oscillations (Covell, 1927). Sir Gordon Covell, the one-time head of the Malaria Institute of India, recorded epidemics of "great intensity" in Murshidabad (Bengal) in 1821. In 1852, one malaria epidemic wiped out the entire village of Ula and then spread across the Bhagirathi River to Hooghly and continued to devastate populations for many years in Burdwan. The development of the Indian railways under the British administration contributed to the spread of malaria. The rather haphazard construction of railway embankments provided a number of breeding sites for the malaria vectors. It is also likely that the labourers who worked on the construction of the railways introduced different strains of the parasite into the areas in which they worked, possibly increasing the incidence of the disease. The year 1908 saw a countrywide pandemic during which huge mortality and morbidity occurred. In the early 1920s, Bengal suffered a severe malaria epidemic which resulted in over 730,000 deaths in 1921 alone. Thereafter, the number of deaths from malaria slowly decreased to within 300,000 to 400,000 per annum. During the Second World War, however, malaria deaths rose again, with Bengal recording over 680,000 deaths in 1943 and 763,220 deaths in 1944. In Bombay Presidency, there were significant outbreaks of malaria during the construction of the Colaba Causeway between 1821 and 1841, as well as during the construction of the Alexander Dock and Hughes Dry Dock. The number of malaria cases in Bombay increased relatively sharply after 1901, during which time the construction of the railway was completed, ensuring an increase in traffic and a greater movement of people.

Among British troops in 1921, over 30% of all hospital admissions were for malaria, and in 1922, 10,552 cases of the disease were reported, with 26 deaths out of a total military force of 60,166. This represents an incidence rate of 175 per 1,000, which increased slightly to 180 per 1,000 in 1923. The number of cases among the Indian troops is broadly similar to that of the British troops, with 173 cases per 1,000 in 1922. However, the mortality rate among Indian soldiers was around three times higher than for British soldiers who possibly did not have access to the same level of medical care and treatment offered to the British soldiers.

22.4.6 Early Malaria Control

Early malaria control efforts tended to concentrate on the removal of breeding sites and later used chemicals such as Paris green and kerosene for larviciding mosquito immatures. One of the first formal operations to control the disease took place at Mian Mir, near the city of Lahore. Mian Mir later became a British military post and the focus of a concerted effort to eradicate malaria. Mian Mir had an intricate system of irrigation canals which provided excellent breeding grounds for the malaria vectors. The incidence of malaria was astonishing, with an average incidence of slightly under 100%, rising to over 300% during epidemic years. The malariologists Dr J.W.W. Stephens and Dr S. R. Christophers, arrived at Mian Mir in 1901 with ambitious plans to remove all the breeding sites, evacuate the infected people, and administer quinine as both a curative and preventative measure. Their programme began in 1901 and eventually developed into a massive effort, with between 400 and 500 soldiers set to work full time at filling in the irrigation canals. The programme of constantly filling in ditches and removing puddles and any other potential breeding sites continued until 1909, when, to their dismay, a serious malaria epidemic broke out, forcing them to abandon the programme mainly on account of high costs and low output. The lesson learnt from the Mian Mir debacle was that mosquitoes, well entrenched, tenacious, and adaptable, are difficult to control. The enemy (mosquito) in Mian Mir proved more formidable than had previously been admitted, especially in its adaptability to attack and its capacity to cover considerable distances and make use of secondary breeding places when those of first choice were closed. It proved above all that for malaria control, every local situation needed to be thoroughly studied and a specific method employed (Harrison, 1978).

Similar failures were conceivable in Bombay, Jhansi, Poona, Meerut, and Secunderabad, as well as at all other military posts where similar activities concurrent to Mian Mir were initiated. In 1917, the Bengal Nagpur Railway (BNR) and the East India Railway formed a separate malaria control organization, specifically to control the disease in and around stations. There, every person suffered three bouts of malaria during a single year. Similar larviciding and breeding pool removal programmes were undertaken during the 1920s on the tea plantations of Assam and in Mysore by the Rockefeller Foundation.

The Rockefeller Foundation began using pyrethrum sprays experimentally in India to great success, and this method of malaria control was recognized as enormously valuable. Pyrethrum was first used for malaria control in Delhi in 1938, following its phenomenally successful usage by William Gorgas in Cuba about two decades prior. The use of pyrethrum was then expanded to Assam in 1942. All of the aforementioned interventions had some impact on the incidences of malaria, yet they were unable to sustain control over the disease.

The availability of DDT, the wonder insecticide, in the late 1940s, however, briefly changed the whole concept of malaria control through vector control.

22.4.7 DDT – The Wonder Insecticide

DDT was first synthesized by Othmar Zeidler in 1874, but the chemical remained in obscurity until 1939, when Dr Paul Muller discovered its insecticidal properties. Its best use was obtained during the Second World War, when ectoparasites like lice felled many troops with the deadly relapse of typhus fever. DDT was doused over people, and it was instantly successful in controlling diseases. In India, DDT was first used by the armed forces in 1944 for the control of malaria and other vector-borne diseases. In 1945, DDT was made available for civilian use in Bombay to control malaria, and shortly afterwards in 1946, pilot schemes using DDT were set up in several areas, including Karnataka, Maharashtra, West Bengal, and Assam. Between 1948 and 1952, WHO set up DDT demonstration teams in Uttar Pradesh, Rayagada, Wynad, and Malnad. The distribution of DDT in Bombay produced some remarkable results within a very short period. On 1 July 1945, the first civilian home was sprayed in India with a 5% solution of DDT mixed in kerosene. This initial spraying ensured that not a single adult *An. fluviatilis* from any house was found either during daytime or nighttime catches. The use of DDT was not only popular among the locals for control of the deadly Anopheles mosquito, but the insecticide also ensured that lizards, cockroaches, scorpions, bedbugs, and ticks all disappeared after a single spraying.

After the spraying in the Kannara district, it was reported that the population began to grow precisely because of a decrease in the death rate as a result of the DDT spraying. Prior to DDT being used, the district reported an average of 50,000 malaria cases every year, which was reduced by around 97% to only 1,500 cases after DDT was introduced. The spray programme in Bombay took place during a period of great upheaval in India and the withdrawal of the British Raj from the Subcontinent. During 1949, it is estimated that over six million people in Bombay were protected from malaria through the use of DDT and that at least half a million cases of malaria were prevented. Along with the control of malaria came the control of a number of other diseases. Incidences of the human plague were completely eliminated, and there were reports of reductions in the incidences of diarrhoea and dysentery. It is likely that this was because the number of insects that contributed to the spread of these diseases were all dramatically reduced. The benefits were not restricted to people; cattle also enjoyed improved health, as the ticks that caused festering sores and infections were all but destroyed.

In the early 1950s, India's population was estimated to be around 360 million, and every year around 75 million people suffered from malaria, and approximately 800,000 died from the disease. The disease put almost the entire population at risk, and the country had, and still has, over 50 different anopheline species. Of these, *An. culicifacies, An stephensi, An. Minimus*, and *An. fluviatilis* were defined as the major vectors of malaria. The apparent health and economic benefits that came from using DDT prompted the formation of the NMCP in 1953. Thel programme first set out to control the disease in the endemic and hyperendemic areas with 125 control units. Each of these control units consisted of between 130 and 275 men, and their plan was to protect approximately 1 million people each. By 1958, the malaria control programme had been increased to protect at least 165 million people from the disease with 160 control units. Controlling the disease became an issue of economic importance. A study conducted in the early 1950s estimated that there would be a net gain of Rs. 200 million if the government spent Rs. 40 million to control malaria.

After consultation with WHO and with the assistance of UNICEF and the US Agency for International Development (USAID), the decision was made to attempt the eradication of malaria. The programme was expanded to cover the entire country with an additional 230 units, and the NMEP was thus born in 1958.

The control units were divided into sub-units, sectors, and sections, where the sections each had the task of protecting around 10,000 people from malaria. This meant that on average, each section would be responsible for spraying 2,000 houses once during the high transmission season (if the area was in a seasonal malaria area) or twice every year in those areas where transmission occurred year-round.

The entire eradication plan was based on the spraying of DDT and the provision of chemotherapy to those in need. Based on the experience of the NMCP, it seemed reasonable to expect that eradication could be achieved in just a few years; originally, the officers did not force DDT on the insect; rather, the insect choose to enter the house, take a blood meal, and then kill itself by landing on the walls. The prescribed manner of spraying DDT was to mix 5% suspension of DDT wettable powder in water into a compression sprayer and to pump it to a pressure of 40 pounds per square inch. The sprayer would then stand about 18 inches away from the wall and spray the mixture onto the wall in a sweeping motion so that the DDT would be left on the walls at a concentration of 2 gm/m^2.

Despite some outstanding results initially, the programme began to encounter setbacks. By the early 1970s, it became apparent that controlling one of the most devastating diseases known to man was not going to be achieved solely with the use of DDT.

22.5 MOSQUITO PHOBIA

The beauty of the mosquito is that, notwithstanding it being infamously known as an ugly, irritating, and dangerous animal, it is also the one most remembered by humans! The sheer feeling of the fall of evening reminds humans about this notorious creature who stalks from close quarters of their own living conditions to allow them neither peaceful outdoor recreation during dusk or sound sleep anywhere at any time. Although malaria, dengue, yellow fever, and encephalitis – all spread by mosquitoes – cause a great concern to human health and economy, perhaps nothing matches the hair-raising, fearsome feeling of the sight of a person suffering with grotesque and disfigured lower limbs due to elephantiasis, a type of extreme manifestation of lymphatic filariasis caused by *Wuchereria* and *Brugia* worm transmitted in Asia by *Culex quinquefasciatus* and certain *Mansonia* species. No one who confronts a patient suffering from this disease ever forgets it. The imprint is so deep, almost indelible, that the mosquito, which is responsible for its transmission, takes on the form of a devil and frightens all, rich and poor. Elephantiasis is one of the most stigmatizing diseases in the world (Spielman and D'Antonio, 2001). It occurs when the flow of lymphatic fluid is blocked and parts of the body – frequently the legs and genitals – swell enormously. Under the constant pressure of swelling, the skin stretches until it becomes thinner, to a hair's breadth. Fluid collects, often forming thin sheaths that flow from a person's hips down his legs. The swollen testicles in males can match the size of a shot-put ball or even larger, dangling between the legs to the utter embarrassment and shock of the patient. In females, the vaginal labia can swell to nearly the same volume. If legs are affected (often with dissimilar impacts), feet and ankles are the most vulnerable sites, though the swelling can envelope more of the leg upwards, giving the patient a semblance of an elephant's leg. The mosquito, *Culex quinquefasciatus*, notwithstanding a ubiquitous presence in all environments of India, is more rampant in urban set-ups.

The situation of 'mosquito phobia' is almost the same in both rural and urban areas, as well as plain (low altitudinal) mountainous and forested regions. Cities and townships are presently worst affected simply because of the evolution of the enormous number and types of breeding habitats supporting different varieties of mosquitoes. Bigger cities and towns today mean bigger mosquitogenic problems, despite bigger shares of financial inputs in sanitizing them. Therefore, whether it is Delhi, Kolkata, Mumbai, Hyderabad, Bangalore, or Chennai – all great cities and the capitals of respective states, in addition to Delhi also being the national capital, the 'mosquito phobia' thrives unchecked.

22.6 THE MOSQUITOGENIC TALE OF TWO INDIAN CITIES

Bigger cities and towns, or metropolises, in India, as well as many other countries both in the New and Old Worlds, are regarded parallelly with the mosquito menace. In India, *Wuchereria bancrofti*–related lymphatic filariasis is principally an urban disease transmitted by the urban black-brown domestic mosquito, *Cx. quiuquefasciatus*. Malaria in Indian urban set-ups, spread mainly by *An. stepheusi*, is presently sharing as high as 40% of the total cases in the country from being virtually negligible in the early twentieth century. Dengue infection, though lately erupting in rural and sylvan environments, is also characteristically associated with an urban style of living. A mosquitogenic cross-section of two major metropolitan cities in the country, the Chennai and Bangalore metropolises, where living is far better, organized, and systematic than in many others, are still heavily punctuated with mosquito-donned havens.

22.6.1 Chennai Metropolis

Chennai (formerly Chennai) is a metropolitan city with a chequered history, but it is also sadly known for a high incidence of malaria and, in recent times, dengue – both transmitted by mosquitoes, the former infection by *Anopheles* mosquitoes and the latter by *Aedes* mosquitoes. Over the years, unorganized and uncontrolled urbanization, population boom, poor quality of housing, and a myriad kinds of unhygienic situations rife with different kinds of mosquito breeding sites have all contributed to the eruption of heavy mosquito density and infections.

The city of Chennai stretches over nearly 172 sq. km. along the western margin, from Thiruvanmiyur in the south to Thiruvttiyur in the north. Chennai is one of the fastest-growing metropolitan cities in the country, both in terms of population and employment. Important buildings and structures such as the Fort, the Marina – the road parallel to the sea – and the fronting public buildings, the High Court, the Moore Market, the Connemara Public Library, and other public buildings are the landmarks of the city.

The Chennai Metropolitan Area (CMA) consisting of the City of Chennai and its environment extends over 1,166.76 sq. km. The estimated population of Chennai's urban agglomeration in 1991 using the exponential growth model was 5.56 million, and using the component method, it was 5.43 million. The population estimates for CMA and the city area estimated by MMDA (Madras Metropolitan Development Authority) were 10.0 million in 2011. Estimates also indicated that between 1991 and 2011, 3.5 million persons would be added to CMA, of which the city share would be 2.0 million.

Three distinct areas are identified within the metropolitan area: (i) first, there is the city itself within the limits of Chennai corporation; (ii) secondly, there are the urban centres and extensions contiguous to the city area; and (iii) thirdly, there is the rural area with a number of rural settlements. Chennai city means the area under the jurisdiction of the city corporation. Chennai urban agglomeration means the area consisting of Chennai city and the urban centres. CMA means the area delineated for planning purposes, which is also the Chennai standard urban area defined by the census.

Until a few years ago, malaria counts in Chennai city surpassed all estimates from rest of the state. Chennai also had the highest annual parasite incidence per 1,000 in India, besides contributing 10% of total cerebral malaria cases caused by the deadly parasite *P. falciparum*. Mosquitoes, which breed in the city's numerous open overhead tanks, wells, drains, and stretches of stagnant water, function to courier the disease pathogens. According to an estimate, the city has 95,023 wells (82,287 uncovered), 89,523 overhead tanks (47,367 uncovered), 23,368 sumps, 72,834 cisterns, 12,3OO cess pools, 15,000 cess pits, 13,000 septic tanks, 700 km of storm water drains, 35 km of minor waterways, and 56 km of major waterways, besides open water-filled tar drums left at construction sites, sewage treatment plants of Metro Water, and trenches dug by government water, power, and communications departments and others – each a potential breeding spot for mosquitoes.

With the advent of independence, industrialization of the state zoomed, with many important industries both in the public and private sectors getting established in and around Chennai. The industrial areas within the city are mainly in the northern and western parts where they are closely mixed with residential developments. In recent times, new industrial establishments have been located outside the city limits; in fact, many important industries are bracing the outskirts. People who came to work in these areas could not afford to rent a house and thus began the growth of slums in Chennai city.

Slums are important places for mosquito proliferation. The vast majority of the people who migrated to the city were attracted by opportunity and comforts offered by modernization. They belonged to the working class and found it difficult to secure accommodation within their means. So, they squatted on every open space available, as near their workplaces as possible, and put up huts with cheap building materials. In this way, slums grew in number and population in Chennai city. The Government of India Slum Areas (Improvement and Clearance) Act of 1954 defines a slum "*as any predominantly residential areas, in which light or sanitary facilities or any combination of these factors are detrimental to the safety, health or morals*." Slums are predominantly residential areas of blight. They are chaotically occupied, unsystematically developed, and generally neglected areas, overpopulated and ill repaired. The other horrors of slum life are insufficient communications, inadequate amenities for physical and social health, and absence of minimum needs and comforts. A slum is taken to mean "*[c]utting areas with huts erected in a haphazard manner without proper access, without protected water supply and drainage arrangements and congested as to allow little free air to get in*." Slums generally present the most unhygienic, ugliest, most nauseating scenes. During the rainy season, the whole area gets flooded, the pathways become swampy, and the entire colony becomes a fertile breeding place for mosquitoes, exposing the slum dwellers in the area to all sorts of diseases and mosquito-bite-related maladies.

Some of the fiercest mosquitogenic situations and or factors complementary to the copious preponderance of mosquito populations in the metropolis are briefly cited next.

22.6.1.1 Slums in Chennai City

The slums are scattered throughout the city. As South Chennai has larger tracts of open space than North Chennai, which is predominantly an industrial area, the number of slums is higher in South Chennai. Due to scarcity of land in North Chennai, huts are built mostly on corporation dumping grounds and even over the drainage and sewage pipelines. It is seen that 56.89% of the slum families are put up on public lands. The other 43.11% of slum families are on private land and on lands of religious and other institutions.

22.6.1.2 Cattle Population

Another menacing problem of environment deterioration is caused by cattle. There are about 70,000 cattle in the city maintained in 2,000 cattle sheds. About 1,900 (76%) of the families depend solely on the cattle. About 350 families maintain cows for their own use.

22.6.1.3 Garbage and Waste Disposal in CMDA

The city generates about 2,400 tonnes of garbage, of which about 1,900 to 2,000 tonnes are cleared every day. The civic body has 302 heavy vehicles, 96 light vehicles, and 150 autorickshaws. Often, in the event of a heavy storm, the city is left behind in garbage and refuse, turning the metropolis of about 5.5 million people into a vast rotting garbage dump.

The rivers that can be treated as waterways of transportation are Coovum, Adyar, and Buckingham Canal. These rivers crisscross the city, meeting at several places. But they are used as no more than open drains to carry the city's sewerage sullage, garbage, and industrial wastes, leaving them hazardous. It is uncontaminated at its entry point to the city, collects the surpluses of 75 small tanks from Kancheepuram where it originates, and runs 16 km from the upper reaches of Chennai until it reaches the Bay of Bengal. The Coovum River runs along the middle of the city, bifurcates on the southern side of the Central Jail, and forms an island; the two arms join again slightly upstream of the Napier Bridge. There are three types of waste discharges that pollute the river.

The first flush of stormwater is a severe pollutant and as bad as raw sewage; in addition nutrients, sediments, and heavy metals are carried in the stormwater. Homeowners throw garbage and other hazardous and non-hazardous wastes into drains leading to Coovum. Further along the Coovum, there are 27 points through which waste and raw sewage enter into Coovum, contributing more than 25 million litres per day. All of this contamination leads to the increased pollution of the Coovum. It is said that the velocity of flow of the river falls far below the self-cleaning velocity, as the terrain is even and flat, and accordingly, the silt and sedimentation are heavy. The problem is much more with the sandbars at the mouth of the Bay of Bengal: stagnation in the estuarine area. The sandbars prevent free water flow into the sea for most parts of the year. The sandbars form a pool of water between the Chetpet Bridge and the river mouth of Coovum. Nearly 10 lakhs to 20 lakhs tonnes of sand deposits collect at the mouth of Coovum per annum. However, during incessant and nonstop rains, the floods carry a substantial quality of Coovum water into the sea, and the increase of population at the estuarian area every year complicates the situation.

22.6.2 Bangalore Metropolis

Bangalore, the garden city of India with the reputation of a 'salubrious environment' and 'peaceful living,' offers a rather glaring example of how the once fine city could virtually turn out to be a haven for blood-sucking creatures – all for the reasons associated with human's needs and activities for prosperity! Incidentally, this is also India's first metropolitan city, which drafted a "Master Plan for Mosquito Control." Bangalore city has lately been plagued by the mosquito nuisance, notwithstanding a good infrastructure for mosquito control and reasonably sufficient funds (> one crore, at early 1980s level). The solution to the metropolis's mosquito nuisance was offered by the Indian Council of Medical Research's Vector Control Research Centre, based at Pondicherry, by practically working out all the reasons underlying the emergence of the mosquito menace in the city and subsequently tabling it in the form of a unique "Mosquito Control Master Plan for Bangalore City" (Rajagopalan et al., 1987). Bangalore city, with all its impeccable glory of being one of India's cleanest, well-organized, and hospitable cities, provides the perfect model for understanding how a nearly mosquito-free environment could be transformed into a veritable paradise for mosquitoes, only to vex its inhabitants and rob them of their hard-earned sleep! More than anything, humans are to bear the blame for this mess. Bangalore, during the 1980s in particular, was rated the fastest-growing metropolitan city in the country when the population boomed from mere 25 lakhs in 1981 to close to 45 lakhs by 1987. A huge influx of humans in search of vocations or in connection with entrepreneurship, coupled with unorganized growth of living conditions like large slums and shanties, added much to convert the area, particularly the peripheral areas, into mosquitogenic situations. The black domestic or house mosquito, *Culex quinquefasciatus*, is the wicked culprit, although *Aedes aegypti*, the dangerous vector for dengue, is also often present.

22.7 REASONS ACCOUNTING FOR MOSQUITOGENIC CONDITIONS AND ERUPTION OF MOSQUITO-BORNE DISEASES

These are aplenty. Almost everything on Earth that collects water is liable to become a viable breeding place for some or another mosquito species, many of which can be serious vectors of human diseases. Some mosquitoes, particularly those belonging to the genus Aedes, require but a little moistened soil in close vicinity to an underlying water source to breed and thrive, obfuscating the 'mosquito phenomenon' a further great deal. Even so, any of these, or all of these, naturally existing breeding conditions hardly ever trigger the onset of a devastating epidemic unless some catastrophic and colossal vicissitudes or interferences with the nature are committed by humans. Often, wars drive people across borders (from hyperendemic areas to non- or hypoendemic areas), and the concentration of refugees in relief camps further adds to the already long list of mosquito breeding conditions and factors effecting spread and exacerbation of a parasitic infection like malaria.

Sometimes drastic climatic variabilities, with a human hand, are culprits: (i) heavy rains in 1990 in the otherwise xeric environment of the Thar Desert caused the onset of an unprecedented malaria epidemic in the State of Rajasthan (Mathur et al., 1992); (ii) prolonged deficient rains or droughts between 1997 and 2000, also in the Thar Desert, forced the migration of local folks, along with their cattle, to other states (with hypoendemic status) and became the cause of a malaria epidemic upon the return of these people, who also brought back the parasite of a non-endemic or hypoendemic home population for a sudden conflagration; (iii) the irreversible changes in physiography in the Thar Desert, with innumerable groundwater-based breeding conditions, developed in an otherwise *terra incognito* for a mosquito species with an infamously high potential for transmitting malaria, like *An. culicifacies*, converting the virgin land into a hotbed for disease epidemics, on one hand, and a virtual paradise for mosquitoes on the other. The countrywide dengue epidemics as recently as mid-2003 rocked the nation once again, as it looked very much like the bearing of a human hand in inviting the problem that infected over 3,500 people and killed 66 others, many of whom were children. A nonstop annually cyclic malaria epidemic in the Thar Desert since 1992, the worst of which occurred in 1994 when 452 deaths were reported on the soil of Thar Desert alone, mostly in the command area covered under Indira Gandhi Nahar Pariyojana (IGNP), is an eye-opener to public health managers and disease control planners in the country. The drastic changes the IGNP brought about in the Thar Desert's physiography, besides the formation of innumerable breeding grounds due to the canal's seepage to several mosquito species, including the dangerous *An. culici- facies*, earlier unknown to the interior of the desert region, have turned the land into one of the 'hotbeds' for malaria outbreaks (Tyagi, 2002, 2003) . These outbreaks and epidemics are a testimony to our constant struggle with mosquito-borne-diseases. The great conundrum, however, in our relationship with the mosquito is that much of the trouble is caused in part by our own deeds: air traffic, sea voyages, and different kinds of commerce *modus operandi* intercontinentally.

22.8 THE MAN-MOSQUITO CONUNDRUM: IS CO-EXISTENCE IMPROBABLE?

Mosquitoes occur worldwide, in most climatic regions and situations, and in many parts, they persist in enormous numbers. They have been collected on mountains and below ground level, with reference to India at altitudes of 14,000 feet or more in Kashmir and, as exemplified by *Culex quinquefasciatus*, 903 feet below sea level in the Kolar Gold Mines. It appears that more outbreaks of mosquito-borne diseases are expected in the future. The issue of mosquitoes is now extremely important with dengue fever on the rise, in addition to the prevailing grim situation of malaria, filariasis, and JE in most parts of India. These insects have long been the bane of human beings. This fascinating little creature has now drawn attention that was never seen before.

Despite the weight of previous failures and the variety of threats that humans endure, they can still learn to live with mosquitoes. To accomplish this, there cannot be a single formula since there are vast differences in economic status between developed countries and those that are still not developed, putting the 'co-existence theory' on a different footing for each group of countries. Those with means and methods – that is, the rich countries – can design their needs, while others with fewer resources – that is, the developing and/or the underdeveloped nations – will have to look for what suits them best.

One thing that is more than sure is that, in the event of a do-or-die conflict between humans and mosquitoes, the latter will likely survive us. Mosquitoes were here on Earth long before humans and certainly have more tricks up their arms (in fact, legs or, may be, proboscis) than the entire human race can conceive to devise with all the modern facilities available. The earlier thought of annihilating vector mosquitoes from the face of Earth, therefore, was derived more by passion and overconfidence than wisdom and logic. Both species can live without one causing harm to other. Man being the thinker has to learn and define how.

All the while, however, researchers may go on finding effective and novel ways to bring mosquitoes under control. One such study is currently embarking on creating transgenic mosquitoes of a given vector mosquito species, and scientists are hopeful it will succeed early. Meanwhile, following experiences obtained through past encounters with humans' deadliest foe, the mosquito, a few time-tested anti-mosquito measures can surely help mitigate the mosquito-borne dangers!

22.8.1 General Living Hygiene Upgradation

Environmental sanitation is a time-tested tool for bringing about long-lasting reduction in mosquito breeding and preponderance. Most of the North American and European countries in the early twentieth century, as well as some of the Asian countries as late as the 1980s, raised their living standards to such an extent that mosquito breeding in and around houses is not only difficult to find, but allowing intra-domestic mosquito breeding is regarded as a punishable offence. Peoples' general awareness is very high, and they work both by themselves as anti-mosquito brigades and in close coordination with government agencies meant to control mosquitoes and/or mosquito-borne diseases in their countries. In such developed nations, humans have adjusted their lifestyles to discourage mosquito blood-feeding by improving quality of house construction, wearing long-sleeve shirts and trousers, and using repellents in evening recreations outdoors. House construction styles play an important role to check on mosquito entry in houses during night. In India, too, regulations governing anti-mosquito breeding are in force in some metropolises, at least, but these are generally insufficiently followed and are overtly ignored in the outgrowth of haphazard and unplanned expansion of cities and towns. A lot of efforts, both at government and people's levels, will have to be meticulously conceived, planned and executed to do what is necessary to check mosquito breeding in and nearby houses.

22.8.2 Improved Housing Conditions

As referred to earlier, improved housing styles make them impregnable to mosquitoes without sacrificing the essential necessities of *fang, sui*, and *chi* (Chinese expression for air, light, and water). To affect not only the design of a house but also the material to be used in making the house in different ecological situations is also important. Adequate databases are presently available to develop the infrastructure indigenously in any country, and it only requires definite regulations and implementing bodies to execute the act.

22.8.3 Obligatory Mosquito Netting of All Entry/Eat Points in Homes

Houses impregnable to mosquito means a proper mesh one on all doors, windows, and exhaust fan points like the ones in kitchens and toilets. In some countries, houses are to be put up invariably with a mosquito netting as per law requirements. This is also found to be a foolproof method for checking mosquito entry into the house, even in the event of some negligence on the part of inhabitants.

In India, houses are properly screened only either under statutory provisions applicable to government quarters in certain services or with the individual awareness of the mosquito menace. In south India, particularly in Kerala, houses are excellently designed and constructed but still mostly without screening, albeit a nearly year-round mosquito menace. Several modes and designs are currently available to place it on doors, windows, and exhaust fan points, but this exercise requires a strong conviction, with a legislative provision to back it up to the people about their long-term benefits.

22.8.4 Compulsory Use of Bed Nets Where Affordable

Mosquito bed nets, though in use in different forms in several countries, including India, have never really had as strong a sense of deployment in personal prophylaxis as in the last three to four decades, especially with the invention of the insecticide-impregnated mode. Use of a bed net is simple, cost-effective, and highly effective against mosquito biting. In fact, in Indian military services, it is mandatory for everyone to use a bed net while sleeping. Anyone found flaunting the instruction has to face punishment. Yet in public life, the usage is generally restricted to either those who sense its significance or the people who are simply overwhelmed by the mosquito nuisance day in and day out. The deployment of bed nets, however, requires a strict discipline since, otherwise, in the wake of improper deployment – which generally is the case – the bed net would be rendered completely useless for the purpose. As a solution to this frequent folly, insecticide-impregnated bed nets provide a good recourse since not only are mosquitoes prevented from attacking humans, but they are also kept at bay even when the bed net is not properly folded under the mattress or, in extreme cases, it is worn out here and there due to excessive and/or faulty use. This method has arisen as a possible alternate to the long-time indoors application of residual insecticides.

22.8.5 Use of Local Indigenous Repellents

While outside the home in the evening hours, for recreation or work, certain repellents are very handy for keeping the mosquito away (e.g., DEET, DMP, DEPA). In recent times, in view of certain physiological disturbances by some of the synthetic repellents, due emphasis has been made to rely on plant-based repellents. Use of repellents bar mosquitoes from making contact with humans and, thereby, prevents not only the painful bite but also the risk of disease transmission.

22.9 MOSQUITOES – THE LITERARY MUSE AND WORK OF ART

22.9.1 Mosquitoes in Poetry

Interestingly, mosquitoes are mentioned in poems and books penned by famous poets and authors. One of earliest famous examples is from Sir Dr Ronald Ross (*vide infra*), who discovered the malaria-mosquito relationship (Nye and Gibson, 1997; Tyagi et al., 2020). A malariologist and polymath, including nurturing innate interest in poetry since childhood (Bhattacharya and Bhattacharya, 2010), Dr Ross naturally dedicated his 2002 Nobel Prize–winning research work to the mosquito in whose body inevitably completes the beast called the *Plasmodium* parasite with the sporogony part of its life cycle. The following are some lines from that immortal verse!

This day designing God
Hath put into my hand
A wondrous thing. And God
Be praised. At his command,
I have found thy secret deeds
Oh million-murdering Death.
I know that this little thing
A million men will save –

Oh death where is thy sting?
Thy Victory, oh grave?

In his famous collection of poems entitled *Birds, Beasts and Flowers: Poems*, D. H. Lawrence (1992) iterated rhetorically about the 'mosquito' as follows:

Yet what an aura surrounds you;
Your evil little aura, prowling, and
casting a numbness on my mind.

That is your trick, your bit of filthy magic:
Invisibility, and the anaesthetic power
To deaden my attention in your direction.

But I know your game now, streaky sorcerer.

Queer, how you stalk and prowl the air
In circles and evasions, enveloping me,
Ghoul on wings
Winged Victory.

Mosquitoes were here on Earth before us and will likely survive us. Needless to reminisce that no other animal on Earth has touched so directly and profoundly the lives of so many human beings as the mosquito. "For all of history and all over the globe she has been a nuisance, a pain, and an angel of death. Mosquitoes have felled great leaders, decimated armies, and decided the fates of nations" (Spielman and D'Antonio, 2001). Bhattacharya (2021) has very elegantly described the mosquito in an ode:

River of faith flowing in the City of Joy,
where I got my first freedom and touched
the divinely inspired soil
Champagne, music and dance
left behind in the white man's town,
Filth, flies, lime traders and noisy crows
in the native's colony,
Unknown enemies flying in the dark,
their singing swarms in a stinging swoop
drunk on wine-red blood murdering millions,
seeking out a crusader alone in an alien land,
who struggled with 'tears and toiling breath'
to unravel the mystery behind the massacre,
with Calcutta as witness and muse.

Whatever be the reason behind such rhetoric about mosquito, it is essentially the fair sex, the *femme fatale*, which hogs the entire limelight, albeit her villainous stardom and human's deadliest foe on the planet. The male mosquito, which is essentially left out unsung, has very rarely been applauded. Tyagi (2023) has extolled the male and emphasized that the entire glory of the female mosquito actually rests on the existence of her subaltern partner.

The female is large and notorious,
And progenitor of a humongous race.

Extolled galore for her prowess unparalleled,
To suck blood and sicken millions human being,
Health, wealth and human intelligence, all,
Appear at the mercy of this 'man's deadliest foe',

To whom do fall countless men and women, and infants, too,

Prey to the singleness of aim of 'the femme fatale'.
And, yet, she fails to see the Sun first,
And depend for all her glory and majesty,
On a singular and much smaller, though promiscuous,
Y – her irreplaceable zaddy, the male mosquito.

22.9.2 Mosquitoes in Body Decoration

Nail art has been in vogue for centuries, but hardly anyone seemed to have chosen mosquitoes on their nails (Figure 22.1a). Surprisingly, the trend has apparently changed, and mosquitoes are increasingly becoming a source fashion art and design on nails for show (YouTube, Bigheadbungo, 2021).

22.9.3 Mosquitoes as Food

What is even queerer is that mosquitoes are now being served as palatable food (Figure 22.1b). These tiny creatures are now being liked on food menus as energetic dishes such as the 'mosquito burger' in the African wilderness (YouTube, BBC News, 2018). It is noteworthy that a wide variety of insects have aroused a lot of interest for their nutritious value, and in many countries, insects are becoming delectable items on menus as loudly advocated through deliberations in the First International Conference on Insects as Human Food and Animal Feed, jointly organized by the Food and Agriculture Organization of the United Nations (FAO) and the Wageningen University and Research Centre (Wageningen UR) in 2013 (Huis et al., 2015).

22.9.4 Mosquitoes in Novels

Mosquitoes, regardless of their frail and slim forms, have not escaped the imagination of either artists, musicians, or novelists (Bedirian 2023). Mosquitoes appear in several novels (Faulkner, 1927), while very recently Tyagi (2025) caricatured about interesting aspects of their behaviours, as "Mosquitoons."

22.9.5 Mosquitoes in Preservation

Most people of all ages still cherish memories of a mosquito preserved in amber on the head of a walking stick that the owner of the Jurassic Park (which was also the name of the movie directed by Steven Spielberg), representing the presence of mosquitoes during the Mesozoic period dominated by dinosaurs (Kelly, 2023; Yong, 2013).

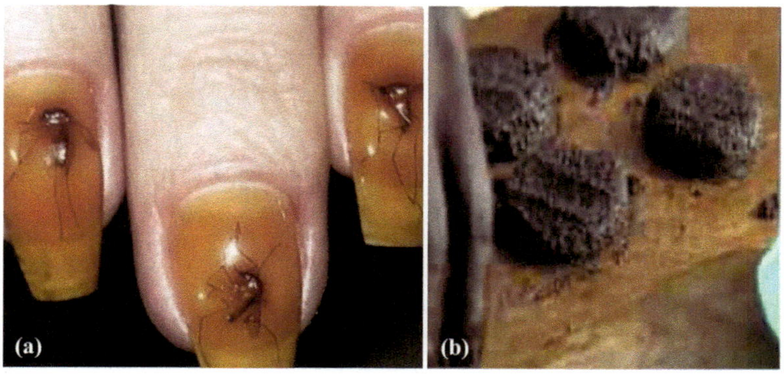

FIGURE 22.1 (a) Mosquito in nail art and (b) mosquito burger on sale in Africa.

Thus, it is clear that mosquitoes have a deeper significance in the lives of humans than is actually understood.

22.10 MOSQUITOES AS BENEFICIAL INSECTS

22.10.1 Appealing Body Ornamentation

There are many mosquitoes that are excellent examples of beauty personified. Though a New World (Central and South Americas) resident, the paddle-legged beauty, *Sabethes cyaneus* or *Sabethes tarsopus*, is universally appreciated, both in its appearance with conspicuously ornamented body topped by shining metallic scales and particularly in context with its courtship manoeuvres (Figure 22.2). Among the whole world of insects, the *Sabethes cyaneus* has few rivals when it comes to looks, though it is only a matter of knowledge that *Sabethes chloropterus* has been found infected with SLE virus and Ilhéus virus, and it transmits yellow fever virus to humans (Harbach, 1994; De Rodaniche and Galindo, 1957). Its colour is predominantly a stunning iridescent blue, marked with contrasting silvery-white spots – especially on the abdomen! Bold colours are not its only unique feature; the feather-like paddles catch the eye as well. These strange paddles, formed by elongated scales located on the tibia of each middle leg, are sported by both males and females.

The mating or courtship behaviour of this mosquito is quite interesting. The vast majority of species do not display any behavioural patterns complex enough to even qualify for the term 'mating.'

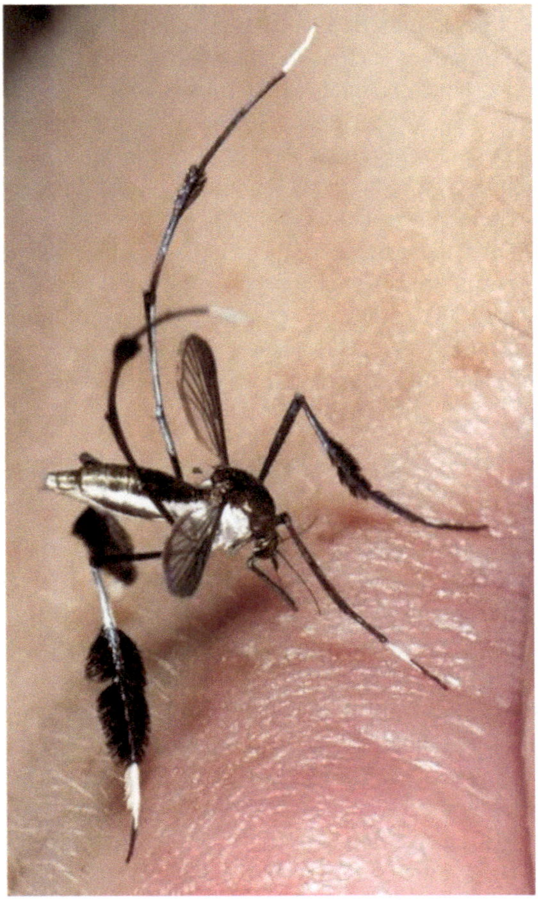

FIGURE 22.2 The paddle-legged beauty, ***Sabethes tarsopus***.

Source: Stephen A. Marshall.

For most mosquitoes, mating is a fast and haphazard occurrence with swarms of males circling around and waiting for the sound of female wing beats. Once detected, the male rushes after the female and immediately attempts copulation without an iota of courtship! In contrast, *Sa. cyaneus* is one notable exception. The male goes through a number of displays and movements in a specific order, and the female has multiple opportunities to accept or reject his advances. The intersexual attraction begins with the female resting on the underside of a horizontal surface, such as a stick or twig. The male, apparently attracted by her stunning beauty, hovers around the female with his leg paddles outstretched. He then places one of these middle legs (midlegs) over her wing, perhaps to prevent her from just flying away but also to be used as a pivot point as he swings around to face his date and simultaneously grasps the twig with his front legs. Once aligned, the true courtship begins. The male waves his free midleg back and forth, apparently showing off his lovely paddle, but eliciting no apparent response from the female. Next, the male will flex his forelegs to create a bout of rhythmic body 'swinging' while simultaneously flicking his proboscis (tube-like mouthpart) downward. If successful, this display will induce the female to lower her abdomen sufficiently for copulation to take place. If unsuccessful, he will return to waving his free midleg, followed by another round of rhythmic body swinging, until she either lowers her abdomen or uses her hind legs to reject him by pushing him away. During copulation, the male gyrates his abdomen while waving his middle and hind legs alternately until mating is completed and the pair disengages!

22.10.2 Mosquitoes as Good Pollinators

Some mosquitoes are good pollinators. It seems strange, but, yes, many do pollinate, albeit they are few in number. Some mosquitoes such as the *Aedes sollicittans, Ae. Provocans*, and *Ae. tritaeniorhynchus* are floral feeder, hence potential pollinators of some selected plant species in certain countries, e.g., Canada. Similarly, *Aedes communis*, known by some as a snow pool mosquito, is an important pollinator of *Platanthera obtusata*, the blunt-leaf orchid, in northern regions. It is most common in the state of New Jersey, although it can also be found throughout much of the northern United States, Canada, and Alaska. It usually inhabits dense forests at high elevations. Its larvae are mostly found a few feet from the shore in deep snow pools, which reside at greater than 1,500 feet in elevation. This snow pool mosquito eats the nectar from the floral spur of the *Platanthera obtusata*, during which time its eye naturally comes into contact with the pollinium, a cluster of pollen. The pollinium sticks to the mosquito's eye even when it flies away. Thus, when it eats from another flower, the pollinium touches the stigma of that flower, and the flower is pollinated!

22.10.3 Mosquitoes Serve as Food to Predators in Both Aquatic and Aerial Environments

Mosquitoes are also delectable food material, as they are easy to catch as well. As larvae, mosquitoes make up a substantial biomass in aquatic ecosystems globally. They abound in bodies of water ranging from ephemeral ponds to tree holes to old tyres, and the density of larvae on flooded plains can be so high that their writhing sends out ripples across the surface. They feed on decaying leaves, organic detritus, and microorganisms. In the absence of their larvae, hundreds of species of fish would have to change their diets to survive, which would not be easy since this feeding behaviour is deeply imprinted genetically in those fish, such as the mosquitofish, Gambusia affinis, which is a specialized predator – so effective at killing mosquitoes that it is stocked in rice fields and swimming pools as pest control – that could go extinct. And the loss of these or other fish could have major effects up and down the food chain. Many species of insect, spider, salamander, lizard, and frog would also lose a primary food source, i.e., mosquitoes. On the contrary, if the mosquito immature aquatic population was sprayed with a microbial mosquito control agent, the birds that are dependent on them would produce, on average, two chicks per nest after the spraying, compared with three for birds at control sites. Most mosquito-eating birds would probably switch to other insects that, post-mosquitoes, might emerge in large numbers to take their place. Other insectivores might not miss them at all: bats feed mostly on moths, and less than 2% of their gut content is mosquitoes.

22.10.4 Mosquitoes Are 'Ecosystem Service Providers'

Mosquito larvae are important members of the tight-knit communities in the 25–100 ml pools inside pitcher plants, *Sarracenia purpurea*, on the east coast of North America. Species of mosquito, *Wyeomyia smithii*, and midge, *Metriocnemus knabi*, are the only insects that live there, along with microorganisms such as rotifers, bacteria, and protozoa. When other insects drown in the water, the midges chew up their carcasses, and the mosquito larvae feed on the waste products, making nutrients such as nitrogen available for the plant. In this case, eliminating mosquitoes might affect plant growth. A stronger argument for keeping mosquitoes might be found if they provide 'ecosystem services' – the benefits that humans derive from nature. Without mosquitoes, thousands of plant species would lose a group of pollinators. Adults depend on nectar for energy (only females of some species need a meal of blood to get the proteins necessary to lay eggs).

Interestingly, a few mosquito species, particularly those belonging to the Sabethine group of genera such as *Malaya*, exhibit a quaint feeding behaviour on the ants belonging to the genus *Cremastogaster* (Figure 22.3a). The Malaya mosquitoes, of which two species occur in India, are totally innocuous to man and other animals, as they never bite due to the presence of a highly specialized dew-seeking proboscis or mouthparts only; the hairy proboscis is peculiar in having a flexible joint and a very much enlarged tip carrying four long hairs. They are small mosquitoes ornamented with brilliant silvery, blue, or violet scales. These mosquitoes imbibe the sugary volume in the oral cavity of ants.

22.10.5 Mosquitoes as Biocontrol Agents Against Disease Vector Mosquitoes

Toxorhynchites spp. are a rare group of mosquitoes, essentially vegetarian (Figure 22.3b). All members of this genus subsist on plant materials: sap, nectar, etc. Elegance nonpareil, these nonbiting mosquitoes make a significant constituent of the culicid biome. The metallic shining of azure on thorax, comparatively gigantic size, elephantine trunk-like sap-sucking proboscis, and spidery long legs are something unseen in any other animal on the planet.

Immatures of *Toxorhynchites* predate the immature stages of other co-inhabiting mosquito species, including dengue vectors. Accordingly, *Toxorhynchites splendens* larvae had been successfully deployed as biological control agents against phytotelmata breeding dengue vector *Aedes aegypti* (Panicker et al., 1981).

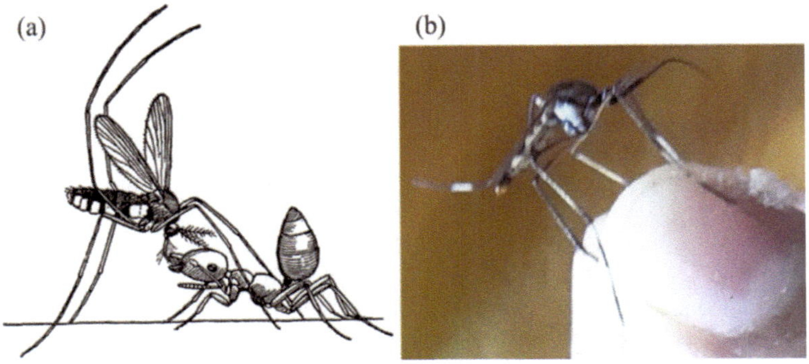

FIGURE 22.3 (a) *Malaya* sp. taking food from the mouth of an ant (*Cremastogaster* sp.) and (b) a member of the nonbiting *Toxorhynchitis* mosquitoes, some of which are time-tested potential mosquito biocontrol agent.

REFERENCES

Alnwick, D. (2000). Roll Back Malaria – What are the prospects? *Bulletin of the World Health Organization* 78:1377.

BBC News (2018). Youtube. https://www.youtube.com/watch?v=LItNFP7icUw. Accessed on June 15, 2023

Bedirian, R. (2023). The national news, arts & culture, arts & design. https://www.thenationalnews.com/arts-culture/art-design/2023/04/01/mahmoud-obaidi-explores-brutality-of-colonialism-in-dubai-exhibition/#:~:text=%E2%80%9CI'm%20trying%20to%20show,You%20see%2C%20I%20hate%20mosquitoes. Accessed on July 19, 2023.

Bhattacharya, S. (2021). *A crusader in my city. Poetry, now and again (Book).* Dey Publications, 60 pp.

Bhattacharya, S. and Bhattacharya, S. (2010). Ronald Ross: Known scientist, unknown man. *Science and Culture* 76(7&8), 253–254.

Bigheadbungo (2021). Youtube, https://www.youtube.com/shorts/vesQHWgLUQA. Accessed on June 15, 2023.

Bynum, W.F. and Overy, C. (1998). *The beast in the mosquito: The correspondence of Ronald Ross and Patrick Manson.* Editions Rodopi B.V., 528 pp.

CDC (2024). https://www.cdc.gov/vector-borne-diseases/about/index.html. Accessed on January 18, 2025.

Christophers, S.R. (1933). *The fauna of British India, including Ceylon and Burma. Diptera: Family Culicidae; Tribe Anophelini. vol. 4.* Taylor & Francis, 271 pp.

Covell, G., 1927. A critical review of the data recorded regarding the transmission of malaria by the different species of Anopheles, with notes on distribution, habits and breeding places. *Indian Medical Research Memoirs*, 7: 87–99.

De Rodaniche, E., Galindo, P. (1957). Isolation of yellow fever virus from Haemagogus mesodentatus, H. equinus and Sabethes chloropterus captured in Guatemala in 1956. *American Journal of Tropical Medicine & Hygiene*, 6(2):232–7. doi: 10.4269/ajtmh.1957.6.232

Faulkner, W. (1927). *Mosquitoes.* Boni & Liveright, 183 pp.

Gubler D.J. (2002) The global emergence/resurgence of arboviral diseases as public health problems. *Archives of Medical Research*, 33, 330–342.

Harbach, R.E. (1994). Review of the internal classification of the genus Anopheles (Diptera: Culicidae): the foundation for comparative systematics and phylogenetic research. *Bulletin of Entomological Research* 84, 331–342.

Harrison, G. (1978). *Mosquitoes, malaria and man: A history of the hostilities since 1880.* Tohn Murray, 314 pp.

Hayes, E.B. (2009) Zika virus outside Africa. *Emerging Infectious Dis*eases, 15, 1347–1350.

Huis, A., Dicke, M. and Joop van Loon (2015). Insects to feed the world. *Journal of Insects as Food and Feed*, 1, 3–5. https://doi.org/10.3920/JIFF2015.x002

Jones, A.V., Tilley, M., Gutteridge, A., Hyde, C., Nagle, M., Ziemek, D., Gorman, D., Fauman, E.B., Chen, X., Miller, M.R., Tian, C., Hu, Y., Hinds, D.A., Cox, P. and Scollen, S. (2017). GWAS of self-reported mosquito bite size, itch intensity and attractiveness to mosquitoes implicates immune-related predisposition loci. *Human Molecular Genetics*, 26(7), 1391–1406.

Kaur S.D. and Singh L. (2017). Identification of mosquitoes, nature of diseases and treatment in early sanskrit literature. *Indian Journal of History of Science*, 52(3), 243–250.

Kelly, B. (2023). Is jurassic park possible using mosquitos preserved in amber? *East End Tick Control.* https://tickcontrol.com/blog/jurassic-park-mosquitos-in-amber/#:~:text=Are%20Amber%20Mosquitoes%20Real%3F,when%20dinosaurs%20walked%20the%20earth. Accessed on June 15, 2023.

Krishnamoorthy, K., Harichandrakumar, K.T., Krishna Kumari, A. & Das, L.K. (2009) Burden of chikungunya in India: Estimates of disability adjusted life years (DALY) lost in 2006 epidemic. *Journal of Vector Borne Diseases*, 46, 26–35.

Lawrence, D. H. (1992). *Birds, beasts and flowers: poems.* David R. Godine Publisher.

Mathur, K.K., Harpalani, G., Kalra, N.L., Murthy, G.G., Narasimham, M.V. (1992). Epidemic of malaria in Barmer district (Thar desert) of Rajasthan during 1990. *Indian J Malariol* 29(1): 1–10.

Nye, E.R. and Gibson, M.E. (1997) *Ronald Ross: Malariologist and polymath – A biography.* Palgrave Macmillan, 336 pp. https://doi.org/10.1057/9780230377547

Panicker, K.N., Bai, M.G., Rao, U.S.B., Viswam, K., Murthy, U.S. (1981).Anopheles subpictus, vector of malaria in coastal villages of South-East India. *Current Science* 50: 694–5.

Rajagopalan, P.K., P.K. Das, M. Kalyanasundaram, B.K. Tyagi and N. Arunachalam, N. Somachary, C.B.S. Reddy and C.M.R. Reddy. 1987. *Bangalore mosquito control project: Master plan.* Vector control Research Centre, pp. 309.

Rao, T.R., (1984). *The Anopheline of India* (Revised Edition). Malaria Research Centre, Indian Council of Medical Research, 518 pp.

Ross, R. (1897). On some peculiar pigmented cells found in two mosquitoes fed on malaria blood. *British Medical Journal* 2: 1786–1788.

Ross, R. (1923). *Memoirs: with a full account of the great malaria problem and its solution.* J. Murray, 547 pp.

Sharma, V.P. (1999). Current scenario of malaria in India. *Parassitologia*, 41(1–3): 349–353.

Sharma, V.P. and Mehrotra, K.N. (1986). Malaria resurgence in India: A critical study. *Social Science & Medicine*, 22, 835–845.

Sinton, J. A. (1935). *What malaria costs India nationally, socially and economically.* Thacker's Press & Directories, 77 pp.

Spielman, A. and M. D'Antonio, (2001). *Mosquito: the story of man's deadliest foe.* Faber & Faber, vi+247 pp.

Tyagi, B.K. (1994). Distribution of arthropod vector-borne communicable diseases and control of their vectors in India. *Indian Review of Life Sciences*, 14, 223–243.

Tyagi, B.K. (2002). *Malaria in the Thar desert: Facts, figures and future.* Agrobios (India). 165 pp.

Tyagi, B.K. (2003). *Handbook of Medical entomology.* Scientific Publishers (India), pp. 262.

Tyagi, B.K. (2004). *The invincible deadly mosquitoes: India's Health and Economy Enemy No. 1.* Scientific Publishers, India, pp 266.

Tyagi, B.K. (2023) *Male mosquito – A zaddy and subaltern sex in the disease epidemiology and control.* Scientific Publishers (India), 232 pp.

Tyagi, B.K. (2025). *Biodiversity of mosquitoes: The ugly, the bold and the beautiful.* M/S Bishen Singh Mahendra Pal Singh, 291pp.

Tyagi, B.K., Bhattacharya, S. and Reddyanaik, B. (2020). *Dr Ronald Ross: Mosquito, malaria, India and the Nobel Prize – An untold story of the First Indian Nobel Laureate.* Scientific Publishers (India), 258 pp.

Tyagi, B.K., and Tilak, R. (2022). Which is the world's deadliest animal: A lion, tiger, shark, snake or mosquito? *Journal of Medical Arthropodology & Public Health*, 2(1), 1–7.

Winegard, T.E. (2020). *The mosquito: A human history of our deadliest predator.* Dutton, 496 pp.

Yong, E. (2013). Blood-filled mosquito is a fossil first. *Nature.* https://doi.org/10.1038/nature.2013.13946

Section VII

Artificial Intelligence and Consilience in Culicidology

23 Overview of Artificial Intelligence (AI) in Mosquito Research, with Special Reference to the Future of Indian Fauna

Alphy Joseph, Krishnadas Ashok and P. Fijy Jose
St Joseph's College, Thrissur, India

E. M. Aneesh
University of Calicut, India

N. R. Mangalambal
St Joseph's College, Thrissur, India

Susanta Kumar Ghosh
Ex-ICMR's National Institute of Malaria Research, Bengaluru, India

Chaitali Ghosh
Tata Institute for Genetics and Society, Bengaluru, India

Abhijit Mazumdar
The University of Burdwan, India

E. Pushpalatha
University of Calicut, Kerala, India

B.K. Tyagi
Chandigarh University, Mohali, India

DOI: 10.1201/9781003327028-30

23.1 INTRODUCTION

As the human population grows and becomes more mobile, as international trade grows, and as the habitats of many vector species expand due to climate change, there is obviously an urgent need for global disease vector identification and surveillance. The foundation of any control approach is determining the present distribution and dynamics of the targeted mosquito species, with effectiveness relying on the early identification of invasive species and the efficient monitoring of the spread. Increased processing power and data volume have made machine learning feasible for environmental applications. For an effective control of vector mosquitoes, culicidologists need development of an efficient automated system for the detection, surveillance, and classification of mosquitoes. Researchers have shown the effectiveness of AI methods in mosquito identification, classification, and prediction (Fuchida et al., 2017; Lee et al., 2016). A branch of convolutional neural networks (CNN) and machine learning (ML) models experienced rapid growth and attention in the early 2010s. Since then, CNNs have consistently beaten human experts in a range of activities. The use of CNN and ML models in mosquito control strategies can help to minimize the development of mosquito-borne diseases (Lorenz et al., 2015; Joshi and Miller, 2021). India is currently making big strides in using deep knowledge in medical science, although, in contrast, certain European, American, and Asian countries have so far taken a lead in systemic AI research dealing with mosquito biology, including species identification, population surveillance and mapping, and control (Genoud, 2020; Kaur et al., 2022; Nayak et al., 2023). In this scenario, the implementation of new techniques in AI to help mosquito research in India will become an emerging field of research. Researchers at IIT Bombay, India, created an AI-enabled model for the diagnosis of malaria parasites by observing the changes in the human blood proteins that aims to eradicate tropical diseases, including malaria, AIDS, and tuberculosis by 2030. For the detection, sampling, and treatment of operations to control malaria vectors in urban areas, a digitally managed larviciding operation is designed by the Zzapp system (Vigodny et al., 2023). AI can facilitate large-scale cost-effective interventions in such operations. The recent popularity of deep learning techniques is mainly due to the availability of large enough image databases and the development in computing power. Section 23.1 deals with the procedures of CNNs. In Section 23.2, we focus on ML techniques for classification and identification of mosquitoes. Section 23.3 gathers the information on recognition of diseases based on deep learning techniques. This study proposes an effective methodology for the classification of mosquitoes using ML and CNN models. To choose the most relevant model based on performance indicators and computational requirements, a comparison analysis of multiple ML and deep learning models has been carried out.

23.2 NEURAL NETWORKS IN MOSQUITO CLASSIFICATION AND PREDICTION

According to wing shape characteristics, Lorenz et al. (2015) introduced an artificial neural network (ANN) technique for the identification and classification of 17 species of the genera *Anopheles, Aedes*, and *Culex*. They conducted research to determine whether classification using ANN is superior to classification using discriminant analysis (DA). A multilayer perceptron classification ANN received its input from 32 wing shape primary components. The generated ANN classified species more effectively than the conventional method of multivariate DA, properly identifying species with accuracy rates ranging from 85.70% to 100%. The outcomes demonstrated ANNs' capacity to identify mosquito species and partially automate taxonomic identification. These results also provided evidence in favour of the theory that wing venation patterns are species-specific and should be included in taxonomic keys (Lorenz et al., 2015).

A prominent empirical model for predicting mosquito abundance, the ANN, was put to the test in 2016 by Lee et al. (2016). They also created a multiple linear regression (MLR) model for comparison. The Digital Mosquito Monitoring System (DMS) was used to make a two-year observation in

Yeongdeungpo-gu, Seoul, and both the ANN and the MLR models were utilized to estimate mosquito abundances. They employed meteorological information such as temperature, wind speed, humidity, and precipitation as input variables. The outcomes demonstrated that the ANN and the MLR models' performances were nearly identical in terms of R and root mean square error (RMSE). In contrast to MLR, the ANN model was able to forecast the high variability. A sensitivity analysis of the ANN model revealed that it adequately explained the relationships between the input variables and mosquito abundances. In summary, ANNs may be able to anticipate changes in mosquito populations (especially the extreme values), and they may be able to do so better than conventional statistical methods (Lee et al., 2016).

Using a dataset of larvae, Sanchez-Ortiz et al. (2017) suggested a method for classifying mosquito larvae based on the VGG16 pre-trained convolution neural network which distinguished between the genus *Aedes* and 'others' genera. The infection rates would be reduced since the proposed method made the larva detection procedure more effective, automatic, and quick than the conventional systems. The outcomes demonstrated the successful identification of *Aedes* larvae, proving the system's viability for use in practical settings (Sanchez-Ortiz et al., 2017). In order to recognize adult mosquitoes of the species *Aedes aegypti, Aedes albopictus*, and *Culex quinquefasciatus* from mosquito photos (Motta et al., 2019) constructed a computer model based on a CNN. They used a dataset with 4,056 mosquito photos to train the CNN to automatically classify mosquitoes based on their morphology. LeNet, AlexNet, and GoogleNet were three of the neural networks used. The accuracy of the mosquito categorization during the validation phase was 57.5% using LeNet, 74.7% using AlexNet, and 83.9% using GoogleNet. During the testing phase, GoogleNet produced the best result (76.2%), whereas LeNet and AlexNet produced 52.4% and 51.2% of the results, respectively. Significantly, accuracy rates of 100% and 90% for the classification of *Aedes* and *Culex*, respectively, were obtained . For *Aedes* females, a classification accuracy of 82% was attained. The findings offered data that is essential for the field-based automatic morphological classification of adult mosquito species. Health professionals and taxonomists can identify some insects that can spread infectious diseases to people by using CNNs, which is a useful and accessible way for autonomous identification (Ocimati et al., 2019).

According to the Centers for Disease Control and Prevention (CDC), a complete solution was presented in the paper by Arista-Jalife et al. (2020) that can identify the Aedes aegypti and Aedes albopictus mosquito in the larval stage, which is easily discardable, restricted to water bodies, and incapable of transmitting diseases. Deep neural networks (DNN) were the foundation of their approach, and they successfully identified larval samples with an accuracy of 94.19%, outperforming other cutting-edge automatic techniques. The capabilities of their proposed DNN also allowed them to automatically crop the region of interest (ROI) with an accuracy of 92.85% and then automatically classify the region as *Aedes* positive or Aedes negative, both without further human intervention and in less than a second, reducing the response time for biological control from days to seconds. The idea incorporated low-cost hardware designs, making it appropriate for remote communities, developing nations, and rural or urban locations (Arista-Jalife et al., 2020). Park et al. (2020) looked at the capacity of cutting-edge deep learning models in classifying mosquito species with high inter-species similarity and intra-species variability. Since there was no pre-made dataset representing the variety of normal field-captured mosquitoes, the researchers created a collection with roughly 3,600 photos of 8 different insect species in varied postures and deformation situations. They looked into the viability of applying general features discovered from a generic dataset to the mosquito classification in order to further solve issues with data scarcity. The outcome showed that, when the right data augmentation approaches are used in combination, classification accuracy of more than 97% may be reached by fine-tuning generic features. They also looked at how deep learning models' usage of discriminative regions could help attain this high classification accuracy. According to their findings, deep learning models exploit morphological features similar to those used by human experts (Park et al., 2020). A CNN model was created by Akter et al. (2020) so that the mosquitoes could be identified from their photos. A local dataset comprising 442 photos, gathered from various sources, was created by them. The proposed CNN model was applied to the gathered dataset and showed a 70% accuracy. However, the accuracy rose to 93% after this dataset was expanded to include 3,600 photos. They also demonstrated some other algorithms, like VGG-16, Random Forest, XGboost, and support vector machine (SVM), compared to the CNN method. The categorization accuracy of the mosquitoes is improved by their suggested CNN approach over these methods. As a result, the study served as an illustration of how humanitarian technology might leverage data science to support mosquito categorization and enable treatment of various mosquito-borne diseases (Akter et al., 2020). In order to test the viability of automatically classifying mosquito sex, genus, species, and strains using whole-body, 2D photos of mosquitoes, Couret et al. (2020) applied CNNs to images of insects. Couret et al. (2020) described a method for processing images, enhancing data, and training and validating a CNN that produced high prediction accuracies of 96.96% for species identification and 98.48% for sex and addressed how these results might be applied in the future to monitor malaria mosquito populations.

The use of two deep CNNs in a side-by-side comparison study to automate the classification task was first introduced by Adhane et al. (2021). They achieved testing accuracy of 94% by using the transfer learning concept to train two state-of-the-art architectures on the data provided by the Mosquito Alert project. The most discriminant areas of the categorized images which correspond with the white band stripes found at the legs, abdomen, and thorax of mosquitoes of the Aedes albopictus species were also visualized using explainable models based on the Grad-CAM method. The model enabled a deeper analysis of the categorization mistakes. They were connected to poor acquisition conditions and significant picture occlusions, according to visual Grad-CAM models. A CNN-based model that can identify mosquito larvae in water was introduced by Saeed et al. (2021). The approach could be applied to portable or robotics applications on both personal computers and small portable onboard computers like Raspberry Pi. The model might therefore be applied to a system for the detection of mosquito larvae in order to remove their habitat. The model was created from scratch and achieved 93.95% training accuracy and 90.18% testing accuracy. Additional testing with 100 images revealed an accuracy of 86.0% and a precision of 92.2%.

ANNs were proposed by Kinney et al. (2021) as a workable replacement for a mechanistic model of mosquito abundance. Additionally, they created a gated recurrent unit network, a long short-term memory recurrent neural network, and a feed-forward neural network. They addressed the effects of supplementing the training data with time series that emphasize particular dynamical behaviours on model performance and assessed the networks' capacity to replicate the spatiotemporal characteristics of mosquito populations predicted by the mechanistic model. They offered a prediction for how such equation-free models might help with vector management or the calculation of disease risk at various spatial scales in their conclusion. A new set of uses for these models can be found when a complex mechanistic model can be replaced with an ANN that is faster and easier to scale up. However, because predictions made by ANNs frequently lose accuracy in 'unfamiliar' circumstances, the training data may need to pay particular attention to certain dynamic characteristics that the modeller believes are significant. A crucial step in the creation of ANN-based models is determining what kind and how much extra data will be required to boost performance.

23.3 MACHINE LEARNING IN CLASSIFICATION AND IDENTIFICATION OF MOSQUITOES

In order to effectively combat dangerous diseases like malaria, filariasis, encephalitis, dengue fever, and others, it is crucial to quickly and accurately identify and classify mosquito species. To create efficient management measures for these diseases, accurate mosquito species identification is a necessary first step. Unfortunately, taxonomic identification of mosquitoes requires experienced professionals and is a time-consuming operation. In this situation, the creation of an automated system for species identification would be a useful tool that non-taxonomists and medical experts could use more easily. In many instances, ML facilitates the classification of citizen reports of disease-carrying mosquitoes. Mosquito Alert (www.mosquitoalert.com/en) is a citizen science platform for monitoring and reducing disease-carrying mosquitoes that have been approved by experts. The use of ML on citizen reports to automatically validate a portion of them was described in (Rodríguez García et al., 2016). It allows entomologists to either handle larger report streams or choose reports from areas with high epidemiological risks or concentrate on more strategic items, such as reporting from new regions. Then, in fact, early warning protocols and the control actions to reduce mosquito populations will be activated.

For a layman, recognizing the species of mosquitoes is challenging. If the vision-based classification of mosquito species from pictures captured by a smartphone or mobile device is available, the results of the classification, however, can be used to inform and educate people about the infectious diseases. This seeks to suggest a vision-based classification scheme for mosquito species. Wang et al. (2012) presented a novel feature extraction approach from the obtained photos of insects. This method uses the SVM and ANN for the order-level categorization of insects from the photographs. This feature extraction approach takes into account the insects' eccentricity, body area ratio, width ratio, upper body length ratio, body shape parameter, and colour complexity characteristics that contribute to their uniqueness. In classification and image retrieval scenarios, both ANN- and SVM-based techniques are employed; nevertheless, it has been found that SVM-based classification may yield more accurate results (Vanajakshi and Rilett, 2004; Byvatov et al., 2003; Wong and Hsu, 2006). By separating the morphological traits from other insects like bees and flies, Fuchida et al. (2017) distinguished mosquitoes from those other insects. In the context of the mosquito classification challenge, it also displays the outcomes of three SVM classifier versions. Hence, the usual approaches of mosquito monitoring and mapping, and sample picture collection are effectively replaced by this vision-based solution to the challenge of mosquito categorization.

Extensive, long-term surveys are needed to assess the effectiveness of mitigation strategies, and they can be expensive and time-consuming. Alternative entomological monitoring techniques require effective classification algorithms to work, which give access to the recent development of optical sensors. Supervised ML algorithms are compared for the identification of mosquitoes through optical signals, which includes linear DA, decision trees, SVM, K-nearest neighbours, and Naïve Bayes (Genoud et al., 2020). These algorithms were trained to perform various classification tasks, including the identification of species, sex, and/or gravidity of mosquitoes present in New Jersey, USA. These algorithms were based on predictor variables derived from the wing beat frequency and optical cross-section of mosquitoes. Results indicate that SVM, which performs on average across all tasks between 0.65 and 7.3%, better than other algorithms, is the most versatile ML algorithm for mosquito identification. ML techniques appear suited to increase the identification accuracy of entomological lidars when dealing with the complexity of interpreting the fluctuations in optical cross-section of mosquitoes. Additionally, ML algorithms are crucial in the forecasting of a disease outbreak brought on by climate change. Prediction of mosquitoes using climate data based on ML is discussed in Hwang et al. (2020). The early diagnosis of such epidemics would give public health organizations time to address the large number of illnesses and possibly even reduce them.

Much worthwhile research that incorporates remote sensing data and ML techniques have been offered recently as a result of the development in the field of ML. In order to create an early warning predictor of the upcoming mosquito population, Tsantalidou et al. (2021) focused on integrating multi-source data to extract environmental-, landscape-, and ecosystem-related information in a uniform, consistent and reliable manner. They created a location and genus agnostic model out of a generic and adaptive framework. As a result, a general framework called MAMOTH (Mosquitoes Abundance Prediction Model autO-calibrated from features pleTHora) was created. MAMOTH does not require human input during the selection of features or during the tuning of the model's hyper-parameters. The designed model makes use of local geospatial data and satellite earth observation. Since MAMOTH is independent of either the place or the species of mosquito, it may be easily copied and applied to several cases without the need for any special parametrization. The model was used in many areas (Italy, Serbia, France, Germany) of interest and with various mosquito genera and species, including *Culex spp.* as potential West Nile virus vectors, Anopheles spp. for malaria and Aedes albopictus for Zika/chikungunya/dengue. The outcomes demonstrate the model's accuracy and consistency across all case studies.

Since mosquitoes are small and fly rapidly it is difficult to detect them manually. ML could be used to categorize the sound of mosquito wing beats in order to find them. An Arduino Nano BLE 33 Sense-based prototype that captures audio data from mosquito wing beats and uses TinyML to intelligently categorize mosquito species was discussed by Trivedi and Shroff (2021). The TinyML system created by Edge Impulse on the basis of the HumBug project mosquito wing beats dataset classifies mosquito types with an accuracy rate of 88.3%. Joshi and Miller (2021) reviewed ML approach to mosquito control in urban areas using geospatial, visual and audio models from various environmental contexts.

A variety of optoelectronic sensors have been created recently and are being utilized with ML methods to classify flying mosquitoes, with promising levels of accuracy (Genoud et al., 2020, 2018; Silva et al., 2015). In recent years, several optoelectronic sensors have been developed and used in conjunction with ML techniques to classify flying mosquitoes, with promising levels of accuracy. The use of auditory and optical sensors in conjunction with ML algorithms to enable an automatic classification of mosquitoes based on their flight characteristics, including wingbeat frequency, is one of the new methods for mosquito surveillance (González-Pérez et al., 2022). The creation and use of these techniques would make it possible to remotely monitor mosquito populations in the field, which would greatly enhance the effectiveness of vector surveillance. This system marks a step forward in mosquito surveillance technology and encourages the adoption of the sensor in distant, real-time mosquito population characterization in the future. *Aedes albopictus, Aedes aegypti*, and *Culex pipiens* were the three species selected for the study because they are important arbovirus vectors, have a considerable influence on public health and are the focus of vector surveillance and control activities in many regions of the world. In order to identify which feature and ML method performed the best at categorizing mosquitoes by genus and sex, a collection of characteristics were extracted from each recording and used to train a variety of algorithms. Therefore, the efficient application of ML to address the problem of mosquito identification has the potential to significantly affect the control of vector-borne diseases.

23.4 DISEASE RECOGNITION BASED ON DEEP LEARNING

The quality of the entire clinical process can be improved by the implementation of an effective clinical decision support system for arboviral diseases, thus increasing the precision of the diagnosis and the accompanying treatment. It should support doctors' decision-making and as a result enhance the use of resources and patients' quality of life. The parasite plasmodium causes malaria, which is transmitted through the bite of the female Anopheles mosquito. Through blood smears, medical professionals identify and assess for malaria infection and determine parasitaemia. However, their validity and accuracy in identifying and classifying malaria infection depends on the smear's quality. Particularly when a large number of diagnoses are made, this type of investigation could produce erroneous and inadequate validation of the results. A useful diagnostic tool will be the application of deep learning algorithms to malaria screening. Militante (2019) evaluated how well various CNN architectures performed in distinguishing and classifying cases of the disease that is unrelated to malaria and cases of the disease that is infected with malaria. Militante (2019) conducted an experimental determination of the optimum CNN model for feature extraction from cell image data.

In the past, pathologists used a microscope to diagnose this condition manually, which is a time-consuming operation in today's computerized world and depends on the pathologists' experience, which is a major issue in rural areas. Therefore, the necessity of detecting malaria using computerized image analysis that has been developed using a dynamic learning mechanism has increased. An image processing-based malaria detection system trained by deep learning is proposed by Kalkan and Sahingoz (2019). In the investigation of multiple deep learning models, Nayak et al. (2019) tracked which of these models offered greater accuracy and quicker resolution than previously employed deep learning models. Deep Learning may prove to be a very useful tool for diagnosing malaria. This provides a quicker and more affordable way to find plasmodium parasites. The primary function of the custom CNN is to discriminate between blood samples that are healthy and those that are infected. (Shah et al., 2020) proposed a model consisting of three convolutional layers. Convolutional layers with multiple filters present in layers yield exceptionally good accuracy. In order to diagnose malaria from microscopic cell images using image classification, a deep learning–based strategy utilizing a CNN is provided by Joshi et al. (2020). The suggested CNN model constructed using a fivefold cross-validation method beats all currently used techniques in terms of accuracy and other evaluation criteria, yielding the best outcomes to date for the diagnosis of malaria using deep learning. Masud et al. (2020) demonstrated how deep learning architecture, like CNN, may be used in real-time malaria detection from input photographs effectively and reliably while reducing human labour via a mobile application. This will make it easier to diagnose malaria by microscopy using a mobile application, thus addressing the problem of reliability of the treatment and a lack of medical expertise. Nabet et al. (2020) evaluated ANNs coupled with matrix-assisted laser desorption ionization-time of flight (MALDI-TOF) mass spectrometry (MS) and examined the impact on the proteome of laboratory-reared *Anopheles stephensi* mosquitoes to predict Anopheles drivers of malaria transmission, such as mosquito age, blood feeding, and *plasmodium* infection. With best prediction accuracies of 73%, 89%, and 78%, respectively, ANNs were sensitive to *Anopheles* proteome alterations and specifically detected spectral patterns linked to mosquito age (0–10 days, 11–20 days, and 21–28 days), blood feeding, and *P. berghei* infection. Nabet et al. (2020) showed how MALDI-TOF MS paired with ANNs can be used to forecast the entomological drivers of malaria transmission, potentially introducing new vector control methods. Similarly, in laboratory settings, near-infrared spectroscopy (NIRS) can distinguish species and age groups with relatively high success. Mid-infrared spectroscopy (MIRS, 4,000–400 cm^{-1}) is an alternative technology for detection of more subtle changes among species or mosquitoes of different ages (Siria et al., 2022). They used the deep learning MIRS (DL-MIRS) model based on geographically and ecologically distinct female mosquitoes in East and West Africa to reconstruct the age structure of semi-field mosquito populations and detect their changes in pre- and post-simulated vector control interventions.

Various image identification tasks have benefited greatly from computer-aided diagnostic (CAD) techniques based on deep learning algorithms, such as CNNs, which carry out an end-to-end feature extraction and classification. Kumar et al. (2022) evaluated the performance of malaria diagnosis using the CADx system. Due to its lower number of parameters and higher computing capacity, the CADx system is significantly more advantageous for diagnosis in isolated and rural locations without access to medical facilities. Deep learning–based disease diagnoses have had a significant impact on the medical community, particularly in recent years. CNN architecture, with 20 weighted layers, is built and proposed by Irmak (2021) as a novel deep learning–based malaria illness detection method to distinguish parasitized microscopic images from uninfected microscopic images. The numerous rising infections and deaths of malaria and acute lymphoblastic leukaemia (ALL) emphasizes the critical need for quick, accurate diagnosis techniques. The framework of artificial intelligence (AI) has recently been used to reduce time-consuming activities, improve the precision and adaptability of clinical diagnoses, and lessen the burden on doctors, diagnosticians and clinical experts. A detection system for malaria and ALL is proposed by Vo et al. (2022) and that utilizes blood smear microscopic images with the aid of deep learning algorithms to identify and classify these two diseases automatically. In short, the effectiveness of mosquito control in lowering outbreaks of diseases carried by mosquitoes is affected by different forms of surveillance data.

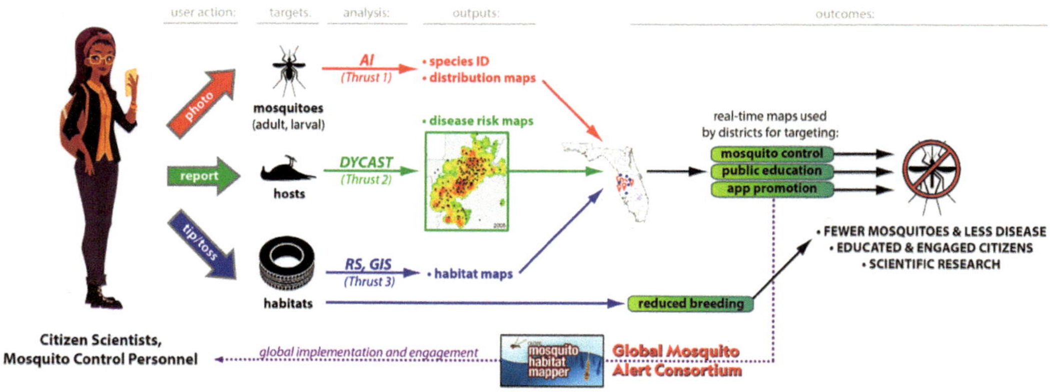

FIGURE 23.1 Citizen science mosquito surveillance (AI = artificial intelligence, DYCAST = dynamic continuous-area space-time system (DYCAST.org), RS = remote sensing, GIS = geographic information systems).

Source: Copyright Dr Ryan Carney (ryancarney.com; permission received).

23.5 CITIZEN SCIENCE IN MOSQUITO SURVEILLANCE

Public community engagement or, more precisely, the citizen science platforms is crucial for mosquito surveillance programmes (Ong and Ahmad, 2022). The citizen science platform in executing mosquito research is a new, novel and effective concept at a national level to display citizen science activities and useful information in their local language (Figure 23.1). Citizen science platforms, for example, Mosquito Alert, GLOBE Habitat Mapper, GLOBE Land Cover, and iNaturalist, designed for mosquito surveillance and control, serve as central tools in research for scientists to access data/observations. The latest AI algorithms has created an image recognition software which will be used to identify mosquito species in uploaded citizen science images (mosquitodashboard.org and mosquitoID.org). The programme will identify the pixels in the image that correspond to the wing, head, legs, and thorax. It will then precisely run the algorithm over each of those anatomical segments to determine the species identification. The following image depicts 'next-generation' mosquito surveillance that involves the novel incorporation of AI and citizen science.

23.6 ETHICAL ISSUE

Recently, the Indian Council of Medical Research (ICMR) has issued Ethical Guidelines for the Application of Artificial Intelligence in Biomedical Research and Healthcare (ICMR, 2023). Whether mosquito research in context with AI will need any ethical permissions will have to be seen in the future.

23.7 CONCLUSION

One of the most promising areas of AI is in the field of healthcare. AI has the potential to greatly benefit society by solving complex problems and addressing global challenges. Since 1900, global mean temperatures have increased by 1.1°C. This is largely driven by green-gas emissions from burning fossil fuels. This has impacted greatly on the vector-borne diseases like dengue, chikungunya, and the Zika viruses and malaria. They could become endemic diseases in many parts of the world, as climate is more likely to be appropriate for mosquito vectors than before (Thomson and Stanberry, 2022). One strategy we use is the application of the Internet of Things concept as soon as possible to boost the accuracy of vector species in time since they appear unexpectedly. The use of ML/deep learning techniques to manage disease vectors is covered in great detail in this chapter.

This study will incorporate current research initiatives in India to simply and precisely monitor the population of disease vectors along with future research prospects of integrated vector management. The long-term objective is to run prediction models, perform risk assessments, issue warnings and make historical analysis based on the proposed model, which can enable local vector surveillance and analysis, and support effective vector control programmes, particularly in remote areas in India with an insufficient number of expert entomologists.

AI uses complex computer algorithms to analyse large datasets. Its use in mosquito research needs reference specimens, and possibly developing through iso-female lines for producing homogenized populations (Ghosh et al., 2023). Another method of DNA barcoding must be aligned for perfect AI modelling (Singh et al., 2023). In this process the correct identification of different mosquito species can be performed at the best perfection for which we need expert mosquito taxonomists.

REFERENCES

Adhane, G., Dehshibi, M.M. and Masip, D. (2021). A deep convolutional neural network for classification of *Aedes albopictus* mosquitoes. *IEEE Access*, 9, 72681–72690.

Akter, M., Hossain, M.S., Ahmed, T.U. and Andersson, K. (2020). Mosquito classification using convolutional neural network with data augmentation. In *International Conference on Intelligent Computing & Optimization, Springer International Publishing*, 865–879.

Arista-Jalife, A., Nakano, M., Garcia-Nonoal, Z., Robles-Camarillo, D., Perez-Meana, H. and Arista-Viveros, H.A. (2020). *Aedes* mosquito detection in its larval stage using deep neural networks. *Knowledge-Based Systems*, 189, 104841.

Byvatov, E., Fechner, U., Sadowski, J. and Schneider, G. (2003). Comparison of support vector machine and artificial neural network systems for drug/nondrug classification. *Journal of chemical information and computer sciences*, 43(6), 1882–1889.

Couret, J., Moreira, D.C., Bernier, D., Loberti, A.M., Dotson, E.M. and Alvarez, M. (2020). Delimiting cryptic morphological variation among human malaria vector species using convolutional neural networks. *PLoS Neglected Tropical Diseases*, 14(12), e0008904.

Fuchida, M., Pathmakumar, T., Mohan, R.E., Tan, N. and Nakamura, A. (2017). Vision-based perception and classification of mosquitoes using support vector machine. *Applied Sciences*, *7*(1), 51.

Genoud, A.P., Basistyy, R., Williams, G.M. and Thomas, B.P. (2018). Optical remote sensing for monitoring flying mosquitoes, gender identification and discussion on species identification. *Applied Physics B*, *124*(3), 1–11.

Genoud, A.P., Gao, Y., Williams, G.M. and Thomas, B.P. (2020). A comparison of supervised machine learning algorithms for mosquito identification from backscattered optical signals. *Ecological Informatics*, *58*, 101090.

Ghosh, C., Kumar, N., Kushwah, R.B.S., Joshi, S.G., Ramanjini, C.K., Alalamath, T., Srinivasan, S., Subramani, S., Kumar, S. and Swain, S., (2023). Enrichment of phenotype among biological forms of *Anopheles stephensi* Liston through establishment of isofemale lines. *Parasites & Vectors*, 16(1):79.

González-Pérez, M.I., Faulhaber, B., Williams, M., Brosa, J., Aranda, C., Pujol, N., Verdún, M., Villalonga, P., Encarnação, J., Busquets, N. and Talavera, S. (2022). A novel optical sensor system for the automatic classification of mosquitoes by genus and sex with high levels of accuracy. *Parasites & Vectors*, *15*(1), 1–11.

Hwang, S.Y., Cha, Y.B., Cha, H.B. and Koh, J. (2020). Prediction of Mosquitoes using Climate Data based on Machine Learning. In *Proceedings of the Korea Information Processing Society Conference, Korea Information Processing Society*, 1031–1033.

Indian Council of Medical Research, 2023. Ethical guidelines for application of Artificial Intelligence in Biomedical Research and Healthcare 2023. https://main.icmr.nic.in/content/ethical-guidelines-application-artificial-intelligence-biomedical-research-and-healthcare

Irmak, E. (2021). A novel implementation of deep-learning approach on malaria parasite detection from thin blood cell images. *Electrica*, 21(2), 216–224.

Joshi, A.M., Das, A.K. and Dhal, S. (2020). Deep learning based approach for malaria detection in blood cell images. In *2020 IEEE Region 10 Conference (TENCON), IEEE*, 241–246.

Joshi, A. and Miller, C. (2021). Review of machine learning techniques for mosquito control in urban environments. *Ecological Informatics*, *61*, 101241.

Kalkan, S.C. and Sahingoz, O.K. (2019). Deep learning based classification of malaria from slide images. In *2019 scientific meeting on electrical-electronics & biomedical engineering and computer science (EBBT), IEEE*, 1–4.

Kaur, I., Sandhu, A.K. & Kumar, Y. (2022). Artificial intelligence techniques for predictive modeling of vector-borne diseases and its pathogens: A systematic review. *Archives of Computational Methods in Engineering* 29, 3741–3771. https://doi.org/10.1007/s11831-022-09724-9

Kinney, A.C., Current, S. and Lega, J. (2021). Aedes-AI: Neural network models of mosquito abundance. *PLoS Computational Biology*, 17(11), e1009467.

Kumar, A., Singh, S.B., Satapathy, S.C. and Rout, M. (2022). MOSQUITO-NET: A deep learning based CADx system for malaria diagnosis along with model interpretation using GradCam and class activation maps. *Expert Systems*, 39(7), e12695.

Lee, K.Y., Chung, N. and Hwang, S. (2016). Application of an artificial neural network (ANN) model for predicting mosquito abundances in urban areas. *Ecological informatics*, 36, 172–180.

Lorenz, C., Ferraudo, A.S. and Suesdek, L. (2015). Artificial Neural Network applied as a methodology of mosquito species identification. *Acta Tropica*, 152, 165–169.

Masud, M., Alhumyani, H., Alshamrani, S.S., Cheikhrouhou, O., Ibrahim, S., Muhammad, G., Hossain, M.S. and Shorfuzzaman, M (2020). Leveraging deep learning techniques for malaria parasite detection using mobile application. *Wireless Communications and Mobile Computing*, 1–15.

Militante, S.V. (2019). Malaria disease recognition through adaptive deep learning models of convolutional neural network. In *2019 IEEE 6th International Conference on Engineering Technologies and Applied Sciences (ICETAS), IEEE*, 1–6.

Motta, D., Santos, A.Á.B., Winkler, I., Machado, B.A.S., Pereira, D.A.D.I., Cavalcanti, A.M., Fonseca, E.O.L., Kirchner, F. and Badaró, R. (2019). Application of convolutional neural networks for classification of adult mosquitoes in the field. *PloS One*, 14(1), e0210829.

Nabet, C., Chaline, A., Franetich, J.F., Brossas, J.Y., Shahmirian, N., Silvie, O., Tannier, X. and Piarroux, R. (2020). Prediction of malaria transmission drivers in *Anopheles* mosquitoes using artificial intelligence coupled to MALDI-TOF mass spectrometry. *Scientific Reports*, 10(1), 1–13.

Nayak, S., Kumar, S. and Jangid, M. (2019). Malaria detection using multiple deep learning approaches. In *2019 2nd International Conference on Intelligent Communication and Computational Techniques (ICCT), IEEE*, 292–297.

Nayak, B., Khuntia, B., Murmu, L.K., Sahu, B., Pandit, R.S., Barik, T.K. (2023). Artificial intelligence (AI): a new window to revamp the vector-borne disease control. *Parasitology Research*, 122(2), 369–379. https://doi.org/10.1007/s00436-022-07752-9

Ocimati, W., Bouwmeester, H., Groot, J.C., Tittonell, P., Brown, D. and Blomme, G. (2019). The risk posed by Xanthomonas wilt disease of banana: Mapping of disease hotspots, fronts and vulnerable landscapes. *PloS one*, 14(4), e0213691.

Ong, S.Q. and Ahmad, H. (2022). An annotated image dataset for training mosquito species recognition system on human skin. *Scientific Data*, 9, 413. https://doi.org/10.1038/s41597-022-01541-w

Park, J., Kim, D.I., Choi, B., Kang, W. and Kwon, H.W. (2020). Classification and morphological analysis of vector mosquitoes using deep convolutional neural networks. *Scientific Reports*, 10(1), 1–12.

Rodríguez García, A., Bartumeus, F. and Gavaldà Mestre, R. (2016). Machine learning assists the classification of reports by citizens on disease-carrying mosquitoes. In: *SoGood 2016: Data Science for Social Good: Proceedings of the First Workshop on Data Science for Social Good co-located with European Conference on Machine Learning and Principles and Practice of Knowledge Discovery in Databases (ECML-PKDD 2016): Riva del Garda, Italy, September 19, 2016*, 1–11.

Saeed, M.S., Nazreen, S.F., Ullah, S.S.S.A., Rinku, Z.F. and Rahman, M.A. (2021). Detection of mosquito larvae using convolutional neural network. In *2021 2nd International Conference on Robotics, Electrical and Signal Processing Techniques (ICREST), IEEE*, 478–482.

Sanchez-Ortiz, A., Fierro-Radilla, A., Arista-Jalife, A., Cedillo-Hernandez, M., Nakano-Miyatake, M., Robles-Camarillo, D. and Cuatepotzo-Jiménez, V. (2017). Mosquito larva classification method based on convolutional neural networks. In *2017 International Conference on Electronics, Communications and Computers (CONIELECOMP), IEEE*, 1–6.

Shah, D., Kawale, K., Shah, M., Randive, S. and Mapari, R. (2020). Malaria parasite detection using deep learning: (Beneficial to humankind). In *2020 4th International Conference on Intelligent Computing and Control Systems (ICICCS), IEEE*, 984–988.

Silva, D.F., Souza, V., Ellis, D.P., Keogh, E.J. and Batista, G.E. (2015). Exploring low cost laser sensors to identify flying insect species. *Journal of Intelligent & Robotic Systems*, 80(1), 313–330.

Singh, U.S., Amdep, F.L., Kshiar, A., Acharya, P., Karumuthil, T., Kale, S., Mishra, S., Khan, N., Kharbisnop, B., Kessler, A. and Carlton, J.M. (2023). Characterisation of *Anopheles* species composition and genetic diversity in Meghalaya, northeast India, using molecular identification tools. *Infection, Genetics and Evolution*, 112:105450. https://doi.org/10.1016/j.meegid.2023.105450

Siria, D.J., Sanou, R., Mitton, J., Mwanga, E.P., Niang, A., Sare, I., Johnson, P.C., Foster, G.M., Belem, A.M., Wynne, K. and Murray-Smith, R.(2022). Rapid age-grading and species identification of natural mosquitoes for malaria surveillance. *Nature Communications*, 13(1), 1501.

Trivedi, K. and Shroff, H. (2021). Identification of deadliest mosquitoes using wing beats sound classification on tiny embedded system using machine learning and Edge Impulse platform. In *2021 ITU Kaleidoscope: Connecting Physical and Virtual Worlds (ITU K), IEEE*, 1–6.

Thomson, M.C. and Stanberry, L.R. (2022). Climate Change and Vectorborne Diseases. *New England Journal of Medicine*, 387(21), 1969–1978.

Tsantalidou, A., Parselia, E., Arvanitakis, G., Kyratzi, K., Gewehr, S., Vakali, A. and Kontoes, C. (2021). MAMOTH: An Earth Observational Data-Driven Model for Mosquitoes Abundance Prediction. *Remote Sensing*, 13(13), 2557.

Vanajakshi, L. and Rilett, L.R. (2004). A comparison of the performance of artificial neural networks and support vector machines for the prediction of traffic speed. In *IEEE Intelligent Vehicles Symposium, IEEE*, 194–199.

Vigodny, A., Ben Aharon, M., Wharton-Smith, W., Fialkoff, Y., Houri-Yafn, A., Braganca, F., Graca, F., Gluck, D., D'Abreu, J. and Ropao, H. (2023). Digitally managed larviciding as a cost-effective intervention for urban malaria: operational lessons from a pilot in São Tomé and Príncipe guided by the Zzapp system. *Malaria Journal*, 22, 114.

Vo, Q.H., Le, X.H. and Le, T.H. (2022). A deep learning approach in detection of malaria and acute lymphoblastic leukemia diseases utilising blood smear microscopic images. *Vietnam Journal of Science, Technology and Engineering*, 64(1), 63–71.

Wang, J., Lin, C., Ji, L. and Liang, A. (2012). A new automatic identification system of insect images at the order level. *Knowledge-Based Systems*, 33, 102–110.

Wong, W.T. and Hsu, S.H. (2006). Application of SVM and ANN for image retrieval. *European Journal of Operational Research*, 173(3), 938–950.

24 Mosquitoes for Novel Functional Systems

B.K. Tyagi
University Institute of Biotechnology, Chandigarh University, Mohali, India

24.1 INTRODUCTION

Mosquito science, or culicidology, is an incredibly exciting discipline which, on one hand, not only constitutes an integral part of public and veterinary health but, on the other hand, also serves as an indispensable resource of novel materials that can pave the development of an array of functional systems to solve various different human infirmities (Clements, 1992, 1999, 2011; Schroeder et al., 2018). Unfortunately, however, since eternity, mosquitoes have been regarded solely as the deadliest *bête noire* of man on our planet (Harrison, 1978; Spielman and D'Antonio, 2001; Tyagi, 2003, 2004, 2020; Winegard, 2019). In a sense, this fact is very well documented too. In 2022, about 250 million cases of human malaria occurred worldwide (World Health Organization, 2023). While it is still considered to be underestimated, the annual dengue burden accounts for 400 million cases and 22,000 deaths worldwide each year (Roy and Bhattacharjee, 2021). Unfortunately, mosquitoes have for ages been exhibited in a villainous form only. This kind of biased and lopsided branding of mosquitoes has inherently overshadowed their phylogenic, biological, ecological, behavioural, physiological, and genetical potentialities to help design, through innovative routes, the integrative biomaterials, medical devices and/or simulating knowledge about system functionalities in enhancing quality of human life. This newly evolving applied aspect of mosquito life portrays unlimited scope of benefits to humans, by way of opening new vistas to combat some of the most nagging health disorders and, of course, not to recall in the least, utility of mosquitoes versus mosquitoes as either biocontrol agents or, using the genetically modified male gender, as the GE mosquito to suppress or replace the vector species' population. There is a very strong argument for viewing mosquitoes "*beyond the parapet*."

24.2 WHY MOSQUITOES?

The birth of mosquitoes is a decisive event in the history of life (Aria, 2022). As insects, mosquitoes are likely the first hematophagous animals to have established themselves on land as well as in water and the air, pervading nearly all ecosystems. For their interesting nature of blood feeding and involvement in the transmission of deadly and/or debilitating infections, mosquitoes have earned sobriquets like "*man's deadliest foe*," "*the deadliest predator of man*," and "*the angel of death*." Even so, much to the lack of public acknowledgement, they have surreptitiously become important pillars of the planet's ecological and food networks. Thus, their wildly successful proliferation across all continents of the earth with an estimated population of 200–300 trillion mosquitoes under more than 3,600 species, mosquitoes have largely dominated and influenced man's own history of evolution. In the process of this long co-evolutionary journey, mosquitoes have garnered an enviable array of diverse adaptations, processes and principles to associate with different habitats, which include adhesives (in eggs), locomotor systems (larvae and pupae), hydrophobic surfaces (*Aedes* species), and sensors (adults) and actuators (iridescent mosquitoes) that transduce mechanical,

DOI: 10.1201/9781003327028-31

acoustic, optical, thermal, and chemical signals. These unique understandings using breakthrough technologies are likely to inspire development of mosquito-inspired solutions to human maladies which often enhance the variety of applicable mechanisms observed in nature and open new pathways for improved technological solutions that are needed in a changing global society.

24.3 DEVELOPMENT OF INTEGRATED MATERIALS FOR FUNCTIONAL LEADS AND SOLUTIONS TO HUMAN MALADIES

In contrast to popular dogma that mosquitoes only cast ill health, the life of these creatures may, in fact, provide an infinite and largely untapped pool of unique problem-solving strategies for bettering human lives. They may offer us unfathomable sources of information through their structure and behaviour using intelligent autonomous artificial intelligence–assisted polypod robots can be built that can exploit their unique biomechanics, morphology, and computational systems to achieve autonomy, adaptability, and energy efficiency. Mosquitoes' structural and behavioural adaptations may create environmental sensing and control, protection, communication, and locomotion. Mosquitoes may become easy attention from a bioinspirational perspective, given their archaic history for survival, astounding abundance, and phenotypic diversity. Therefore, it is considered commensurate to comprehend, along with man's viewpoint, the mosquito's basis of evolution in sustaining a common planetary niche – the global village ecosystem!

24.3.1 Brain Research Gets a Boost from Mosquitoes

The human brain is responsible for our remarkable capabilities from seeing to singing, from remembering to running that originate in the electrical activities of neurons. Neurons interact with each other forming circuits, which process sensory information and drive appropriate behaviours. An understanding of how these circuits encode innate behavioural preferences is a matter of deep research. Using the mosquito olfactory system (sense of smell) as a model, it is likely to improve understanding in detail about human brain functioning. Singh et al. (2023), using a variety of techniques including *in vivo* preparation for patch-clamp electrophysiology, histology, behavioural observations, genetic manipulations, and computational modelling, have studied the neural responses in the brain of a mosquito, *Aedes aegypti*, which is capable to detect the attractive odour of a human or the repulsive odour of a repellent cream. It is being considered a likelihood to decode neural circuits control similarities and differences across individuals of the same mosquito species. Their results have provided for the first time a detailed description of the second-order olfactory neurons in the central nervous system of mosquitoes and lay a foundation for understanding the neural basis of their olfactory behaviours (Figure 24.1).

Can a protein found in a mosquito lead to a better understanding of the workings of our own brains? Scientists have discovered new leads to understand light-sensitive proteins that could illuminate the dark corners of our brain's communication pathways (Mahn et al. 2021). Researchers have made a giant stride in the medical sciences by identifying a light-sensitive protein derived from mosquitoes and used it to devise an improved method for investigating the messages that are passed from neuron to neuron in the brains of mice. This method can potentially help scientists solve age-old cerebral mysteries that could pave the way for new and improved therapies to treat neurological and psychiatric conditions. This has been facilitated by developing the so-called optogenetic methods – research techniques that allow them to 'reverse engineer' the activity of specific brain circuits in order to better understand their function. Optogenetics uses proteins known as rhodopsins to control the activity of neurons in the mouse brain. Rhodopsins are light-sensing proteins – they are most known for their role in organs like the retina rather than in the dark inner reaches of the body. But in the new study the rhodopsins in the mice enabled researchers to control the activity of specific neurons when they shone a minuscule beam of light into the mouse's brain. They hope to detect the

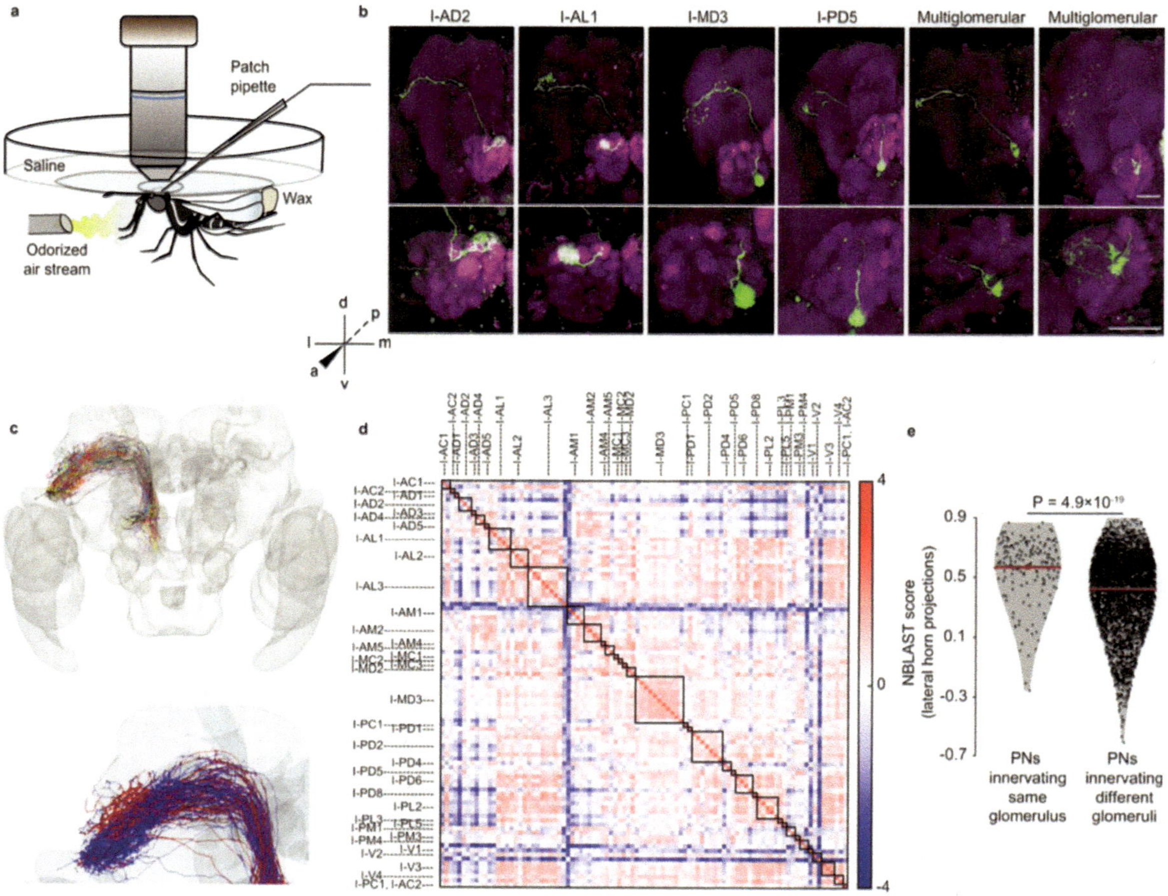

FIGURE 24.1 Morphological characterization of projection neurons (PNs) in *A. aegypti:* (a) Schematic showing the experimental preparation for in vivo patch-clamp recording. The animal was fixed with glue and wax to a recording chamber placed under the objective lens of an upright, fixed-stage microscope. Odours were delivered to the olfactory organs below the recording chamber and removed using a vacuum tube (not shown) placed behind the animal. Recordings were obtained from cell bodies present in the dorsal and the lateral regions around the antennal lobe (AL) using a glass pipette through a hole in the head cuticle. (b) Examples of some morphological sub-classes of PNs, including 4 uniglomerular (out of 201 obtained) and 2 multiglomerular (out of 7 obtained) PNs. For each PN, the top image shows the complete morphology in a maximum-intensity projection of the image stack (the region displayed includes ipsilateral AL, mushroom body, and protocerebrum); the bottom image shows the dendritic innervations in a maximum-intensity projection from the AL. Co-ordinate axes: dorsal-ventral (d–v), anterior-posterior (a–p), and medial-lateral (m–l). Green: biocytin or lucifer yellow used to fill the cell; magenta: Dncad (neuropil marker). Scale bars, 50 µm. (c) Top image shows the axonal reconstructions of 94 PNs registered to an adult female Aedes aegypti brain template. Each PN sub-class is shown in a different colour. The bottom image zooms into the LH (lateral horn) projections of PNs innervating dorsomedial glomeruli (red) and other glomeruli (blue). (d) The matrix shows the NBLAST (Nucleotide Basic Local Alignment Search Tool) scores (after z-score normalization) indicating the morphological similarity in the protocerebral projections of pairs of PNs. PNs are ordered by glomerular identity, marked by the boxes along the diagonal and the labels along the axes. e NBLAST scores, indicating morphological similarity of projections to the LH, were higher for pairs of PNs belonging to the same glomerulus (n = 171) than for pairs of PNs belonging to different glomeruli (n = 4,200). P value from two-sided signed-rank test is displayed. Red lines: means, error bars: s.e.m. Source data for (d, e) are provided as a source data fil.)

Source: Singh et al. 2023.

presence of the various neurotransmitters since optogenetics is enabling to not only see the 'ink,' but really to decipher the 'message.' The outputs of optogenetics in mice rhodopsins can be finicky.

To evaluate the efficacy of the new mosquito-derived tool, the researchers tested their method against a drug that is known to reduce the strength of the communication between neurons in the brain. They found that the interference was just as effective, and much more stable with the mosquito rhodopsin. More than that, unlike a conventional drug that affects numerous parts of the brain and is hard to control, the researchers found that since only neurons that produce the mosquito sensor are affected by the light, the modulatory effect on the brain's synapses can be precisely controlled in both space and time – just by switching the light on or off in specific brain regions. They then validated the utility of the new tool by using it to block the release of the neurotransmitter dopamine on one side of the brain only: illuminating the hemisphere expressing the mosquito rhodopsin with green light led to a one-sided bias in the behaviour of these mice. In other words, they had created a tool that was precise, selective, and controllable. Researchers now emphasize that one of the major advantages of the mosquito rhodopsin is that it is bistable, i.e., it does not need refreshing, and it is potentially very specific, so that they can control just the precise synapses in which they are interested. Not only this, but the scientists are of the opinion that mosquito protein could open the way to developing a whole family of new optogenetic tools for use in neuroscience research.

24.3.2 Mosquito Biting is Sprouting Medically Important New Products

Mosquito-based innovations can be unique in functionalities because their characteristics and behaviour are the cornerstones that drive the urge for the development of select biomaterials and medical devices, such as the following.

24.3.2.1 Microneedles

Mosquito-inspired microneedles have been fabricated from a variety of materials, including biocompatible metals and polymers, to mimic the mouthparts that some mosquitoes use to bite a host with minimal injury during blood collection.

24.3.2.2 Anaesthetics

The salivary components that these mosquitoes use to reduce the clotting of blood extracted during the biting process provide a rich source of anticoagulants that could potentially be integrated into blood-contacting biomaterials or administered in therapeutics to reduce the risk of thrombosis.

24.3.2.3 Behaviour-Based Technologies

Mosquito movement, vision, and olfaction are other behaviour that also have the potential for inspiring the development of medically relevant technologies, such as (a) viscoelastic proteins: these facilitate mosquito movement and are being investigated for use in tissue engineering and drug delivery applications, (b) non-wetting eye-lens: nanostructure of a mosquito eye has inspired the creation of a robust superhydrophobic surface coating that shows promise for biomaterial and drug delivery applications, and (c) biosensors: biosensors incorporating mosquito olfactory receptors have been built to detect disease-specific volatile organic compounds.

24.3.3 Understanding Mosquito's Infrared Detection System Could Lead to More Effective Traps

The mosquito integrates multiple cues to locate a potential blood meal host such as, for example, multiple cues, such as CO_2 from breath, skin odour, and visual cues, at different ranges. A new factor, infrared (IR), has now been added to the mosquito's repertoire of detection senses which is located, at the tip of the antennae (Chandel et al., 2024; Figure 24.2). Likely a temperature-sensitive protein, TRPA1 is there to facilitate this phenomenon.

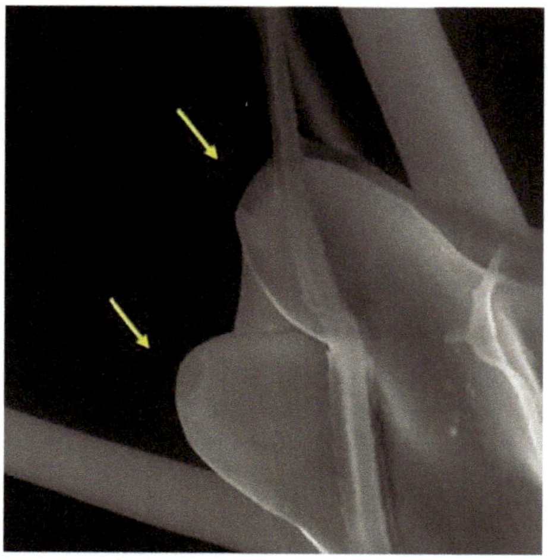

FIGURE 24.2 Pits at the end of the mosquito's antennae shield the peg-like structures that detect thermal IR.

Source: DeBeaubien and Chandel et al.; permissible with credit given. Original Source: https://news.ucsb.edu/2024/021587/mosquitoes-sense-infrared-body-heat-help-track-humans-do.

The tip of each antenna has peg-in-pit structures that are well adapted to sensing radiation. The pit shields the peg from conductive and convective heat, enabling the highly directional IR radiation to enter and warm the structure. This new information can be explored to develop new potential strategies for better controlling the transmission of mosquito-borne diseases. *Aedes aegypti*, a deadly vector for dengue, yellow fever, chikungunya fever etc., can detect heat rising from our skin within about 10 cm. It can then directly sense the temperature of human skin once it lands. These two senses correspond to convection, heat carried away by a medium such as air, and conduction, which is heat via direct touch. These are two kinds of heat transfer. Energy from heat can also travel longer distances when converted into electromagnetic waves, generally in the IR range of the spectrum. The IR can then heat whatever it hits. Mosquitoes, specifically *Ae. aegypti*, are probably able to detect thermal IR as a more reliable directional cue.

Aedes aegypti, it appears, is able detect thermal IR as an important mid-range cue to home in on warm-blooded hosts, and therefore, can be extensively exploited for further research.

24.3.4 Controlling Mosquito Populations Through Genetic Breeding May Help Reduce Global Disease Spread

Hybridization between two species of mosquitoes, one of which is a deadly vector for various diseases, is a newfound way to identify genetic targets useful for controlling mosquito populations, potentially offering an alternative to insecticides. Research recently made out of hybridization between *Aedes aegypti*, a major global arboviral disease vector, and *Aedes mascarensis*, a sibling species of the former in the Indian Ocean, leads to disruption of male sex determination. This can be a strong reason for disrupting population buildup by the vector species due to species incompatibility. When *Ae. aegypti* crosses with *Ae. mascarensis*, with a subsequent back crossing of its F1 offspring with one parent (*Ae. aegypti*), it was analysed that about 10% of the progeny became intersex and were unable to reproduce. Understanding the sex determination pathway and its disruptions in mosquitoes is critical for the effective control of disease vectors through genetic manipulations based on sex separation. Further, analysis of the sex-specific splicing of double sex and fruitless transcripts demonstrated the presence of both male and female splice variants, indicating that sex determination is disrupted. A comparative transcriptomic analysis revealed similar expression levels of most female-associated genes in reproductive organs and carcasses between intersexual males and normal females. It is safely concluded now that evolving hybrid incompatibilities between

Ae. aegypti and *Ae. mascarensis* involve disruption of sex determination and are accompanied by changes in gene expression associated with sexual differentiation. Intersexes can serve as a valuable model for the discovery of genetic factors involved in sex determination, sexual differentiation, mating, host-seeking, and blood-biting behaviours in mosquitoes. Finding new sex determination pathway genes and their disruptions in mosquitoes will contribute to the effective control of disease vectors through genetic manipulations based on sex separation.

24.3.5 Genetically Modifying Wild Populations with Less Risk

Genetic engineering could address major global challenges by altering the genetic make-up of certain wild populations, e.g., to combat mosquito-borne illnesses such as malaria, or stop the spread of environmentally harmful invasive pests like cane toads. But there is genuine public concern about introducing genetic modification into wild populations and many regulatory constraints. Traditional methods of genetic modification can also see a rapid spread of engineered genes within a population, raising both ecological and ethical questions, according to the research team. In response, many regulatory frameworks have been introduced to address genetic modification, presenting further challenges. Many in society believe that modified organisms could contain foreign DNA that could cause unpredictable ecological consequences over time. The answer to this frequently asked question is the Allele Sail, which is different from other genetic modification techniques like gene drives, which quickly spread engineered genes through a population. In contrast, the Allele Sail introduces genetic changes using an editor that remains at low levels and could potentially be removed from the population completely. The Allele Sail offers a way to change the traits and fates of wild populations in ways that may be more acceptable, as the genetically modified part is introduced at low frequencies and usually won't last forever. The Allele Sail system uses a gene editor to make specific DNA changes to an organism's genome. But unlike other genetic modification methods, the editor is inherited normally and does not increase in the population, except that the edits it creates can spread rapidly, potentially making desirable traits common.

On testing this hypothesis through computer models, it is understood that the new technique could address some of the regulatory challenges and public concerns associated with existing genetic modification methods.

24.3.6 Skin Microbiome with 'Living Repellent' Reduces Mosquito Attraction

A microbiome-based mosquito repellent is being developed by engineering common human skin commensals to produce much less of a form of lactic acid known to attract mosquitoes (Liu et al., 2024). The gold-standard synthetic repellent, N,N-diethyl-m-toluamide (DEET), protects against mosquitoes for just four to eight hours. More lasting protection would be invaluable. Not only would it save us from untold numbers of annoying mosquito bites, but it would also reduce the spread of mosquito-borne diseases such as dengue fever, Zika, and malaria. This new means of keeping mosquitoes at bay is the genetically engineered version of the common human skin commensals *Staphylococcus epidermidis* and *Corynebacterium amycolatum*. These 'living repellents' produce much less l-(+)-lactic acid, a form of lactic acid that is known to attract mosquitoes. In experiments with mice, protection lasted 7–11 days, longer than the 4 to 8 hours common with DEET. In other words: less spray, more stay.

The aforementioned research has suggested the feasibility of creating a living and long-lasting engineered microbiome-based mosquito repellent.

24.4 CONCLUSION

In conclusion, it becomes absolutely necessary to understand the positive dimension of the life of mosquitoes in a broader perspective of fathomless utility to humans rather than just the intermediate hosts to some parasites. In an evolutionary sense, mosquitoes naturally inherit molecules that prevent blood from clotting and enzymes that degrade pesticides – traits that make them invaluable. Milking the potential of the unseen part of mosquitoes' lives allies with combatting diseases instead of abhorring them for their nefarious acts of transmitting deadly diseases and marks the dawn of a new era in modern medicine.

REFERENCES

Aria, C. (2022) The origin and early evolution of arthropods. *Biological Reviews of the Cambridge Philosophical Society* 97(5):1786–1809. https://doi.org/10.1111/brv.12864

Chandel, A., DeBeaubien, N.A., Ganguly, A. et al. (2024). Thermal infrared directs host-seeking behaviour in *Aedes aegypti* mosquitoes. *Nature* https://doi.org/10.1038/s41586-024-07848-5

Clements, A.N., 1992. *The biology of mosquitoes. Vol. 1. Development, nutrition and reproduction.* CABI Publishing, 571 pp.

Clements, A.N., 1999. *The biology of mosquitoes. Vol. 2. Sensory reception and behaviour.* CABI Publishing, 752 pp.

Clements, A.N., 2011. *The biology of mosquitoes. Vol. 3. Transmission of viruses and interactions with bacteria.* CABI Publishing, 592 pp.

Harrison, G., 1978. *Mosquitoes, malaria and man: A history of the hostilities since 1880.* Tohn Murray, 314 pp.

Liu, F., Coutinho-Abreu, I.V., Raban, R., Nguyen, T.T.D., Dimas, A.R., Merriman, J.A. and Akbari, O.S. (2024). Engineered skin microbiome reduces mosquito attraction to mice, *PNAS Nexus* 3(7):267. https://doi.org/10.1093/pnasnexus/pgae267

Mahn, M., Saraf-Sinik, I., Patil, P., Pulin, M., Bitton, E., Karalis, N., Bruentgens, F., Palgi, S., Gat, A., Dine, J., Wietek, J., Davidi, I., Levy, R., Litvin, A., Zhou, F., Sauter, K., Soba, P., Schmitz, D., Lüthi, A., Rost, B.S., Wiegert, J.S. and Yizhar, O. (2021).Efficient optogenetic silencing of neurotransmitter release with a mosquito rhodopsin. *Neuron* 109(10):1621–1635. https://doi.org/10.1016/j.neuron.2021.03.013.

Roy, S.K. and Bhattacharjee, S. (2021). Dengue virus: Epidemiology, biology, and disease aetiology. *Canadian Journal of Microbiology* 67:687–702.

Schroeder, T.B.H., Houghtaling, J., Wilts, B. D. and Maye, M. (2018). It's not a bug, it's a feature: Functional materials in insects. *Advanced Materials* 30:1–48.

Singh, P., Goyal, S., Gupta, S., Garg, S., Tiwari, A., Rajput, V., Bates, A.S., Gupta, A.K. and Gupta, N. (2023). Combinatorial encoding of odors in the mosquito antennal lobe. *Nature Communications* 14:3539. https://doi.org/10.1038/s41467-023-39303-w

Spielman, A. and D'Antonio, M. 2001. *Mosquito: The story of man's deadliest foe.* Faber & Faber, vi+247 pp.

Tyagi, B.K., 2003. *Handbook of medical entomology.* Scientific Publishers, India, pp. 262.

Tyagi, B.K. 2004. *The invincible deadly mosquitoes: India's health and economy enemy no. 1.* Scientific Publishers, India, pp 266.

Tyagi, B.K., 2020. *Mosquito hunters: A history of hostilities against man's deadliest Foe – The mosquito – Since 1881.* Scientific Publishers, India, 474 pp.

Winegard, T.C. (2019). *The mosquito: A human history of our deadliest predator.* Allen Lane, pp. 496.

World Health Organization (2023) *World malaria report 2023.* World Health Organization.

Appendix I
Global Resources on Mosquito Research and Major Organizations and Programs Engaged in Providing Information, Training, and Capacity Building

Resource Documents

- Mosquitoes – information on preventing bites and mosquito-borne viruses
- Vector-Borne Disease (DVBD) – resources and guidance from the Division of Vector-Borne Disease
- Climate Change Increases the Risk of Vector-Borne Diseases pdf icon [PDF – 111 KB]
- Mosquito and Mosquito Habitat Science (https://observer.globe.gov/do-globe-observer/mosquito-habitats/science)
- CDC's DPDx – Laboratory Identification of Parasites of Public Health Concern (https://www.cdc.gov/dpdx/index.html)
- CDC's Global Malaria Activities (https://www.cdc.gov/malaria/php/public-health-strategy/global-activities.html)

Global Programmes

- The World Mosquito Program (https://www.worldmosquitoprogram.org/)
- GLOBE Mosquito Project (https://www.globe.gov/web/globe-mosquito-project/overview/resources-for-students-and-teachers)
- Mosquito Taxonomic Inventory (https://mosquito-taxonomic-inventory.myspecies.info/)

Training and Capacity Building

- Biosafety for human health and the environment in the context of the potential use of genetically modified mosquitoes. Training manual. A tool for biosafety training (CD and internet version) based on courses in Africa, Asia and Latin America, 2008–2011 I. World Health Organization. (https://iris.who.int/bitstream/handle/10665/180388/9789241549271_eng.pdf?sequence=1)
- Guidance framework for testing of genetically modified mosquitoes, second edition (https://iris.who.int/bitstream/handle/10665/341370/9789240025233-eng.pdf?sequence=1)

Appendix II
Glossary

(Terms, Abbreviations, and Explanations)

%Pf	Slides with *P. falciparum* as % of all positive cases
ABER	Annual blood examination rate. Slides examined as % of population over a one-year period
Abiotic factors	The climatic characteristics (temperature, humidity, saturation deficit, rainfall, light, wind) and edaphic characteristics (type of soil, soil texture) of a region or habitat
ACD/PCD	Active/passive case detection
Active ingredient	The component of an insecticide formulation that kills or controls pests
Aedeagus	A finger-like evagination of the ventral body wall of male mosquito enclosing the terminal section of the ejaculatory duct
Aestivation	The survival of mosquitoes during hot summer periods
Al Nina	A phenomenon opposite of El Nino
Allergy	A state of hypersensitivity to a given material (called an allergen), acquired through exposure; re-exposure to the allergen shows altered reactivity
Anal papillae	The soft fleshy lobes found on the anal segment of mosquito larvae. These lobes are associated with the physiological regulation of water balance in larvae. Species adapted to saline habitats have very small papillae to avoid the excessive loss of water; freshwater species have large papillae to avoid excess uptake of water
Antenna	One of the paired sense organs on the heads of insects
Antennae	Paired appendage of head of insects and other arthropods., usually with many segments. Antennae carry chemosensory organs
Anterior	The front end
Anterodorsal	The front of the upper surface
Anteroventral	The front of the under surface
Anthropophilic	Man-biting (literally: liking man)
API	Annual parasite incidence. The number of parasite-positive blood slides per thousand population
APICAL	At the tip or end of (APEX)
Arbovirus	Any virus that is associated with an arthropod
Area protective measures	Measures applied, usually by a mosquito control agency to protect the population of a specific area from the impact of vectors and the parasites they transmit.
Arthropod	An animal belonging to the phylum Arthropoda in the animal kingdom, having a hard jointed exoskeleton and paired, jointed legs. The phylum includes, among other classes, the Arachnida and the Insecta, many of which are important medically as parasites or as vectors of organisms capable of causing disease in humans

Attack phase	(of a control programme) The phase in which all houses are treated, regardless of whether or not the house was infested
Autogenous	Mosquitoes that demonstrate the trait of autogeny
Autogeny	The production of viable eggs by mosquitoes and other blood-sucking insects
Bacteria	A group of single-celled microorganisms. Some are pathogens (cause disease) bacteria, and certain minerals. The Environmental Protection Act recognizes three categories of biorational insecticides
Basal	Refers to the base or the position opposite the apex of an insect structure; the base of the proboscis
Biological control	Control measures by means of living organisms
Bionomics	The (quantitative) relation of the development, reproduction and survival of an organism or population of organisms to factors in the environment
Biopesticides	Another term applied to biorational pesticides and insecticides
Biorational insecticides	Insecticides derived from natural materials such as animals or plants
Biotic factors	Factors connected directly with the organisms living in a given environment, e.g., nature and availability of hosts, water quality, and plant environment
Biotype	A small topographic unit, including the biotic community
Breeding place	A place where mosquito larvae develop. No breeding takes place there in BSC/BSE – blood slides collected/examined
Bti	*Bacillus thuringiensis* subsp. *israelensis*, a bacterium well known for its toxicity against vector/pest mosquito larvae
California encephalitis	A disease of humans first reported in California caused by the California encephalitis virus; it cannot be separated into other substances by a process that does not involve a chemical and cannot be displayed in a readily recognizable form
CDC	Centers for Disease Control and Prevention
CE	California encephalitis virus, the cause of the disease California. CE was the first virus to be classified in the serogroup
Chemical control	The control of pests, including mosquitoes, by the use of chemicals
CHIKV	Chikungunya virus
Cholinesterase	An enzyme present at nerve junctions in animals and necessary for proper transmission of stimuli
Chromosome	A linear structure within the nucleus of a cell composed of proteins and DNA, which holds the genetic material or genes of an organism, the arrangement of which is unique to each organism
CI	Cytoplasmic incompatibility
Classification	The process of arranging a series of some kind into groups according to specific characters
Complete metamorphosis	The process of insect development that includes the egg; complete metamorphosis includes eggs, larvae, pupae, and adults
Control	(of insects [or other undesirable animals]) The restriction of the population density of such insects to a level below that at which they can be harmful to the interests of humans
CRISPR	Clustered regularly interspaced short palindromic repeats
Cuticle	The outer covering (or skin) of an arthropod

Cytospecies	Species that, because of their morphological similarity, have been identified primarily by cytological characters, such as the banding patterns of the chromosomes
Cytotaxonomy	The description of species based on a study of the banding patterns of the chromosomes
DDC	Drug distribution centre
DDT	Dichlorodiphenyltrichloroethane
DEET	The common name for the insect repellent *N,N*-Diethyl-*meta*-toluamide
Density (of flies)	Number of flies per given area or another unit
Density-dependent	A term used to describe factors that vary according to the population density and provide a form of 'negative feedback' to stabilize population numbers. All living populations must be limited by density-dependent factors operating on either their birth rate or death rate (or both); otherwise, their populations would increase forever. In some cases, however, populations can be limited by density-independent factors, before density-dependent factors come into operation
DENV	Dengue virus
DENV-1	Dengue 1 virus
DENV-2	Dengue 2 virus
DENV-3	Dengue 3 virus
DENV-4	Dengue 4 virus
Diapause	In insects, including mosquitoes, an altered physiological state in which certain animals do not resort to perform many of the physiological activities and are dormant
Diptera	An order of insects containing the two-winged flies that have only the anterior pair of wings well developed
Disease	Any departure from normal health in an organism
Dispersion	The movements of flies in and around a locality or area in which they emerged or are found at a given time
Diurnal	Daytime – usually relating to a cycle of activity. The opposite of nocturnal
DNA	Deoxyribonucleic acid
Dorsal	Refers to the uppermost surface of an organism
Dorsal	Upper surface or back
Drug resistance in parasite	The ability of a parasite strain to survive and/or multiply despite the administration and absorption of a drug in doses equal to higher than those usually recommended but within the limits of tolerance of the subject
Ecology	The study of the interrelationship between organisms and their environment (including other organisms)
Ecosystem	A type of biological organization made up of all the organisms in a given area
Ectoparasite	A parasite that lives on the external surface of an animal, e.g., fleas, lice, mites, ticks
El Nino	A term used to describe anomaly in the flow of ocean waters along the west coast of South America. This anomaly is the result of the nutrient-rich cold water of the coastal Humboldt Current being replaced by eastward flowing warm ocean water (which is nutrient poor) from the equatorial Pacific

EIR	Entomological inoculation rate. The number of infectious mosquito bites a person is exposed to in a certain time, typically a year. The EIR is the product of the human biting rate (HBR) and sporozoite rate
ELISA	Enzyme-linked immunosorbent assay
Encephalitis	A disease characterized by inflammation of the brain
Endemic disease	A disease that is caused by factors that are constantly present within a given geographical area or population group
Endophilic insects	Insects that regularly enter houses and spend at least part of their life indoors
ENSO	The 'El Nino' Southern Oscillation (SO) is a large-scale atmospheric 'seesaw' entered over the equatorial Pacific Ocean. The variation in pressure is accompanied in surrounding areas by fluctuations in wind strengths, ocean currents, sea surface temperatures, and precipitation. The SO and the warm waters of the El Nino are part of the same climate phenomenon referred to as ENSO (El Nino Southern Oscillation) in southern Asia, part of Australia, and parts of Africa, and heavy rainfall and flooding in arid areas of Southern America have been observed during El Nino years, while the Indian summer monsoon sometimes weakens and winters in western Canada and parts of the northern USA become milder. Overall, disasters triggered by drought are twice as frequent worldwide during El Nino years
Environment	The sum total, at a given moment, of all external influences and conditions to which an organism or object is subjected, including all physical, chemical, and biological factors
Epidemic	A period of increased prevalence of a given disease in a population (usually, but not necessarily, as a result of factors that are not constantly present in the populations)
Epidemiology	The study of the factors that determine the frequency and distribution of disease and of other health-related states in populations
Eradication	Complete removal or destruction (e.g., of an infectious agent in a given geographical area)
Exophilic insects	Insects that normally do not enter buildings
Exotic	Refers to anything that is from some other place
Facultative	Refers to phenomena such as diapause or autogeny where the phenomenon is timed and temporary
Family	A taxonomic grouping of organisms containing one or more genera
Fauna	All the species of animals that are present in a given area, e.g., the fauna of Chile
Femur	The third segment of an insect's leg, distal to the trochanter
FF/HF	Fully engorged/partially engorged female mosquitoes; abdomen of female mosquito showing state of engorgement due to blood fed on
FISH	Fluorescence in-situ hybridization
FTD	Fever treatment depot
G/SG	Gravid/semi-gravid females. Females with fully or partially developed state of eggs in ovaries
Ganglion	A nerve mass in insects, including mosquitoes
GE	Genetically engineered
Genetic	A hereditary factor controlled by genes of the previous generation

Genetic control	A modern approach to mosquito control involving altering the genetic configuration
Genitalia	The external genital organs of the insect; the gonads and their ducts and all associated accessory organs
Genus	(plural, genera) a taxonomic category within a family, consisting of one or more generally similar species
Gonotrophic cycle	The events between successive egg-laying, or the time interval between them
Gravid	Swollen and full of or nearly full of eggs
Habitat	The physical place (home) where an organism lives
Haematophagous	The term applied to an insect that feeds exclusively on blood; in mosquitoes, only females blood feed
Hazard	In toxicology, the risk of poisoning when a material is used
Head	The anterior region of an insect, which bears the mouthparts, eyes, and antennae and houses the brain
HEGs	Homing endonuclease genes
Herbicide	A type of pesticide designed to kill weeds
Hibernation	An altered state of some kind by which insects survive the winter
Host	The animal parasitized by a parasite or disease organism
Humidity or relative humidity	The content of water vapour in the air measured as a percentage of the content of saturated air (100% RH or relative humidity)
Hyperendemic	Heavily endemic
IGR	Insect growth regulator
IIBN	Insecticide impregnated bed nets. A bed net that is treated with a fast-acting and long-standing insecticide such as deltamethrin, permethrin, or lambdacyhalothrin that is safe for humans
IIT	Incompatible insect technique
Immunity	Resistance to parasites or other noxious agents or organisms in a given time, e.g., a year
Incidence	Occurrence of new cases in a population over time
Indigenous	Refers to an organism native to an area. The opposite of exotic
Infection	The invasion of a host by pathogenic microorganisms (establishing a host-parasite relationship) and the subsequent multiplication of the latter within the body of the host
Infectious diseases	Diseases of animals or plants caused by infectious pathogens
Infestation	The invasion of the surface of the body of the host by parasites or the harbouring of parasites on the surface of the body
Inorganic	Materials not containing carbon atoms
Insect development inhibitors (IDI)	Chemicals that inhibit the development of larvae or pupae
Insect growth regulator	A type of biorational insecticide that kills insects by interfering with the developmental process of the target stage
Insecticide	A chemical substance or a mixture of substances used to kill insects. It may be applied as a liquid, powder, fine spray, or vapour. The term 'larvicide' is usually applied to an insecticide that kills immature stages of insects, and the term 'adulticide' applies to an insecticide that kills mature or adult forms
Insecticide resistance	A characteristic that renders a given population of an insect species resistant to a given insecticide, although the species is normally susceptible to that being no longer controlled by the insecticide in the

	area concerned or the ability of an insect to withstand the lethal effects of an insecticide
Insecticide resistance in vector	Development of strength in a population of a given vector species to counteract the effect of the conventional or even higher dose of a given insecticide
Instar	The stage of an insect's life between successive moults; for example, the first instar is between hatching from the egg and the first moult
Integrated control	Applied pest control that contains and integrates chemical, physical and biological control measures, including the use of natural enemies of pests
IRS	Indoor residual spraying
IVM	Integrated vector management
Larva	An immature form occurring in some animals after hatching, differing markedly from the adult form
Larvicide	A insecticide used to kill larvae, usually of insects
LC50	A toxicological term used in pesticide testing that means the concentration is required to kill by 50%
LD50	A toxicological term used in pesticide testing that means the dose required to kill lengthened by days or warm temperatures
Life cycle	Stages of development through which an organism passes between fertilization of the egg of one generation and the same stage in the subsequent generation. The term is sometimes used loosely to mean the number of days between egg deposition and the attainment of sexual maturity
Malaria	A disease of humans and other animals caused by protozoan in humans by *Plasmodium* spp.
Mesonotum	The notum (tergum) of the mesothorax
Metamorphosis	Changes that an insect goes through during its life cycle. In insects, the series of changes through which the insect passes in its development from the larval through the pupal to the adult form. If the pupa is inactive and does not feed, metamorphosis is said to be 'complete'; if there is no pupa, or if the pupa is active and feeds, it is said to be 'incomplete'
Mortality	The action of death on populations; more specifically, the number of deaths in a given population in a given period of time
Multivoltine	Many generations per year. *Culex pipiens* is a mosquito that is multivoltine
NAMP	National Anti-Malaria Programme
Natural host	A host in which a given pathogenic microorganism (or parasite) is commonly found and in which the pathogen can complete its development. The term 'natural host' implies that the host is the usual one; it has the same meaning as 'typical host'
NMEP	National Malaria Eradication Programme
Nocturnal	Active during nighttime. The opposite of diurnal
Nomenclature	The scientific (Latin) names given to species, genera, families, etc., of animals and plants
Non-target organism	Any organism in an environment that is not the intended target of the control
NP	Nulliparous female mosquitoes, which have as yet not had an opportunity to oviposit eggs even once

Order	A taxonomic group of organisms containing one or more families
Organic	Chemical substances containing carbon
Organochlorines	A class of insecticides that contains chlorine groups, includes DDT
Organophosphates	A class of insecticides that contains phosphate groups, includes malathion
Oviposition	The laying of eggs by an insect, including mosquitoes
PAHO	Pan American Health Organization
Parasite	An organism that lives on, and at the expense of another organism (called the host). Or an organism that derives its nutrition by competing with its host, often with deleterious results to the host or by feeding upon its tissues
Parasitemia	The presence of circulating parasites in the blood of a host
Pathogens	Microorganisms (germs), such as viruses, bacteria, protozoa, eggs, and cysts of worms, that can cause disease if a person is infected
PCR	Polymerase chain reaction. A method for amplifying DNA in vitro, involving the use of oligonucleotide primers complementary to nucleotide sequences in a target gene and the copying of the target sequences by the action of DNA polymerase
Pest	Any mosquito species that is harmful in any way, biting or otherwise, or present in sufficient numbers to be considered a nuisance by humans
Pesticide	An agent or a substance or a mixture of substances used to kill species regarded as pests by humans
Phylum	A taxonomic group of organisms containing one or more orders
Physical control	The management or alteration of physical features of the environment
Phytotelmata	Bodies of water held by plants. Mosquito larvae often develop in such habitats
Plasmodium	The generic name for the parasite causing human malaria
PMH	Per work-hour density. Average number of vectors collected in human or animal dwellings (or even outdoors shelters) by hand collection done by standard methods by searches over fixed periods of time
Polytene	Chromosomes that have divided many times but have not separated, thereby forming thick bundles of up to a thousand parallel strands. After staining, their banding pattern can be seen under moderate magnification, and they are therefore convenient for examination. They are found in the nuclei of the cells of a few specialized tissues only
Population	A large group of organisms of the same species living in a geographic area
Population density	The number of individuals of a given population per unit area or volume
Posterior	Situated behind. The opposite of anterior
Potential	Existing and ready for action, but not yet active
Precipitin testing	A serological technique whereby the source of a blood meal can be identified by testing it with antibodies to blood from known host species
Predator	An organism that devours another organism for food.

Preparatory phase (of a control programme) The planning phase, when areas are mapped and surveyed for infestations, prior to the attack phase

Prevalence Existing cases in a population (point prevalence: cases at a certain point of time; period prevalence: cases during a period of time). The proportion of individuals in a population having malaria parasites in the blood

Proboscis In mosquitoes, a bundle of individual structures called stylets that are bound within a case called proboscis or the mouthparts of mosquito, which form a tube containing the stylets that penetrate the host tissues in order to suck up the host fluids

Pupa An insect form that occurs after the final larval stage and before appearance of the adult or the free living, highly mobile but not feeding stage of development of a mosquito that occurs between the larval and adult forms

Pv *Plasmodium vivax*

Pyrethrin The insecticidally active chemical component of pyrethrum insecticides

Salivary glands A set of glands located in the thorax of larval and adult mosquitoes

SFR Slide falciparum rate. The number of blood slides found with *P. falciparum* as a percentage of those examined

Sibling species Good species that do not interbreed but are difficult to separate purely on morphological basis

SIT Sterile insect technique

Species The smallest unit of classification commonly used, i.e., the group whose members have the greatest mutual resemblance, are able to interbreed (if not separated in space or time) but not breed with organisms of other groups (species). Species may be divided into subspecies that differ in certain genetic characters but are able to interbreed

Species complex A species complex consists of a number of species with almost identical morphological characteristics but with differences in certain aspects of their biology, behaviour and distribution. In most cases, the distinction has to be made using modern diagnostics, such as cytological tools or the PCR techniques

Spermathecae Structures in the abdomens of female mosquitoes in which sperm is stored

SPR Slide positivity rate. The number of blood slides found with *Plasmodium* parasites as a percentage of those examined

Subspecies A geographical group of local populations, differing taxonomically from the other groups of the same type forming part of the species under consideration

Surveillance Observation of a thing, an animal, or a person. The term also implies supervision or inspection or the monitoring, or close watch, over something

Survey To examine for some specific purpose, to inspect or consider carefully, to review in detail

Symptom A feeling of unhealthiness that can be expressed by a person. It may represent a *symptom*

Taxonomy	Classification of animals and plants divided into species, genera, families, etc.
Thorax	Flies and other insects are divided into three regions: head, thorax, and abdomen The thorax carries the legs and wings
Tibia	The fourth segment of an insect's leg, distal to the femur
Trachea	In insects, one of the major tubes that conduct air throughout the body of an insect
Tracheole	In insects one of the fine tubules that branch off at the end of tracheae
Transovarial transmission	The transmission of microorganisms from parent to offspring
ULV	Ultra-low-volume concentrate
Vector	An arthropod or another animal that carries a parasite from one host to another host. The vector may or may not be essential for the completion of the life cycle or the parasite. If it is not essential, it is referred to as a 'mechanical vector'
Vectorial capacity	The number of infective bites a person receives in a given period. A function of relative populations of mosquitoes and people, feeding frequency of the mosquito, duration of the latent period in the vector and the survival of the vector
Viremia	The presence of circulating virus in the blood of a host
Virus	A microorganism that can grow and reproduce only in living cells of other organisms
Wettable powder	A water dispersible but insoluble, formulation of an insecticide, in which the active ingredient is bound to an inert insoluble carrier material such as talcum powder. Wettable powder formulations are mixed with water to be sprayed and usually have greater residual activity than liquid formulations of the same insecticide
WHO	World Health Organization
Xenobiotic	Any substance foreign or strange to life, like synthetic insecticides, such as yellow fever virus
Zoogeography	The study of the geographic distribution of animals, including mosquitoes
Zoonosis	A disease of animals that may be transmitted under natural conditions to humans
Zoophilic	Feeding on animals other than humans

Index

Pages in *italics* refer to figures and pages in **bold** refer to tables.

L

M

N

O

P

Q

R

S

T

U

V

W

Y

Z